Handbook of Plant Biotechnology

Handbook of Plant Biotechnology

Editor: Zoe Eastwood

R CALLISTO
REFERENCE

www.callistoreference.com

Callisto Reference,
118-35 Queens Blvd., Suite 400,
Forest Hills, NY 11375, USA

Visit us on the World Wide Web at:
www.callistoreference.com

ISBN: 978-1-64116-111-4 (Hardback)

Trademark Notice: Registered trademark of products or corporate names are used only for explanation and identification without intent to infringe.

Cataloging-in-Publication Data

Handbook of plant biotechnology / edited by Zoe Eastwood.
    p. cm.
Includes bibliographical references and index.
ISBN 978-1-64116-111-4
1. Plant biotechnology. 2. Plant genetic engineering. 3. Agricultural biotechnology.
I. Eastwood, Zoe.
SB106.B56 H36 2019
631.523 3--dc23

Table of Contents

Permissions

List of Contributors

Index

Preface

Plant biotechnology is involved with the modification of plants in order to produce desired characteristics. These characteristics include higher yield, disease resistance, drought tolerance and higher adaptability. It employs the scientific principles and techniques of genetic engineering and molecular biology in order to make desired modifications in plants for applications in agriculture. Some of the widely used techniques of this field include reverse breeding, marker assisted selection, genetic recombination and double haploidy among many others. This book studies, analyses and upholds the pillars of plant biotechnology and its utmost significance in modern times. It aims to shed light on some of the unexplored aspects and the recent researches in this field. Students, researchers, experts and all associated with plant biotechnology will benefit alike from this book.

This book has been the outcome of endless efforts put in by authors and researchers on various issues and topics within the field. The book is a comprehensive collection of significant researches that are addressed in a variety of chapters. It will surely enhance the knowledge of the field among readers across the globe.

It gives us an immense pleasure to thank our researchers and authors for their efforts to submit their piece of writing before the deadlines. Finally in the end, I would like to thank my family and colleagues who have been a great source of inspiration and support.

Editor

A phosphate starvation-driven bidirectional promoter as a potential tool for crop improvement and *in vitro* plant biotechnology

Oropeza-Aburto Araceli[1], Cruz-Ramírez Alfredo[2], Mora-Macías Javier[1] and Herrera-Estrella Luis[1,*]

[1]*Metabolic Engineering Laboratory, Unidad de Genómica Avanzada – LANGEBIO CINVESTAV, Irapuato, Guanajuato, Mexico*
[2]*Molecular and Developmental Complexity Laboratory, Unidad de Genómica Avanzada – LANGEBIO CINVESTAV, Irapuato, Guanajuato, Mexico*

*Correspondence
email lherrerae@cinvestav.mx

Keywords: phosphate starvation, crop improvement, bioengeneering, roots, enhancer.

Summary

Phosphate (Pi)-deficient soils are a major limitant factor for crop production in many regions of the world. Despite that plants have innovated several developmental and biochemical strategies to deal with this stress, there are still massive extensions of land which combine several abiotic stresses, including phosphate starvation, that limit their use for plant growth and food production. In several plant species, a genetic programme underlies the biochemical and developmental responses of the organism to cope with low phosphate (Pi) availability. Both protein- and miRNA-coding genes involved in the adaptative response are transcriptionally activated upon Pi starvation. Several of the responsive genes have been identified as transcriptional targets of PHR1, a transcription factor that binds a conserved cis-element called PHR1-binding site (P1BS). Our group has previously described and characterized a minimal genetic arrangement that includes two P1BS elements, as a phosphate-responsive enhancer (*EZ2*). Here, we report the engineering and successful use of a phosphate-dependent bidirectional promoter, which has been designed and constructed based on the palindromic sequences of the two P1BS elements present in *EZ2*. This bidirectional promoter has a potential use in both plant *in vitro* approaches and in the generation of improved crops adapted to Pi starvation and other abiotic stresses.

Introduction

One of the major limitations for sustained plant growth in most soils is the scarcity of inorganic phosphate (Pi). A large fraction of Pi in soils is present as diverse organic and inorganic chemical forms that are not readily available for plant uptake. As this macronutrient is needed for the synthesis of vital molecules such as nucleic acids and phospholipids and for important metabolic processes, plants have evolved a complex and multifactorial strategy to adapt and grow in soils with low Pi abundance (Lynch, 1995; Plaxton, 2004; Raghothama, 1999). Such combined strategy involves the action of Pi as a signal for triggering a signalling cascade to respond to the internal and external level of Pi (Lin *et al.*, 2009; Shen *et al.*, 2011).

Cells of roots and rhizoids are in charge of Pi uptake and are the first organs involved in sensing Pi availability in the rhizosphere. In Pi-scarce soils, a still unclear signalling pathway is triggered in the most external cell layers of the roots, and in rhizoids, such pathway activates a transcriptional machinery that induces the expression of mRNA and miRNAs that are involved in the myriad of biochemical and morphological mechanisms that allow the plant to: (i) reconfigure the root system architecture (RSA) and release Pi from the organic compounds in the soil, (ii) increase Pi uptake and transport and (iii) recycle Pi from cellular organic sources (Baker *et al.*, 2015; Chiou and Lin, 2011; Rouached *et al.*, 2010; Vance *et al.*, 2003; Williamson *et al.*, 2001).

In Arabidopsis and other plant species, it has been well characterized that under Pi starvation, the expression of several miRNAs and mRNAs involved in the adaptive mechanisms mentioned above is positively regulated by the transcription factor PHOSPHATE STARVATION RESPONSE 1 or PHR1 (Bari *et al.*, 2006; Lundmark *et al.*, 2010; Nilsson *et al.*, 2007; Rubio *et al.*, 2001). PHR1 is a MYB-CC transcription factor with several homologues in diverse plant species, including PSR1 (Phosphate Starvation Response 1), the homolog of PHR1 in the chlorophyte algae *Chlamydomonas reinhardtii* (Wykoff *et al.*, 1999). PHR1 homologues share the conserved function of orchestrating the transcriptional programme triggered in response to low Pi availability and activate the transcription of its target genes by binding to P1BS (PHR1-binding sites), a conserved DNA motif, with the consensus sequence GNATATNC, which is present in the promoters of many phosphate-responsive genes (Franco-Zorrilla *et al.*, 2004; Müller *et al.*, 2007; Sobkowiak *et al.*, 2012). We have previously reported the finding and characterization of a conserved enhancer element that regulates the expression of the Arabidopsis *PHOSPHOLIPASE D-Z2* (*PLDZ2*) gene in response to Pi availability (Oropeza-Aburto *et al.*, 2012). This enhancer element, denominated *EZ2*, is present in the promoter of *PLDZ2* orthologues and in diverse Pi-responsive genes conserved along plant lineages (Acevedo-Hernández *et al.*, 2012). *EZ2* is composed by two P1BS motifs that are not identical (sequence of the upstream one is GAATATTC and the other GGATATTC) with a spacer sequence of between 21 and 28 bp in average, depending on the plant species (Figure 1a). It also has a conserved motif in the spacer region with the consensus sequence GCAYCAAA and a motif in its 5′

region with the sequence TTTGG or TTTGC. We modified the *EZ2* native enhancer and found that when the two P1BS elements have the same GAATATTC motif, the induction by Pi starvation is stronger than the native enhancer, indicating that the Arabidopsis PHR1 displays a higher affinity for this Modified *EZ2* (*M-EZ2*) version of the enhancer (Figure 1a). A recent study demonstrates that when the two P1BS motifs in the promoter of *OsPHF1* were replaced by the same sequence we used in *M-EZ2*, there is a drastic increase in the affinity of OsPHR2, the PHR1 ortholog in rice, for this type of cis-regulatory enhancer (Ruan *et al.*, 2015).

In this work, we describe the design and functional characterization of a bidirectional promoter based on the *M-EZ2* enhancer, which responds with high sensitivity to Pi concentrations. Our

results show that this novel tool has a potential use in the generation of crops with improved capacity to deal not only with Pi starvation, but also other abiotic stresses associated with Pi-scarce soils. Moreover, *M-EZ2*-based bidirectional promoter can be used to drive gene expression for *in vitro* plant biotechnology, where Pi availability can be used to modulate expression of diverse proteins as reporters and/or resistance markers, in a reversible manner.

Results

In-tandem arrangement of *M-EZ2*

In a previous work (Oropeza-Aburto *et al.*, 2012), we identified an enhancer element in the regulatory region of the Arabidopsis

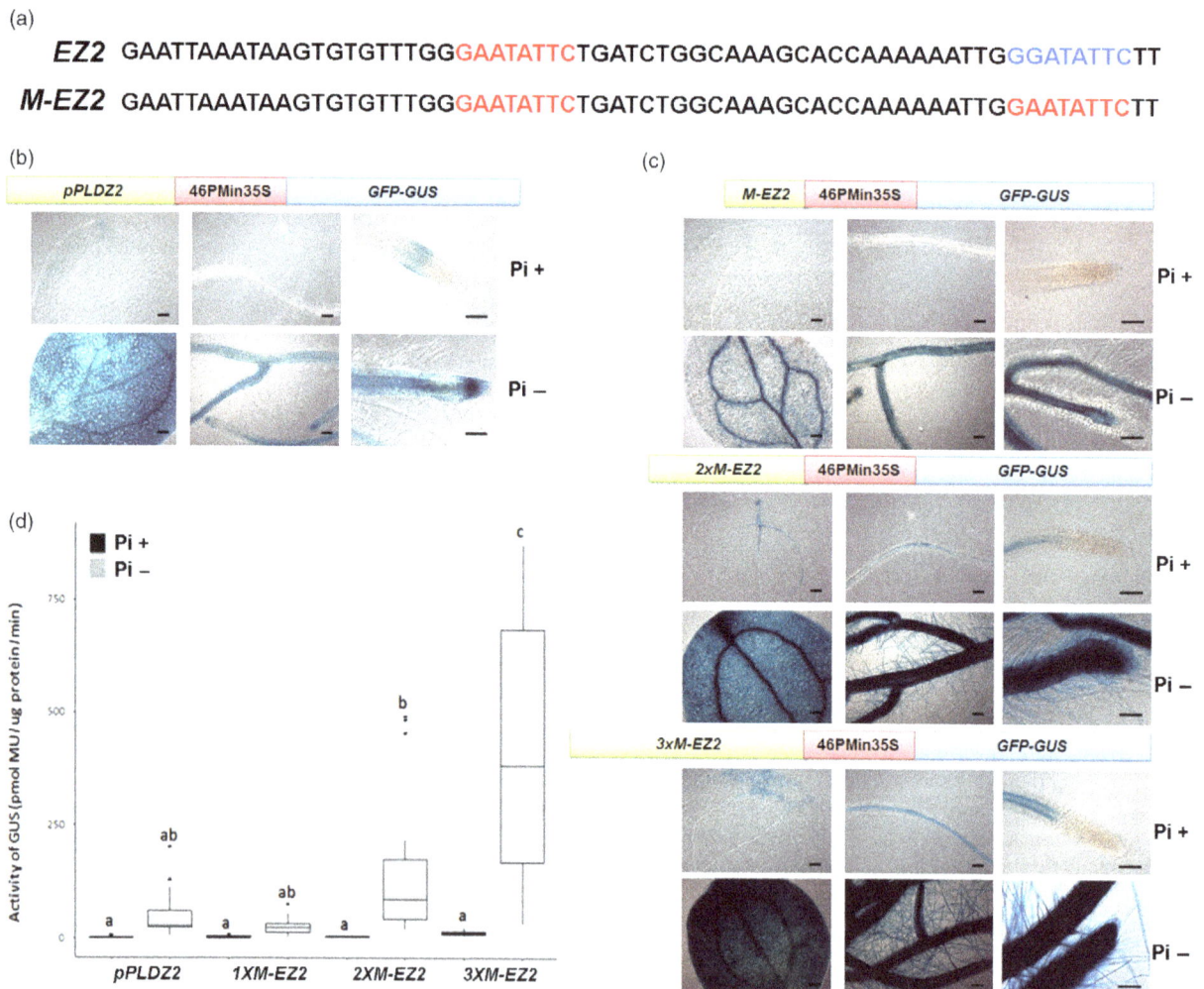

Figure 1 Structure of *EZ2* and *M-EZ2* and level expression of tandem enhancers. (a) The native upstream sequence of the *PLDZ2* (locus At3g05630) between position −782 and 717 pb relative to the start codon (top sequence), and the modified sequence containing the duplicated P1BS motif (red) to obtain the *M-EZ2* enhancer (bottom sequence) are shown. (b) Structure and expression of *pPLDZ2-GUS::GFP*, Arabidopsis plants carrying the *PLDZ2* promoter driving the gene *GUS* were grown in P+ (1 mM) and P− (0 mM) medium for 10 dag. Plants were stained for GUS activity, and cotyledons, lateral roots and root meristematic region were photographed using Nomarsky optics. (c) Structure and expression of chimeric promoters composed of the sequence of modified enhancer once (*1XM-EZ2*), twice (*2XM-EZ2*) and three times (*3XM-EZ2*) inserted upstream of the −46 S minimal 35S promoter driving the expression of *GUS*. Seedlings were grown for 10 dag in medium Pi+ (1 mM) or Pi− (0 mM) and stained for GUS activity. Cotyledons, lateral roots and root apical meristems of representative plants from each line and for each condition (Pi+ and P−) were photographed using Nomarsky optics. Bars, 100 μm. (d) Results of the fluorometric assay of eight independent lines for each construction are shown. Values indicate the mean obtained for two biological replicates and four independent technical replicates for each biological sample. Different letters above bars indicate statistically significant differences based on an ANOVA test.

PLDZ2 gene. The sequence and arrangement of this enhancer is described in detail in Figure 1a. A modified version of this enhancer when fused to the −46 cauliflower mosaic virus 35S minimal promoter (46PMin35S) was shown to drive high levels of expression of both UidA (GUS) and GFP reporter genes in a low Pi-dependent manner (Oropeza-Aburto et al., 2012). The modified promoter, named M-EZ2, turns on transcription in response to Pi starvation. To determine whether a promoter with more than one copy of M-EZ2 could increase the responsiveness to Pi starvation, we generated synthetic promoters with two or three copies of the M-Z2 enhancer (2XM-EZ2 and 3XM-EZ2 versions), fused to the 46PMin35S and a GUS-GFP double reporter gene construct (Figure 1c). Arabidopsis transgenic lines were obtained by transforming plants independently with 1XM-EZ2-GUS::GFP, 2XM-EZ2-GUS::GFP or 3XM-EZ2-GUS::GFP and its responsiveness to Pi availability tested (Figure 1c).

The number of M-EZ2 copies in the promoter is directly proportional with the strength in GFP and GUS expression

As a first approach to qualitatively determine the transcriptional effect of different number of M-EZ2 copies, seeds from several independent transgenic 1XM-EZ2-GUS::GFP, 2XM-EZ2-GUS::GFP, 3XM-EZ2-GUS::GFP and pPLDZ2-GUS::GFP lines were germinated and grown either in P+ (1 mM) or P− (0 mM) medium. Seedlings of 10 days after germination (dag) were stained for GUS activity, and representative images of each line are shown (Figure 1b and c). As previously reported (Oropeza-Aburto et al., 2012), under Pi-sufficient conditions, pPLDZ2-GUS::GFP seedlings only stain for GUS in the root meristematic region, whereas in Pi-limiting conditions there is a clear increase in GUS activity in cotyledons and all root tissues (Figure 1b). We observed that for 1XM-EZ2-GUS::GFP lines there is no GUS in Pi+ conditions, while in Pi− conditions 1XM-EZ2-GUS::GFP seedlings showed a similar GUS pattern to that observed for pPLDZ2-GUS::GFP lines grown in P− media, although less intense (Figure 1b and c). Seedlings of 2XM-EZ2-GUS::GFP and 3XM-EZ2-GUS::GFP lines grown in Pi+ conditions showed weak GUS staining in the vascular tissue of both cotyledon and root, but never the root meristematic zone, even when plants were incubated overnight for staining (Figure 1c). However, in Pi− conditions seedlings showed a dramatic increase in GUS activity in all root tissues, including root hairs, a strong was also observed in the cotyledons. In summary, the number of copies of the M-Z2 enhancer determines the spatial–temporal patterns of GUS expression and the strength of the response to Pi deprivation (Figure 1c).

To determine in a quantitative manner the increase in GUS expression, we first performed fluorometric GUS assays to quantify the activity of Beta-Glucuronidase (the protein coded in the UidA gene or GUS) in seedlings of eight independent lines for each of the synthetic promoters that differ in the number of M-EZ2 enhancer elements (Figure 1c–d and Table S1). Seedlings were germinated and grown in medium with phosphate (1 mM) or without phosphate (0 mM) for 10 days after germination (dag) and then GUS activity determined for each line. In P− conditions, the 1XM-EZ2-GUS::GFP seedlings showed an average activity of 25.249 pmol MU/μg protein min, which is only the half of the average value of the full Arabidopsis PLDZ2 promoter (pPLDZ2-GUS::GFP). The average activity for the 2XM-EZ2-GUS::GFP line was 153.609 pmol MU/μg protein min, which represents 2.9 times higher than full PLDZ2 promoter, whereas the 3XM-EZ2-GUS::GFP lines showed the highest average value of expression,

433.1 pmol MU/μg protein min, this is 8.321 times more than that showed by the PLDZ2-GUS::GFP lines. Statistical analyses (ANOVA) showed that between the 1XM-EZ2-GUS::GFP and the complete PLDZ2-GUS::GFP there is no significant difference in the level of GUS expression in seedlings grown under Pi-limiting conditions; by contrast, there is significant differences between the expression directed by the 2XM-EZ2-GUS::GFP and 3XM-EZ2-GUS::GFP synthetic promoters and that observed for the complete PLDZ2 promoter. Furthermore, in Pi+ media there is no difference in the level of expression directed between the synthetic promoters with that observed for the complete PLDZ2 promoter (Figure 1d). The fold change of GUS activity between seedling grown in Pi+ media compared to Pi− media for 1XM-EZ2-GUS::GFP, 2XM-EZ2-GUS::GFP and 3XM-EZ2-GUS::GFP lines was 10.11, 79.44 and 35.01, respectively. The lower fold change induction in 3XM-EZ2-GUS::GFP seedlings with respect to 2XM-EZ2-GUS::GFP is due to a higher basal activity when they are grown in Pi supply media (Table S1). Also in the 3XM-EZ2-GUS::GFP lines in which GUS activity was as high as 848.45 pmol MU/μg protein min, this represents 16.3 times higher than that quantified for the PLDZ2-GUS::GFP line and it was comparable or even higher than the GUS expression observed in constructs driven by the CAMVS35S promoter (Table S1).

These results indirectly show that the transcriptional activation in response to Pi availability, as indicated by the level of GUS activity, increases as the number of copies of M-EZ2 also increases.

To directly measure the effect of the number of copies of M-EZ2 on gene transcription, we performed both semiquantitative reverse transcription PCR (RT-PCR) and quantitative reverse transcription PCR (qRT-PCR) assays to determine the levels of GUS transcripts. For these experiments, seedlings were germinated and grown for 7 days in Pi+ solid medium (1 mM) and transferred to Pi− liquid medium (0 mM). For RNA extraction, samples were frozen and processed at different time points: 12, 24, 48, 72 and 96 h post-transference (hpt). Our results showed that GUS mRNA levels, when transcription is under the control of the unmodified (native) EZ2, were almost absent in seedlings grown in P+ media and a weak gradual response to Pi-starvation conditions was observed when compared to the RT-PCR products observed when GUS transcription was driven under the control of the full PLDZ2 promoter and the 1X version of MEZ2 (Figure 3a). Also, in these RT-PCR assays we were able to observe that the 2XM-EZ2 promoter version gradually increases the GUS gene transcription upon time in response to Pi starvation in a similar fashion to that observed for the full PLDZ2 promoter (Figure 3a). Interestingly, the 3XM-EZ2 promoter version responded earlier and stronger than any other of the promoter versions, including the full PLDZ2 promoter (Figure 3a).

In order to quantify the higher sensitivity of 3XM-EZ2 promoter in comparison with that of PLDZ2, the GUS transcripts levels were determined by a qRT-PCR. Our results showed that in response to low Pi, the 3XM-EZ2 promoter induces the expression of the GUS transcripts 48.2-fold 24 h after seedlings were submitted to the stress, whereas for the PLDZ2 promoter, GUS transcript levels at this time point were 7.08-fold higher than the controls in Pi+ media. This behaviour was similar in the following time points of the treatment; by 48, 72 and 96 h, the GUS transcript levels under the 3XM-EZ2 promoter were, respectively, 207.63, 395.92 and 710.86 while those of the PLDZ2 promoter in the corresponding times were 69.19, 203.45 and 250.29 (Figure 3b and c). These results correlate well with the qualitative analyses of GFP expression in *PLDZ2-GUS::GFP* and *3XM-EZ2-GUS::GFP* lines, by analysing the expression pattern of the transgene in a similar conditions and time point used for *GUS* q-RT-PCR analysis. Our results clearly show a higher GFP expression, especially in the vascular tissue of *3XM-EZ2-GUS::GFP* roots, when compared with those of *PLDZ2-GUS::GFP* (Figure 2).

M-EZ2 displays a high sensitivity to internal and external Pi concentrations

It has been shown that phosphite (Phi), a nonassimilable source of phosphorus for plants, a structural analogue of Pi that it is transported to the cytoplasm via the same transport system as Pi in plant cells (Danova-Alt *et al.*, 2008), is perceived by the plant as Pi altering some of the responses to Pi deficiency, mainly delaying the initial response triggered by external and internal reduction of Pi levels (Ticconi *et al.*, 2001; Varadarajan *et al.*, 2002). Taking advantage of this knowledge, we tested whether Phi addition to the media could alter the intensity and/or timing of the response of *PLDZ2* and *M-EZ2* promoters to Pi starvation. For this, *PLDZ2-*

GUS::GFP and *3XM-EZ2-GUS::GFP* seedlings were transferred from P+ medium either to Phi+ Pi− or Pi− medium, and samples were taken at different time points (12, 24, 48, 72 and 96 hpt) for RNA extraction and qRT-PCR evaluation of *GUS* transcripts. Ours results showed that the presence of Phi decreases the *GUS* transcripts at each time point after plants are subjected to low Pi availability. This decrease indirectly shows that Phi delays the response of *PLDZ2* along the time course experiment. For instance, while in plants transferred to Pi− medium there is a clear increase in *GUS* transcripts by 48 hpt (Figure 3b), in plants transferred to Phi+ Pi− medium this increase was observed until 96 hpt and the final level of induction was significantly lower in seedlings transferred to media containing Phi than those transferred into media lacking Pi (Figure 3b). *3XM-EZ2-GUS::GFP* seedlings also showed this delay, as plants transferred to media containing Phi had a detectable decrease with respect to the signal detected for seedlings grown in Pi− media. However, the expression is significantly higher than that observed for the *PLDZ2* promoter at all tested time points (Figure 3c). This suggests that

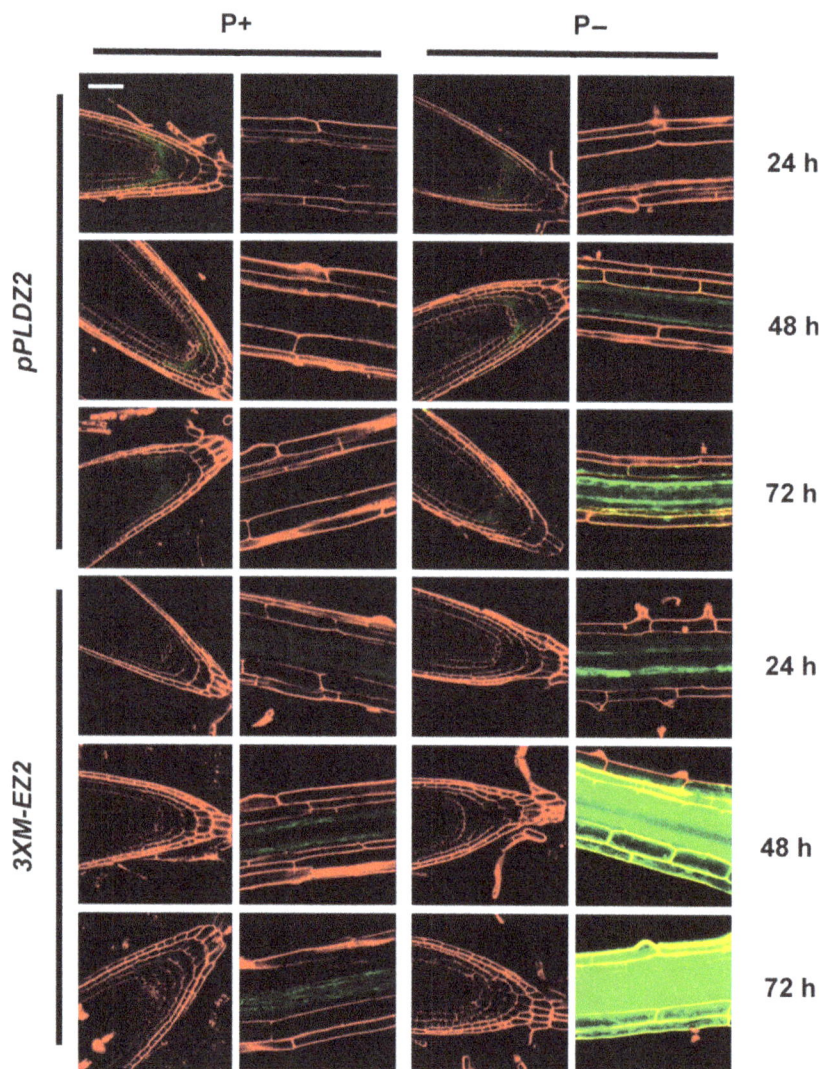

Figure 2 Pi-starvation-induced expression of GFP in Arabidopsis plants. Transgenic *pPLDZ2-GUS::GFP* and *3XM-EZ2-GUS::GFP* lines were grown in Pi-sufficient conditions for 7 dag and induced in Pi-deficient medium (P−) for 24, 48 and 72 h. Plants grown in P+ medium as control are shown. Photographs of root apical meristem and root elongation zone were taken using Zeiss confocal optics in a LSM510 META microscope. Bar in the upper left picture = 50 μm. All the other pictures have the same scale.

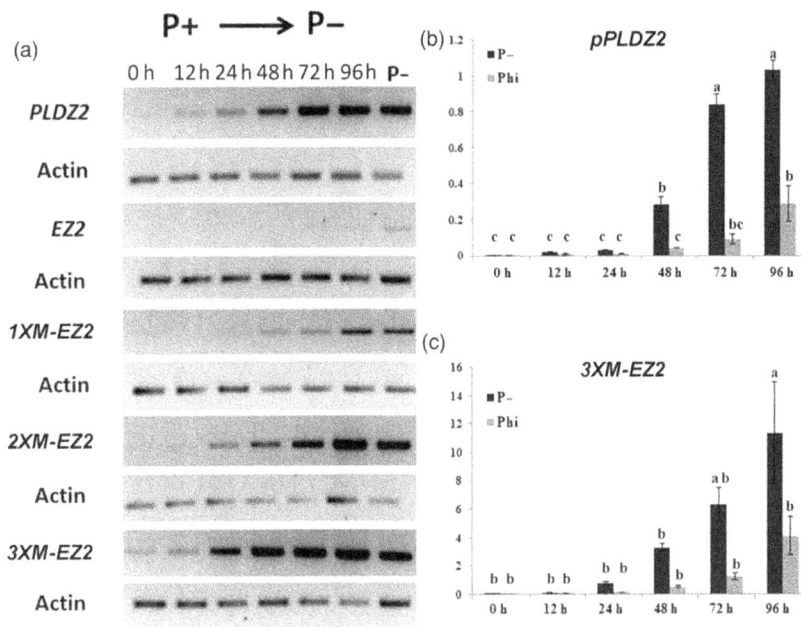

Figure 3 Comparative *GUS* expression analysis among the native *PLDZ2* promoter, *EZ2* native and the *M-EZ2* enhancer versions upstream of the −46 S minimal 35 promoter in response to Pi deficiency and Phi supply. (a) Semiquantitative RT-PCR assays of *GUS* transcripts driven by either *pPLDZ2*, *EZ2*, *1XM-EZ2*, *2XM-EZ2* or *3XM-EZ2* were carried out for plants grown for 7 dag in P+ (1 mM) and transferred to P− (0 mM) medium for 12 h, 24 h, 48 h, 72 h and 96 h. Transcript levels for plants grown continuously in P− are also indicated. (b) Quantitative reverse transcription PCR analysis for *GUS* expression of plants carrying *pPLDZ2-GUS::GFP* and (c) *3XM-EZ2-GUS::GFP*, grown in P+ (1 mM) for 7dag and transferred to P− (0 mM) or phosphite (Phi, 1 mM) are shown. A time course indicating the transcript level of the reporter gene is shown. Values are reported as a relative quantification [$2^{(-\Delta\ C_T)}$] between the reporter gene and the *ACTIN2* gene for each condition. Data are the mean of two biological replicas, and letters above bars indicate statistically significant differences supported by ANOVA test.

three copies of *M-EZ2* increased the sensitivity of the *3XM-EZ2* synthetic promoter to Pi availability in such a manner that even in the presence of a molecule that mimics Pi presence, as Phi does, the promoter is still able to strongly respond to Pi scarcity. These results suggest that Phi acts in Arabidopsis plants via the P1BS DNA motif, which is the only regulatory element in the *3XM-EZ2* synthetic promoter that is known to respond to low phosphate availability and, therefore, it must be by promoting the interaction of PHR1 and SPX1, as previously shown under in vitro binding conditions (Puga *et al.*, 2014).

Reversible behaviour of *M-EZ2*

To determine whether the expression directed by the EZ2 synthetic promoters is reversible after the transcriptional activation by Pi starvation, we transferred seedling grown in Pi− media for 7 dag into Pi+ media and then determine *GUS* transcript levels. We could clearly observe that the expression of the chimeric promoters was repressed upon transfer from Pi− media into Pi+ media (Figure 4a). This behaviour was quantified by qRT-PCR (Figure 4b and c) comparing the complete promoter and the *3X* enhancer version. The transcription levels were reduced 36.04% and 39.99% in the first 30 min in *pPLDZ2* and the *3XM-EZ2* seedlings, respectively. 1, 3, 6 and 12 h post-treatment the level of transcription for *pPLDZ2* progressively declined in 59.04%, 86.84%, 90.33% and 92.5%. In the case of *3XM-EZ2* seedlings, this percentage was reduced in 40.16%, 86.83%, 95.56% and 98.82% at the respective time points. These results confirm that the activity of the *M-EZ2* enhancer is directly regulated by internal levels of Pi.

Generation and function of a bidirectional Pi-responsive promoter

In nature, the presence of genes located on opposite DNA strands, which transcription is under the control of a bidirectional promoter has been reported in several cases (Banerjee *et al.*, 2013; Dhadi *et al.*, 2009; Liu and Han, 2009; Wang *et al.*, 2009). The capacity of bidirectional promoters to control the expression of two opposite open reading frames (ORFs) is many times dependent on the presence of enhancer elements that can affect the expression of proximal promoters located upstream and downstream of the enhancer element (Chaturvedi *et al.*, 2006). Based on the fact that enhancers can modulate transcription independent of their orientation, genetic engineering strategies have been designed to develop minimal bidirectional promoters that confer a specific spatio-temporal transcriptional pattern (Mehrotra *et al.*, 2011; Venter, 2007). As the P1BS sequence is palindromic (GAATATTC) and *EZ2* acts an enhancer, it is predictable that the activity of the triple enhancer element could act in both orientations. Therefore, we tested the capacity of the *3XM-EZ2* element to function bidirectionally by fusing a minimum nopaline synthase promoter (PMinNOS) at its 5′ end in opposite direction to 46PMin35S (Figure 5a, Figure S1). The *PMinNOS-3XM-EZ2-46PMin35S* bidirectional promoter recombined in the pBGWFS7 plasmid to generate the *PMinNOS-3XM-EZ2-46PMin35S-GUS::GFP* transcriptional fusion. Then, the gene coding regions for hygromycin resistance (hygromycin B phosphotransferase or HYGBP) and *GUS-GFP* were placed

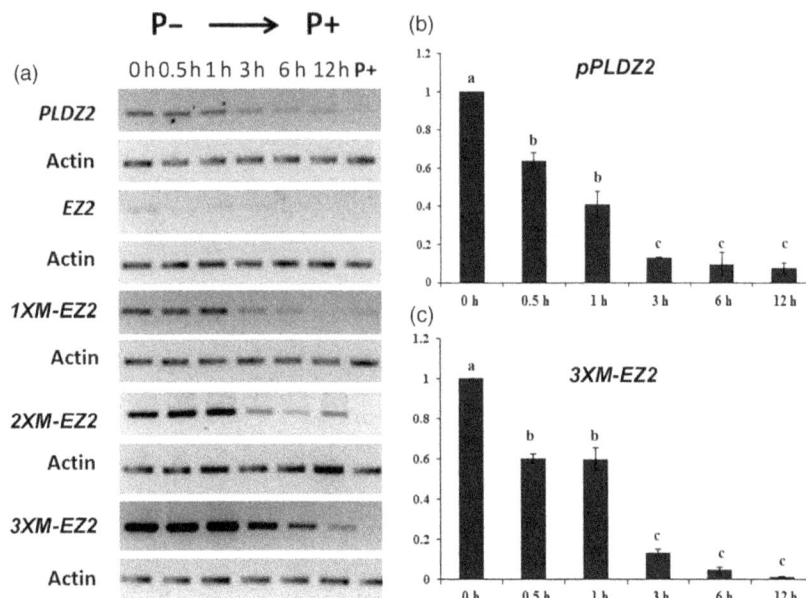

Figure 4 Comparative *GUS* expression analysis among the native *PLDZ2* promoter, *EZ2* native and the *M-EZ2* enhancer versions upstream of the −46 S minimal 35 promoter in response to Pi supply. (a) Plants grown for 7 dag in medium P− (0 mm) and then transferred to medium P+ (1 mm) were analysed by semiquantitative RT-PCR assay. *GUS* transcript levels from the *pPLDZ2*, *EZ2* native, *1XM-EZ2*, *2XM-EZ2* and *3XM-EZ2* constructs are shown at 0.5, 1, 3, 6 and 12 h. GUS transcript levels of plants grown continuously in P+ are also indicated. (b) Real-time qRT-PCR analysis of *pPLDZ2::GUS:GFP* and c) *3XM-EZ2:: GUS:GFP* from plants grown 7 dag in P− (0 mm) and transferred to P+ (1 mm) are shown. Values are reported as a relative quantification [$2^{(-\Delta\Delta\,C_T)}$] in which was considered the sample P− 0 h as calibrator and as an endogenous control *ACTIN2* gene. Values are the mean of two biological replicas, and letters above bars indicate statistically significant differences resulted from ANOVA test of the data.

downstream of the PMinNOS and 46PMin promoters, respectively. The binary vector carrying the dual transcriptional phusion *HYGBP::PMinNOS-3XM-EZ2-46PMin35S-GUS::GFP* (Figure 5a) between the T-DNA borders was used to transform Arabidopsis plants, and the resulting homozygous lines were tested for GUS and HYGBP activity. To test the capacity of *HYGBP::PMinNOS-3XM-EZ2-46PMin35S-GUS::GFP* (*3XM-EZ2Bi*) lines to express the Hygromycin resistance gene, seedlings were germinated and grown in solid medium lacking Pi for 6 days and then transferred to either Pi+ or Pi− mediums, both added with 15 mg/L of hygromycin. Seedlings transferred to Pi+/Hyg medium died in a similar manner to hygromycin sensible controls (HYGS in Figure 5b), whereas seedlings transferred to Pi−/Hyg medium were able to survive, developing green cotyledons and leaves similar to that observed for seedlings harbouring the Hygromycin resistance gene under the 35S promoter (HYGR in Figure 5b, left panel). These results correlate well with GUS patterns of *3XM-EZ2Bi* seedlings transferred to Pi−/Hyg subjected to the staining protocol which show a strong blue after the reaction (Figure 5b, right panel), while the seedlings grown in Pi+ medium do not show detectable blue staining, indicating that they do not express *GUS* gene at all (Figure 5b, right panel).

We performed a quantitative analysis of *GUS* and *HYGBP* transcripts in *3XM-EZ2 Bi* seedlings grown 7dpg in Pi+ solid medium and then transferred to Pi− medium for 72 h. Seedlings of six independent lines showed a transcriptional *GUS* behaviour similar to the unidirectional *3XM-EZ2* lines (Figure 5c). In the case of the *HYGBP* gene, transcript levels were also induced in seedlings transferred to Pi− when compared to the control. Our results also evidenced that, although enough to confer hygromycin resistance to plants, the *HYGBP* transcripts levels are 10

times lower than those of *GUS* transcripts, suggesting that either the strength with which *M-EZ2* enhancer influences the transcription of minimal promoters is orientation dependent or that the nature of the minimal promoter determines the strength of the effect of this enhancer element; in this case, *PMinNOS* can influence the level of induction conferred by the *3XM-EZ2* enhancer construct (Figure 5d).

Discussion

In this study, we built a promoter by arranging in tandem three copies of a modified enhancer from the promoter region of the Arabidopsis *PLDZ2* gene, which strongly responds to Pi starvation (*M-EZ2*; Oropeza-Aburto et al., 2012). We found that the *3XM-EZ2* promoter is highly sensitive to Pi scarcity, showing that the number of copies of the enhancer drastically influences the strength of the response. It has been previously shown the importance of P1BS as a *cis*-acting motif in the Pi-starvation response. However, the presence of P1BS in a given promoter is not enough to confer Pi-responsiveness to the gene, the specific sequence of the motif and the distribution of one or more motifs in the sequence context define the strength of the transcriptional response to the stress (Bustos et al., 2010; Oropeza-Aburto et al., 2012; Ruan et al., 2015). We also show that the composition of the *M-EZ2* element confers an optimal sequence and spatial arrangement to increase the sensitivity to the action of the transcriptional activator PHR, a sensitivity that increases in function of the number of *M-EZ2* copies.

The P1BS, as a target sequence of PHR1 in a natural or artificial promoter, depends on the amount of this TF and its activity. PHR1 capacity to act on P1BS depends, in turn, on the direct binding among SPX1 and PHR1 proteins, a Pi-dependent interaction which

(a)

3XM-EZ2-Bi | HYGBP | PMinNOS | 3XM-EZ2 | 46PMin35S | GFP-GUS |

(b)

(c)

UidA

(d)

HygB Phosphotransferase

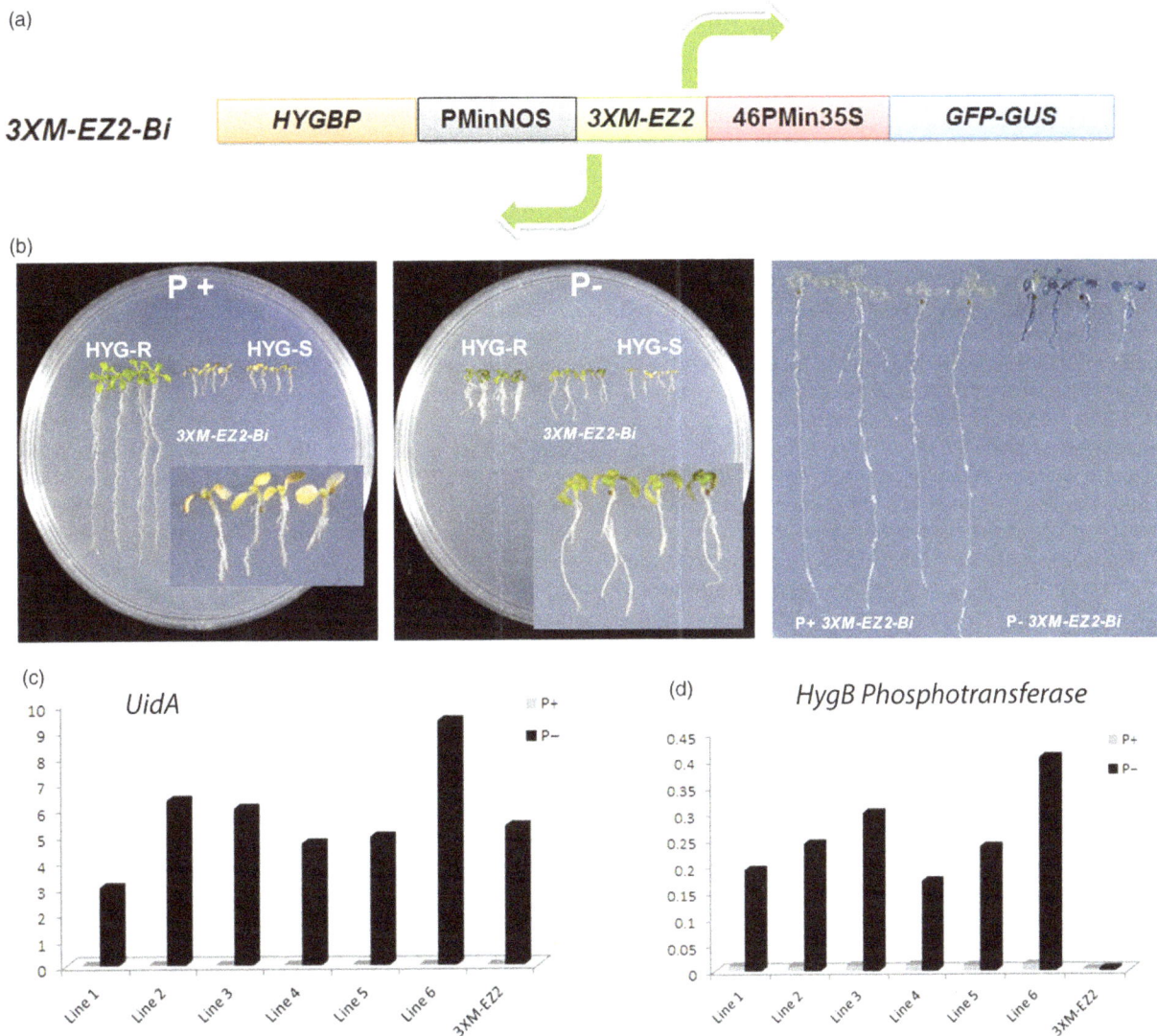

Figure 5 Pi-starvation responsiveness of bidirectional *3XM-EZ2* enhancer. (a) Diagram showing the structure of *3XM-EZ2 Bi*, harbouring at the core the *M-EZ2* enhancer, at the 5′ end the minimal *NOS* promoter and the *HYG B* phosphotransferase gene in opposite directions, and at the 3′ end the −46 S minimal 35 S promoter followed by the *GFP* and *uidA* genes. (b) Expression analysis of the two reporter genes in Arabidopsis plants. Hygromycin B resistance of Arabidopsis seedlings growing in induction medium without Pi (5 μM) and then transferred to Hyg medium (15 mg/L) with (1 mM) or without (5 μM) Pi. Hygromycin B-resistant and hygromycin B-sensitive plants were grown as controls (two left panels). The right panel shows GUS activity of seedlings grown for 10 dag in medium Pi+ (1 mM) and Pi− (0 mM) and then stained and photographed. (c) Quantitative reverse transcription PCR assay of plants carrying *3XM-EZ2 Bi*. Six independent lines were analysed for (c) *GUS* and (d) *HYG B* phosphotransferase expression, and a representative line carrying *3XM-EZ2* enhancer was analysed as control. Values are the result of a relative quantification [$2^{(-\Delta C_T)}$] between the reporter gene and *ACTIN2* gene for each condition.

inhibits PHR1 DNA-binding. Puga *et al.* (2014) showed that in the presence of Pi the SPX1 binds PHR1, inhibiting the transcriptional activation of PHR1 targets. They also reported that phosphite (Phi) was able to interfere with the interaction between PHR1 and SPX using an *in vitro* binding assay. As the *M-EZ2* enhancer only contains the P1BS-binding site as a Pi–responsive element, the observation that the *3XM-EZ2* is still influenced by Phi provides *in vivo* support of the notion that Phi directly acts by promoting the interaction between PHR1 and SPX1. However under Pi-starvation conditions, low Pi concentrations decrease the interaction between SPX1 and PHR1 complex, releasing PHR1 which activates its transcriptional targets, among which is *SPX1* gene itself. Indeed, *SPX1* transcription is repressed in the presence of Phi and absence of Pi, in both shoots and roots (Jost *et al.*, 2015).

An important observation in this study is that the *3XM-EZ2* promoter responds to Pi limitation, even in the presence of Phi. This suggests that the sensitivity of the *3XM-EZ2* promoter to Pi absence is high enough to discriminate among Phi and Pi; in other words, the promoter detects specifically the scarcity of the ion orthophosphate as a signal even in the presence of Phi. However, more and detailed experiments should be performed in the future to fully confirm or discard such hypothesis.

Moreover, we found that the induction of transcription in response to Pi scarcity, driven by the *3XM-EZ2* promoter, can be reversed by resupplying Pi to the medium. Therefore, our designed promoter behaves as a switch that can be turned On or Off in function of the Pi concentration in the medium. Such features of the *3XM-EZ2* promoter prompted us to construct a bidirectional version of the promoter by fusing two unidirectional minimal promoters in opposite direction driven by *3XM-EZ2* cis-regulatory elements (*3XM-EZ2Bi*). Our results demonstrated that *3XM-EZ2Bi* was able to efficiently induce the transcription of two reporter transgenes in Arabidopsis seedlings in response to Pi availability. In the past years, the need for simultaneous multiple gene expression for biotechnological application in plants has led to the generation of bidirectional promoters base on plant cis-regulatory regions (Chaturvedi *et al.*, 2006; Frey *et al.*, 2001; Li *et al.*, 2004; Mitra *et al.*, 2009; Xie *et al.*, 2001). The latest attempts have improved the design of bidirectional promoters avoiding gene silencing, and this is the case of a light-induced 'natural' bidirectional promoter of Arabidopsis, which was used to express simultaneously two transgenes (Mitra *et al.*, 2009). Although bidirectional promoters occur naturally in plants, very few have been shown to activate the expression of two genes by the same environmental stimuli and with a similar tissue-specific pattern of expression. Besides Mitra *et al.* (2009) work, to date no natural or designed bidirectional promoters triggered by environmental factors have been reported. To our knowledge, this study describes the first designed bidirectional promoter which can be turned on by phosphate deficiency. Moreover, it behaves in a reversible manner. Therefore, the *3XM-EZ2Bi* design represents a novel tool for multiple gene expression modulated first by Pi availability and second by the nature of the opposite promoters used. The fact that *3XM-EZ2* is able to turn On and Off in a high sensitive fashion, depending on the Pi concentrations in plant roots, points to very promising potential uses of this tool not only for *in vitro* research and biotechnology applications, but also for crop improvement approaches to allow plants grow in soils with low phosphate amounts and other stresses, as the bidirectionality of the promoter will allow the expression of genes to cope with more than one abiotic stress.

Experimental procedures

Plant material and growth conditions

Arabidopsis thaliana transgenic lines were generated in Col0 ecotype background. Seeds were disinfected with 20% (v/v) bleach in water, followed by several rinses of sterile distilled water. The medium used for germination and all other experiments was MS 0.1X supplemented with NaH_2PO_4 (1 mM) or without NaH_2PO_4 (0 mM), 0.5% sucrose and 10 g/L agar.

Protein extraction and fluorometric GUS assays

GUS activity was determined in plantlets grown 10dag under limiting (0 mM) or sufficient (1 mM) phosphate. The plantlets were ground in protein extraction buffer containing 50 mM KPO_4, pH 7.0, 10 mM EDTA, 0.1% Triton X-100, 0.1% Sarkosyl and 10 mM β-mercaptoethanol, and protein was quantified by Bio-Rad Protein Assay. One μg of plant protein extract from the phosphate-limiting or phosphate-sufficient conditions was incubated with 2 mM of 4-methylumbelliferyl-b-D-glucuronide in protein extraction buffer for 90 min and 240 min, respectively. GUS activity was measured fluorometrically using a high-performance multilabel plate reader TECAN Infinite M1000.

Histochemical GUS

Seedlings grown for 10 dag on solid medium with (1 mM) or without (0 mM) phosphate were incubated in GUS reaction buffer (0.5 mg/mL of 5-bromo-4-chloro-3-indolyl-b-D-glucuronide in 100 mM sodium phosphate, pH 7.0) overnight at 37 °C. After, tissues were cleared (Malamy and Benfey, 1997) and representative plants were photographed using Nomarski optics in a Leica DMR microscope.

Real-time quantitative analysis

Plants carrying the *pPLDZ2-GUS::GFP* and *3XM-EZ2-GUS::GFP* were grown for 7 dag in Pi-sufficient medium and then transferred to liquid medium without Pi for 12, 24, 48, 72 and 96 h, or liquid medium with Phi (1 mM); plants were then collected, frozen and ground to isolate total RNA using the TRIzol reagent method (Invitrogen). The same lines were grown in Pi-limiting medium and then transferred to Pi-sufficient liquid medium for 0.5, 1, 3, 6 and 12 h. cDNA was synthesized with SuperScript III reverse transcriptase (Invitrogen) using 30 μg of total RNA for each sample. The qPCR was performed in a Real-time PCR ABI PRISM 7500 sequence detection system (Applied Biosystems), using SYBR Green PCR Master Mix (Applied Biosystems) and specific primers (Table S2). The PCR conditions were as follows: 10 min at 95 °C and 40 cycles at 95 °C for 30 s, 60 °C for 30 s and 72 °C for 40 s. Relative transcript abundance was determined using the *ACTIN2* transcript as control. At least three independent PCRs were performed for each sample. The same procedure for qRT-PCR assay was used to analyse plants carrying the *3XM-EZ2Bi* grown in medium P+ (1 mM) for 7 dag and transferred for 72 h to P− (0 mM) liquid medium.

Semiquantitative RT-PCR assay

RNA was isolated from plants carrying the *EZ2-GUS::GFP*, *pPLDZ2-GUS::GFP*, *1XM-EZ2-GUS::GFP*, *2XM-EZ2-GUS::GFP* and *3XM-EZ2-GUS::GFP* constructs. Plants were subjected to transfer experiments from Pi sufficient to Pi limiting, and Pi limiting to Pi sufficient as in real-time assay. cDNA synthesis was performed using 100 ng of total RNA with SuperScript III (Invitrogen) following the manufacturer instructions. The PCR amplification conditions for *GUS* transcripts were 94 °C for 3 min and 26 cycles at 94 °C for 30 s, 58 °C for 30 s and a final extension step at 72 °C for 40 s. PCR amplification for *ACTIN2* were 94 °C for 3 min and 25 cycles of 94 °C for 30 s, 60 °C for 30 s and 72 °C for 30 s.

The *1X*, *2X* and *3XM-EZ2* enhancer element construct

We used the previously reported construct *EZ2P1BS4(2X)-GUS::GFP* in plasmid pKGWFS7 (Oropeza-Aburto *et al.*, 2012) as a backbone to generate new constructs containing two and three times this modified enhancer element (*M-EZ2*) driving *GUS-GFP* expression.

We designed single-stranded oligonucleotides containing the sequence of the DNA chain with one (*M-EZ2*) or two modified enhancer elements and *Hind* III in both extreme sites (Table S2). We then synthesized the complementary chain with the DNA polymerase Klenow Fragment using a reverse specific primer (Table S2). These double-stranded fragments were restricted with *Hind* III and cloned in the *Hind* III site of the backbone plasmid pKGWFS7 in which the modified enhancer element *EZ2P1BS4 (2X)-GUS::GFP* had been cloned previously (Oropeza-Aburto *et al.*, 2012).

These constructs were introduced in *Agrobacterium tumefaciens* and then used to transform *Arabidopsis thaliana* by the floral dip method (Martinez-Trujillo *et al.*, 2004).

The *3XM-EZ2* bidirectional hygromycin and GUS-GFP construct

A bidirectional enhancer *3XM-EZ2* containing the *EZ2* enhancer element three times, the −46 minimum promoter of 35S at the 3′ and the minimum promoter of the nopaline synthase gene in opposite direction at the 5′ end, was synthesized (Figure S1).

This fragment carrying the *3XM-EZ2* bidirectional enhancer was cloned by recombination in pDONR221 plasmid using a GATEWAY BP kit (Invitrogen), and then subcloned in the destiny plasmid pBGWFS7 using the GATEWAY LR kit (Invitrogen).

A DNA fragment containing the hygromycin B phosphotransferase gene and the NOS terminator was obtained by XbaI restriction from plasmid pWRG1515 and isolated. This fragment was cloned in the unique Xba I site of plasmid pBGWFS7 carrying the *3XM- EZ2* bidirectional enhancer, verifying that it had been cloned in the correct orientation.

This construction was introduced in *Agrobacterium tumefaciens* by electroporation, and this strain was used for Arabidopsis transformation by the floral dip method (Martinez-Trujillo *et al.*, 2004).

Acknowledgements

This work was supported by the HHMI grant 55007646 to LHE.

References

Acevedo-Hernández, G., Oropeza-Aburto, A. and Herrera-Estrella, L. (2012) A specific variant of the PHR1 binding site is highly enriched in the Arabidopsis phosphate-responsive phospholipase DZ2 coexpression network. *Plant Signaling & Behavior*, **7**, 914–917.

Baker, A., Ceasar, S.A., Palmer, A.J., Paterson, J.B., Qi, W., Muench, S.P. and Baldwin, S.A. (2015) Replace, reuse, recycle: improving the sustainable use of phosphorus by plants. *J. Exp. Bot.* **66**, 3523–3540.

Banerjee, J., Sahoo, D.K., Dey, N., Houtz, R.L. and Maiti, I.B. (2013) An intergenic region shared by At4g35985 and At4g35987 in Arabidopsis thaliana is a tissue specific and stress inducible bidirectional promoter analyzed in transgenic Arabidopsis and tobacco plants. *PLoS ONE*, **8**, e79622.

Bari, R., Pant, B.D., Stitt, M. and Scheible, W.-R. (2006) PHO2, microRNA399, and PHR1 define a phosphate-signaling pathway in plants. *Plant Physiol.* **141**, 988–999.

Bustos, R., Castrillo, G., Linhares, F., Puga, M.I., Rubio, V., Pérez-Pérez, J., Solano, R. *et al.* (2010) A central regulatory system largely controls transcriptional activation and repression responses to phosphate starvation in Arabidopsis. *PLoS Genet.* **6**, e1001102.

Chaturvedi, C.P., Sawant, S.V., Kiran, K., Mehrotra, R., Lodhi, N., Ansari, S.A. and Tuli, R. (2006) Analysis of polarity in the expression from a multifactorial bidirectional promoter designed for high-level expression of transgenes in plants. *J. Biotechnol.* **123**, 1–12.

Chiou, T.-J. and Lin, S.-I. (2011) Signaling network in sensing phosphate availability in plants. *Annu. Rev. Plant Biol.* **62**, 185–206.

Danova-Alt, R., Dijkema, C., De Waard, P. and Koeck, M. (2008) Transport and compartmentation of phosphite in higher plant cells–kinetic and 31P nuclear magnetic resonance studies. *Plant, Cell Environ.* **31**, 1510–1521.

Dhadi, S.R., Krom, N. and Ramakrishna, W. (2009) Genome-wide comparative analysis of putative bidirectional promoters from rice, Arabidopsis and Populus. *Gene*, **429**, 65–73.

Franco-Zorrilla, J.M., González, E., Bustos, R., Linhares, F., Leyva, A. and Paz-Ares, J. (2004) The transcriptional control of plant responses to phosphate limitation. *J. Exp. Bot.* **55**, 285–293.

Frey, P.M., Schärer-Hernández, N.G., Fütterer, J., Potrykus, I. and Puonti-Kaerlas, J. (2001) Simultaneous analysis of the bidirectional African cassava mosaic virus promoter activity using two different luciferase genes. *Virus Genes*, **22**, 231–242.

Jost, R., Pharmawati, M., Lapis-Gaza, H.R., Rossig, C., Berkowitz, O., Lambers, H. and Finnegan, P.M. (2015) Differentiating phosphate-dependent and phosphate-independent systemic phosphate-starvation response networks in Arabidopsis thaliana through the application of phosphite. *J. Exp. Bot.* **66**, 2501–2514.

Li, Z.T., Jayasankar, S. and Gray, D. (2004) Bi-directional duplex promoters with duplicated enhancers significantly increase transgene expression in grape and tobacco. *Transgenic Res.* **13**, 143–154.

Lin, W.-Y., Lin, S.-I. and Chiou, T.-J. (2009) Molecular regulators of phosphate homeostasis in plants. *J. Exp. Bot.* **60**, 1427–1438.

Liu, X. and Han, B. (2009) Evolutionary conservation of neighbouring gene pairs in plants. *Gene*, **437**, 71–79.

Lundmark, M., Kørner, C.J. and Nielsen, T.H. (2010) Global analysis of microRNA in Arabidopsis in response to phosphate starvation as studied by locked nucleic acid-based microarrays. *Physiol. Plant.* **140**, 57–68.

Lynch, J. (1995) Root architecture and plant productivity. *Plant Physiol.* **109**, 7.

Malamy, J.E. and Benfey, P.N. (1997) Organization and cell differentiation in lateral roots of Arabidopsis thaliana. *Development*, **124**, 33–44.

Martinez-Trujillo, M., Limones-Briones, V., Cabrera-Ponce, J.L. and Herrera-Estrella, L. (2004) Improving transformation efficiency of Arabidopsis thaliana by modifying the floral dip method. *Plant Mol. Biol. Rep.* **22**, 63–70.

Mehrotra, R., Gupta, G., Sethi, R., Bhalothia, P., Kumar, N. and Mehrotra, S. (2011) Designer promoter: an artwork of cis engineering. *Plant Mol. Biol.* **75**, 527–536.

Mitra, A., Han, J., Zhang, Z.J. and Mitra, A. (2009) The intergenic region of Arabidopsis thaliana cab1 and cab2 divergent genes functions as a bidirectional promoter. *Planta*, **229**, 1015–1022.

Müller, R., Morant, M., Jarmer, H., Nilsson, L. and Nielsen, T.H. (2007) Genome-wide analysis of the Arabidopsis leaf transcriptome reveals interaction of phosphate and sugar metabolism. *Plant Physiol.* **143**, 156–171.

Nilsson, L., Müller, R. and Nielsen, T.H. (2007) Increased expression of the MYB-related transcription factor, PHR1, leads to enhanced phosphate uptake in Arabidopsis thaliana. *Plant, Cell Environ.* **30**, 1499–1512.

Oropeza-Aburto, A., Cruz-Ramírez, A., Acevedo-Hernández, G.J., Pérez-Torres, C.-A., Caballero-Pérez, J. and Herrera-Estrella, L. (2012) Functional analysis of the Arabidopsis PLDZ2 promoter reveals an evolutionarily conserved low-Pi-responsive transcriptional enhancer element. *J. Exp. Bot.* **63**, 2189–2202.

Plaxton, W.C. (2004) *Plant Response to Stress: Biochemical Adaptations to Phosphate Deficiency. Encyclopedia of Plant and Crop Science*, pp. 976–980. New York: Marcel Dekker.

Puga, M.I., Mateos, I., Charukesi, R., Wang, Z., Franco-Zorrilla, J.M., de Lorenzo, L., Irigoyen, M.L. *et al.* (2014) SPX1 is a phosphate-dependent inhibitor of PHOSPHATE STARVATION RESPONSE 1 in Arabidopsis. *Proc. Natl. Acad. Sci.* **111**, 14947–14952.

Raghothama, K. (1999) Phosphate acquisition. *Annu. Rev. Plant Biol.* **50**, 665–693.

Rouached, H., Arpat, A.B. and Poirier, Y. (2010) Regulation of phosphate starvation responses in plants: signaling players and cross-talks. *Molecul. Plant*, **3**, 288–299.

Ruan, W., Guo, M., Cai, L., Hu, H., Li, C., Liu, Y., Wu, Z. *et al.* (2015) Genetic manipulation of a high-affinity PHR1 target cis-element to improve phosphorous uptake in Oryza sativa L. *Plant Mol. Biol.* **87**, 429–440.

Rubio, V., Linhares, F., Solano, R., Martín, A.C., Iglesias, J., Leyva, A. and Paz-Ares, J. (2001) A conserved MYB transcription factor involved in phosphate starvation signaling both in vascular plants and in unicellular algae. *Genes Dev.* **15**, 2122–2133.

Shen, J., Yuan, L., Zhang, J., Li, H., Bai, Z., Chen, X., Zhang, W. *et al.* (2011) Phosphorus dynamics: from soil to plant. *Plant Physiol.* **156**, 997–1005.

Sobkowiak, L., Bielewicz, D., Malecka, E., Jakobsen, I., Albrechtsen, M., Szweykowska-Kulinska, Z. and Pacak, A.M. (2012) The role of the P1BS element containing promoter-driven genes in Pi transport and homeostasis in plants. *Front. Plant Sci.* doi: 10.3389/fpls.2012.00058.

Ticconi, C.A., Delatorre, C.A. and Abel, S. (2001) Attenuation of phosphate starvation responses by phosphite in Arabidopsis. *Plant Physiol.* **127**, 963–972.

Vance, C.P., Uhde-Stone, C. and Allan, D.L. (2003) Phosphorus acquisition and use: critical adaptations by plants for securing a nonrenewable resource. *New Phytol.* **157**, 423–447.

Varadarajan, D.K., Karthikeyan, A.S., Matilda, P.D. and Raghothama, K.G. (2002) Phosphite, an analog of phosphate, suppresses the coordinated expression of genes under phosphate starvation. *Plant Physiol.* **129**, 1232–1240.

Venter, M. (2007) Synthetic promoters: genetic control through cis engineering. *Trends Plant Sci.* **12**, 118–124.

Wang, Q., Wan, L., Li, D., Zhu, L., Qian, M. and Deng, M. (2009) Searching for bidirectional promoters in Arabidopsis thaliana. *BMC Bioinform.* **10**, 1.

Williamson, L.C., Ribrioux, S.P., Fitter, A.H. and Leyser, H.O. (2001) Phosphate availability regulates root system architecture in Arabidopsis. *Plant Physiol.* **126**, 875–882.

Wykoff, D.D., Grossman, A.R., Weeks, D.P., Usuda, H. and Shimogawara, K. (1999) Psr1, a nuclear localized protein that regulates phosphorus metabolism in Chlamydomonas. *Proc. Natl. Acad. Sci.* **96**, 15336–15341.

Xie, M., He, Y. and Gan, S. (2001) Bidirectionalization of polar promoters in plants. *Nat. Biotechnol.* **19**, 677–679.

2

The genetic architecture of amino acids dissection by association and linkage analysis in maize

Min Deng, Dongqin Li, Jingyun Luo, Yingjie Xiao, Haijun Liu, Qingchun Pan, Xuehai Zhang, Minliang Jin, Mingchao Zhao and Jianbing Yan*

National Key Laboratory of Crop Genetic Improvement, Huazhong Agricultural University, Wuhan, China

*Correspondence
email yjianbing@mail.hzau.edu.cn

Summary

Amino acids are both constituents of proteins, providing the essential nutrition for humans and animals, and signalling molecules regulating the growth and development of plants. Most cultivars of maize are deficient in essential amino acids such as lysine and tryptophan. Here, we measured the levels of 17 different total amino acids, and created 48 derived traits in mature kernels from a maize diversity inbred collection and three recombinant inbred line (RIL) populations. By GWAS, 247 and 281 significant loci were identified in two different environments, 5.1 and 4.4 loci for each trait, explaining 7.44% and 7.90% phenotypic variation for each locus in average, respectively. By linkage mapping, 89, 150 and 165 QTLs were identified in B73/By804, Kui3/B77 and Zong3/Yu87-1 RIL populations, 2.0, 2.7 and 2.8 QTLs for each trait, explaining 13.6%, 16.4% and 21.4% phenotypic variation for each QTL in average, respectively. It implies that the genetic architecture of amino acids is relative simple and controlled by limited loci. About 43.2% of the loci identified by GWAS were verified by expression QTL, and 17 loci overlapped with mapped QTLs in the three RIL populations. GRMZM2G015534, GRMZM2G143008 and one QTL were further validated using molecular approaches. The amino acid biosynthetic and catabolic pathways were reconstructed on the basis of candidate genes proposed in this study. Our results provide insights into the genetic basis of amino acid biosynthesis in maize kernels and may facilitate marker-based breeding for quality protein maize.

Keywords: amino acid, Quality Protein Maize (QPM), co-expression network, GWAS, linkage mapping, metabolism.

Introduction

Maize (*Zea mays*) is one of the most widely grown crops worldwide. It is not only a staple food for people and animals, but also an important industrial material for fuel and other applications. Typically, the maize endosperm is ~10% protein, and seed storage proteins supply nitrogen for the germinating seedling and are also an important protein source for humans and animals. The amino acid composition and quantity of seed storage proteins are related to the nutritional quality of seeds (Mandal and Mandal, 2000; Young and Pellett, 1994). However, the maize cultivars widely planted usually have insufficient levels of essential amino acids, such as lysine and tryptophan (Misra *et al.*, 1972). In order to facilitate breeding for balanced amino acid composition, it is important to identify the genes controlling amino acid content in the maize kernel.

Although more than 180 amino acids have been discovered in nature, only 20 amino acids constitute proteins. Many amino acids, such as homoserine, homocysteine, ornithine and citrulline, play important roles in growth and development (Dunlop *et al.*, 2015), defence against insect herbivores (Huang *et al.*, 2011). Amino acids are also important signalling molecules regulating several signal pathways related to the growth and development of both animals and plants. Some studies have found that aspartate plays an important role in human cell proliferation (Birsoy *et al.*, 2015; Sullivan *et al.*, 2015). Proline could maintain cellular osmotic homoeostasis, as well as redox balance and

energy status (Krishnan *et al.*, 2008). Proline also may function as a molecular chaperone to protect proteins from denaturation (Mishra and Dubey, 2006; Sharma and Dubey, 2005), an antioxidant to scavenge ROS, a singlet oxygen quencher (Matysik *et al.*, 2002; Smirnoff and Cumbes, 1989), or a regulator of the cell cycle in maize (Wang *et al.*, 2014).

The amino acid metabolism pathways, including biosynthesis, degradation and regulation, are well studied in microorganisms (Miflin and Lea, 1977; Umbarger, 1969, 1978). Studies of the model plant *Arabidopsis thaliana* have focused on the roles of amino acids in nitrogen nutrition (Crawford and Forde, 2002), N-assimilation (Coruzzi, 2003), metabolism and regulation (Hell and Wirtz, 2011; Ingle, 2011; Jander and Joshi, 2009; Tzin and Galili, 2010a; Verslues and Sharma, 2010). Some key genes regulating free amino acid content have been identified in *Arabidopsis* (Angelovici *et al.*, 2013), tobacco (Maloney *et al.*, 2010), soya bean (Ishimoto *et al.*, 2010; Takahashi *et al.*, 2003), rapeseed (Moulin *et al.*, 2000, 2006), rice (Kang *et al.*, 2005; Zhou *et al.*, 2009) and maize (Mertz *et al.*, 1964; Muehlbauer *et al.*, 1994; Shaver *et al.*, 1996; Wang *et al.*, 2001, 2007). Opaque2 (O2) is an endosperm-specific transcription factor belonging to the bZIP family, whose mutation could increase free lysine levels and enhance the overall nutritional value of grain by reducing the 22-kD α- and β-zein transcripts and proteins in maize (Hunter *et al.*, 2002; Kodrzycki *et al.*, 1989; Mertz *et al.*, 1964). Due to the lysine content in *o2* mutant maize kernels being 70% higher than wild type, it has become a subject of intense research over the past several decades (Wu and Messing, 2014). However, the *o2* gene has not been widely used for breeding high-nutrition maize lines because its pleiotropic effects are negatively associated with agronomic performance (Loesch *et al.*, 1976; Nass and Crane, 1970; Zhang *et al.*, 2016). Identification of more favourable genes and increasing the understanding of the underlying amino acid biosynthetic pathways are the key steps for breeding maize with high-quality protein (Ufaz and Galili, 2008).

With the rapid development of DNA and RNA-sequencing technologies, high-density genotyping with SNPs became easily accessible, enabling genomewide association studies (GWAS). This method became a powerful tool for complex trait dissection in plants (Xiao *et al.*, 2016; Yan *et al.*, 2011). Many GWAS were performed in plants including maize (Li *et al.*, 2013; Xiao *et al.*, 2016), rice (Huang *et al.*, 2010, 2012), canola (Liu *et al.*, 2016d; Luo *et al.*, 2015), sorghum (Morris *et al.*, 2013), foxtail millet (Jia *et al.*, 2013), *Arabidopsis* (Atwell *et al.*, 2010) and others. Recently, the expression data of 28 769 genes and 1.03 million high-quality SNPs were obtained by deep RNA-sequencing of the immature seeds at 15 days after pollination of 368 diverse maize inbred lines (Fu *et al.*, 2013). These data were used for studies of maize quality traits, including oil concentration (Li *et al.*, 2013), vitamin E content (Li *et al.*, 2012b) and metabolites (Wen *et al.*, 2014). They provide a valuable resource for studying the genetic architecture of maize quantitative traits.

To better understand the genetic components underlying the natural variation and the metabolism of amino acids in the maize kernel, we used an automatic amino acid analyser to quantify the total amino acids of mature maize kernel from a diversity association panel of 513 lines (Yang *et al.*, 2011, 2014) and three RIL populations (Pan *et al.*, 2016). GWAS and linkage mapping were combined to dissect the genetic architecture of amino acids in the maize kernel. Many previously known and unknown genes directly or indirectly involved in amino acid metabolism were identified, which has helped to ascertain the amino acid metabolism network. Some of the candidate genes were validated by multiple approaches, including expression QTL mapping, QTL fine mapping, bioinformatics, and further confirmed by genetic transformation. These results provide new insights for understanding amino acid biosynthesis and thus enhancing the breeding of high-nutrition maize.

Results

Natural variation of amino acids in maize kernel

Using an automatic amino acid analyzer L-8800 (L-8800, Hitachi Instruments Engineering, Tokyo, Japan), we assessed the variation in total amino acid content in dry matured maize kernels, which included an association panel (513 inbred lines) harvested from two environments and three RIL populations (169, 152, 146 lines for B73/BY804 (BB), KUI3/B77 (KB) and ZONG3/YU87-1 (ZY), respectively). The concentrations of seventeen amino acids (Ala, Arg, Asx, Glx, Gly, Lle, Leu, Lys, Met, Pro, Phe, Val, Tyr, His, Cys, Thr and Ser in mg/g dry maize kernel) and total amino acid concentration (sum of the seventeen amino acids) were calculated. Forty-seven derived compositional traits were also calculated (detailed in methods). The level of each amino acid-related trait varied widely in both the association panel and three RIL populations (Figure S1). Variation ranged from a 1.2-fold difference in Phe/PT to 14.9-fold difference in Cys/Total, and 1.1-fold difference in GT/Total and Glx/GT to 5.7-fold difference in Met/Total in association and linkage mapping populations,

respectively (Tables S1, S2). For the average total lysine content, the maximum ratio of 3.1-fold difference was found in the KB population (1.72–5.37 mg/g). The skewness, kurtosis and other detailed information for each amino acid are shown in Tables S1 and S2.

Loci associated with amino acid content identified by GWAS and linkage mapping

GWAS was performed using an association panel including 513 maize diverse inbred lines (Yang *et al.*, 2011, 2014) and 1.25 million high-quality single nucleotide polymorphisms (SNPs) with minor allele frequency (MAF) >0.05 (Fu *et al.*, 2013; Liu *et al.*, 2016a). In total, 247 and 281 associated loci were identified in AM1 and AM2 at $P \leq 2.04 \times 10^{-6}$, with an average of 5.1 and 4.4 loci for each trait, respectively (Table 1, Figure S2, Table S3). The phenotypic variation explained by each locus for each amino acid trait ranged from 5.21% (Ala/AT in AM2) to 19.74% (Leu/Total in AM1), with an average of 7.44% for AM1 and 7.90% for AM2 (Figure S3, Table S3). Ten loci with effects greater than 15% were identified in two environments. For each trait, the total phenotypic variation explained by all the identified loci was 23.3% (ranged from 5.6% to 66.3%) and 19.3% (ranged from 5.4% to 49.5%) in AM1 and AM2, respectively.

Three RIL populations (BB, KB and ZY) were genotyped with high-density SNP array (Pan *et al.*, 2016) and were used for QTL mapping for the amino acid traits. At least one QTL was identified for 45, 56, 59 among 65 measured traits in BB, KB and ZY RIL populations, respectively. In total, 89, 150, and 165 QTLs were identified for BB, KB, and ZY populations with an average of 2.0, 2.7 and 2.8 QTLs for each trait, respectively (Table 1, Figure S2, Table S4). For the same trait, only 15 QTLs were detected in more than one population, implying that different low-frequency QTL existed in different genetic backgrounds (Xiao *et al.*, 2016). Each QTL explained the phenotypic variation of 6.40%–14.88% (BB), 3.42%–16.96% (KB), and 5.87%–23.32% (ZY), with an average of 9.03%, 9.39% and 10.15%, respectively (Figure S3, Table S4). Thirteen QTLs with effects greater than 15% were identified in the three RIL populations. For each trait, all the identified QTLs on average explained 13.6% (ranged from 7.2% to 32.6%), 16.4% (ranged from 4.9% to 32.4%) and 21.4% (ranged from 8.5% to 49.9%) of the total phenotypic variance in BB, KB and ZY RIL population, respectively.

Table 1 Summary of significant loci–trait associations identified by GWAS and QTL by linkage mapping

Population*	BB	KB	ZY	AM1	AM2
Number of Traits with QTL[†]	45	56	59	48	64
Number of Loci[‡]	89	150	165	247	281
Average loci per trait[§]	2.0 ± 1.2	2.7 ± 1.5	2.8 ± 1.6	5.1 ± 6.9	4.4 ± 3.3

*BB, KB, ZY represent three linkage populations B73/By804, Kui3/B77, Zong3/Yu87-1, respectively; AM1, AM2 represent the two environments.

[†]Number of traits with QTLs identified. 65 amino acids traits were analysed in each population.

[‡]Number of significant loci detected on the association panel ($P \leq 2.04 \times 10^{-6}$, MLM) and a uniform threshold for significant QTLs was determined by 500 permutations ($P = 0.05$).

[§]Average number of significant loci (or QTL) detected per trait ± S.D.

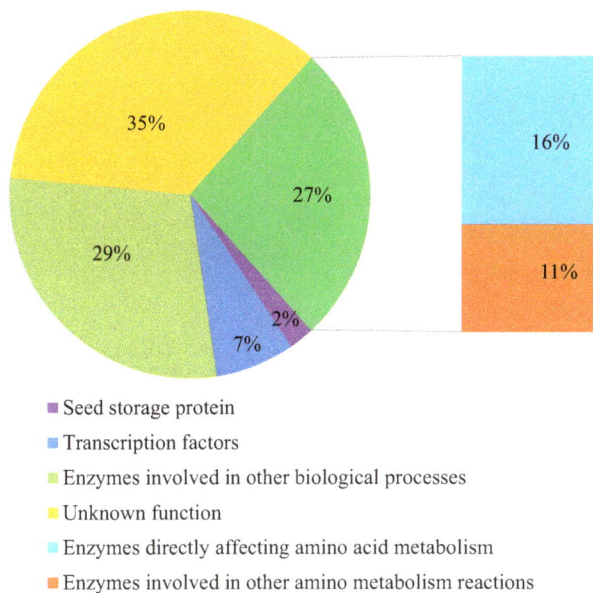

Figure 1 Functional category annotations for 308 candidate genes and their respective percentages identified via GWAS as significantly associated with amino acid traits in maize kernels.

Candidate genes and QTL hotspots

Subsequently, limited overlaps were found between the loci (17/528) identified by GWAS and the QTLs identified by linkage mapping for the same trait in the present study. A total of 308 unique candidate genes corresponding to 528 trait–locus associations identified in two experiments were annotated, and other potential candidate genes within 200 kb (100 kb upstream and downstream of the lead SNPs) of the 528 loci were also listed in Table S3. Among the candidate genes, those encoding enzymes or other protein directly or indirectly affecting amino acid metabolism accounted for 27%, the enzymes involved in other biological processes accounted for 29%, and the functions were unknown for 35%, based on the current database (Figure 1). Gene Ontology (GO) term analysis revealed significant enrichment in terms relating to cellular nitrogen metabolism, amine metabolism, amino acid and derivative metabolism, organic acids and other processes (Figure S4). Expression QTLs (eQTL, $n = 368$) were identified for a plurality of these candidate genes (43.2%, or 133/308) using the previous RNA-sequencing data of immature kernels (Fu *et al.*, 2013). Significant correlations ($P < 0.05$, $n = 295–326$) between the expression level of the candidate genes with eQTLs identified and the phenotypic variation of the corresponding amino acid traits were found in 50 cases (16.2%) (Table S3), which suggests that some of these loci affect phenotypic variation via transcriptional regulation.

QTLs were not distributed evenly on the chromosomes, based on 1000-time permutation tests at the level of 0.05, and eight QTL hotspots were observed on chromosomes 1, 3, 7, 8 (Figure 2, Tables S3, S4). These QTLs were often shared by biologically related amino acids. For example, the QTLs affecting Leu, Val, and Ile contents or derived traits were enriched on chromosome 7. The candidate genes underlying these QTL hotspots could include regulators of the metabolic pathway, and influence the rate-limiting reactions. Interestingly, two QTL

hotspots (on chromosome 3 and 7) overlapped with the metabolite QTL hotspot identified in a previous study using three different tissues from the BB population in (Wen *et al.*, 2015), which helps identify the underlying genes and their regulating pathway.

Amino acid metabolic network involving identified genes and their co-expression genes

We reconstructed a maize amino acid metabolism network based on the published results in *Arabidopsis* (Coruzzi, 2003; Hell and Wirtz, 2011; Ingle, 2011; Jander and Joshi, 2009; Tzin and Galili, 2010a,b; Verslues and Sharma, 2010) and data obtained from this study. Notably, 23 candidate genes involved in amino acid anabolism and catabolism were identified by GWAS (Figure 3, Table 2). Five of 23 genes have been reported previously in maize, including isocitrate dehydrogenase (*IDH*) (Curry and Ting, 1976; Zhang *et al.*, 2010), phenylalanine ammonia-lyase (*PAL*) (Havir, 1971) tryptophan synthase (*TS*) (Wright *et al.*, 1992), asparagine synthase (*AS*) (Chevalier *et al.*, 1996; Schmidt *et al.*, 1987) and aconitate hydratase (*ACO*) (Wendel *et al.*, 1988). The remaining candidate genes identified in this study may be involved in amino acid biosynthetic pathways, based on the available database annotation and comparative genomic approaches although the functions have not been fully explored in maize (Table 2).

A Pearson correlation was calculated between the expression level of the 23 candidate genes (source genes) and 28 769 genes analysed by RNA-sequencing from immature kernels (Fu *et al.*, 2013). A total of 6641 directed edges connected 14 of the 23 source genes (big red nodes) and were involved in 4670 target genes ($P \leq 1 \times 10^{-20}$, $r \geq 0.5$, Figure 4). Among these 4670 genes, 49 genes (including five source genes) were identified by GWAS (big yellow nodes) as well. Another 140 annotated genes (big green nodes), including 33 transcription factors (big blue nodes), were identified to be directly or indirectly associated with amino acid metabolism. GO term analysis of the 4670 co-expressed genes revealed significant enrichment in terms relating to metabolism, including amine metabolism, cellular processes, developmental processes and biological regulation (Figure S5, Table S5). In addition, we found that four candidate genes (GRMZM2G147191, GRMZM2G009808, GRMZM2G119482, GRMZM2G178826) were related in glycolytic pathway and TCA cycle based their annotation in this co-expression network (Figure 4).

Functional validation of candidate genes

A strong signal ($P = 1.05 \times 10^{-8}$, $n = 393$) was identified on the short arm of chromosome 7 (Figure 5a), associated with Lys/Total, which could explain 8.5% of the phenotypic variation. The *O2* (GRMZM2G015534) gene is located about 98Kb downstream of the lead SNP chr7.S_10695002 (Figure 5b–d). *O2* is a bZIP transcription factor that regulates the expression of various genes during maize kernel development, particularly abundant endosperm storage protein genes like encoding the 22-kD α-and β-zein (Li *et al.*, 2015). The lead SNP was strongly associated with the *O2* expression level ($P = 2.25 \times 10^{-10}$, $R^2 = 11.96\%$, $n = 318$) and phenotypic trait ($P = 2.92 \times 10^{-17}$, $R^2 = 16.71\%$, $n = 393$). Subsequently, a strong *cis*-eQTL was detected for *O2* ($P = 1.04 \times 10^{-10}$, MLM, $n = 368$, Figure 5e), and the expression level of *O2* was significantly negatively correlated with the level of Lys/Total ratio ($r = -0.448$, $P = 2.24 \times 10^{-15}$, $n = 283$, Figure 5f-g, Table S6). In addition, the significant correlations

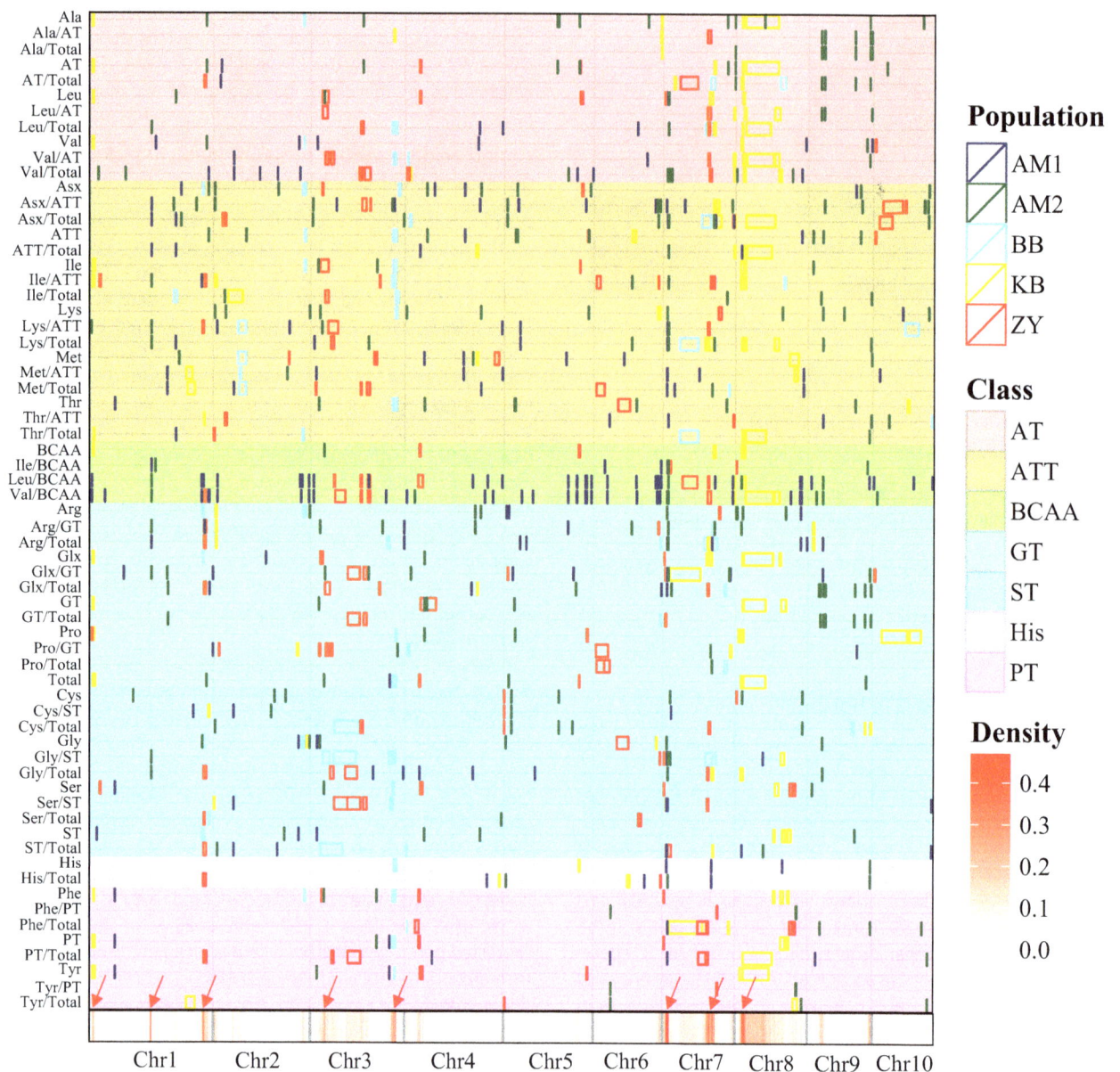

Figure 2 Chromosomal distribution of amino acids loci and QTLs identified in this study. QTL regions (represented by the confidence interval for linkage mapping and the 100 kb up- and downstream of the lead SNP for association mapping) across the maize genome responsible for amino acid levels from the different populations are shown as midnight blue (BB), green (AM1), cyan (AM2) gold (KB) and red (ZY) boxes, respectively. The class represents different amino acid families. AT, pyruvate-derived amino acid family related traits; ATT, aspartate-derived amino acid family related traits; BCAA, branched-chain amino acid family related traits; GT, glutamate-derived amino acid family related traits; PT, phenylalanine-derived amino acid family related traits; ST, serine-derived amino acid family related traits; His, histidine family related traits. The x-axis indicates the genetic positions across the maize genome in Mb. Heatmap under the x-axis illustrates the density of amino acid loci and QTLs across the genome. The red arrows show the QTL hotspots. The detailed information of all detected loci and QTLs is shown in Tables S3 and S4. Amino acid traits from different derived families are marked by distinct colours as shown on the right.

between the expression levels of *O2* and many other genes were found. The top 2% of genes (575) with the lowest *P*-value ($P < 1.0 \times 10^{-15}$) were retained for further analysis including nine genes identified by present GWAS affecting different amino acid traits (Tables S3 and S7). And 22 of 575 genes were also identified by ChIP-Seq and RNA-sequencing in *o2* mutant and wide type (Li *et al.*, 2015; Table S7). Another 40 genes involved in amino acid

metabolism were in the relevant pathways but were not detected by GWAS (Figure S6). These results confirm the importance of *O2* for regulating the amino acid biosynthesis pathway, and the novel candidate genes may help to identify the *o2* modifiers or regulators and to expand the known regulation pathway.

A major QTL on chromosome 7 (LOD = 7.38, R^2 = 14.88%) affecting Lys/Total was identified in BB RIL population (Figure 6a)

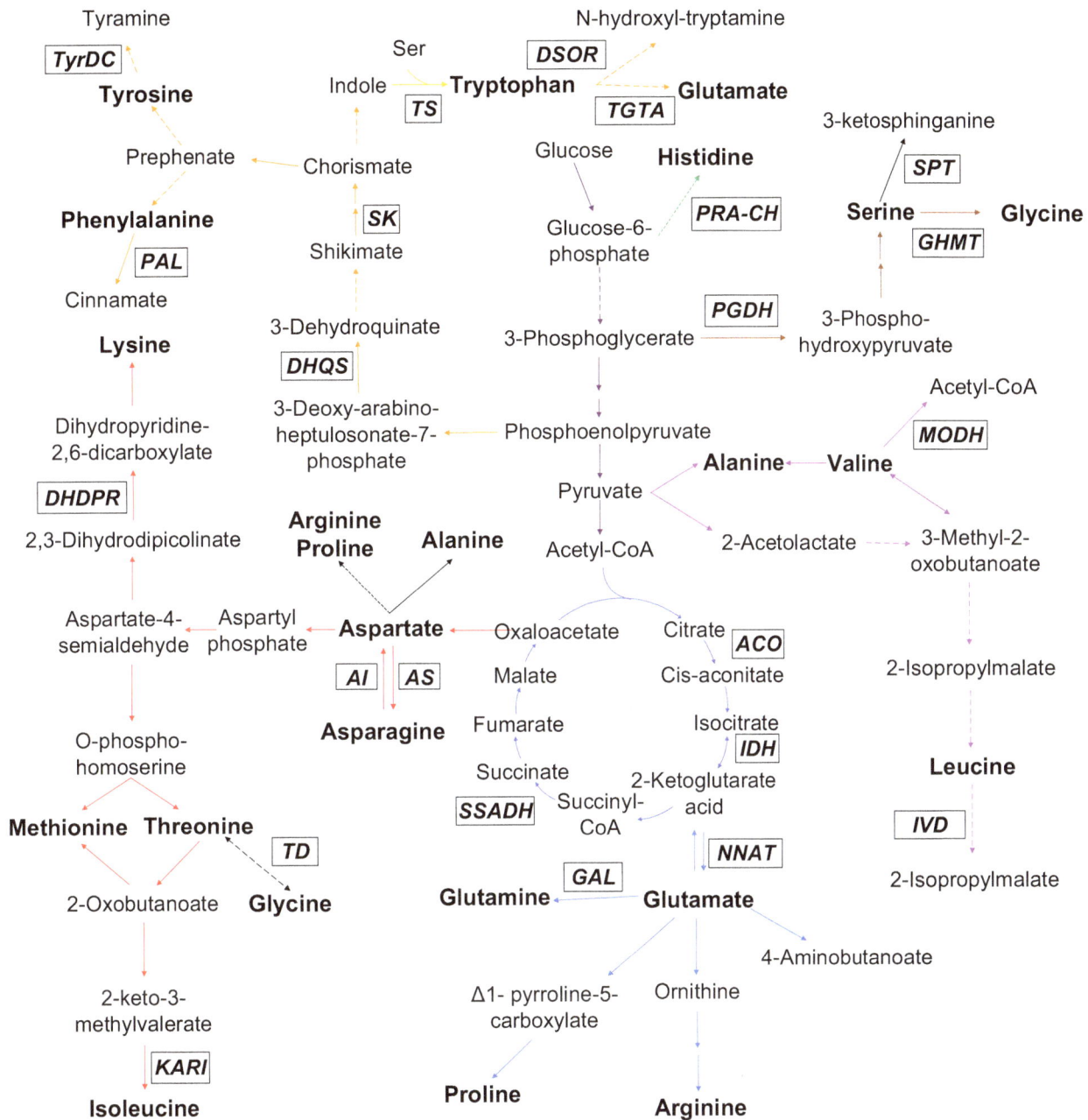

Figure 3 A maize amino acids network involving key genes identified in this study by GWAS. The different colours represent the different amino acids families. The purple, sky-blue, red, brown, dark green, orange lines represent the metabolism pathway of pyruvate-derived, glutamate-derived, aspartate-derived, serine-derived, Histidine, phenylalanine-derived amino acids, respectively. The blue lines represent the TCA cycle. Candidate genes identified in this study by GWAS are shown in the respective pathway. KARI, Ketol-acid reductoisomerase; GHMT, Glycine hydroxymethyltransferase; SPT, Serine palmitoyltransferase; PGDH, Phosphoglycerate dehydrogenase; IDH, Isocitrate dehydrogenase; TS, Tryptophan synthase; TGTA, Tryptophan Glutamate transaminase; DSOR, Disulphide oxidoreductase; MODH, 3-methyl-2-oxobutanoate dehydrogenase; IVD, Isovaleryl-CoA dehydrogenase; DHDPR, Dihydrodipicolinate reductase; TD, L-threonine 3-dehydrogenase; AS, Asparagine synthase; AI, Asparaginase; SK, Shikimate kinase; PAL, Phenylalanine ammonia-lyase; GAL, Glutamate-ammonia ligase; NNAT, Nicotianamine aminotransferase; ACO, Aconitate hydratase; SSADH, Succinate semialdehyde dehydrogenase; DHQS, 3-dehydroquinate synthase; TyrDC, Tyrosine decarboxylase; PRA-CH, Phosphoribosyl-AMP cyclohydrolase.

with a confidence interval greater than 10 cM (70.4–81.1 cM), and physical length greater than 27 Mb (102.65–129.86 Mb) (Table S4). This QTL was validated in a heterogeneous inbred family (HIF) covering the target region (Figure 6b). Four geno-typed and phenotyped progeny families were obtained, which helped to narrow the location of this QTL to a 5.7 Mb region (115.7–121.4 Mb) (Figure 6c). A GWAS signal was detected within the QTL interval located at 120.57 Mb ($P = 6.26 \times 10^{-6}$, $n = 393$, Figure 6d). Ten candidate genes were obtained within the 400Kb region around the peak including one zp27 (GRMZM 2G138727), two ARID-transcription factors (GRMZM2G138976 and GRMZM5G873335), one AP2-EREBP-transcription factor (GR

Table 2 SNPs and candidate genes significantly associated with amino acid traits and were used in the amino acids network analysis

Candidate Gene*	Lead SNP	Chromosome	Position (bp)[†]	Allele	MAF[‡]	P value[§]	R^2 (%)[¶]	P value (eQTL)[ǁ]	Correlation (Phenotype vs expression)**	Annotation
GRMZM2G373859	chr1.S_210471253	1	210471253	T/C	0.241	5.35E-08	9.59	NS		Glutamine dumper
GRMZM2G081886	chr2.S_5014818	2	5014818	C/A	0.169	1.75E-06	5.15	2.17E-11	−0.206	Phosphoglycerate dehydrogenase
GRMZM2G139463	chr2.S_20803501	2	20803501	C/T	0.201	1.34E-06	8.56	6.00E-34	−0.096	L-asparaginase
GRMZM2G118345	chr2.S_28051836	2	28051836	G/T	0.395	1.61E-06	6.70	1.76E-21	−0.151	Phenylalanine ammonia-lyase
GRMZM2G161868	chr3.S_202363139	3	202363139	A/C	0.208	7.05E-07	7.07	NS		Ketol-acid reductoisomerase
GRMZM2G006480	chr4.S_3888039	4	3888039	G/A	0.412	9.23E-07	6.95	NS		Nicotianamine aminotransferase
GRMZM2G169593	chr4.S_35917250	4	35917250	C/T	0.085	3.39E-07	6.78	NS		Tryptophan synthase
GRMZM2G090241	chr4.S_53259978	4	53259978	C/T	0.093	1.71E-06	5.96	NS		Dihydrodipicolinate reductase
GRMZM2G036464	chr4.S_167077316	4	167077316	C/T	0.084	1.11E-06	5.98	4.15E-17	−0.051	Glutamate-ammonia ligase
GRMZM2G119482	chr4.S_236213884	4	236213884	T/A	0.052	9.56E-07	11.34	NS		Succinate semialdehyde dehydrogenase
GRMZM2G381051	chr6.S_27117945	6	27117945	A/G	0.109	1.10E-06	5.74	NS		Isovaleryl-CoA dehydrogenase
GRMZM2G009400	chr6.S_158268310	6	158268310	C/G	0.083	1.71E-06	10.84	NS		Tyrosine decarboxylase
GRMZM5G829778	chr6.S_165625349	6	165625349	C/T	0.181	5.34E-07	5.91	3.23E-19	−0.114	Isocitrate dehydrogenase (NADP(+))
GRMZM2G015534	chr7.S_10695002	7	10695002	C/G	0.105	2.36E-07	7.09	1.58E-08	−0.270	Opaque 2
GRMZM2G138727	chr7.S_120252509	7	120252509	G/T	0.498	3.43E-09	8.61	1.40E-13	0.428	Glutelin-2 Precursor (27 kDa zein)
GRMZM2G082214	chr8.S_2198542	8	2198542	T/C	0.081	1.48E-06	7.42	1.78E-12	0.069	Phosphoribosyl-AMP cyclohydrolase
GRMZM2G127308	chr8.S_16853869	8	16853869	G/A	0.185	1.33E-06	8.61	NS		Tryptophan transaminase
GRMZM2G010202	chr8.S_159961615	8	159961615	G/A	0.197	5.72E-07	5.79	1.24E-20	−0.134	Serine palmitoyltransferase
GRMZM2G004824	chr9.S_31098627	9	31098627	G/A	0.454	1.80E-06	5.16	NS		Glycine hydroxymethyltransferase
GRMZM2G078472	chr9.S_137749795	9	137749795	T/C	0.051	1.41E-06	9.32	NS		Asparagine synthase (glutamine-hydrolysing)
GRMZM2G009808	chr9.S_151665914	9	151665914	G/A	0.361	1.60E-06	6.36	NS		Aconitate hydratase
GRMZM2G091819	chr10.S_16572309	10	16572309	C/G	0.116	1.86E-06	5.48	NS		Disulphide oxidoreductase
GRMZM2G147191	chr7.S_126350017	7	126350017	T/C	0.057	6.88E-07	6.527	NS		L-threonine 3-dehydrogenase
GRMZM2G139412	chr5.S_151841984	5	151841984	G/A	0.116	2.51E-07	12.338	NS		Shikimate kinase
GRMZM2G037614	chr2.S_233675625	2	233675625	G/A	0.059	9.88E-07	5.6551	5.75E-10	−0.102	3-methyl-2-oxobutanoate dehydrogenase
GRMZM2G178826	chr9.S_151366977	9	151366977	A/G	0.417	6.06E-09	9.0918	1.76E-22	0.060	3-dehydroquinate synthase

*A plausible biological candidate gene in the locus or the nearest annotated gene to the lead SNP.

[†]Position in base pairs for the lead SNP according to version 5b.60 of the maize reference sequence.

[‡]Minor allele frequency of the lead SNP.

[§]P value of the corresponding metabolic trait calculated by MLM.

[¶]The phenotypic variance explained by the corresponding locus.

[ǁ]P value of the expression QTL of the candidate gene. The P value is the lead SNP of eQTL rather than the GWAS lead SNP. NS, not significant; ND, the expression of the candidate gene is not detected. P value was calculated by MLM, the sample size $N = 368$.

**Pearson correlation between the expression amount and the phenotypic data of the corresponding metabolic trait.

MZM2G052667) and six unknown genes. GRMZM2G123018 was not detected in RNA-sequencing of 15 DAP (Fu et al., 2013) (Figure 6e, white arrow shown). eQTLs were identified for seven of the nine expressed genes (except GRMZM2G700198 and GRMZM2G003225, Figure 6e). Lys/Total was significantly correlated with the expressions of five of the seven genes (Figure 6e and Table S8) which were then considered as candidate genes. Recently, a QTL (qγ27) designated o2 modifier1 in bin 7.02 affecting the expression of 27-kDa γ-zein was cloned and co-localizes with our present locus (Liu et al., 2016b). qγ27 resulted

from a 15.26 kb duplication at the 27-kDa γ-zein locus contained four genes (GRMZM2G138727, GRMZM2G565441, GRMZM 2G138976, and GRMZM5G873335) which overlap with our proposed candidate genes (Figure 6e). We used the primer pair (0707) reported in previous study (Liu et al., 2016b) to genotype the association panel and the parents of the BB RIL population. The results showed that this duplication significantly influenced the Lys/Total level ($P = 2.97 \times 10^{-3}$, $R^2 = 2.18\%$, $n = 402$) and the expression level of the four candidate genes (Figure 6f, Figure S7, $P = 1.35 \times 10^{-27}$, $n = 333$). That included this

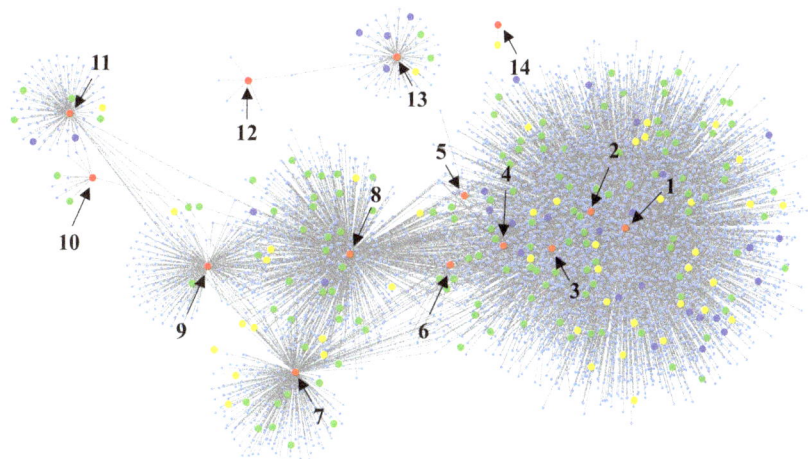

Figure 4 A co-expression network of the amino acids metabolism. The red nodes represent the 14 candidate genes from GWAS. The yellow nodes represent the co-expressed genes overlapping with candidate genes of GWAS. The green nodes represent that genes directly or indirectly related to amino acids metabolism. The blue nodes represent the transcription factors. 1, GRMZM2G147191; 2, GRMZM2G009808; 3, GRMZM2G119482; 4, GRMZM2G178826; 5, GRMZM2G010202; 6, GRMZM5G829778; 7, GRMZM2G081886; 8, GRMZM2G090241; 9, GRMZM2G082214; 10, GRMZM2G161868; 11, GRMZM2G169593; 12, GRMZM2G006480; 13, GRMZM2G127308; 14, GRMZM2G036464.

duplication not only influenced the 27-kDa γ-zein level, but also influenced the Lys/total level. Surprisingly, a QTL was identified in BB RIL population, but the B73 and By804 did not contain the duplication. This implies that other causal variants may exist within the target gene, in addition to the duplication. Haplotype analysis identified four major haplotypes at GRMZM2G138727 (Figure S8) and a significant difference was observed between B73-like (GAT) and By804-like (TAT) haplotypes, both for Lys/Total level (R^2 = 1.96%, P = 8.55 × 10^{-3}, n = 352) and expression (P = 1.05 × 10^{-3}, n = 286) (Figures 6g, S7). To exclude the possible influence of the duplications, we compared the difference between B73-like and By804-like haplotypes within the lines without duplications. Significant association was still observed for Lys/Total level (R^2 = 3.68%, P = 0.014, n = 164) but not for expression (P = 0.681, n = 127) (Figure 6h), although the sample size was more than halved. Low-linkage disequilibrium (r^2 = 0.1) was found between the duplication and the two haplotypes which implies that they were two independent variants and that the gene may affect the phenotype, but not gene expression. Combining effects of the two variants was much greater (R^2 = 3.74%, P = 6.96 × 10^{-3}, n = 322) than single variant that provided beneficial information for high-quality maize breeding.

ALS, Acetolactate synthase 1 (GRMZM2G143008), located on chromosome 5 and involved in branched-chain amino acid metabolism, catalyses the first step of Val and Leu biosynthesis. *ALS* was found to associate with Leu/Total (P = 3.59 × 10^{-6}, R^2 = 6.84%, n = 394), and the lead SNP (chr5.S_163943054) was located about 41 kb upstream of the *ALS* gene (Figure 7a–c). Two eQTLs including one strong *cis*-eQTL (P = 1.91 × 10^{-9}, MLM, n = 368, Figure 7d, Tables 2, S3) and one *trans*-eQTL (P = 3.8 × 10^{-10}, MLM, n = 368) were detected for *ALS*. The *trans*-eQTL was *O2*, which implies that *O2* may regulate the expression of *ALS*. In addition, the aforementioned co-expression analysis of *O2* and the difference in the expression of genes between *o2* mutant and wide type (Li *et al.*, 2015) both identified *ALS* that was regulated by *O2*. *ALS* may affect the trait by regulating the gene expression as the expression level of *ALS* was positively correlated with Leu/Total (r = 0.178, P = 2.20 × 10^{-3},

n = 295, Figure 7e) based the phenotype and RNA-sequencing data of association panel, and this process may be regulated by *O2*, as discussed above. Consequently, we overexpressed *ALS* in rice and a significant difference was observed between the transgenic (Figure 7f) and nontransgenic plants for a number of traits including Leu/Total, Val/BCAA, Val/Total, Val/TA and others involved in the branched-chain amino acids pathway (Figure 7g). The nontransgenic plants had higher Leu/BCAA, Leu/AT and Leu/Total level than the transgenic ones, but the Val/BCAA, Val/Total and Val/TA involved in the same metabolic pathway increased in transgenic plants. According to the previous study (Binder, 2010), the *ALS* catalyses the first step in the parallel pathway towards Val/Leu and Ile in Arabidopsis. Here, we observed a significant difference in Val, Leu, Ala, and Met between transgenic and nontransgenic lines, but not in Ile. More studies are still required to fully explore the biosynthesis of branched-chain amino acids.

Discussion

Amino acids provide essential building blocks for proteins and act as signalling molecules during plant germination, growth, development and reproduction. Grain proteins are the major source of essential amino acids in food and feed. Amino acid biosynthesis is not fully elucidated in higher plants as compared to bacteria (Umbarger, 1969, 1978) and most of the information has been from model plant Arabidopsis (Coruzzi, 2003; Hell and Wirtz, 2011; Ingle, 2011; Jander and Joshi, 2009; Tzin and Galili, 2010a,b; Verslues and Sharma, 2010). In this study, GWAS and linkage mapping were used to dissect the genetic basis of amino acid content in mature maize kernel. We identified 528 loci and 404 QTLs through GWAS and linkage mapping, respectively. Most of the identified loci or QTLs had moderate effects, explaining between 5% and 15% of the phenotypic variation (Figure S3, Tables S3, S4). Similar results have also been reported in other metabolite studies in maize (Riedelsheimer *et al.*, 2012; Wen *et al.*, 2014, 2015, 2016). It is only a few QTLs (15/404) could be identified in multiple RIL populations, implying that QTLs affecting amino acid composition were genetic background dependent. On average, 5.1 and 4.4 loci per trait were

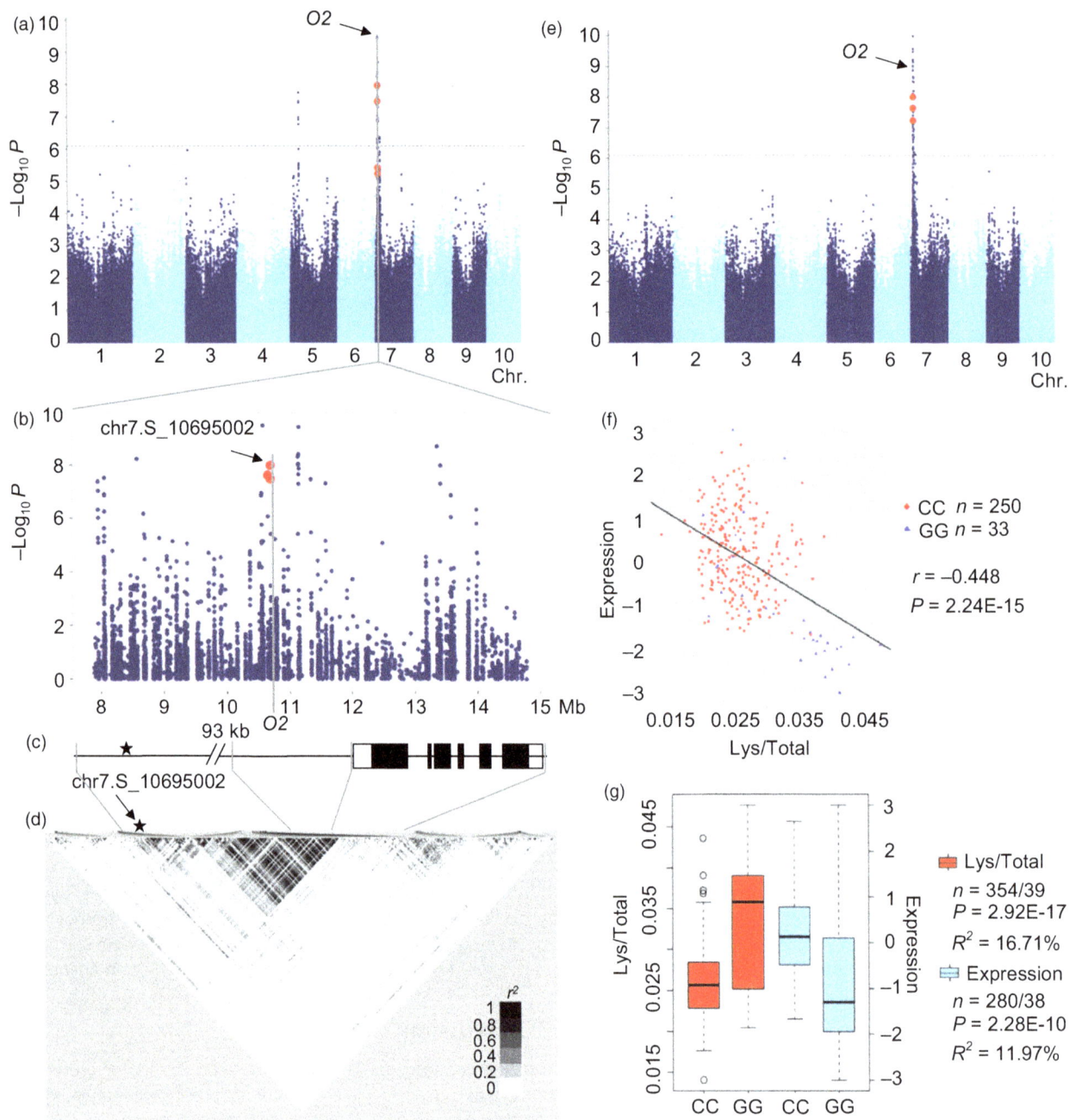

Figure 5 GWAS for Lys/Total with significant SNP-trait association in this study. (a) Manhattan plot displaying the GWAS result of the Lys/Total level. (b) Regional association plot for locus O2. The SNPs in the promoter and gene body of O2 were shown in red. (c) Gene structure of O2. Filled black boxes represent exons, and filled white ones represent UTRs. (d) A representation of the pairwise r^2 value among all polymorphic sites in O2, where the colour of each box corresponds to the r^2 value according to the legend. (e) Manhattan plot shows the association between expression level of O2 and genomewide SNPs. Significant signals are mapped to SNPs within O2, indicating a *cis* transcriptional regulation of this gene. (f) Plot of correlation between the Lys/Total level (red) and the normalized expression level (sky blue) of the O2. The r value is based on a Pearson correlation coefficient. The P value is calculated using the Student's-*t* test. (g) Box plot for Lys/Total level (red) and expression of O2 (sky blue).

identified using GWAS in AM1 and AM2, respectively, and some of them were located within the identified QTLs (17/528). It appears that the genetic basis of amino acid content in the maize kernel is relatively be simple and controlled by few genes compared with other complex quantitative traits, including agronomic traits (Xiao *et al.*, 2016). A co-expression network was constructed based on the genes identified by GWAS and gene expression data in kernel of 15 DAP (Figure 4) and novel

genes involved were found. These genes are enriched in different metabolic processes and may function as downstream and/or upstream regulators. Further studies are required to fully explore the genetic control of amino acid biosynthetic pathways.

QTLs were not randomly distributed on the chromosomes, with eight QTL hotspots observed (Figure 2) on four different chromosomes. The underlying genes were not identified for most of the QTL hotspots. This kind of QTL clustering was also observed in

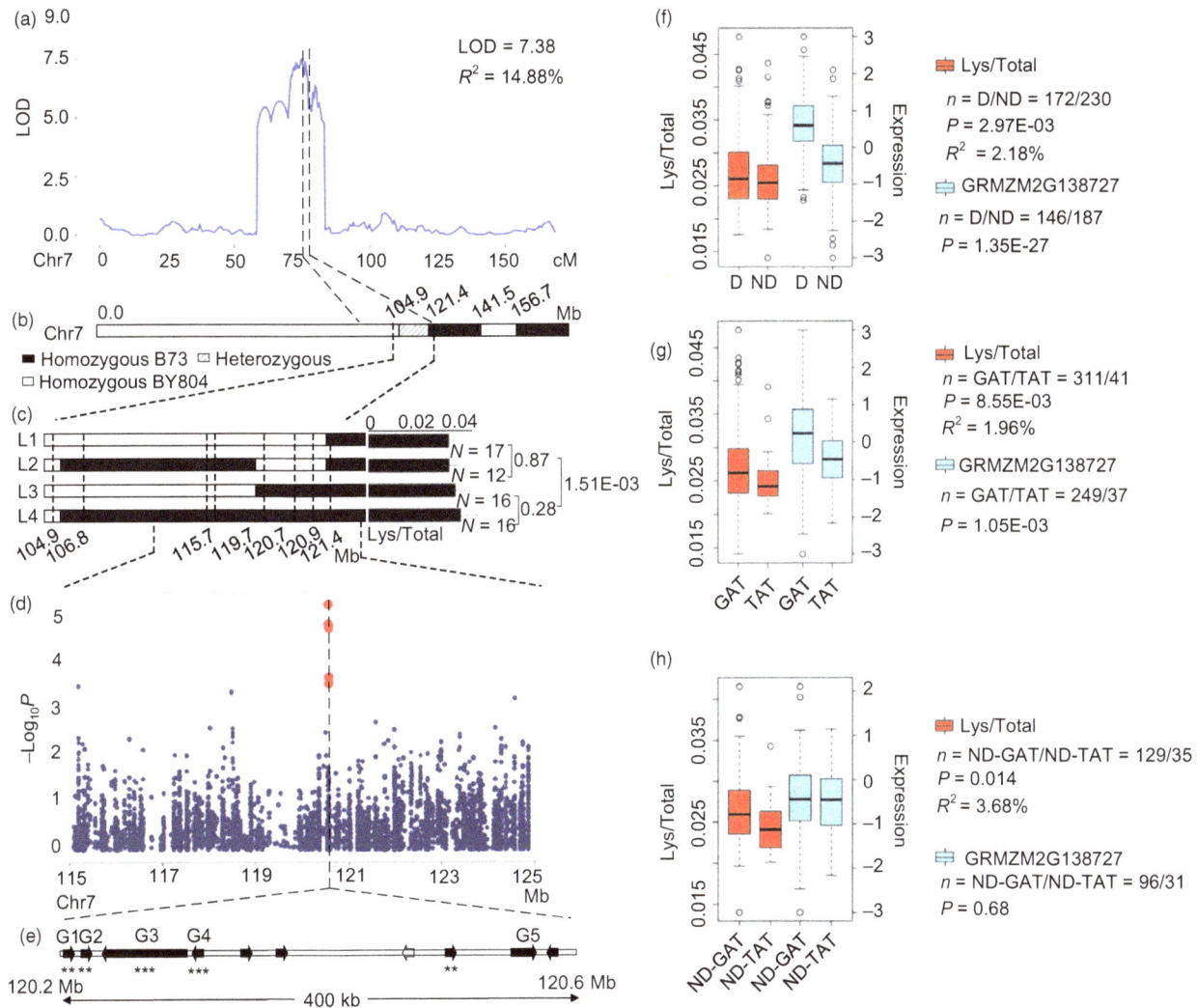

Figure 6 Validation of association analysis using QTL Interval and progeny test. (a) LOD curves of QTL mapping for level of Lys/Total in maize kernels on chromosome 7. (b) Bin map of a heterogeneous inbred family with a heterozygous region on chromosome 7. (c) Progeny test using four progeny families derived from the residual heterozygous line. (d) Scatterplot of association results between SNPs in the confidence interval and the level of Lys/Total. Association analysis was performed using the mixed linear model controlling for the population structure (Q) and kinship (K). (e) The candidate genes of 400 kb in the confidence interval. G1 to G5 represent GRMZM2G138727, GRMZM2G565441, GRMZM2G138976, GRMZM5G873335 and GRMZM2G446625, respectively. *** and *** indicate significant correction between the Lys/Total and the normalized expression levels of candidate genes at $P < 0.01$ and $P < 0.001$. (f) Box plot for Lys/Total (red) and expression of GRMZM2G138727 (skyblue) based on duplication (D) and no duplication (ND). (g) Box plot for Lys/Total (red) and expression of GRMZM2G138727 (skyblue) based on B73 (GAT) and By804 (TAT) like haplotype. (h) Box plot for Lys/Total (red) and expression of GRMZM2G138727 based on B73 (GAT) and By804 (TAT) like haplotype within no duplication (ND).

other maize studies (Riedelsheimer *et al.*, 2012; Wen *et al.*, 2015; Zhang *et al.*, 2015) and in other plants: tomato (Causse *et al.*, 2002; Schauer *et al.*, 2008), rice (Chen *et al.*, 2014; Gong *et al.*, 2013; Matsuda *et al.*, 2012) and *Arabidopsis* (Lisec *et al.*, 2008). This could be explained by the joint effects of closely linked genes (in local LD) (Bergelson and Roux, 2010) or by pleiotropy. Two QTL hotspots that affect many different phenotypic traits was identified on chromosome 7 (Figure 2). *O2* is located in one of the two QTL hotspots and appears to regulate many other genes, as identified by co-expression analysis (Figure S6). In a recent study, up to 35 *O2*-modulated target genes were identified by RNA-sequencing and ChIP-sequencing based on the *o2* mutant (Li *et al.*, 2015), some of which overlapped with our findings (Figure S6, Table S7). *o2* mutants have higher lysine content but usually worse agronomic performance, limiting their commercial

utility. The materials used in the present study are all elite inbred lines with normal field performance, differing in amino acid content, including lysine, implying that natural genetic variation in *O2* and other genes existing in the maize germplasm could be used for the improvement of amino acid composition in the future. Identification of the favourable alleles affecting amino acid composition for enhancing high nutritional maize breeding is an important priority.

The quality protein maize (QPM) was developed by introducing the *o2 modifier(s)* into *o2* maize (Lopes *et al.*, 1995) and has normal phenotype and yield, but the high lysine content of the *o2* mutant. However, the breeding process is time-consuming, and the mechanism and genetic architecture of *o2 modifiers* is poorly understood. Seven *o2 modifiers* have been located using a F_2 population (Holding *et al.*, 2008). More recently, one of the

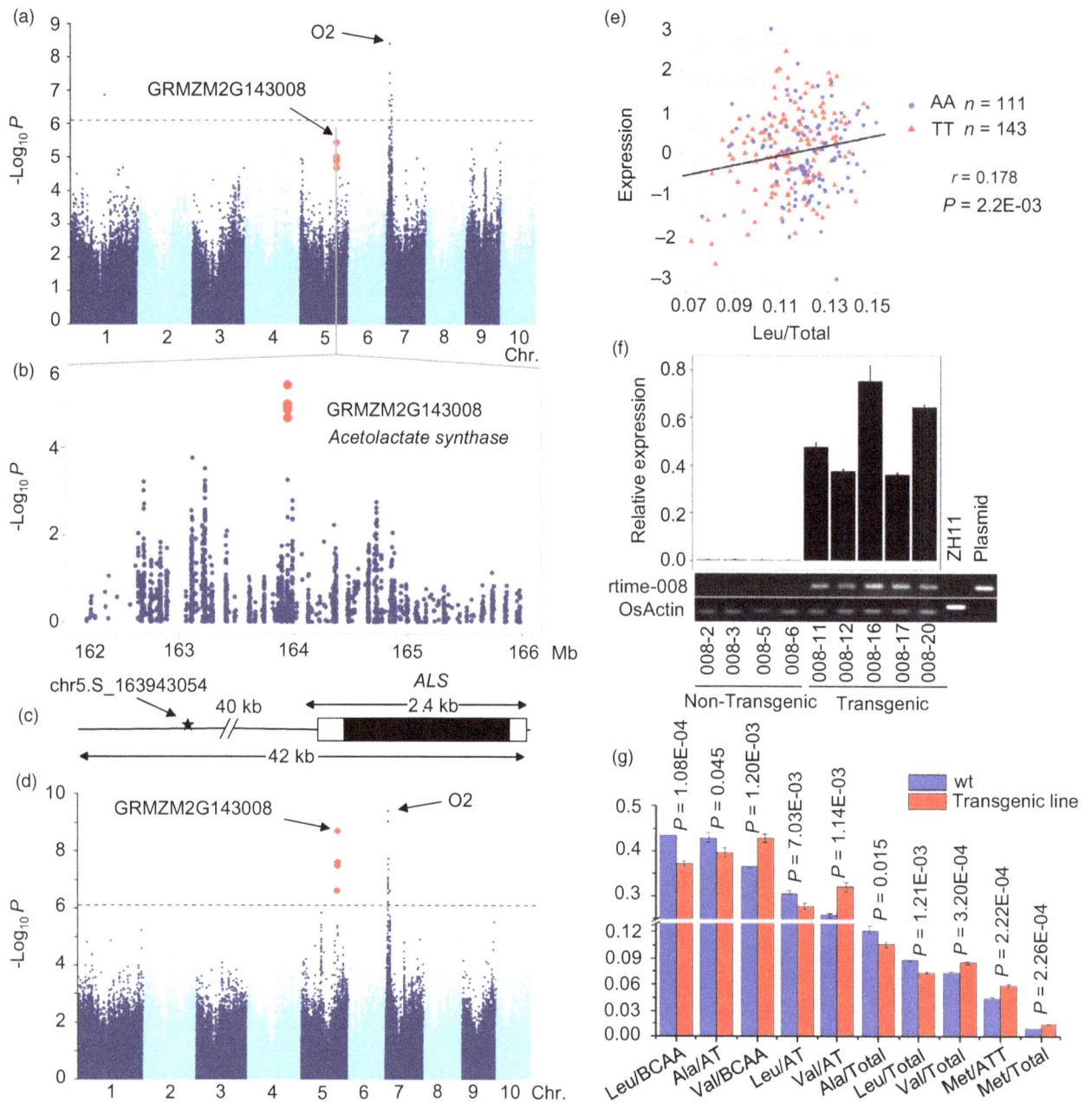

Figure 7 GWAS for Leu/Total with significant SNP-trait association in this study. (a) Manhattan plot displaying the GWAS result of the Leu/Total level. (b) Regional association plot for locus Acetolactate synthase (GRMZM2G143008). (c) Gene structure of ALS. (d) Manhattan plot shows the association between expression level of GRMZM2G143008 and genomewide SNPs. (e) Plot of correlation between the Leu/Total level (red) and the normalized expression level of the Acetolactate synthase gene (skyblue). The *r* value is based on a Pearson correlation coefficient. The *P* value is calculated using the Student's-*t* test. (f) The relative expression of GRMZM2G143008 in transgenic and non-transgenic plants. ZH11 was DNA as the positive control, and plasmid was the over-expression construct as the negative control. (g) Bar plot for amino acid traits in rice transgenic lines relative to wide type.

modifiers, *qγ27,* was cloned and gene duplication was found to increase the expression of 27-kDa γ-zein, affecting protein content (Liu *et al.,* 2016b). It was confirmed that this duplication is also present in our diverse maize inbred collections and affects the Lys/Total level and lysine content. It is interesting that a QTL was also identified in the BB RIL populations, whose parents did not contain this duplication. Additional causal variation exists within *qγ27* and was not in linkage disequilibrium with the duplication may provide new alleles for future quality protein maize breeding.

Materials and methods

The association panel and RIL populations

Genetic materials used in this study included an association panel of 513 diverse maize inbred lines for GWAS (Li *et al.,* 2012b; Yang *et al.,* 2011, 2014) and three recombinant inbred line (RIL) populations B73/BY804 (BB), ZONG3/YU87-1 (ZY) and KUI3/B77 (KB) for linkage analysis (Pan *et al.,* 2016; Xiao *et al.,* 2016). The association panel was composed of tropical, subtropical and temperate materials representing global maize diversity; details were described in previous studies (Li *et al.*, 2012b; Yang *et al.*, 2011, 2014). Field trials for the association panel were conducted in two environments: Yunnan (N 24 25', E 102 30') in 2011 and Chongqing (N 29 25', E 106 50') in 2012. RIL populations were phenotyped in three environments. The 197 RILs from BB were planted in Hainan (N 18 25', E 109 51') in 2011, and the 197 RILs from ZY and 177 RILs from KB were planted in Yunnan (N 24 25', E 102 30') in 2011. An incompletely randomized block design was used for the field trials of all the inbred lines including the association panel and three RIL populations, and a single replicate was conducted in each location. All lines were self-pollinated and five ears were harvested from each plot at maturity and were air-dried and shelled. A mixture of kernels from five self-pollinated ears was used to measure the amino acids.

Genotypes

The association panel was genotyped using Illumina MaizeSNP50 BeadChip (Ganal *et al.*, 2011) and a genotyping by sequencing method (Elshire *et al.*, 2011). Kernels from five immature ears of 368 maize inbred lines were collected at 15 days after self-pollination for RNA extraction. 1.03 million high-quality SNPs and the expression data of 28 769 genes were obtained by RNA-sequencing, (Fu *et al.*, 2013; Li *et al.*, 2013). Affymetrix Axiom Maize 600K array (Unterseer *et al.*, 2014) was used to genotype 153 lines of the association panel. After strict quality controls for each dataset, the genotypes from four different genotyping platforms were merged and 1.25M SNPs with a MAF> = 5% were used for further studies (Liu *et al.*, 2016a). The three RIL populations were also genotyped by Illumina MaizeSNP50 BeadChip and high-density linkage maps were constructed with 2496, 3071, and 2126 unique bins for BB, ZY and KB, respectively (Pan *et al.*, 2016; Xiao *et al.*, 2016).

Amino acids analysis

The amino acid concentrations of the matured maize kernel from the association panel and the three RIL populations were determined using an automatic amino acid analyzer L-8800 (L-8800, Hitachi Instruments Engineering, Tokyo, Japan). About 50–70 mg per sample of seed powder was used for the total amino acids analysis. Each sample was solubilized in 10 mL 6 M HCl at 110° for 22 h. To remove the insoluble materials, all samples were filtered into a 50-mL volumetric flask, then deionized water was added to 50 mL and mixed well. 750 μL mix of each sample was transferred to a 2-mL tube and evaporated. The dried materials were then re-dissolved in 750 μL 0.02N HCl. Subsequently, 20 μL of the re-dissolved materials were injected into an automatic amino acid analyser and the raw data was analysed with L-8800 software ASM (Zhou *et al.*, 2009). Finally, the levels of seventeen amino acids of mature maize kernel (Ala = Alanine, Arg = Arginine, Asx = Aspartic acid and Asparagine, Glx = Glutamine and Glutamic acid, Gly = Glycine, Ile = Isoleucine, Leu = Leucine, Lys = Lysine, Met = Methionine, Pro = Proline, Phe = Phenylalanine, Val = Valine, Tyr = Tyrosine, His = Histidine, Cys = Cysteine, Thr = Threonine and Ser = Serine in mg/g dry maize kernel) and the total amino acid content (sum of the seventeen amino acids) were obtained using this method. Forty-seven derived traits were determined: aspartate-derived amino acid (abbreviated ATT, included Lys, Asx, Met, Ile and Thr), pyruvate-derived amino acid (abbreviated AT, included Ala, Leu and Val), the branched-chain amino acid (abbreviated BCAA, included Ile, Leu and Val), serine-derived amino acid

(abbreviated ST, Ser, Gly and Cys), phenylalanine-derived amino acid (abbreviated PT, Phe and Tyr), glutamate-derived amino acid (abbreviated GT, included Glx, Pro and Arg) (Table S1). Each amino acid content was expressed as a percentage of the total amino acid, and the ratio of each relative amino acid content to the sum of corresponding derived amino acids were the derived traits, including Ala/Total, Arg/Total, Asx/Total, Glx/Total, Gly/Total, Ile/Total, Leu/Total, Lys/Total, Met/Total, Pro/Total, Phe/Total, Vla/Total, Tyr/Total, His/Total, Cys/Total, Thr/Total, Ser/Total, Lys/ATT, Asx/ATT, Met/ATT, Ile/ATT, Thr/ATT, Ala/AT, Leu/AT, Val/AT, Ile/BCAA, Leu/BCAA, Val/BCAA, Ser/ST, Gly/ST, Cys/ST, Phe/PT, Tyr/PT, Glx/GT, Pro/GT and Arg/GT.

Genomewide association study

A genome wide association study (GWAS) was conducted for maize kernel amino acid traits. To test the statistical associations between genotype and phenotype, a mixed linear model was used for accounting for the population structure and relative kinship (Li *et al.*, 2013; Yu *et al.*, 2006). Considering the maker number in present study is 1.25 million and many of them should be in linkage disequilibrium. The effective number of independent marker (N) was calculated using the GEC software tool (Li *et al.*, 2012a). Suggestive ($1/N$) P value thresholds were set to control the genomewide type 1 error rate. The suggestive value was 2.04E-06 for whole population and used as the cut-offs. The P value of each SNP was calculated using Tassel3.0. For all traits, the lead SNP (SNP with the lowest p value) at an associated locus and its corresponding candidate genes in or near (within 100 kb up-, downstream of the lead SNP) known genes were reported (Table S3). If the associated SNPs were not in or near an annotated amino acid metabolism gene, the closest of the lead SNP candidate gene was considered the most likely candidate gene. The physical locations of the SNPs were based on the B73 RefGen_v2.

QTL mapping

The linkage mapping was conducted using Composite Interval Mapping (CIM) implemented in Windows QTL Cartographer V2.5 (Wang *et al.*, 2006; Zeng *et al.*, 1999) for all amino acid traits measured in maize kernel of the three RIL populations. The methods followed the Windows QTL Cartographer V2.5 user manual. Zmap (model 6) with a 10-cM window and a walking speed of 0.5 cM was used. For each trait, a uniform threshold for significant QTLs was determined by 500 permutations ($P = 0.05$). The parameter was set as default. 2.0 LOD–drop confidence interval was used for each QTL as described.

In total, 13 progeny families were derived from one heterogeneous inbred line that were identified for the major QTL on chromosome 7 and planted at Wuhan in the summer of 2014 for QTL validation and cloning. Six families with enough seeds ($n = 10$ to 25 rows, 11 individuals per row for each family) were planted at Hainan in the winter of 2014. Two families ($n = 29$ and 32 individuals for each family) with enough recombinant individuals were measured for amino acids with one replicate. Primers used for linkage analysis were listed in Table S9.

eQTL mapping

Expression mapping (eQTL) analysis (SNP vs. gene expression level) used the same method described above for GWAS. The association analysis between the genomewide SNPs and the identified candidate gene expression level was performed. Only those genes expressed in more than 50% of 368 lines and for which at least 10 reads were available were used in this analysis (Liu *et al.*, 2016a).

Co-expression network

In order to construct the co-expression network of chosen genes, we calculated pairwise relative expression coefficients in R (https://www.r-project.org/) and used these coefficients and *P*-values to filter the genes. The filtered co-expression genes were used to construct the co-expression network. The pairwise relative expression coefficients shown the relationship between genes. The program Cytoscape was used to draw the network with only the most highly connected genes (http://www.cytoscape.org/). The Gene Ontology term analysis was conducted at AGRiGO (http://bioinfo.cau.edu.cn/agriGO/).

Plasmid construction and rice transformation

The overexpression vector pCAMBIA1300nu was provided by Dr. Yongjun Lin, Huazhong Agricultural University, Wuhan, China. To generate the GRMZM2G143008 over-expression construct, the open reading frame of GRMZM2G143008 was amplified from the cDNA of maize inbred line B73 developing kernel by PCR using the gene-specific primers DMp008Os-F and DMp008Os-R, which contained a 20-bp fragment complementary with pCAMBIA1300nu. The PCR product was cloned into pCAMBIA1300nu with a homologous recombination clone kit (Vazyme, China). The target gene was driven by a maize ubiquitin promoter. Then the correct clone was selected by sequencing the construct. These constructs were introduced into *japonica* rice cultivar ZhongHua 11 (ZH11) by *Agrobacterium tumefaciens*-mediated transformation (Lin and Zhang, 2005). Primers used in this study were listed in Table S9.

Expression analysis of transgenic plant

Total RNA was prepared from leaves using a Quick RNA Isolation kit (HUAYUEYANG, Beijing). For RT-PCR, the first-strand cDNA was synthesized from 1.5 mg total RNA using the TransScript One-Step gDNA Removal and cDNA Synthesis SuperMix kit (TransGen, China). Semi-quantitative PCR was performed for gene expression analysis using gene-specific (DMp008Os-F and DMp008Os-R) and rice ACTIN (OsrActin-F and OsrActin-R) primers. Real-time PCR was performed on an optical 96-well plate in a BIO-RAD CFX96 Real-Time system using TransStart Tip Green qPCR SuperMix (TransGen, China). Actin was used as an endogenous control. Primers used in this study were listed in Table S9.

Acknowledgements

This research was supported by the National Natural Science Foundation of China (31525017) and the National key research and development programme of China (2016YFD0101003). The authors declare no conflict of interest.

References

Angelovici, R., Lipka, A.E., Deason, N., Gonzalez-Jorge, S., Lin, H., Cepela, J., Buell, R. *et al.* (2013) Genome-wide analysis of branched-chain amino acid levels in *Arabidopsis* seeds. *Plant Cell*, **25**, 4827–4843.

Atwell, S., Huang, Y.S., Vilhjalmsson, B.J., Willems, G., Horton, M., Li, Y., Meng, D. *et al.* (2010) Genome-wide association study of 107 phenotypes in *Arabidopsis Thaliana* inbred lines. *Nature*, **465**, 627–631.

Bergelson, J. and Roux, F. (2010) Towards identifying genes underlying ecologically relevant traits in *Arabidopsis Thaliana*. *Nat. Rev. Genet.* **11**, 867–879.

Binder, S. (2010) Branched-chain amino acid metabolism in *arabidopsis thaliana*. *Arabidopsis Book*, **8**, e0137.

Birsoy, K., Wang, T., Chen, W.W., Freinkman, E., Abu-Remaileh, M. and Sabatini, D.M. (2015) An essential role of the mitochondrial electron transport chain in cell proliferation is to enable aspartate synthesis. *Cell*, **162**, 540–551.

Causse, M., Saliba-Colombani, V., Lecomte, L., Duffe, P., Rousselle, P. and Buret, M. (2002) QTL analysis of fruit quality in fresh market tomato: a few chromosome regions control the variation of sensory and instrumental traits. *J. Exp. Bot.* **53**, 2089–2098.

Chen, W., Gao, Y.Q., Xie, W.B., Gong, L., Lu, K., Wang, W.S., Li, Y. *et al.* (2014) Genome-wide association analyses provide genetic and biochemical insights into natural variation in rice metabolism. *Nat. Genet.* **46**, 714–721.

Chevalier, C., Bourgeois, E., Just, D. and Raymond, P. (1996) Metabolic regulation of asparagine synthetase gene expression in maize (*Zea mays* L.) root tips. *Plant J.* **9**, 1–11.

Coruzzi, G.M. (2003) Primary N-assimilation into amino acids in *Arabidopsis*. *Arabidopsis Book*, **2**, e0010.

Crawford, N.M. and Forde, B.G. (2002) Molecular and developmental biology of inorganic nitrogen nutrition. *Arabidopsis Book*, **1**, e0011.

Curry, R.A. and Ting, I.P. (1976) Purification, properties, and kinetic observations on the isoenzymes of NADP isocitrate dehydrogenase of maize. *Arch. Biochem. Biophys.* **176**, 501–509.

Dunlop, R.A., Main, B.J. and Rodgers, K.J. (2015) The deleterious effects of non-protein amino acids from desert plants on human and animal health. *J. Arid Environ.* **112**, 152–158.

Elshire, R.J., Glaubitz, J.C., Sun, Q., Poland, J.A., Kawamoto, K., Buckler, E.S. and Mitchell, S.E. (2011) A robust, simple genotyping-by-sequencing (GBS) approach for high diversity species. *PLoS ONE*, **6**, e19379.

Fu, J.J., Cheng, Y.B., Linghu, J.J., Yang, X.H., Kang, L., Zhang, Z.X., Zhang, J. *et al.* (2013) RNA sequencing reveals the complex regulatory network in the maize kernel. *Nat. Commun.* **4**, 2832.

Ganal, M.W., Durstewitz, G., Polley, A., Berard, A., Buckler, E.S., Charcosset, A., Clarke, J.D. *et al.* (2011) A large maize (*Zea mays* L.) SNP genotyping array: development and germplasm genotyping, and genetic mapping to compare with the B73 reference genome. *PLoS ONE*, **6**, e28334.

Gong, L., Chen, W., Gao, Y.Q., Liu, X.Q., Zhang, H.Y., Xu, C.G., Yu, S.B. *et al.* (2013) Genetic analysis of the metabolome exemplified using a rice population. *Proc. Natl Acad. Sci. USA*, **110**, 20320–20325.

Havir, E.A. (1971) L-Phenylalanine ammonia-lyase (maize): evidence for a common catalytic site for L-Phenylalanine and L-Tyrosine. *Plant Physiol.* **48**, 130–136.

Hell, R. and Wirtz, M. (2011) Molecular biology, biochemistry and cellular physiology of cysteine metabolism in *Arabidopsis thaliana*. *Arabidopsis Book*, **9**, e0154.

Holding, D.R., Hunter, B.G., Chung, T., Gibbon, B.C., Ford, C.F., Bharti, A.K., Messing, J. *et al.* (2008) Genetic analysis of *opaque2* modifier loci in quality protein maize. *Theor. Appl. Genet.* **117**, 157–170.

Huang, X.H., Wei, X.H., Sang, T., Zhao, Q., Feng, Q., Zhao, Y., Li, C.Y. *et al.* (2010) Genome-wide association studies of 14 agronomic traits in rice landraces. *Nat. Genet.* **42**, 961–967.

Huang, T., Jander, G. and de Vos, M. (2011) Non-protein amino acids in plant defense against insect herbivores: representative cases and opportunities for further functional analysis. *Phytochemistry*, **72**, 1531–1537.

Huang, X.H., Zhao, Y., Wei, X.H., Li, C.Y., Wang, A., Zhao, Q., Li, W.J. *et al.* (2012) Genome-wide association study of flowering time and grain yield traits in a worldwide collection of rice germplasm. *Nat. Genet.* **44**, 32–39.

Hunter, B.G., Beatty, M.K., Singletary, G.W., Hamaker, B.R., Dilkes, B.P., Larkins, B.A. and Jung, R. (2002) Maize opaque endosperm mutations create extensive changes in patterns of gene expression. *Plant Cell*, **14**, 2591–2612.

Ingle, R.A. (2011) Histidine biosynthesis. *Arabidopsis Book*, **9**, e0141.

Ishimoto, M., Rahman, S.M., Hanafy, M.S., Khalafalla, M.M., El-Shemy, H.A., Nakamoto, Y., Kita, Y. *et al.* (2010) Evaluation of amino acid content and nutritional quality of transgenic soybean seeds with high-level tryptophan accumulation. *Mol. Breeding*, **25**, 313–326.

Jander, G. and Joshi, V. (2009) Aspartate-derived amino acid biosynthesis in *Arabidopsis thaliana*. *Arabidopsis Book*, **7**, e0121.

Jia, G.Q., Huang, X.H., Zhi, H., Zhao, Y., Zhao, Q., Li, W.J., Chai, Y. *et al.* (2013) A haplotype map of genomic variations and genome-wide association studies of agronomic traits in foxtail millet (*Setaria Italica*). *Nat. Genet.* **45**, 957–961.

Kang, H.G., Park, S., Matsuoka, M. and An, G. (2005) White-core endosperm floury endosperm-4 in rice is generated by knockout mutations in the C-type pyruvate orthophosphate dikinase gene (*OsPPDKB*). *Plant J.* **42**, 901–911.

Kodrzycki, R., Boston, R.S. and Larkins, B.A. (1989) The *opaque-2* mutation of maize differentially reduces zein gene transcription. *Plant Cell*, **1**, 105–114.

Krishnan, N., Dickman, M.B. and Becker, D.F. (2008) Proline modulates the intracellular redox environment and protects mammalian cells against oxidative stress. *Free Radical. Bio. Med.* **44**, 671–681.

Li, M.X., Yeung, J.M., Cherny, S.S. and Sham, P.C. (2012a) Evaluating the effective numbers of independent tests and significant *P*-value thresholds in commercial genotyping arrays and public imputation reference datasets. *Hum. Genet.* **131**, 747–756.

Li, Q., Yang, X.H., Xu, S.T., Cai, Y., Zhang, D.L., Han, Y.J., Li, L. *et al.* (2012b) Genome-wide association studies identified three independent polymorphisms associated with alpha-tocopherol content in maize kernels. *PLoS ONE*, **7**, e36807.

Li, H., Peng, Z.Y., Yang, X.H., Wang, W.D., Fu, J.J., Wang, J.H., Han, Y.J. *et al.* (2013) Genome-wide association study dissects the genetic architecture of oil biosynthesis in maize kernels. *Nat. Genet.* **45**, 43–50.

Li, C., Qiao, Z., Qi, W., Wang, Q., Yuan, Y., Yang, X., Tang, Y. *et al.* (2015) Genome-wide characterization of *cis*-acting DNA targets reveals the transcriptional regulatory framework of opaque2 in maize. *Plant Cell*, **27**, 532–545.

Lin, Y.J. and Zhang, Q.F. (2005) Optimising the tissue culture conditions for high efficiency transformation of indica rice. *Plant Cell Rep.* **23**, 540–547.

Lisec, J., Meyer, R.C., Steinfath, M., Redestig, H., Becher, M., Witucka-Wall, H., Fiehn, O. *et al.* (2008) Identification of metabolic and biomass QTL in *Arabidopsis thaliana* in a parallel analysis of RIL and IL populations. *Plant J.* **53**, 960–972.

Liu, H.J., Wang, F., Xiao, Y.J., Tian, Z.L., Wen, W.W., Zhang, X.H., Chen, X. *et al.* (2016a) MODEM: multi-omics data envelopment and mining in maize. *Database (Oxford)*, **7**, 2016 pii: baw117.

Liu, H.J., Luo, X., Niu, L.Y., Xiao, Y.J., Chen, L., Liu, J., Wang, X.Q. *et al.* (2016b) Distant eQTLs and non-coding sequences play critical roles in regulating gene expression and quantitative trait variation in maize. *Mol Plant*, **10**, 414–426.

Liu, H., Shi, J., Sun, C., Gong, H., Fan, X., Qiu, F., Huang, X. *et al.* (2016c) Gene duplication confers enhanced expression of 27-kDa gamma-zein for endosperm modification in quality protein maize. *Proc. Natl Acad. Sci. USA*, **113**, 4964–4969.

Liu, S., Fan, C.C., Li, J.N., Cai, G.Q., Yang, Q.Y., Wu, J., Yi, X.Q. *et al.* (2016d) A genome-wide association study reveals novel elite allelic variations in seed oil content of *Brassica napus*. *Theor. Appl. Genet.* **129**, 1203–1215.

Loesch, P., Foley, D. and Cox, D. (1976) Comparative resistance of *opaque-2* and normal inbred lines of maize to ear-rotting pathogens. *Crop Sci.* **16**, 841–842.

Lopes, M.A., Takasaki, K., Bostwick, D.E., Helentjaris, T. and Larkins, B.A. (1995) Identification of two *opaque2* modifier loci in quality protein maize. *Mol. Gen. Genet.* **247**, 603–613.

Luo, X., Ma, C.Z., Yue, Y., Hu, K.N., Li, Y.Y., Duan, Z.Q., Wu, M. *et al.* (2015) Unravelling the complex trait of harvest index in rapeseed (*Brassica napus* L.) with association mapping. *BMC Genom.* **16**, 379.

Maloney, G.S., Kochevenko, A., Tieman, D.M., Tohge, T., Krieger, U., Zamir, D., Taylor, M.G. *et al.* (2010) Characterization of the branched-chain amino acid aminotransferase enzyme family in tomato. *Plant Physiol.* **153**, 925–936.

Mandal, S. and Mandal, R. (2000) Seed storage proteins and approaches for improvement of their nutritional quality by genetic engineering. *Curr. Sci. India*, **79**, 576–589.

Matsuda, F., Okazaki, Y., Oikawa, A., Kusano, M., Nakabayashi, R., Kikuchi, J., Yonemaru, J. *et al.* (2012) Dissection of genotype-phenotype associations in rice grains using metabolome quantitative trait loci analysis. *Plant J.* **70**, 624–636.

Matysik, J., Alia Bhalu, B. and Mohanty, P. (2002) Molecular mechanisms of quenching of reactive oxygen species by proline under stress in plants. *Curr. Sci. India*, **82**, 525–532.

Mertz, E.T., Bates, L.S. and Nelson, O.E. (1964) Mutant gene that changes protein composition and increases lysine content of maize endosperm. *Science*, **145**, 279–280.

Miflin, B. and Lea, P. (1977) Amino acid metabolism. *Annu. Rev. Plant Physiol.* **28**, 299–329.

Mishra, S. and Dubey, R.S. (2006) Inhibition of ribonuclease and protease activities in arsenic exposed rice seedlings: role of proline as enzyme protectant. *J. Plant Physiol.* **163**, 927–936.

Misra, P.S., Jambunathan, R., Mertz, E.T., Glover, D.V., Barbosa, H.M. and McWhirter, K.S. (1972) Endosperm protein synthesis in maize mutants with increased lysine content. *Science*, **176**, 1425–1427.

Morris, G.P., Ramu, P., Deshpande, S.P., Hash, C.T., Shah, T., Upadhyaya, H.D., Riera-Lizarazu, O. *et al.* (2013) Population genomic and genome-wide association studies of agroclimatic traits in sorghum. *Proc. Natl Acad. Sci. USA*, **110**, 453–458.

Moulin, M., Deleu, C. and Larher, F. (2000) L-Lysine catabolism is osmo-regulated at the level of lysine-ketoglutarate reductase and saccharopine dehydrogenase in rapeseed leaf discs. *Plant Physiol. Bioch.* **38**, 577–585.

Moulin, M., Deleu, C., Larher, F. and Bouchereau, A. (2006) The lysine-ketoglutarate reductase-saccharopine dehydrogenase is involved in the osmo-induced synthesis of pipecolic acid in rapeseed leaf tissues. *Plant Physiol. Bioch.* **44**, 474–482.

Muehlbauer, G.J., Gengenbach, B.G., Somers, D.A. and Donovan, C.M. (1994) Genetic and amino-acid analysis of two maize threonine-overproducing, lysine-insensitive aspartate kinase mutants. *Theor. Appl. Genet.* **89**, 767–774.

Nass, H. and Crane, P. (1970) Effect of endosperm mutants on germination and early seedling growth rate in maize (*Zea mays* L.). *Crop Sci.* **10**, 139–140.

Pan, Q., Li, L., Yang, X., Tong, H., Xu, S., Li, Z., Li, W., Muehlbauer, G.J., Li, J. and Yan, J. (2016) Genome-wide recombination dynamics are associated with phenotypic variation in maize. *New Phytol.* **210**, 1083–1094.

Riedelsheimer, C., Lisec, J., Czedik-Eysenberg, A., Sulpice, R., Flis, A., Grieder, C., Altmann, T. *et al.* (2012) Genome-wide association mapping of leaf metabolic profiles for dissecting complex traits in maize. *Proc. Natl Acad. Sci. USA*, **109**, 8872–8877.

Schauer, N., Semel, Y., Balbo, I., Steinfath, M., Repsilber, D., Selbig, J., Pleban, T. *et al.* (2008) Mode of inheritance of primary metabolic traits in tomato. *Plant Cell*, **20**, 509–523.

Schmidt, R.J., Burr, F.A. and Burr, B. (1987) Transposon tagging and molecular analysis of the maize regulatory locus opaque-2. *Science*, **238**, 960–963.

Sharma, P. and Dubey, R.S. (2005) Modulation of nitrate reductase activity in rice seedlings under aluminium toxicity and water stress: role of osmolytes as enzyme protectant. *J. Plant Physiol.* **162**, 854–864.

Shaver, J.M., Bittel, D.C., Sellner, J.M., Frisch, D.A., Somers, D.A. and Gengenbach, B.G. (1996) Single-amino acid substitutions eliminate lysine inhibition of maize dihydrodipicolinate synthase. *Proc. Natl Acad. Sci. USA*, **93**, 1962–1966.

Smirnoff, N. and Cumbes, Q.J. (1989) Hydroxyl radical scavenging activity of compatible solutes. *Phytochemistry*, **28**, 1057–1060.

Sullivan, L.B., Gui, D.Y., Hosios, A.M., Bush, L.N., Freinkman, E. and Vander Heiden, M.G. (2015) Supporting Aspartate Biosynthesis is an essential function of respiration in proliferating Cells. *Cell*, **162**, 552–563.

Takahashi, M., Uematsu, Y., Kashiwaba, K., Yagasaki, K., Hajika, M., Matsunaga, R., Komatsu, K. *et al.* (2003) Accumulation of high levels of free amino acids in soybean seeds through integration of mutations conferring seed protein deficiency. *Planta*, **217**, 577–586.

Tzin, V. and Galili, G. (2010a) The biosynthetic pathways for shikimate and aromatic amino acids in *Arabidopsis thaliana*. *Arabidopsis Book*, **8**, e0132.

Tzin, V. and Galili, G. (2010b) New insights into the shikimate and aromatic amino acids biosynthesis pathways in plants. *Mol. Plant*, **3**, 956–972.

Ufaz, S. and Galili, G. (2008) Improving the content of essential amino acids in crop plants: goals and opportunities. *Plant Physiol.* **147**, 954–961.

Umbarger, H.E. (1969) Regulation of amino acid metabolism. *Annu. Rev. Biochem.* **38**, 323–370.

Umbarger, H.E. (1978) Amino acid biosynthesis and its regulation. *Annu. Rev. Biochem.* **47**, 532–606.

Unterseer, S., Bauer, E., Haberer, G., Seidel, M., Knaak, C., Ouzunova, M., Meitinger, T. *et al.* (2014) A powerful tool for genome analysis in maize: development and evaluation of the high density 600 k SNP genotyping array. *BMC Genom.* **15**, 823.

Verslues, P.E. and Sharma, S. (2010) Proline metabolism and its implications for plant-environment interaction. *Arabidopsis Book*, **8**, e0140.

Wang, X., Stumpf, D.K. and Larkins, B.A. (2001) Aspartate kinase 2. A candidate gene of a quantitative trait locus influencing free amino acid content in maize endosperm. *Plant Physiol.* **125**, 1778–1787.

Wang, S., Basten, C. and Zeng, Z. (2006) *Windows QTL Cartographer 2.5.* Department of Statistics, North Carolina State University, Raleigh, NC, 2010.

Wang, X., Lopez-Valenzuela, J.A., Gibbon, B.C., Gakiere, B., Galili, G. and Larkins, B.A. (2007) Characterization of monofunctional aspartate kinase genes in maize and their relationship with free amino acid content in the endosperm. *J. Exp. Bot.* **58**, 2653–2660.

Wang, G., Zhang, J.S., Wang, G.F., Fan, X.Y., Sun, X., Qin, H.L., Xu, N. *et al.* (2014) Proline responding1 plays a critical role in regulating general protein synthesis and the cell cycle in maize. *Plant Cell*, **26**, 2582–2600.

Wen, W.W., Li, D., Li, X., Gao, Y.Q., Li, W.Q., Li, H.H., Liu, J. *et al.* (2014) Metabolome-based genome-wide association study of maize kernel leads to novel biochemical insights. *Nat. Commun.* **5**, 3438.

Wen, W., Li, K., Alseekh, S., Omranian, N., Zhao, L., Zhou, Y., Xiao, Y. *et al.* (2015) Genetic determinants of the network of primary metabolism and their relationships to plant performance in a maize recombinant inbred line population. *Plant Cell*, **27**, 1839–1856.

Wen, W., Liu, H., Zhou, Y., Jin, M., Yang, N., Li, D., Luo, J. *et al.* (2016) Combining quantitative genetics approaches with regulatory network analysis to dissect the complex metabolism of the maize kernel. *Plant Physiol.* **170**, 136–146.

Wendel, J.F., Goodman, M.M., Stuber, C.W. and Beckett, J.B. (1988) New isozyme systems for maize (*Zea mays* L.): aconitate hydratase, adenylate kinase, NADH dehydrogenase, and shikimate dehydrogenase. *Biochem. Genet.* **26**, 421–445.

Wright, A.D., Moehlenkamp, C.A., Perrot, G.H., Neuffer, M.G. and Cone, K.C. (1992) The maize auxotrophic mutant orange pericarp is defective in duplicate genes for tryptophan synthase beta. *Plant Cell*, **4**, 711–719.

Wu, Y.R. and Messing, J. (2014) Proteome balancing of the maize seed for higher nutritional value. *Front Plant Sci.* **5**, 240.

Xiao, Y., Tong, H., Yang, X., Xu, S., Pan, Q., Qiao, F., Raihan, M.S. *et al.* (2016) Genome-wide dissection of the maize ear genetic architecture using multiple populations. *New Phytol.* **210**, 1095–1106.

Yan, J.B., Warburton, M. and Crouch, J. (2011) Association mapping for enhancing maize (*Zea mays* L.) genetic improvement. *Crop Sci.* **51**, 433–449.

Yang, X.H., Gao, S.B., Xu, S.T., Zhang, Z.X., Prasanna, B.M., Li, L., Li, J.S. *et al.* (2011) Characterization of a global germplasm collection and its potential utilization for analysis of complex quantitative traits in maize. *Mol. Breeding*, **28**, 511–526.

Yang, N., Lu, Y.L., Yang, X.H., Huang, J., Zhou, Y., Ali, F., Wen, W.W. *et al.* (2014) Genome wide association studies using a new nonparametric model reveal the genetic architecture of 17 agronomic traits in an enlarged maize association panel. *PLoS Genet.* **10**, e1004573.

Young, V.R. and Pellett, P.L. (1994) Plant proteins in relation to human protein and amino acid nutrition. *Am. J. Clin. Nutr.* **59**, 1203S–1212S.

Yu, J.M., Pressoir, G., Briggs, W.H., Bi, I.V., Yamasaki, M., Doebley, J.F., McMullen, M.D. *et al.* (2006) A unified mixed-model method for association mapping that accounts for multiple levels of relatedness. *Nat. Genet.* **38**, 203–208.

Zeng, Z.B., Kao, C.H. and Basten, C.J. (1999) Estimating the genetic architecture of quantitative traits. *Genet. Res.* **74**, 279–289.

Zhang, N., Gur, A., Gibon, Y., Sulpice, R., Flint-Garcia, S., McMullen, M.D., Stitt, M. *et al.* (2010) Genetic analysis of central carbon metabolism unveils an amino acid substitution that alters maize NAD-dependent isocitrate dehydrogenase activity. *PLoS ONE*, **5**, e9991.

Zhang, N., Gibon, Y., Wallace, J.G., Lepak, N., Li, P., Dedow, L., Chen, C. *et al.* (2015) Genome-wide association of carbon and nitrogen metabolism in the maize nested association mapping population. *Plant Physiol.* **168**, 575–583.

Zhang, Z.Y., Zheng, X.X., Yang, J., Messing, J. and Wu, Y.R. (2016) Maize endosperm-specific transcription factors O2 and PBF network the regulation of protein and starch synthesis. *Proc. Natl Acad. Sci. USA*, **113**, 10842–10847.

Zhou, Y., Cai, H.M., Xiao, J.H., Li, X.H., Zhang, Q.F. and Lian, X.M. (2009) Over-expression of aspartate aminotransferase genes in rice resulted in altered nitrogen metabolism and increased amino acid content in seeds. *Theor. Appl. Genet.* **118**, 1381–1390.

3

A single point mutation in *Ms44* results in dominant male sterility and improves nitrogen use efficiency in maize

Tim Fox, Jason DeBruin, Kristin Haug Collet, Mary Trimnell, Joshua Clapp, April Leonard, Bailin Li, Eric Scolaro, Sarah Collinson, Kimberly Glassman, Michael Miller, Jeff Schussler, Dennis Dolan, Lu Liu, Carla Gho, Marc Albertsen, Dale Loussaert and Bo Shen*

DuPont Pioneer, Johnston, IA, USA

*Correspondence
email Bo.Shen@pioneer.com

Summary

Application of nitrogen fertilizer in the past 50 years has resulted in significant increases in crop yields. However, loss of nitrogen from crop fields has been associated with negative impacts on the environment. Developing maize hybrids with improved nitrogen use efficiency is a cost-effective strategy for increasing yield sustainably. We report that a dominant male-sterile mutant *Ms44* encodes a lipid transfer protein which is expressed specifically in the tapetum. A single amino acid change from alanine to threonine at the signal peptide cleavage site of the Ms44 protein abolished protein processing and impeded the secretion of protein from tapetal cells into the locule, resulting in dominant male sterility. While the total nitrogen (N) content in plants was not changed, *Ms44* male-sterile plants reduced tassel growth and improved ear growth by partitioning more nitrogen to the ear, resulting in a 9.6% increase in kernel number. Hybrids carrying the *Ms44* allele demonstrated a 4%–8.5% yield advantage when N is limiting, 1.7% yield advantage under drought and 0.9% yield advantage under optimal growth conditions relative to the yield of wild type. Furthermore, we have developed an *Ms44* maintainer line for fertility restoration, male-sterile inbred seed increase and hybrid seed production. This study reveals that protein secretion from the tapetum into the locule is critical for pollen development and demonstrates that a reduction in competition between tassel and ear by male sterility improves grain yield under low-nitrogen conditions in maize.

Keywords: *Ms44*, lipid transfer protein, male sterility, grain yield, NUE, maize.

Introduction

Nitrogen (N) is an essential nutrient for plant growth and development and is an important factor determining maize grain yield. Unfortunately, only about 1/3 of applied nitrogen is utilized by crops, while 2/3 is lost to the environment (McAllister *et al.*, 2012; Raun and Johnson, 1999; Xu *et al.*, 2012). Excessive application of nitrogen fertilizer is associated with negative impacts on the environment including eutrophication of lakes and rivers, and toxic levels of nitrates in underground water for human consumption (Good and Beatty, 2011; Liu *et al.*, 2013). In addition, nitrogen fertilizers account for ~25% of the total input costs (seeds, fertilizers and pesticides) in maize production. Improving nitrogen use efficiency (NUE) through genetic engineering in major crops is still in its infancy and far from ready for commercialization. Multiple studies have reported an improvement in NUE in *Arabidopsis*, rice, wheat and maize using transgenic approaches, but the NUE of the transgenic lines has not been validated in field trials using elite germplasm (McAllister *et al.*, 2012; Xu *et al.*, 2012). On the other hand, through plant breeding efforts, newer maize hybrids developed for North America have shown significant improvements in NUE compared to older hybrids. Grain yields have increased steadily in the United States in the past 30 years, while N fertilizer application rate has not been increased significantly (Ciampitti and Vyn, 2014; Han *et al.*, 2015). Further improvements in maize NUE will not only reduce the environmental footprint of nitrogen fertilizer and cost to the grower, but also increase grain yields at current N application rates. In the majority of African countries, improving NUE is critical for food security as farmers cannot afford to apply more N fertilizer.

Maize is a monoecious plant with separate male and female flowers on the same plant. To produce pure hybrid seed, detasseling of the female inbred parent is employed widely by the seed industry. Although mechanical detasseling is effective in maize hybrid seed production, it is time-consuming and labour-intensive. In addition, damage to the top leaves during detasseling reduces hybrid seed yield. Male sterility is the most efficient way to ensure cross-pollination (Chen and Liu, 2014; Kempe and Gils, 2011; Perez-Prat and van Lookeren Campagne, 2002). Cytoplasmic male sterility (CMS) has been used to produce hybrid seed in both maize and rice. CMS is based on mutations in mitochondrial DNA, and male fertility can be restored by specific nuclear restorer genes. However, CMS is difficult to develop as it may not work across germplasms and sterility and fertility restoration may not be stable in certain environments. A CMS system requires a male-sterile line, a maintainer line and a fertility restoration line, which involves a complex integration in the breeding process (Chen and Liu, 2014; Weider *et al.*, 2009). Alternatively, nuclear genetic male sterility is stable in different germplasms and growth environments. Over 40 genetic male-sterile mutants in maize have been identified and characterized with the majority of the mutations being recessive (Skibbe and Schnable, 2005). However, it is problematic to use genetic male-sterile mutants in hybrid seed production because the male-sterile female inbred cannot be self-pollinated. Recently, *ms45*, a recessive genetic male-sterile mutant, was cloned and a hybrid seed production technology (SPT) was developed (Wu *et al.*, 2015). The SPT construct contained a wild-type *Ms45* gene for fertility restoration, an α-amylase gene to disrupt pollination and a seed colour marker for seed sorting. The maintainer line containing the SPT construct was fertile and shed 50% nontransgenic *ms45* pollen and 50% transgenic *ms45* pollen. Pollen harbouring the SPT transgene were unable to germinate due to depletion of starch by α-amylase, whereas pollen that did not carry the SPT transgene were able to fertilize homozygous mutant plants. The resulting homozygous *ms45* seeds were nontransgenic, with respect to the SPT transgene, and could be used for hybrid seed production. Other transgenic male sterility systems have also been reported, such as the split-gene system for hybrid wheat (Kempe *et al.*, 2014) and the SeedLink for rapeseed from Bayer Crop Science (Newhouse *et al.*, 1996).

During the early phases of reproductive growth (initiation of the ear shoot and tassel) in maize, tassel development outcompetes ear development for nitrogen and other nutrients, especially under stress conditions when resources are limited. We hypothesize that a reduction in tassel apical dominance through male sterility may promote greater ear development at anthesis and improve grain yield under stress conditions. *Ms44* is a dominant male-sterile mutant identified from an EMS mutagenized population (Albertsen and Trimnell, 1992) and can be used easily to produce male-sterile hybrids. In this study, we cloned the *Ms44* gene which encodes a putative lipid transfer protein expressed specifically in the tapetum. A single amino acid change from alanine to threonine at the predicted secretory signal sequence cleavage site was responsible for dominant male sterility. Expression of an artificial microRNA targeting and silencing the *Ms44* gene restored male fertility fully in an *Ms44* mutant background. *Ms44* male-sterile plants showed a reduced tassel growth and enhanced ear growth, resulting in an average of 9.6% increase in kernel number per ear. Hybrids carrying the *Ms44* allele showed a 4%–8.5% yield increase under N-limited conditions. These findings indicated that protein secretion from tapetal cells into the locule was essential for proper pollen development and that a reduction in competition between tassel and ear by male sterility improved grain yield under low-N conditions.

Results

Ms44 map-based cloning and characterization

Ms44 was previously mapped to chromosome 4 (Albertsen and Trimnell, 1992). To understand the molecular basis of dominant male sterility, *Ms44* was isolated by map-based cloning (Figure 1a). In a large population of 2686 individuals, *Ms44* was validated as mapping to chromosome 4 and narrowed down to a 109-kb region between markers 9212_4 and 2221_3. There were five annotated genes in this interval, and AC225127.3_FGT003 was selected as the apparent candidate based on its male preferred expression pattern (Wright *et al.*, 1993). Sequence analysis of AC225127.3_FGT003 from the *Ms44* mutant revealed a single base change in the gene as compared to its original W23 parent. To verify that this single base pair change was the causal *Ms44* mutation, the allele was transformed into wild-type maize. Expression of the candidate

Ms44 genomic clone that included 1.2 kb upstream and 0.8 kb downstream from the coding sequence resulted in a male-sterile phenotype (Figure 1b), confirming that AC225127.3_FGT003 was the *Ms44* gene.

The *ms44* coding region is comprised of two exons and contains a small 101-bp intron. The first exon of *ms44* encodes all but the last two amino acids of the predicted protein. The *Ms44* encodes a lipid transfer protein containing a domain found in the superfamily of bifunctional inhibitor/plant lipid transfer protein/seed storage helical domain proteins. The *Ms44* protein contains eight conserved cysteine residues, important for secondary structure, and belongs to the type III or type C nonspecific lipid transfer protein (nsLTP), based on sequence analysis (Figure 1c). Homology searches with the *Ms44* protein revealed a large number of male-expressed plant-specific proteins, all of which have homology to ns-LT proteins. Shown in Figure 1c is an alignment of the N-terminal region of these proteins which show a higher variability than the C-terminal region, but reveal that all contain a predicted secretory signal sequence (SSS), which is predicted to be processed in the endoplasmic reticulum (ER) for extracellular secretion (Von Heijne, 1986).

Spatiotemporal expression and protein localization of *Ms44*

Ms44 is an anther-specific gene. Expression was first detected during meiosis and persists through the quartet and uninucleate stages of microspore development, but was not found after the first pollen mitosis (Figure S1a–b). To assess spatial expression of the *ms44* gene, transgenic constructs were made using the *ms44* promoter driving the expression of the fluorescent protein Zs-Green. As shown in Figure S1c, fluorescence was detected only in the tapetal cell layer of the anther. The signal was detectable but weak during meiosis and increased during quartet and microspore release stages of development, consistent with the temporal expression data.

The SNP found in the *Ms44* allele (G to A) causes an amino acid change from an alanine to a threonine at amino acid 37. The mutation is 14 amino acids downstream of the predicted SSS cleavage site, which is predicted to cleave between G23 and G24. However, comparing predicted SSS sites between the other male specific nsLTPs in Figure 1c reveals a discrepancy between the predicted sites. Half of the compared proteins have a predicted SSS cleavage site equivalent to the G23/G24 position, while the others are predicted to cleave at the comparable A37/Q38 position. Notably, the barley protein had roughly equivalent SSS scores at both positions. Additionally, the comparable A37/Q38 residues are conserved in these sequences except for the Lily sequence which has a S/Q, whereas the G/G site is not conserved. Therefore, there is a question as to where SSS cleavage occurs in Ms44, and whether the mutation from an alanine to a threonine at amino acid 37 affects this process. To investigate whether the A37/Q38 are critical for signal sequence cleavage, point mutations were made around the A37 amino acid. If cleavage occurs between A37 and Q38, A37 can be considered to be at the −1 position, with the Q38 being at the +1 position or the first amino acid in the mature form of the protein. Amino acid changes were made at both the −1 and +1 positions and were chosen using consensus amino acid patterns near signal sequence cleavage sites (Von Heijne, 1986). At the −1 position, variant clones were made having A37G or A37V, with a single variant at the +1 position, Q38P. The A37G was the only variant that would be predicted to cleave.

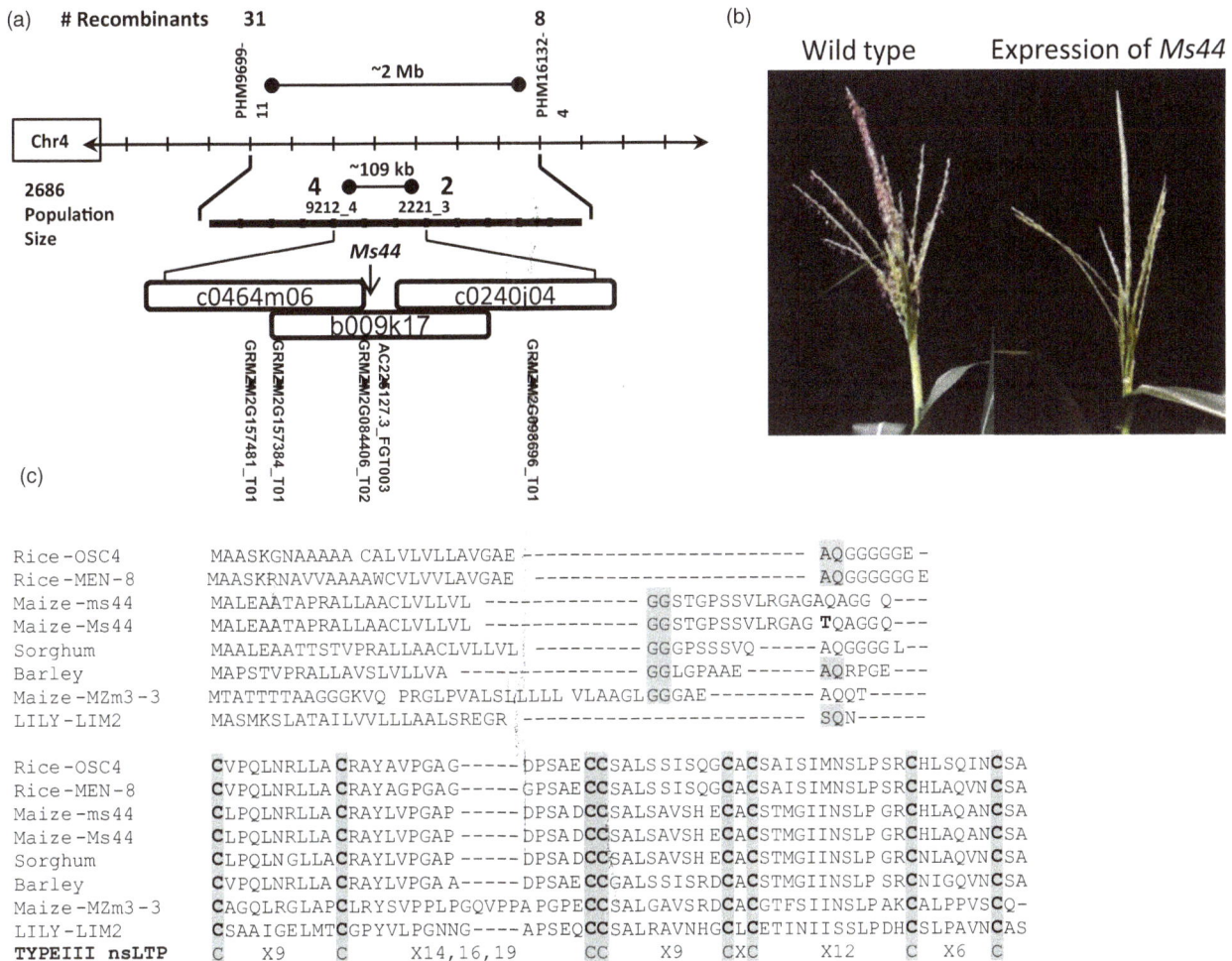

Figure 1 Ms44 map-based cloning. (a) Fine mapping of Ms44. (b) Expression of Ms44 genomic in wild-type maize. Ms44 genomic fragment was isolated from Ms44 mutant and introduced into wild type. Wild-type plant with expression of Ms44 dominant allele showed male-sterile phenotype. (c) Protein sequence alignment of Ms44 with other plant LTPs. The N-terminal sequence up to the first conserved cysteine is hand-aligned. SSS sites predicted by SignalP are shaded grey. The barley sequence had two highly predicted SSS sites. The threonine mutation in Ms44 is red bold. The remaining protein is a CLUSTALW alignment. Conserved cysteine residues are shaded bold and compared to consensus type III nsLTPs. The accession numbers for the LTPs are as follows: LILY-LIM2 (Q43534); Sorghum (XP_002445754); Barley (BAK05897); Rice-OSC4 (BAD09233); Rice-MEN-8 (XP_006660357); and Maize-MZm3-3 (NP_001105123).

A eukaryotic cell-free in vitro synthesis system containing the machinery required for signal peptide processing was employed to measure the effect of Ms44 amino acid changes on protein processing. As shown on the protein blot in Figure 2a, only the wild-type A37 and the A37G variant had the majority of the detectable Ms44 protein processed to a lower molecular weight form. The A37T, A37V and Q38P showed little to no protein processing, having the majority of Ms44 protein uncleaved. This indicates that A37 and Q38 are critical for signal peptide cleavage and confirms that the T37 found in the dominant Ms44 mutant abolishes this processing. Ultimately to demonstrate that a lack of SSS processing in the Ms44 protein is responsible for the dominant male sterility phenotype, we used the amino acid variant clones of ms44 described above and transformed them into maize plants. Single-copy plants were selected and grown to maturity where male fertility phenotypes were evaluated. As shown in Figure 2b–d, transgenic plants containing cleavable A37G variant were male fertile, while variants that disrupt Ms44 signal peptide cleavage, A37V and Q38P, were completely male-sterile. These results confirm that a lack of Ms44 signal peptide cleavage results in a disruption of pollen development leading to dominant male sterility in maize.

To confirm that Ms44 mutation affects protein processing and secretion, both wild-type and mutant Ms44 genes were fused with AcGFP, and separately introduced into plants to assess spatial localization of the two Ms44 protein forms. Anthers were harvested for analysis using both light and confocal microscopy. As shown in Figure 2e, the Ms44 mutant protein fusion was confined only to the tapetal cell layer of the anther (Figure S2). In contrast, the wild-type ms44 protein fusion was found mainly in the locule. This result indicated that in the absence of proper protein processing through cleavage of the secretory signal peptide, secretion of the Ms44 protein into the locule was blocked.

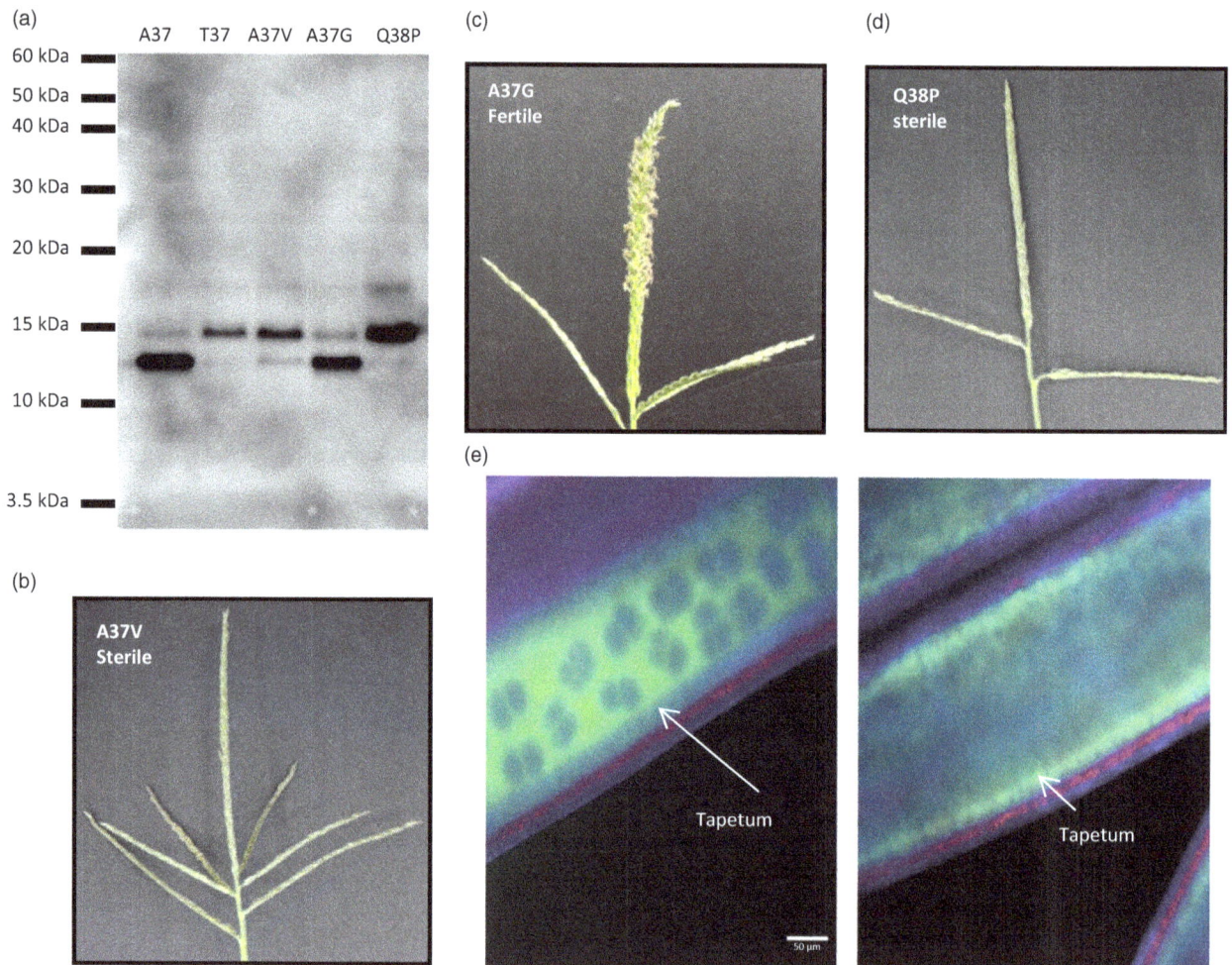

Figure 2 Disruption of signal peptide processing by *Ms44* mutation. (a) *In vitro* protein processing of Ms44 variants. (b) Transgenic tassel of Ms44 A37-to-V37 change. (c) Transgenic tassel of Ms44 A37-to-G37 change. (d) Transgenic tassel of Ms44 Q38-to-P38 change. (e) Localization of MS44::AcGFP fusion proteins in anthers at dyad-to-tetrad stages of microspore development by wide-field fluorescence microscopy. Wild-type ms44::AcGFP fusion (left) and mutant Ms44::AcGFP fusion (right) were expressed under the *ms44* promoter. Red signal indicates autofluorescence of chloroplasts in the endothecium. Tapetum cell layer is indicated by arrow.

Ms44 plants increase N partitioning from tassel to ear and improve ear development

Tassel competes with ear for nutrients during early reproductive stages. To determine whether *Ms44* dominant male sterility reduces competition between tassel and ear growth, we measured shoot, tassel, ear biomass and total N content at early reproductive stages from V9 to V17. Shoots from *Ms44* sterile plants did not show a significant difference in biomass or total N content compared to wild type, but the male sterility reduced the accumulation of biomass and N in tassels with an increase in ear biomass and total ear N (Table S1; Figure 3). At V17, male-sterile plants showed a 64.7% reduction in tassel biomass with a 24.6% increase in ear biomass (Figure 3). More N shifted from tassel to ear growth in sterile plants compared to wild type. As a result, *Ms44* sterile plants produced an average of 9.6% more kernels per ear (35 kernels ear^{-1}) than wild-type plants (Figure 4).

Ms44 male-sterile hybrids increase grain yield under low-N conditions

To determine whether an increase in kernel number by *Ms44* male sterility translated into higher grain yield, four *Ms44* elite male-sterile inbreds were each pollinated by 4–5 male testers to produce hybrid seeds for yield trials. Hybrid seed harvested from the heterozygous *Ms44* female plants segregated 50% fertile and 50% male-sterile (named *Ms44* hybrid). Hybrid seed harvested from wild-type female plants were 100% fertile and were used as control. We conducted field trials in multiple locations to determine whether the *Ms44* hybrid resulted in a yield advantage compared to fertile sibling controls under limited N, drought and optimal growth conditions. When N was limiting, *Ms44* hybrids increased yield by an average of 4% (352 kg ha^{-1}), ranging from 3.3% to 5.7% in 17 hybrids tested under low-N conditions, and an average of 8.5% (579 kg ha^{-1}), ranging from 6.7% to 10.6% in 10 hybrids tested under ultralow-N conditions, compared to the fertile control. All hybrids showed higher yield increases under extreme low-N conditions compared to mild-N stress conditions (Tables 1, S2). When water was limiting during the grain filling period, *Ms44* hybrids increased yield significantly by an average of 1.6% (126 kg ha^{-1}). *Ms44*, in three of 17 hybrids, showed a significant increase in yield but did not show a significant yield increase across all 17 hybrids under drought stress during the flowering period (Tables 2, S2). Under optimal conditions, *Ms44* hybrids increased grain yield significantly by an average of 0.9%

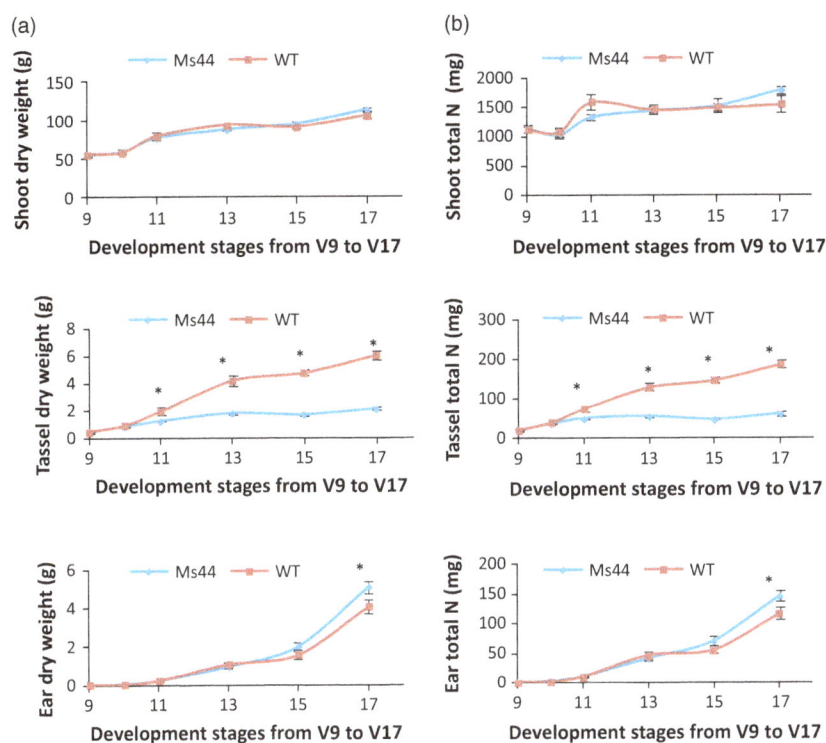

Figure 3 Effect of *Ms44* male sterility on growth and total N content at early reproductive stages. (a) Biomass accumulation of shoot, tassel and ear in *Ms44* male-sterile plant (Ms44) relative to wild-type fertile plant (WT) at early reproductive stages from V9 to V17. Tassel is visible fully at V17. Ear weight includes husk. Data are mean ± SEM; five plants were sampled from eight replicated plots for each data point. (b) Total N content of shoot, tassel and ear in *Ms44* male-sterile plant relative to wild-type fertile plant at early reproductive stages from V9 to V17. Data are mean ± SEM; five plants were sampled from eight replicated plots for each data point. * indicates a significant difference at $P < 0.01$.

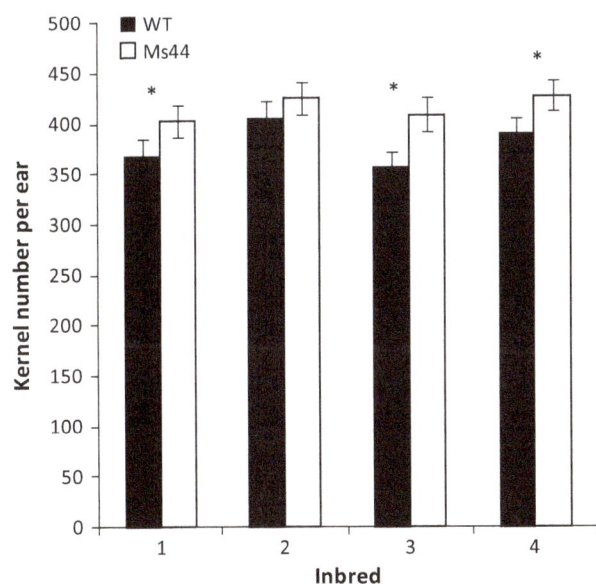

Figure 4 Kernel number per ear of *Ms44* inbred relative to wild-type inbred. Ears harvested from male-sterile plants and fertile wild-type plants were digitally imaged to determine kernel number. * indicates that the kernel number per ear from the *Ms44* sterile inbred was significantly greater than the wild-type inbred at the $P < 0.05$ confidence level.

(126 kg ha^{-1}; Tables 1, S2). *Ms44* hybrids showed no yield penalty in any of the tested environments, which is desirable for commercially viable products.

Transgenic *Ms44* maintainer line development

Dominant male-sterile mutants such as *Ms44* pose additional challenges for seed propagation as the plants carrying the *Ms44*

mutation cannot be self-pollinated. Given that *Ms44* male-sterile hybrids outperform their fertile siblings under N-limited conditions, we have modified the previously described seed production technology (SPT; Wu *et al.*, 2015) to create a transgenic maintainer line. Instead of complementing a recessive male-sterile mutation with a wild-type allele, *Ms44* fertility can be restored by silencing *Ms44* expression using an artificial miRNA (amiRNA). The pollen nonviability and seed colour marker cassettes in the maintainer construct (*AG533*) were identical to the previously described SPT (Figure 5a). The *AG533* construct was transformed into a heterozygous *Ms44* inbred line. As expected, silencing the expression of *Ms44* in mutant plants fully restored male fertility (Figure 5b). Because the amiRNA was directed against the coding sequence of *Ms44*, both the dominant sterility allele and the wild-type allele were targeted and silenced. Interestingly, the wild-type plants containing the *AG533* construct remained fertile, suggesting that the *ms44* gene is not absolutely required for male fertility. Expression of α-amylase using the pollen-specific *PG47* promoter reduced starch content in the transgenic pollen grains and resulted in those being nonviable (Figure 5c). Seeds expressing red fluorescent protein using an aleurone-specific Ltp2 promoter were visually distinguishable from nontransgenic yellow seed under appropriate illumination (Figure 5d). Lines with the *AG533* insertion segregating independently from the native *Ms44* locus were selected for advancement. T1 transgenic lines (*Ms44/−*; *AG533/−*) were self-pollinated to obtain *Ms44* homozygous mutant (*Ms44/Ms44*; −/−) and sibling maintainer lines (*Ms44/Ms44*; *AG533/−*). *Ms44* homozygous mutant plants clearly showed larger ears with more silks and were shorter when compared to the fertile maintainer plants, but ear height was similar between the two (Figure 5b). For the maintainer line, the *Ms44* allele was present in all pollen, with half of that harbouring the unlinked *AG533* construct. Because *AG533* pollen grains were not viable due to α-amylase activity (Figure 5c), the transmission of the

Table 1 Grain yield of *Ms44* hybrid and wild type under optimal growth conditions and nitrogen-limited conditions

	Optimal growth conditions				Low-nitrogen conditions				Ultralow-nitrogen conditions			
Hybrid	Isogenic hybrid yield prediction Mg/ha	Ms44 hybrid yield prediction Mg/ha	% change	N	Isogenic hybrid yield prediction Mg/ha	Ms44 hybrid yield prediction Mg/ha	% change	N	Isogenic hybrid yield prediction 1 Mg/ha	Ms44 hybrid yield prediction Mg/ha	% change	N
1	14.09 ± 0.19	14.23 ± 0.19	1.0*	33	9.27 ± 0.24	9.68 ± 0.24	4.4**	15	6.69 ± 0.31	7.20 ± 0.31	7.7**	10
2	14.13 ± 0.18	14.29 ± 0.18	1.1*	25	9.13 ± 0.19	9.52 ± 0.19	4.3**	24	6.97 ± 0.32	7.54 ± 0.32	8.2**	10
3	14.42 ± 0.19	14.54 ± 0.19	0.8	19	10.03 ± 0.22	10.38 ± 0.22	3.5**	19	8.10 ± 0.31	8.64 ± 0.31	6.7**	10
4	14.21 ± 0.19	14.30 ± 0.19	0.6	20	9.79 ± 0.22	10.16 ± 0.22	3.8**	19	7.34 ± 0.32	8.01 ± 0.32	9.2**	10
5	14.21 ± 0.19	14.35 ± 0.19	1.0	19	9.49 ± 0.22	9.82 ± 0.22	3.5**	20	7.35 ± 0.31	7.89 ± 0.31	7.4**	10
6	13.96 ± 0.18	14.15 ± 0.18	1.4*	24	8.76 ± 0.19	9.12 ± 0.19	4.1**	24	6.63 ± 0.31	7.33 ± 0.31	10.7**	10
7	13.76 ± 0.19	13.86 ± 0.19	0.7	36	8.94 ± 0.24	9.24 ± 0.24	3.3**	14	6.84 ± 0.32	7.40 ± 0.31	8.2**	10
8	13.94 ± 0.18	14.05 ± 0.18	0.8	36	8.83 ± 0.24	9.16 ± 0.24	3.7**	14	7.30 ± 0.31	7.86 ± 0.31	7.6**	10
9	13.93 ± 0.19	14.13 ± 0.19	1.4*	18	8.80 ± 0.22	9.20 ± 0.22	45**	20	6.52 ± 0.31	7.12 ± 0.31	9.2**	10
10	13.56 ± 0.18	13.66 ± 0.18	0.7	24	8.29 ± 0.19	8.59 ± 0.19	3.7**	24	6.35 ± 0.31	6.94 ± 0.31	9.3**	10
11	13.98 ± 0.19	14.09 ± 0.19	0.7	21	9.47 ± 0.22	9.80 ± 0.22	3.4**	19				
12	13.87 ± 0.22	14.03 ± 0.22	1.2*	10	8.61 ± 0.24	9.00 ± 0.24	4.4**	15				
13	13.41 ± 0.18	13.58 ± 0.18	1.2*	23	7.73 ± 0.19	8.17 ± 0.19	5.7**	23				
14	13.59 ± 0.21	13.71 ± 0.21	0.9	13	8.10 ± 0.24	8.44 ± 0.24	4.1**	16				
15	13.30 ± 0.19	13.40 ± 0.19	0.8	21	8.03 ± 0.22	8.37 ± 0.22	4.2**	20				
16	13.89 ± 0.21	14.01 ± 0.21	0.8	11	8.70 ± 0.24	9.03 ± 0.24	3.7**	16				
17	13.47 ± 0.19	13.55 ± 0.19	0.6	20	8.27 ± 0.22	8.55 ± 0.22	3.4**	19				

Yield trials and statistical analysis were conducted as described in Methods. *Ms44* hybrid was segregating 50% male-sterile plants and 50% fertile wild-type plants. Grain yield of *Ms44* hybrid was compared to the wild-type hybrid with 100% fertile plants. The LN treatment targeted a yield reduction of 30% through N limitation, and the ULN treatment (in which only 10 hybrids were tested) targeted a 50% yield reduction through N limitation. Data shown are the predicted yield ±SE. *N* is the total number of plots tested.
*Significant at $P < 0.05$.
**Significant at $P < 0.01$.

AG533 construct occurred only through the female gametes. Thus, crossing the homozygous *Ms44* male-sterile inbred with pollen from the maintainer resulted in only nontransgenic yellow seed progeny.

In practice, the maintainer line can be increased by self-pollination followed by the selection of red seeds (Figure S3a). The homozygous *Ms44* male-sterile inbred can be increased with pollen from the sibling maintainer. The progeny seed should be yellow in colour and not contain the transgenic *AG533* construct (Figure S3b). In hybrid seed production, the female inbred is completely male-sterile (*Ms44/Ms44*), and any cross with a male parent (*ms44/ms44*) will result in F1 hybrid seed that also will be male-sterile (*Ms44/−*) due to dominance. To enable adequate pollination in the grower's field, blending of the male-sterile hybrid with a male fertile hybrid will be required at a certain percentage. This can be accomplished by adding a step after the initial maintainer line cross during inbred increase. Pollination of the male-sterile *Ms44* homozygous inbred with the wild-type inbred will generate heterozygous *Ms44* male-sterile female inbred plants. Subsequent crosses with any male inbred during hybrid production will result in 50% male-sterile hybrid progeny (Figure S3c).

Discussion

Plant nonspecific lipid transfer proteins (nsLTPs) are small proteins with eight conserved cysteine residues and unique secondary protein structures. All LTPs have an N-terminal ER-targeting signal peptide for processing in the secretory pathway. This family of protein involved in numerous biological processes, such as lipid transfer, pathogen defence, abiotic stress response and anther development (Liu *et al.*, 2015). The maize genome contains 63 nsLTP genes which can be divided into five types (Wei and Zhong, 2014). *Ms44* encodes a lipid transfer protein containing eight conserved cysteine residues and has a secondary protein structure that falls within the type III or alternatively named type C (Figure 1; Boutrot *et al.*, 2008; Edstam *et al.*, 2011; Wei and Zhong, 2014) of nsLTPs, due to its intron position 1 bp past the last conserved cysteine (C8) codon and the presence of twelve amino acid residues between C6 and C7, which is specific for this class. The type III or type C is found only in plants that produce pollen and seeds (Boutrot *et al.*, 2008; Edstam *et al.*, 2011). LTP transcripts have been shown to be highly abundant in male organs, comprising 8% of the total transcripts found in rice anthers and over 10% of tapetal transcripts in *Arabidopsis*, and may play an important role in anther development (Huang *et al.*, 2009). In *Arabidopsis* and rice, type III LTPs have tapetal specific expression coupled with an ER-targeting signal sequence, and are secreted from the tapetum via the ER–Golgi system and then become components of the microspore exine (Huang *et al.*, 2013; Zhang *et al.*, 2010). The maize *ms44* gene expresses specifically in tapetal cells. Fusions of the wild-type ms44 gene with GFP (ms44-GFP), as shown in Figure 2e, demonstrate that the Ms44 protein is also secreted into the locule via the ER–TGN. In contrast, the *Ms44* mutant protein is confined to tapetal cells and is not secreted into the locules (Figures 2e, S2). A single amino acid change at the signal peptide cleavage site disrupts cleavage of the signal peptide and blocks secretion of Ms44

Table 2 Grain yield of *Ms44* hybrid and wild type under drought stress conditions

Hybrid	Drought stress during flowering time				Drought stress during grain filling			
	Isogenic hybrid yield prediction Mg/ha	Ms44 hybrid yield prediction Mg/ha	% change	N	Isogenic hybrid yield prediction Mg/ha	Ms44 hybrid yield prediction Mg/ha	% change	N
1	7.70 ± 0.20	7.92 ± 0.20	2.8**	55	8.62 ± 0.27	8.78 ± 0.27	1.9	22
2	7.78 ± 0.25	7.97 ± 0.25	2.4*	15	8.92 ± 0.26	9.06 ± 0.26	1.5	19
3	8.25 ± 0.26	8.37 ± 0.27	1.5	9	8.97 ± 0.28	9.08 ± 0.28	1.2	14
4	7.92 ± 0.26	8.03 ± 0.26	1.5	10	8.73 ± 0.28	8.89 ± 0.28	1.8	11
5	7.86 ± 0.26	7.99 ± 0.26	1.6	12	8.81 ± 0.28	9.02 ± 0.28	2.4*	14
6	7.55 ± 0.25	7.68 ± 0.25	1.7	14	8.57 ± 0.26	8.73 ± 0.26	1.8	19
7	7.50 ± 0.20	7.58 ± 0.20	1.1	55	8.50 ± 0.27	8.64 ± 0.27	1.7	23
8	7.18 ± 0.20	7.25 ± 0.20	0.9	55	8.24 ± 0.27	8.38 ± 0.27	1.8	23
9	7.51 ± 0.26	7.70 ± 0.26	2.5*	12	8.43 ± 0.28	8.58 ± 0.28	1.8	13
10	7.33 ± 0.25	7.45 ± 0.25	1.6	15	8.30 ± 0.26	8.43 ± 0.26	1.6	18
11	7.84 ± 0.26	7.93 ± 0.26	1.1	11	8.41 ± 0.28	8.55 ± 0.28	1.7	14
12	7.55 ± 0.28	7.71 ± 0.28	2.1	6	8.60 ± 0.30	8.70 ± 0.30	1.2	10
13	6.73 ± 0.25	6.86 ± 0.25	2.0	15	8.10 ± 0.26	8.25 ± 0.26	1.8	17
14	6.92 ± 0.28	7.04 ± 0.28	1.7	6	8.19 ± 0.30	8.30 ± 0.30	1.4	10
15	6.66 ± 0.26	6.76 ± 0.26	1.5	11	7.67 ± 0.28	7.82 ± 0.28	2.0	14
16	7.50 ± 0.28	7.58 ± 0.28	1.2	7	8.49 ± 0.30	8.59 ± 0.30	1.2	8
17	6.83 ± 0.26	6.88 ± 0.26	0.7	12	7.96 ± 0.28	8.02 ± 0.28	0.7	15

Yield trials, drought treatment and statistical analysis were conducted as described in Methods. *Ms44* hybrid was segregating 50% male-sterile plants and 50% fertile wild-type plants. Grain yield of *Ms44* hybrid was compared to the wild-type hybrid with 100% fertile plants. Data shown are the predicted yield ±SE. *N* is the total number of plots tested.
*Significant at $P < 0.05$.
**Significant at $P < 0.01$.

protein into the locule, resulting in dominant male sterility (Figure 2). Interestingly, the maize wild-type *ms44* gene is not required for fertility as plants with silencing of the wild-type *ms44* gene showed normal pollen development. The A9 gene in *Brassica* was shown not to be required for male fertility by an antisense knockdown experiment and no phenotype was found for RNA interference knockdown (RNAi) of individual type III LTPs in *Arabidopsis*, although RNAi targeting two *Arabidopsis* type III LTPs did have some effect on intine morphology but not overall male fertility (Huang *et al.*, 2013; Turgut *et al.*, 1994). In contrast, silencing of OsC6 reduced pollen fertility in rice (Zhang *et al.*, 2010). This may not be surprising given the abundance of anther-expressed LTPs. Maize has three closely related type III LTP genes (Wei and Zhong, 2014). Clearly, the Ms44 protein itself is not critical for proper pollen development, but the single amino acid change at the secretory signal cleavage site results in dominant male sterility phenotype, indicating that tapetum secretion is a critical step for the dominant phenotype. It would be interesting to determine whether secretion of other proteins from tapetal cells is impeded in the *Ms44* mutant, or whether blocking protein secretion by chemical BFA treatment would also lead to dominant male sterility.

Maize hybrid seed production requires the crossing of two inbred parent lines to produce the F1 hybrid seed sold to a grower. The female inbred parent needs to be prevented from shedding pollen to ensure pure hybrid seed production. Relative to other hybrid seed production systems, *Ms44* technology has four advantages (Chen and Liu, 2014; Kempe and Gils, 2011; Perez-Prat and van Lookeren Campagne, 2002). First, when compared to detasseling, *Ms44* inbred lines show an average of 9.4% increase ($P < 0.05$) in kernel numbers per ear (34.7 kernels

ear^{-1}) over wild type (Figure 4). Hybrid seed yield from the female inbred is important for the cost of hybrid seed production. An increase in female inbred yield, especially in the number of kernels, reduces hybrid seed production costs. Although effective, mechanical detasseling is time-consuming and labour-intensive and can be imperfect. It also reduces seed yields due to damage to the upper leaves and stalk. Second, compared to other sterility technologies (Feng *et al.*, 2014; Weider *et al.*, 2009), such as cytoplasmic male sterility (CMS) or Roundup hybridization system (RHS), *Ms44* is a native genetic male-sterile mutant that has shown to be stable across different germplasms and broad environments. CMS technology is restricted to certain germplasms and may result in poor sterility stability in inbreds or poor fertility restoration in hybrids depending on growth environments. The transgenic RHS is based on tissue specific expression of a 5-enolpyruvylshikimate 3-phosphate synthase (EPSPS). Application of glyphosate herbicide at V8-to-V13 growth stages prevents pollen development and eliminates the production of viable pollen in the female inbred parent during the hybrid production phase. Large-scale spraying of chemicals across all production field acreages is required at a narrow developmental window for this technology to be effective. Hybrid production costs increase, and hybrid purity and production itself may be at risk if complete sterility is not achieved in the female parents. This could be a consequence of field variability as well as environmental variability. Third, although *Ms44* inbred seeds are propagated using a transgenic maintainer line, the progeny do not inherit the transgenic insertion; thus, commercial hybrid seeds and commodity grain are nontransgenic, with respect to the maintainer transgenes. Finally, the dominant *Ms44* hybrid production system is an elegant and simple method to produce

(a)

| Pollen expression of α-amylase | Seed expression of Ds-RED | Ms44 silencing |

(b)

(c)

(d)

Ms44/Ms44
mutant

Ms44/Ms44;
AG533/–
maintainer

Figure 5 Development of transgenic Ms44 maintainer line. (a) AG533 construct map. The Maintainer construct containing three expression cassettes was transformed into a male-sterile inbred which was heterozygous for Ms44. (b) Ms44 homozygous mutant plant (left) and maintainer plant (right) with fertility restored by expressing AG533. (c) Transgenic pollen fertility ablated by overexpression of α-amylase in pollen. Pollen grains were stained for starch granules using potassium iodide solution (I2KI). The transgenic pollen containing AG533 were not viable for pollination due to starch breakdown by α-amylase (brown stained pollen). The wild-type pollen (dark stained) were viable for pollinating male-sterile inbred. (d) Expression of DsRed2 in seed as a marker for seed sorting. Transgenic maintainer line seed is red, while nontransgenic seeds are yellow (lower panel is under fluorescence light).

blends of male-sterile F1 hybrids. Recessive ms45 male-sterile mutant needs to be bred into both the male and female inbred parents to produce sterile hybrid F1 plants, while the dominant Ms44 only needs to be in the female inbred parent, reducing breeding efforts by half. Another benefit of the Ms44 dominant system is that sterility is 100% maintained in the hybrid production field even with contamination during inbred increase, whereas, for ms45 recessive sterility, any contamination in the inbred increase field results in fertile females in the hybrid production field.

Maize hybrid yields have been increased over the past 50 years, in part, by indirectly selecting for reduced tassel size (Duvick and Cassman, 1999). Reduction in tassel dominance through a dominant genetic male sterility mechanism with a concomitant yield increase is consistent with the indirect selection for smaller

tassels performed by maize breeders. Multiple studies have indicated that detasseling and cytoplasmic male-sterile hybrids improved grain yield under higher population densities for some cultivars (Chinwuba et al., 1961; Duvick, 1958). We report here that dominant Ms44 male-sterile hybrid improved grain yield under low-N conditions across all 17 hybrids tested. The efficient use of N can be improved either by increasing N uptake and assimilation or by increasing N partitioning to grain production. Numerous studies have shown that manipulation of candidate genes involved in N uptake, assimilation, root development and N signalling improved NUE using transgenic approaches, but none have demonstrated a significant improvement of NUE in field trails in elite germplasm (McAllister et al., 2012; Xu et al., 2012). The most advanced lead is alanine aminotransferase. Expression of a barley alanine aminotransferase in canola and rice increased biomass and seed yield under low-N conditions (Good et al., 2007; Shrawat et al., 2008). To our knowledge, this is the first report demonstrating that male sterility caused by a single point mutation in Ms44 improves N use for grain production. The blends of Ms44 male-sterile F1 hybrids increase maize yield under stress conditions in elite hybrids, especially under nitrogen-limited conditions across multiple location field trials (Table 1). While Ms44 male sterility does not change the total N content, the sterility improves N utilization efficiency by reducing the N use in tassel and pollen development and portioning more N to immature ear development, resulting in an increase in kernel number (Figures 3, 4). Maize is the most widely cultivated crop in sub-Saharan Africa and provides 70% of the total human caloric intake. Almost 80% of African countries are challenged with nitrogen scarcity in the soil with an average maize yield of 1.6 Mg/ha. On average, farmers apply less than 20 kg/ha N fertilizer in Africa (Potter et al., 2010; Vitousek et al., 2009). Ms44 has the greatest potential to increase maize grain yield without increasing N application in countries where farmers cannot afford to apply more N fertilizer but more food production is needed (Edmonds et al., 2009; Liu et al., 2010). Further testing is needed to validate whether Ms44 increases grain yield in local germplasms in African low-N soil. In addition, the Ms44 hybrid seed production system provides a cost-effective method to produce high-quality hybrid seed, which can promote the adoption of hybrid maize in this region. Currently, only 20% of the total maize in Africa is hybrid (Gaffney et al., 2016). In North America, although farmers apply sufficient amounts of nitrogen, more than 10% of maize fields are still confronted with nitrogen stress due to nitrate leaching and run-off. Furthermore, N levels can vary significantly across a single field and parts of the field may experience yield loss due to N stress. Ms44 male-sterile hybrids provide an insurance towards stabilizing grain yields if nitrogen stresses occur. The successful use of Ms44 hybrid seed production technology would not only reduce maize hybrid seed production costs and protect yield under low-nitrogen conditions, but would also provide benefits to the environment resulting from increased nitrogen use efficiency in maize.

Experimental procedures

Map-based cloning and gene characterization

Ms44, a dominant male-sterile mutant, was initially identified from an EMS mutagenized W23 population and mapped to Chr4 (Albertsen and Trimnell, 1992). A map-based cloning approach was used to identify the gene and mutation responsible for the sterile phenotype. A BC6-F2 segregating population was generated by backcrossing *Ms44* into B73 and then crossed to Mo17. Single nucleotide polymorphism (SNP) markers were selected across the previously identified region of Chr4, and 414 individuals were used to confirm the *Ms44* interval. Subsequently, 2686 individuals were genotyped and the *Ms44* interval was defined between PHM9699-11 (~193 701 964 bp, B73Ref_v3) and PHM16132-4 (~195 800 812 bp, B73Ref_v3), with 39 recombinants within the interval. Additional SNP markers were developed and the Ms44 interval was further delimited between two CAPS markers, 9212_4 (~195 191 010 bp, B73Ref_v3, 9212_4 forward primer CAGTCCTGCTCGGAG CTTGCTT/reverse primer ACCGAAGGATGCCTGGGAAT) and 2221_3 (~195 300 664 bp, B73Ref_v3, 2221_3 forward prime AGTTGTTGTGCTTGAAGTACTTGGG/reverse primer GGTCA-TAGGCTTTCAAGTGTACACA).

Vector construction, plant transformation and transgene expression analysis

The *Ms44* ('MS44DOM') gene was PCR-cloned from maize genomic DNA based on the original mapping data. It comprises the 413-bp coding sequence (two exons interrupted by a 101-bp intron), 1221 bp of upstream promoter sequence (including the 5′ UTR) and 722 bp of downstream genomic sequence (including the 3′ UTR). The complete gene was ligated into a vector containing *Agrobacterium* T-DNA RB and LB components, plus a plant selectable marker (CAMV35S PRO::PAT). This vector was mobilized into *Agrobacterium tumefaciens* LBA4404 cells carrying extra copies of several VIR genes (pSB1 (JT reference)) to create plant transformation vector PHP42163 which was introduced into wild-type maize. The tassel phenotype was recorded at T0 plants in glasshouse.

An artificial microRNA was designed to silence specifically the *Ms44* allele, using the scaffolding of the maize miR396 h pri-miRNA as template. The endogenous miR396 h targeting sequence was replaced with TCTTATTCCTCTCCCCTCCTG and the endogenous star sequence replaced with CAGGAGGG-CAGAGGAATAAGA. The resulting artificial microRNA was synthesized by GenScript and placed under the control of the MS44 DOM promoter and terminator sequences described above. Gateway™ homologous recombination-based cloning was used to move this expression cassette into a binary transformation vector and stack it with the pollen-specific α-amylase cassette and the *DsRed2* colour marker gene previously described (Wu *et al.*, 2015). The final vector AG533 was introduced into *Ms44* heterozygous inbred line to generate maintainer lines.

Maize transformation was carried out as described previously (Unger *et al.*, 2001). Plasmid was co-integrated into a pSB1 vector in *Agrobacterium tumefaciens* strain LBA4404 and then used to transform maize embryos from a proprietary inbred. Multiple lines were generated for each construct. Single-copy T-DNA integration lines that expressed the transgene were selected for advancement to glasshouse or field test.

Fluorescence microscopy, pollen staining and seed sorting

Male spikelets were harvested and anthers were dissected out of the florets and mounted in 1× PBS on glass slides. Light microscopy images were taken with a Leica (Wetzlar, Germany) DMRXA epifluorescence microscope with a mercury light source using fluorescence, bright-field and differential interference contrast (DIC) optics. The fluorescent filter sets used were from

Chroma Technology (Bellows Falls, VT): Alexa 488 #MF-105 (exc. 486–500, dichroic 505LP, em. 510–530), DAPI #31013 (exc. 360–370, dichroic 380LP, em. 435–485) and Cy3 #C-106250 (exc. 541–551, dichroic 560LP, em. 565–605). Images were captured with a Hamamatsu (Hamamatsu City, Japan) ORCA-Flash4.0 LT digital CMOS camera and images manipulated by Molecular Devices (Downingtown, PA) MetaMorph imaging software. Confocal images were taken with the Leica (Wetzlar, Germany) TCS SPE using the solid-state 405-nm laser line with the DAPI setting (exc. 350 and em. 461) and the 488-nm laser with the GFP setting (exc. 489 and em. 508) using the Leica LAS X software.

Pollen staining and seed sorting were carried out as described previously (Wu *et al.*, 2015).

Materials, yield trials, field experiment design and statistical analysis

The dominant male-sterile allele *Ms44* was backcrossed into four elite inbred maize lines. Two were dose 6 (BC5) and two were dose 4 (BC3) for the numbers of crosses to the recurrent parent. Four to five ear sources were selected for increase based upon positive marker calls for a donor insertion site, decreased insertion size and overall recurrent parent percentage. We harvested dose 5 (BC4) and dose 7 (BC6) inbred seed respectively to be used for trial hybrid production. These inbred lines segregated 1 : 1 for male sterility and male fertility. Molecular markers were utilized to assure genetic purity of the final inbred lines. It has approached 97% for the dose 5 seed and ~99% for the dose 7. The four converted *Ms44* female inbred parental lines were crossed to 4–5 fertile male inbred parental lines to produce 17 hybrids. The resulting F1 hybrid seed segregated 1 : 1 for male sterility and was used for yield trials. Male fertile sibs of female inbred lines were crossed to the same male inbred lines and used as 100% fertile controls to compare to *Ms44* hybrids segregating 1 : 1 for male sterility in yield trials.

During 2014, *Ms44* hybrids, segregating 1 : 1 for male sterility, were submitted to yield trials in a number of North American locations in which stress conditions (drought and reduced nitrogen fertility) were imposed. The experiment was planted at six optimal growth locations (two replications at each location) where N was applied at rates greater than 224 kg ha^{-1} and water was not limiting (Marion, IA; San Jose, IL; Sciota, IL; York, NE; and a second location in York, NE), at four low-N locations (four replications at each location) where N was applied at rates ranging from 45 to 78 kg/ha to target yield reduction by approximately 30% but water was not limiting (Johnston, IA; Marion, IA; Sciota, IL; and Woodland, CA), at two flowering stress locations (four replications per location) where N was not limiting and applied at rates greater than 168 kg ha^{-1} but water was withheld during flowering (Woodland, CA; and Plainview, TX), and at three grain filling stress locations (three replications per location) where N was not limiting and applied at rates greater than 168 kg ha^{-1} but water was withheld during the grain fill period (Fruitland, IA; Garden City, KS; and Plainview, TX). In addition, there were two locations that targeted an ultralow-nitrogen (ULN) treatment (in which only 10 hybrids were included) that targeted a 50% yield reduction through N limitation. One location with an ULN treatment was Woodland, CA, where 22 kg N ha^{-1} was applied at planting and an additional 45 kg N ha^{-1} in the irrigation water may be applied. At the second ULN location, Fruitland, IA, 112 kg N ha^{-1} was applied prior to planting. This location has a high percentage of sand, is irrigated frequently and has a high propensity for leaching N during the season. All low-N and ULN locations were on sites previously depleted of nitrogen. Urea–ammonium nitrate (UAN) was used as N fertilizer. At all locations, phosphorus and potassium levels were within the optimal range for maize production based on either soil testing results or application of fertilizer to ensure N was the main nutrient limiting crop yield.

An additional field trial was conducted in South America near Viluco, Chile, during the 2014–2015 growing season. The experiment was grown in a N-depleted location where N was limiting (0 kg N ha^{-1} applied at planting and 72 kg N ha^{-1} was applied via irrigation water), a location where water was withheld during the flowering period (total N applied during the season was 77 kg N ha^{-1}), a location where water was withheld during the grain filling period (total N applied during the season was 69 kg N ha^{-1}) and a final location where N and water were both adequate (total N applied during the season was 106 kg N ha^{-1}). The soils at the experiment location Viluco, CH, provide a large proportion of required N through mineralization and do not require large inputs of fertilizer N to produce grain yields in excess of 12.5 Mg ha^{-1}.

The experimental design was a randomized complete block design in a split-plot arrangement. The main plot consisted of hybrid pedigree and the subplot was 100% male fertile or *Ms44* hybrid segregating for 50% male-sterile plants. Experimental units in North American and South American trials were all four rows with 76 cm row spacing and a row length from 4.3 to 5.1 m depending on the testing location. At harvest, a combine was used to collect grain weight and grain moisture data from the centre two rows. Yield was calculated and adjusted to a standard moisture of 155 g kg^{-1}.

Each location, or field × irrigation × density × experiment, was classified as optimal growth condition, low-N condition, ultralow-N condition, flowering drought stress or grain filling drought stress. In each location, hybrids were randomized in incomplete blocks in the main plots, and sterile plants (100% fertile control and the segregating 50 : 50 treatment) were randomized in the subplots. Yield data were modelled using the ASReml-R package within R (Butler, 2009; R core Team, 2015). The model can be specified as:

$$y = t + l + s + l*s + r + h + h*l + h*r + h*s + h*s*l + \xi + \eta$$

where *y* denotes yield, *t* denotes location classification, *l* denotes location, *s* denotes sterility, *r* denotes block–location combination, *h* denotes hybrid, ξ denotes location-specific residual errors and η denotes location-specific spatial correlations. Location classification, location, sterility and location × sterility interaction were treated as fixed effects. All the other effects were treated as random effects. The mixed model with spatial adjustment allows the reduction of noise caused by field variability while preserving the genetic signal. Yield for sterility cross hybrids and yield for sterility cross hybrids within each location classification were estimated using best linear unbiased estimator (BLUE). Yield for sterility within hybrid was predicted using best linear unbiased predictor (BLUP), as hybrid effect was treated as random. The BLUPs are not strictly equal to the average of the BLUEs because they have been adjusted to reflect hybrid variation. Differences between the 100% fertile control and the *Ms44* segregating 50 : 50 treatment were considered significant at the 5% confidence level.

Physiological analysis of *Ms44* male-sterile plants in field

Ms44 male-sterile plants produced by expression of *Ms44* genomic DNA were planted in Johnston field in 2013 to determine whether male sterility affects tassel and ear biomass and N accumulation at early reproductive stages. The experimental design was a multistage complete randomized design; eight plots of sterile and eight plots of fertile treatments were completely randomized within each developmental stage. Five plants were harvested from eight replicates and separated into developing ear, tassel and shoots at V9 (with nine leaves at this stage), V10, V11, V13, V15 and V17 (when tassel is visible fully). Samples were dried at 70 °C for 3 days. Dry weight was recorded. Total N content was determined by the combustion method. Averaged data were modelled using the ASReml-R package within R (Butler, 2009; R Core Team, 2015). The model can be specified as: y=v+s+v*s+ ε, where y denotes averaged trait, v denotes developmental stage, s denotes sterility and ε denotes residual. Stage, sterility and stage × sterility interaction were treated as fixed effects. Sterility within each stage was estimated using best linear unbiased estimator (BLUE).

Four inbred lines carrying the dominant *Ms44* allele were tested to determine the effect of male sterility on kernel number per ear. In the North American trial, a 100% fertile line and a segregating line for each inbred were grown at five locations (Champaign, IL; Macomb, IL; Miami, MO; Princeton, IN; York, NE) in 2014. In South America near Viluco, Chile, the same four segregating inbred lines were tested at four locations. At physiological maturity, 10 ears were harvested from the tagged sterile plants from the segregating row. Ten ears were harvested from fertile plants in the 100% fertile row as the control. From both the North American and South American experiments, harvested ears were prepared for imaging by removing all dry silks. A digital image was collected of sterile and fertile ears and processed to determine kernel number per ear.

Acknowledgements

We thank Hua Mo for field experiment design and statistical analysis of field data, Brian Loveland and Nick Mongar for technical support, Laura Church for glasshouse support and Katherine Thilges and Mark Chamberlin for microscopy support. We acknowledge and thank the many DuPont Pioneer staff members at testing and seed production locations for their contributions. We also thank Tom Greene, Michael Lassner, Dave Warner and Mark Cooper for their organizational leadership and helpful input and Jim Gaffney for critical review of the manuscript.

Reference

Albertsen, M.C. and Trimnell, M. (1992) Linkage between Ms44 and C2. *Maize Genet. Coop. Newslett.* **66**, 49.

Boutrot, F., Chantret, N. and Gautier, M.F. (2008) Genome-wide analysis of the rice and Arabidopsis non-specific lipid transfer protein (nsLtp) gene families and identification of wheat nsLtp genes by EST data mining. *BMC Genom.* **9**, 86. doi:10.1186/1471-2164-9-86.

Butler, D. (2009) *ASReml: Asreml () fits the linear mixed model.* R package version 3.0-1. Hemel Hempstead, UK: VSN International Ltd. http://www.vsni.co.uk.

Chen, L. and Liu, Y.G. (2014) Male sterility and fertility restoration in crops. *Annu. Rev. Plant Biol.* **65**, 579–606.

Chinwuba, P.M., Grogan, C.O. and Zuber, M.S. (1961) Interaction of detasseling, sterility, and spacing on yields of maize hybrids. *Crop Sci.* **1**, 279–280.

Ciampitti, I.A. and Vyn, T.J. (2014) Understanding global and historical nutrient use efficiencies for closing maize yield gaps. *Agron. J.* **106**, 2107–2117.

Duvick, D.N. (1958) Yields and other agronomic characteristics of cytoplasmically pollen sterile corn hybrids, compared to their normal counterparts. *Agron. J.* **50**, 121–125.

Duvick, D.N. and Cassman, K.G. (1999) Post-green revolution trends in yield potential of temperate maize in the North-Central United States. *Crop Sci.* **39**, 1622–1630.

Edmonds, D.E., Abreu, S.L., West, A., Caasi, D.R., Conley, T.O., Daft, M.C., Desta, B. *et al.* (2009) Cereal nitrogen use efficiency in sub Saharan Africa. *J. Plant Nutr.* **32**, 21070–22122.

Edstam, M.M., Viitanen, L., Salminen, T.A. and Edqvist, J. (2011) Evolutionary history of the non-specific lipid transfer proteins. *Mol. Plant* **4**, 947–964.

Feng, P.C., Qi, Y., Chiu, T., Stoecker, M.A., Schuster, C.L., Johnson, S.C., Fonseca, A.E. *et al.* (2014) Improving hybrid seed production in corn with glyphosate-mediated male sterility. *Pest Manag. Sci.* **70**, 212–218.

Gaffney, J., Anderson, J., Franks, C., Collinson, S., MacRobert, J., Woldemariam, W. and Albertsen, M. (2016) Robust seed systems, emerging technologies, and hybrid crops for Africa. *Glob. Food Security*, **9**, 36–44.

Good, A.G. and Beatty, P.H. (2011) Fertilizing nature: a tragedy of excess in the commons. *PLoS Biol.* **9**, e1001124. doi:10.1371/journal.pbio.1001124.

Good, A.G., Johnson, S.J., De Pauw, M., Carroll, R.T., Savidov, N., Vidmar, J., Lu, Z. *et al.* (2007) Engineering nitrogen use efficiency with alanine aminotransferase. *Can. J. Bot.* **85**, 252–262.

Han, M., Okamoto, M., Beatty, P.H., Rothstein, S.J. and Good, A.G. (2015) The Genetics of nitrogen use efficiency in crop plants. *Annu. Rev. Genet.* **49**, 269–289.

Huang, M.D., Wei, F.J., Wu, C.C., Hsing, Y.I. and Huang, A.H. (2009) Analyses of advanced rice anther transcriptomes reveal global tapetum secretory functions and potential proteins for lipid exine formation. *Plant Physiol.* **149**, 694–707.

Huang, M.D., Chen, T.L. and Huang, A.H. (2013) Abundant type III lipid transfer proteins in *Arabidopsis* tapetum are secreted to the locule and become a constituent of the pollen exine. *Plant Physiol.* **163**, 1218–1229.

Kempe, K. and Gils, M. (2011) Pollination control technologies for hybrid breeding. *Mol. Breed.* **27**, 417–437.

Kempe, K., Rubtsova, M. and Gils, M. (2014) Split-gene system for hybrid wheat seed production. *Proc. Natl Acad. Sci. USA*, **111**, 9097–9102.

Liu, J., You, L., Aminid, M., Obersteiner, M., Herreroe, M., Zehnderf, A.J.B. and Yang, H. (2010) A high-resolution assessment on global nitrogen flows in cropland. *Proc. Natl Acad. Sci. USA*, **107**, 8035–8045.

Liu, X., Zhang, Y., Han, W., Tang, A., Shen, J., Cui, Z., Vitousek, P. *et al.* (2013) Enhanced nitrogen deposition over China. *Nature*, **494**, 459–463.

Liu, F., Zhang, X., Lu, C., Zeng, X., Li, Y., Fu, D. and Wu, G. (2015) Non-specific lipid transfer proteins in plants: presenting new advances and an integrated functional analysis. *J. Exp. Bot.* **66**, 5663–5681.

McAllister, C.H., Beatty, P.H. and Good, A.G. (2012) Engineering nitrogen use efficiency crop plants: the current status. *Plant Biotechnol. J.* **10**, 1011–1025.

Newhouse, K., Fischer, J. and Gobel, E. (1996) Pollination control with SeedLink™. In *51st Annual Corn and Sorghum Ind. Research Conference*

(Wilkinson, D., ed.), pp. 227–234. Washington, DC: American Seed Trade Association (ASTA).

Perez-Prat, E. and van Lookeren Campagne, M.M. (2002) Hybrid seed production and the challenge of propagating male-sterile plants. *Trends Plant Sci.* **7**, 199–203.

Potter, P., Ramankutty, N., Bennett, E.M. and Donner, S.D. (2010) Charactering the spatial patterns of global fertilizer application and manure production. *Earth Interact.* **14**, 1–21.

R Core Team. (2015) *R: A Language and Environment for Statistical Computing.* Vienna, Austria: R Foundation for Statistical Computing. http://www.R-project.org.

Raun, W.R. and Johnson, G.V. (1999) Improving nitrogen use efficiency for cereal production. *Agronomy J.* **91**, 357–363.

Shrawat, A.K., Carroll, R.T., DePauw, M., Taylor, G.J. and Good, A.G. (2008) Genetic engineering of improved nitrogen use efficiency in rice by the tissue-specific expression of alanine aminotransferase. *Plant Biotechnol. J.* **6**, 722–732.

Skibbe, D.S. and Schnable, P.S. (2005) Male sterility in maize. *Maydica*, **50**, 367–376.

Turgut, K., Barsby, T., Craze, M., Freeman, J., Hodge, R., Paul, W. and Scott, R. (1994) The highly expressed tapetum-specific A9 gene is not required for male fertility in *Brassica napus*. *Plant Mol. Biol.* **1**, 97–104.

Unger, E., Betz, S., Xu, R. and Cigan, A.M. (2001) Selection and orientation of adjacent genes influences DAM-mediated male sterility in transformed maize. *Transgenic Res.* **10**, 409–422.

Vitousek, P.M., Naylor, R., Crews, T., David, M.B., Drinkwater, L.E., Holland, E., Johnes, P.J. *et al.* (2009) Nutrition imbalances in agricultural development. *Science*, **324**, 1519–1520.

Von Heijne, G. (1986) A new method for predicting signal sequence cleavage sites. *Nucleic Acids Res.* **11**, 4683–4690.

Wei, K. and Zhong, X. (2014) Non-specific lipid transfer proteins in maize. *BMC Plant Biol.* **14**, 281. doi:10.1186/s12870-014-028-8.

Weider, C., Stamp, P., Christovc, N., Hüskend, A., Foueillassare, X., Campb, K.H. and Munsch, M. (2009) Stability of cytoplasmic male sterility in maize under different environmental conditions. *Crop Sci.* **49**, 77–84.

Wright, S.Y., Suner, M.M., Bell, P.B., Vaudin, M. and Greenland, A.J. (1993) Isolation and characterization of male flower cDNAs from maize. *Plant J.* **3**, 41–49.

Wu, Y., Fox, T.W., Trimnell, M.R., Wang, L., Xu, R.J., Cigan, A.M., Huffman, G.A. *et al.* (2015) Development of a novel recessive genetic male sterility system for hybrid seed production in maize and other cross-pollinating crops. *Plant Biotechnol. J.* **14**, 1046–1054.

Xu, G., Fan, X. and Miller, A.J. (2012) Plant nitrogen assimilation and use efficiency. *Annu. Rev. Plant Biol.* **63**, 153–182.

Zhang, D., Liang, W., Yin, C., Zong, J., Gu, F. and Zhang, D. (2010) OsC6, encoding a lipid transfer protein, is required for postmeiotic anther development in rice. *Plant Physiol.* **154**, 149–162.

4

Extensive homoeologous genome exchanges in allopolyploid crops revealed by mRNAseq-based visualization

Zhesi He[1], Lihong Wang[1], Andrea L. Harper[1], Lenka Havlickova[1], Akshay K. Pradhan[2], Isobel A. P. Parkin[3] and Ian Bancroft[1],*

[1]*Department of Biology, University of York, Heslington, York, UK*
[2]*Department of Genetics and Centre for Genetic Manipulation of Crop Plants, University of Delhi, New Delhi, India*
[3]*Agriculture and Agri-Food Canada, Saskatoon, SK, Canada*

*Correspondence
email ian.bancroft@york.ac.uk

Keywords: crop genomes, genome structural evolution, mRNAseq.

Summary

Polyploidy, the possession of multiple sets of chromosomes, has been a predominant factor in the evolution and success of the angiosperms. Although artificially formed allopolyploids show a high rate of genome rearrangement, the genomes of cultivars and germplasm used for crop breeding were assumed stable and genome structural variation under the artificial selection process of commercial breeding has remained little studied. Here, we show, using a repurposed visualization method based on transcriptome sequence data, that genome structural rearrangement occurs frequently in varieties of three polyploid crops (oilseed rape, mustard rape and bread wheat), meaning that the extent of genome structural variation present in commercial crops is much higher than expected. Exchanges were found to occur most frequently where homoeologous chromosome segments are collinear to telomeres and in material produced as doubled haploids. The new insights into genome structural evolution enable us to reinterpret the results of recent studies and implicate homoeologous exchanges, not deletions, as being responsible for variation controlling important seed quality traits in rapeseed. Having begun to identify the extent of genome structural variation in polyploid crops, we can envisage new strategies for the global challenge of broadening crop genetic diversity and accelerating adaptation, such as the molecular identification and selection of genome deletions or duplications encompassing genes with trait-controlling dosage effects.

Introduction

Polyploid organisms have multiple sets of chromosomes, and genome studies indicate that the evolutionary flexibility endowed by polyploidy has shaped the genomes of most if not all eukaryotes (Comai, 2005). Stable polyploidy occurs in fish and frogs (Gregory and Mable, 2005), although plants offer the best systems for its study as this is where it is most widespread. Indeed, it has been considered a predominant factor in the evolution and success of the angiosperms (Leitch and Bennett, 1997; Wendel, 2000). As part of the genome stabilization process termed 'diploidization' (Hillier et al., 2004; Wang et al., 2005), newly formed polyploid genomes undergo rapid structural evolution, including gene copy number variation (CNV) (Adams and Wendel, 2005). As CNVs are frequently associated with genetic traits (Beckmann et al., 2007), they are likely to be of crucial importance to crop science as many important crops (e.g. bread wheat, cotton, soybean, potato and rapeseed) are recently formed polyploids. Inferring the mechanisms involved in the diploidization process by comparative genomics of extant species has become a key aim of plant genomics. However, the lack of cost-effective analysis tools means that there has been relatively little analysis of genome structural variation in commercial varieties of polyploid crops.

Bread wheat (*Triticum aestivum*) is an allohexaploid comprising three genomes: A, B and D (Chantret et al., 2005). Less than 800 000 years ago, a hybridization between the A genome progenitor, *Triticum urartu*, and the B genome progenitor (a close relative of *Aegilops speltoides*) formed the AABB allotetraploid emmer wheat *T. turgidum*. Finally, less than 400 000 years ago, hybridization between emmer wheat and the D genome progenitor, *Aegilops tauschii* (Marcussen et al., 2014), formed the AABBDD allohexaploid bread wheat, *T. aestivum*.

The cultivated *Brassica* species are the group of crops most closely related to *Arabidopsis thaliana*, which was the first plant for which a high-quality genome sequence was available (The Arabidopsis Genome Initiative, 2000). The species *Brassica rapa* and *Brassica oleracea*, which contain the *Brassica* A and C genomes, respectively, last shared a common ancestor *ca.* 3.7 Mya (Inaba and Nishio, 2002). *Brassica napus* is an allopolyploid, arising from the hybridization of A and C genome progenitors (U.N., 1935), and the related (homoeologous) regions of the genomes are clearly discernible (Bancroft et al., 2015). A diverse range of *B. napus* crop types have already been developed, including oilseed rape, fodder types, leafy vegetables (kale types) and root vegetables (swede or rutabaga), underlining the phenotypic plasticity of recently formed polyploids such as *B. napus*. Analysis of the genome sequence reported for an oilseed rape *B. napus* variety (Chalhoub et al., 2014) provided clear evidence for homoeologous exchanges (HEs). These were characterized by the loss of a chromosomal region that was replaced by a duplicate copy of the corresponding homoeologous region of the other genome. This indicates that HEs occurred at some stage in the formation of the particular crop variety sequenced.

Genome resequencing can be used to study HEs. Indeed, this approach was used to identify putative HEs in six varieties of *B. napus* crops (Chalhoub *et al.*, 2014). Even using NGS sequencing technology, however, this is an expensive approach. In contrast, transcriptome-based studies using mRNAseq are particularly rapid and cost-effective, with a suite of methodologies developed and applied in *B. napus* for SNP discovery (Trick *et al.*, 2009), linkage mapping and genome characterization (Bancroft *et al.*, 2011), transcript quantification (Higgins *et al.*, 2012) and association genetics (Harper *et al.*, 2012).

In this study, we exploited a method we developed recently for assessing genome dominance in polyploid species (Harper *et al.*, 2016) to visualize HEs in allopolyploid crops and assess the extent to which these occur in cultivars.

Results

mRNAseq-based visualization of homoeologous genome exchanges in *Brassica napus*

We produced leaf mRNAseq data from a panel of 27 *B. napus* varieties, in four biological replicates, and quantified gene expression by mapping the reads to an ordered pan-transcriptome resource that ensured correct allocation of *B. napus* gene sequences to either the *Brassica* A or C genome (He *et al.*, 2015) (Data S1). We identified homoeologous gene pairs from the ordered pan-transcriptome resources. The result was a set of 32 904 homoeologous pairs (Data S2), which define the homoeology relationships between the genomes as shown in Figure 1. Analysis of differential gene expression into the nine categories

indicative of genome dosage changes as listed in Table 1 (Data S3, S4 and S5) revealed numerous clearly defined blocks in the genome where many nearby genes showed the same directionality of one genome over-expressed with the other genome under-expressed, suggestive of potential blocks of homoeologous exchange. To better visualize regions of the genome involved in homoeologous exchanges, we used Transcriptome Display Tile Plots (TDTPs), as used previously for visualizing genome dominance in wheat (Harper *et al.*, 2016). This involves the following: (i) assigning quantitative transcript abundance for each member of each gene pair a value in CMYK colour space where the contributions from the A and C genome copies were coded to cyan and magenta channels, respectively; (ii) displaying the results using tile plots, with order based on the genome order of one member of the gene pair. The result is shown in Figure 2. Most of the genome displays blue, indicating approximately equal contributions to the transcriptome from each member of the homoeologous gene pairs. However, the predominant colour of several regions of the genome was cyan or magenta, indicative of HEs with increased copy number of the A and C genome segments, respectively. To confirm that the HEs identified were the result of structural exchanges rather than gene expression dominance effects, we undertook genome resequencing on one replicate from each of the 27 varieties and visualized the results in the equivalent manner. As shown in Figure 3, the results were identical, confirming the validity of using the more cost-effective approach of undertaking such analyses based on mRNAseq rather than genome resequencing. We included in our experiment the swede variety Sensation NZ, which had been analysed previously by genome resequencing and mapping of reads to the Darmor-*bzh* genome assemblies (Chalhoub *et al.*, 2014). The TDTPs improved the analysis, showing clearer distinction between the balanced and exchanged regions. In addition, our approach

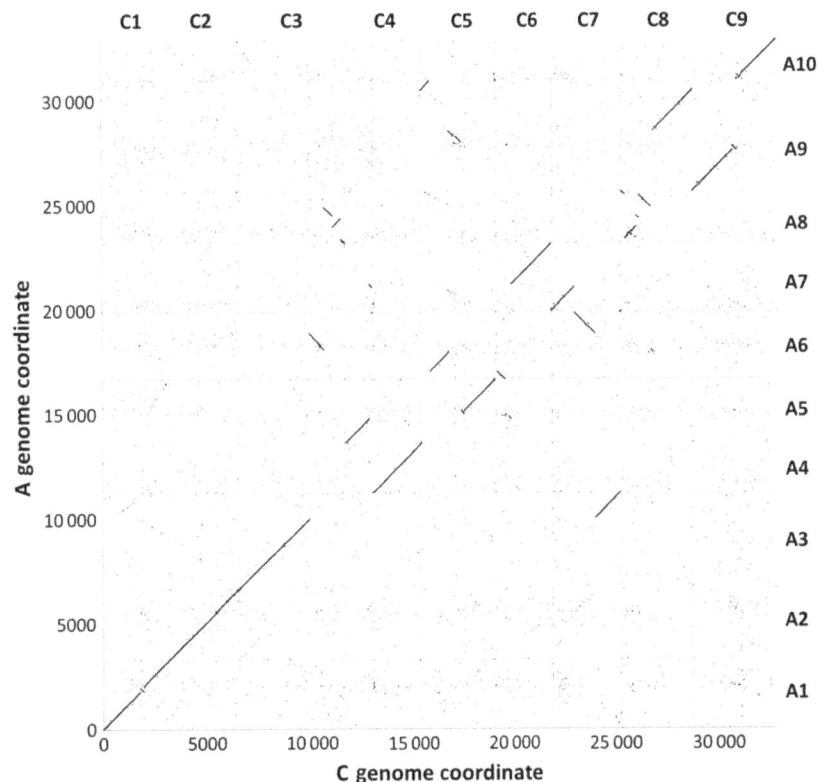

Figure 1 Homoeology relationships between *Brassica* A and C genomes. 32 904 pairs of homoeologous genes in the *Brassica* A and C genomes are plotted by their order in the respective genomes.

Table 1 Categories of genome dosage changes

Expression of samples from one accession on A genome compared to all samples	Expression of samples from one accession on C genome compared to all samples	Gene dosage inference	Differential expression inference
High	High	Duplication A and C	One or both genome over-expression
High	No change (null)	Duplication A	
No change (null)	High	Duplication C	
High	Low	Exchange C → A	One genome over-expressed
Low	High	Exchange A → C	with the other genome under-expressed
			relative to the mean of all samples
Low	No change (null)	Deletion A	One or both genome under-expression
Low	Low	Deletion A and C	
No change (null)	Low	Deletion C	
No change (null)	No change (null)	No difference	No difference

T-test analysis performed between the expression profiles of a single accession (with four replicated samples) on one genome (A or C) and mean of all samples. The expression profiles of each homoeologous CDS model can be allocated into nine categories of genome dosage inferences.

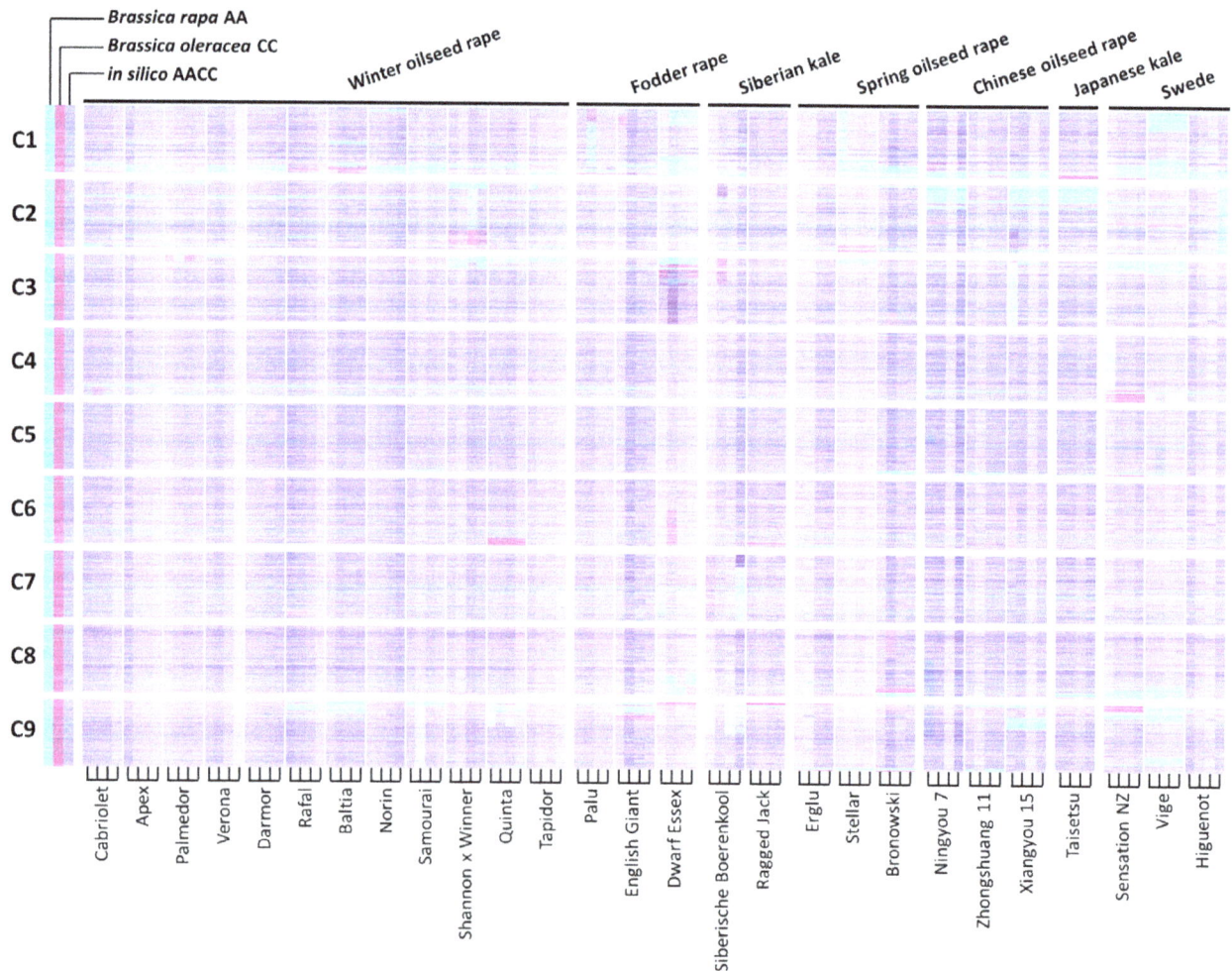

Figure 2 Visualization of homoeologous genome exchanges in *Brassica napus* using Transcriptome Display Tile Plots based on mRNAseq data. The relative transcript abundance of A and C genome homoeologous gene pairs is represented in CMYK colour space, with cyan component representing transcript abundance of the *Brassica* A genome copy and magenta component representing transcript abundance of the *Brassica* C genome copy. The pairs are plotted in *Brassica* C genome order (chromosomes denoted C1 to C9) for four biological replicates of each of 27 accessions of *B. napus* and controls comprising parental species and their *in silico* combination.

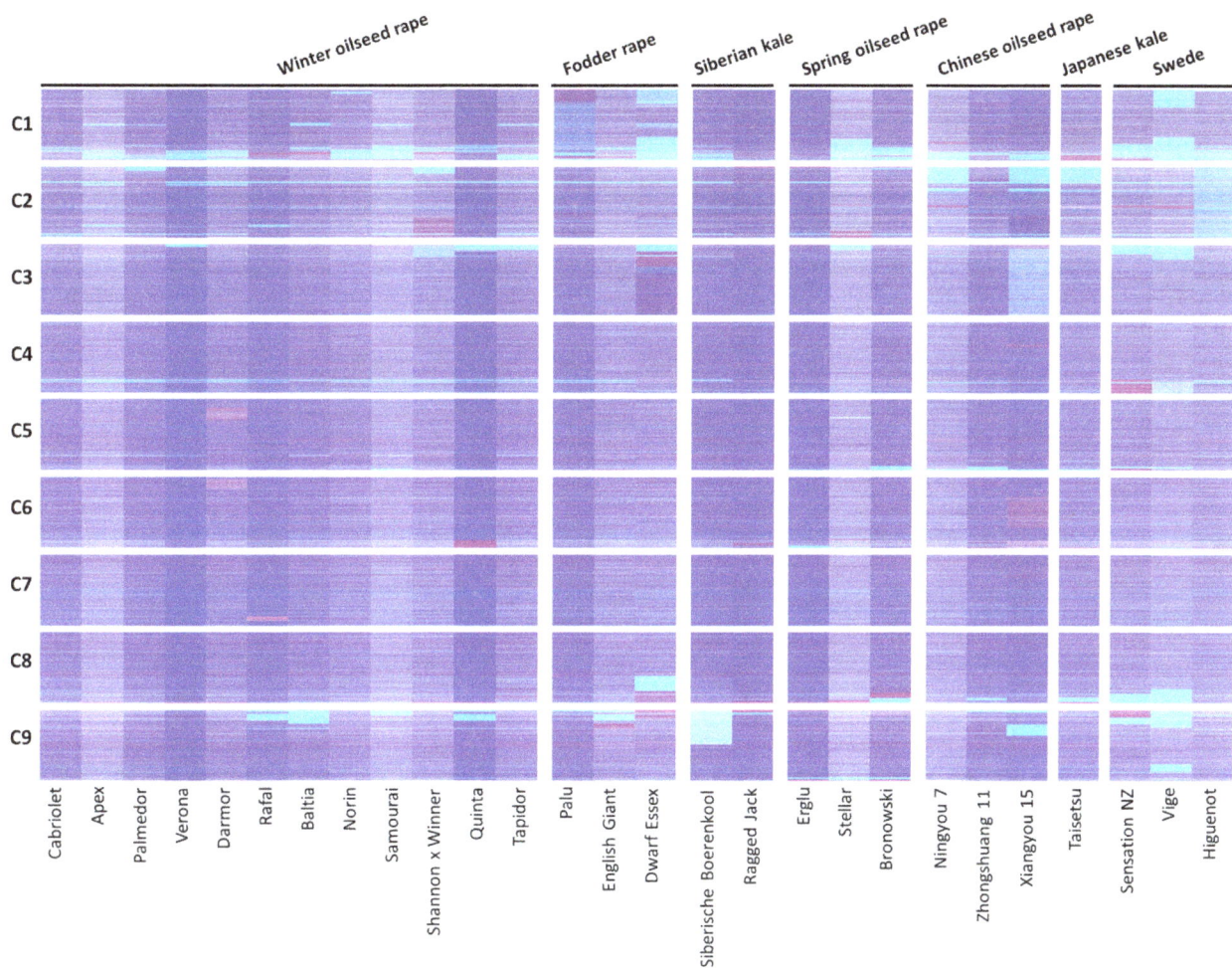

Figure 3 Visualization of homoeologous genome exchanges based on DNA resequencing. The relative redundancy of coverage of A and C genome homoeologous gene pairs is represented in CMYK colour space, with cyan component representing coverage of the *Brassica* A genome copy and magenta component representing coverage of the *Brassica* C genome copy. The pairs are plotted in Brassica C genome order (chromosomes denoted C1 to C9).

identified an exchange affecting the bottom of chromosome C4 (and the homoeologous region of A4) that had not been detectable by mapping reads to the Darmor-*bzh* genome assemblies (Chalhoub *et al.*, 2014). The mRNAseq-based analysis and TDTPs are very consistent between replicates and with genome resequencing results, presumably because of the relative nature of colour rendition within each homoeologue pair, meaning that biological replication is unnecessary for the identification of large HEs. All of the 27 cultivars analysed showed at least one HE, with some showing multiple exchanges.

Homoeologous genome exchanges in *Brassica napus* germplasm

To assess how representative the observations may be of the germplasm used by breeders, we produced TDTPs for a widely shared *B. napus* genetic diversity panel. This used recently produced mRNAseq data from 383 accessions comprising the 'RIPR' diversity panel (Data S6), which overlaps extensively with the 'ASSYST' diversity panel (Bus *et al.*, 2011). The overview plots are shown in Figure 4. Remarkably, they show that all of the *B. napus* accessions examined contain identifiable segmental HEs. The sizes of these vary greatly. A few instances of whole chromosomes being duplicated or lost are observed, but this appears rare.

Assessing the basis of variation for homoeologous genome exchanges in *Brassica* polyploids

The distribution of HEs across the genome in the *B. napus* genetic diversity panel appears not to be random. Analysis of the frequency of occurrence of HEs across the panel (Data S7) confirms that the distribution of HEs is highly skewed towards certain regions of the genome. Alignment of the genome segments frequently involved in such HEs shows these relate to the set of instances where the corresponding regions in both genomes extend to telomeres. Alignment of genome regions rarely involved in such HEs shows these correspond to the set of instances where the regions of the A genome are (relative to the C genome) rearranged, with the sequences homoeologous to C genome telomeres being internal to A genome chromosomes. This suggests that HEs occur most frequently where homoeologues are able to pair over long regions extending to telomeres. To test this hypothesis we identified, a suitable polyploid species related to *B. napus* that contains genomes with less collinearity: *Brassica juncea* (an allotetraploid formed by hybridization of *B. rapa*, contributing the A genome and *Brassica nigra*, contributing the B genome), which is the principal oil crop in India (mustard rape). We identified, based on available *B. rapa* and

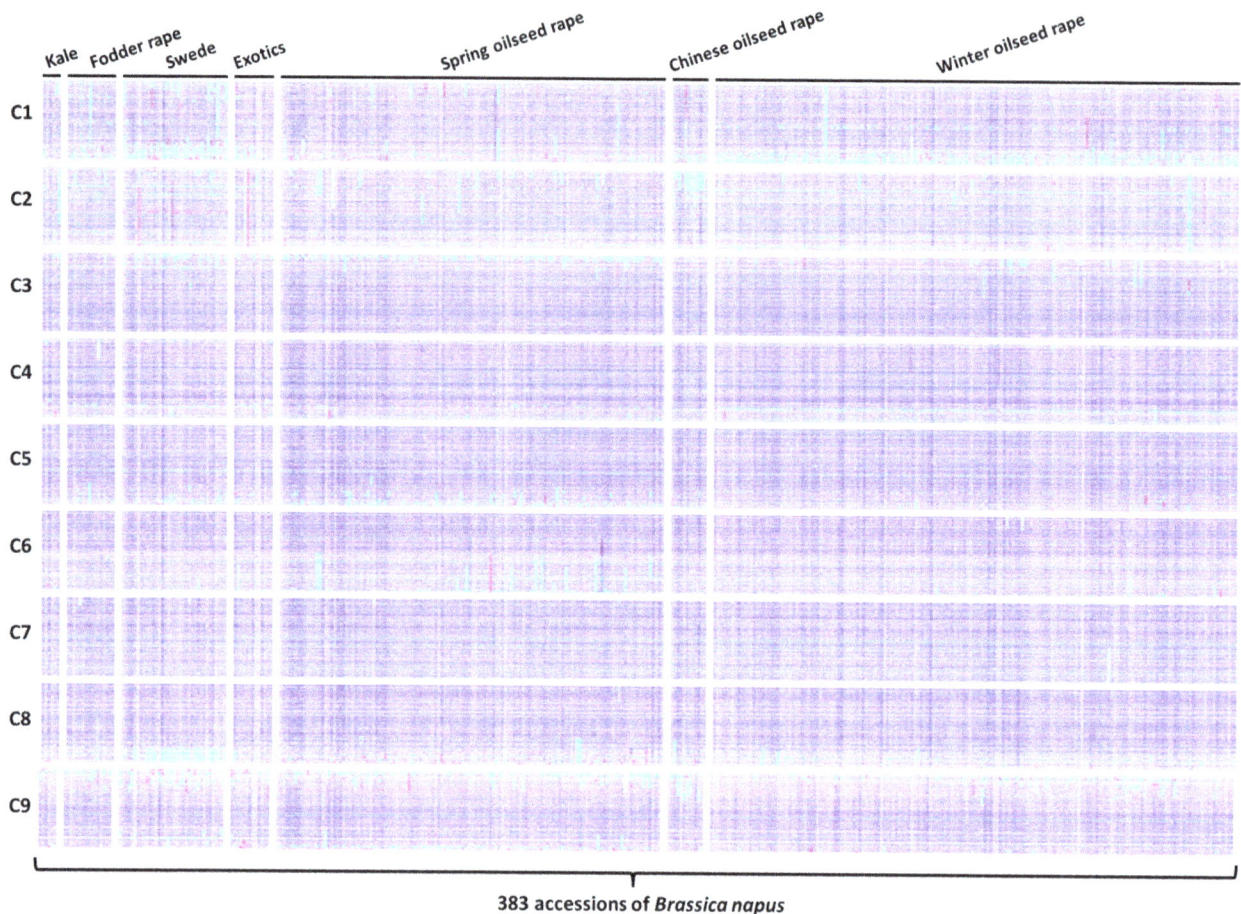

Figure 4 Visualization of homoeologous genome exchanges in *Brassica napus* RIPR diversity panel using Transcriptome Display Tile Plots based on mRNAseq data. The relative transcript abundance of A and C genome homoeologous gene pairs is represented in CMYK colour space, with cyan component representing transcript abundance of the *Brassica* A genome copy and magenta component representing transcript abundance of the *Brassica* C genome copy. The pairs are plotted in *Brassica* C genome order (chromosomes denoted C1 to C9).

B. nigra genome sequence resources, a set of 25 167 homoeologous gene pairs (Data S8), which define the homoeology relationships between the genomes, as shown in Figure 5, confirming that they exhibit less collinearity than those of *B. napus* (as shown in Figure 1). To assess the germplasm used by breeders, we produced TDTPs for the 205 accession 'CGAT' *B. juncea* genetic diversity panel (Data S9). The results are shown in Figure 6 and reveal a much lower incidence of HEs than had been observed in the *B. napus* germplasm and, again in contrast to *B. napus*, those that are present are mainly small segments internal to chromosomes. The lower rate of HEs extending to telomeres is consistent with the hypothesis that HEs occur most frequently where homoeologues are able to pair over long regions extending to telomeres.

The breeding of *Brassica* crops sometimes includes the production of doubled haploid (DH) plants (Möllers and Iqbal, 2009). The process involves colchicine treatment (to induce chromosome doubling) and results in completely homozygous plants. However, colchicine treatment inhibits meiotic telomere clustering (Cowan and Cande, 2002), so could affect the rate of HEs as collinearity extending to telomeres appears to be important. DH formation was used for the production of the *B. juncea* linkage mapping population VHDH (Pradhan *et al.*, 2003). We therefore analysed this population for HEs to assess whether the

process of DH production might increase the frequency of HEs occurring in this species despite the relatively lack of collinearity between its genomes. The results are shown in Figure 7 and reveal that numerous HEs are present in the population, in addition to clear segregation of an HE already represented in one of the parents, suggesting that DH production may indeed result in an elevated rate of HEs.

The identification of homoeologous genome exchanges in bread wheat

To assess the applicability of the approach to the identification of genome structural changes in further polyploid crops, we used it to analyse the genome of bread wheat. We identified homoeologous triplets (15 527 in total) and assigned transcript abundance to colour space using mRNAseq reads: A genome represented by cyan, B genome by magenta and D genome by yellow, as described previously for the analysis of potential genome dominance effects (Harper *et al.*, 2016). We analysed a linkage mapping population derived by single seed decent (six generations) from a cross between cultivars Chinese Spring (CS) and Paragon. The results of the analysis of 47 lines, plus the parent cultivars, are shown in Figure 8. This analysis revealed segregation in the population of a small homoeologous exchange present in the parent line CS (near the bottom of group 7

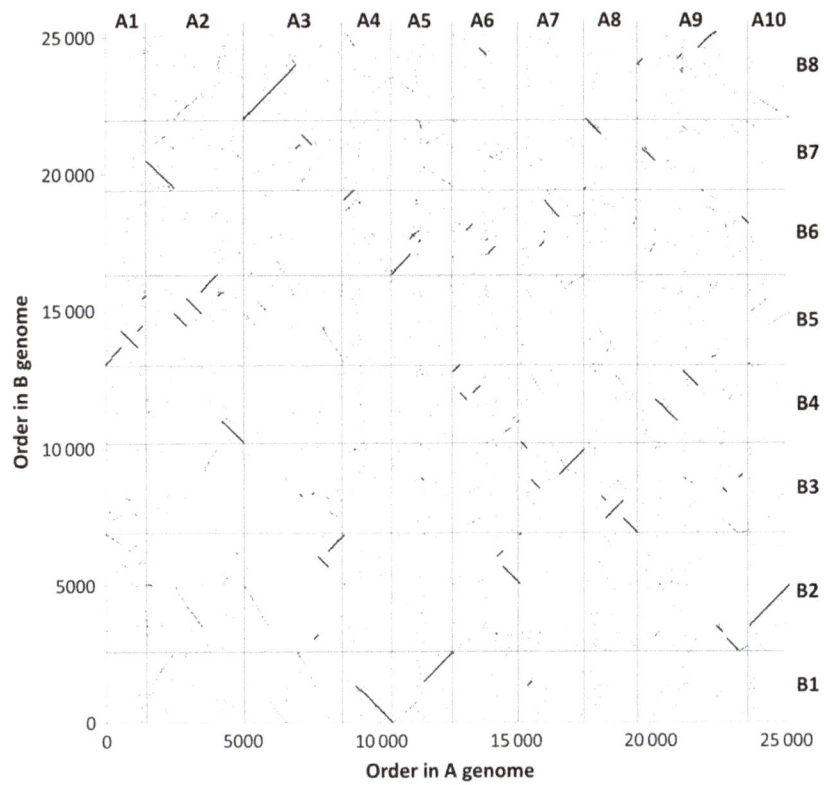

Figure 5 Homoeology relationships between *Brassica* A and B genomes. 25 167 pairs of homoeologous genes in the *Brassica* A and B genomes are plotted by their order in the respective genomes.

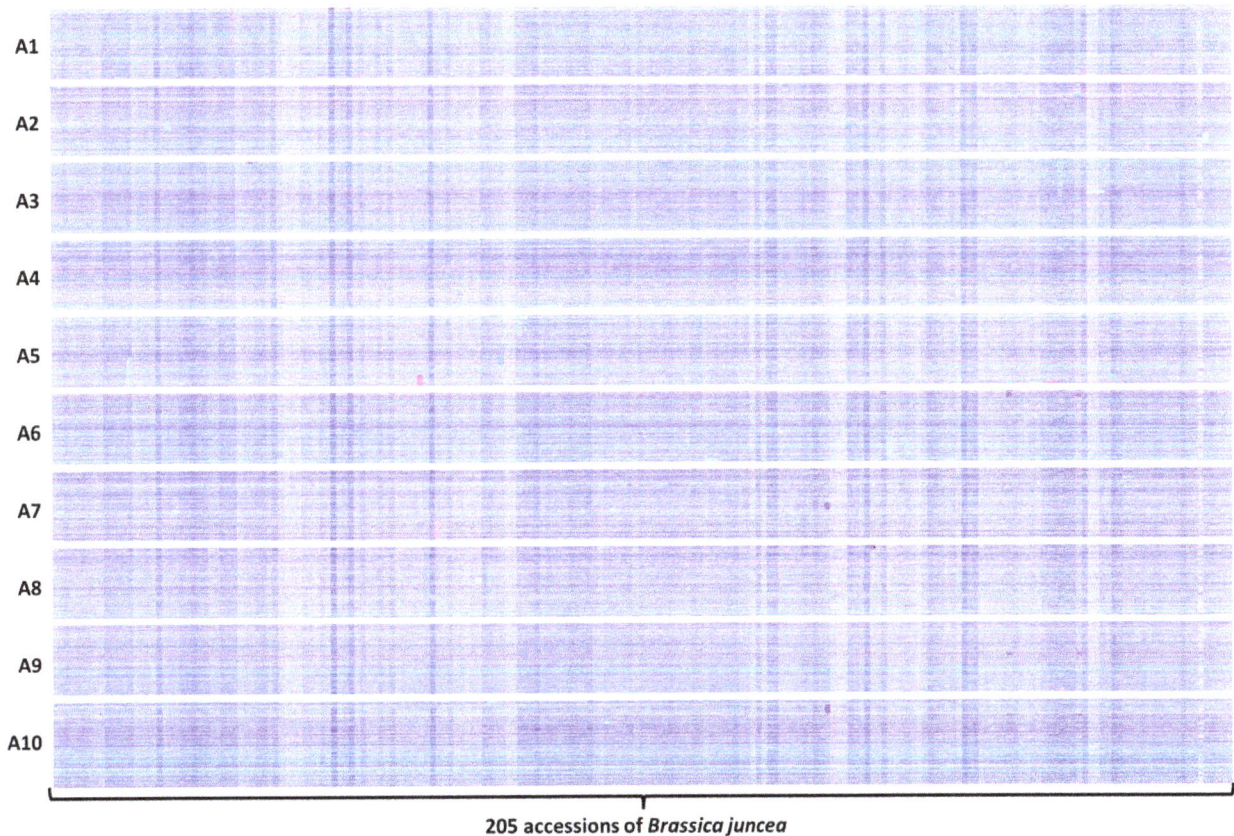

Figure 6 Visualization of homoeologous genome exchanges in *Brassica juncea* CGAT diversity panel using Transcriptome Display Tile Plots based on mRNAseq data. The relative transcript abundance of A and B genome homoeologous gene pairs is represented in CMYK colour space, with cyan component representing transcript abundance of the *Brassica* A genome copy and magenta component representing transcript abundance of the *Brassica* B genome copy. The pairs are plotted in *Brassica* A genome order (chromosomes denoted A1 to A10).

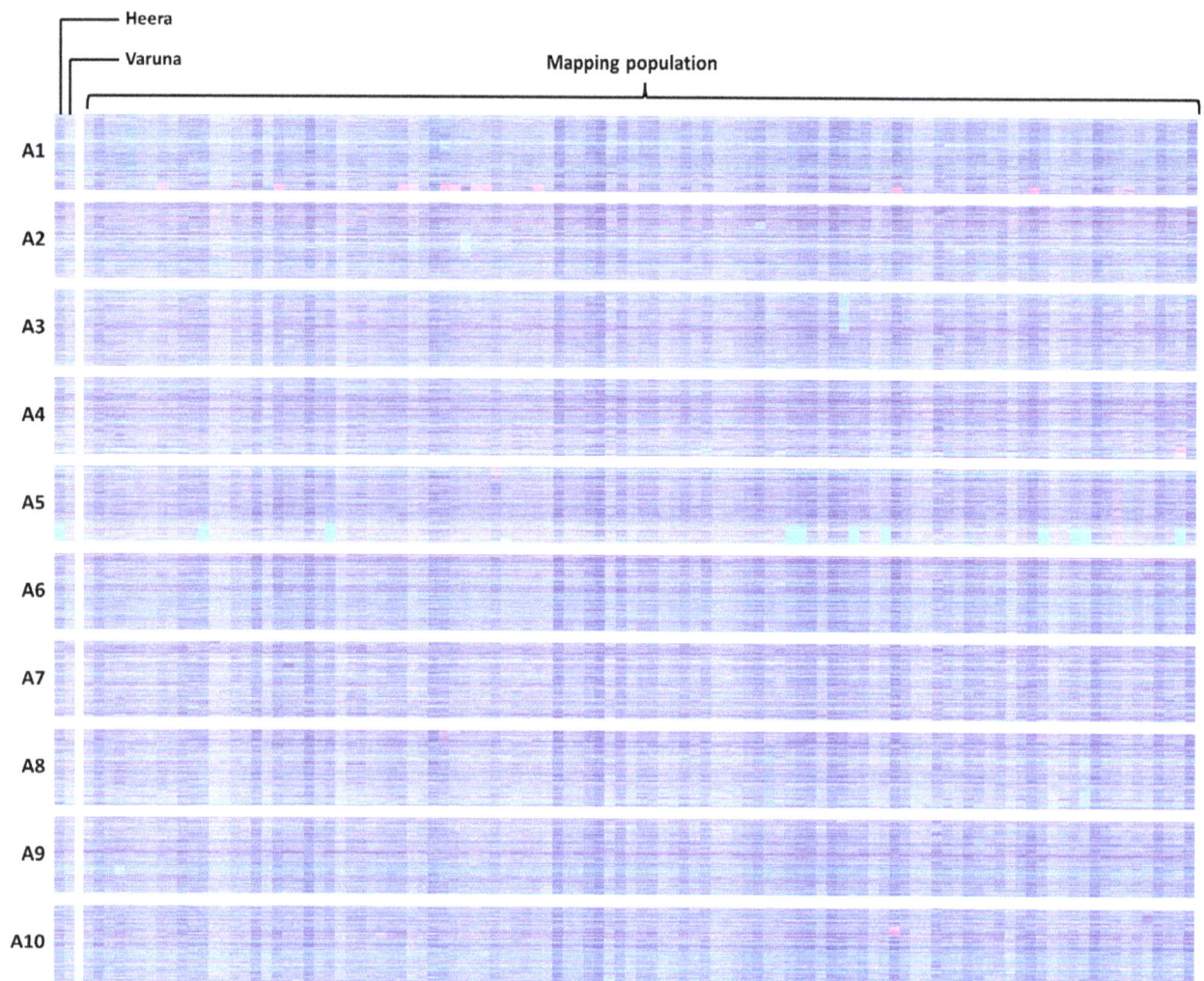

Figure 7 Visualization of homoeologous genome exchanges in VHDH population. The relative transcript abundance of A and B genome homoeologous gene pairs is represented in CMYK colour space, with cyan component representing transcript abundance of the *Brassica* A genome copy and magenta component representing transcript abundance of the *Brassica* B genome copy. The pairs are plotted in *Brassica* A genome order (chromosomes denoted A1 to A10).

chromosomes) and large, newly arising exchanges affecting groups 2 (CSxP67) and 6 (CSxP78) chromosomes. An additional copy of chromosome 4B was identified in CSxP11, and chromosome 4A has been lost from CSxP38. This analysis not only confirms the applicability of the method beyond *Brassica* species, but also shows that in this key polyploid crop, even in lines developed without treatment with colchicine, structural genome change occurs on the timescale of a few generations.

Discussion

The repurposing of TDTPs for the visualization of homoeologous genome exchanges in allopolyploid crops makes tractable the analysis of germplasm collections used for breeding. Compared with the methods available previously, the mRNAseq-based approach is rapid and inexpensive. There will be limits. For example, sequencing-based approaches will not be able to discriminate between the genomes of autopolyploid species, although quantitative analysis of transcript abundance may still be indicative of dosage (copy number) variation. Strong genome dominance effects, for example where one genome of an

allopolyploid is transcriptionally silenced, would permit observation of doubling of the expressed genome as intensified signals, but copy number of the silenced genome could not be determined. A great advantage, however, is that genomics resources enabling the hypothetical ordering of genes are necessary only for one of the genomes being analysed to generate the TDTPs.

mRNAseq-based detection of genome structural variation arising, in polyploid species, from homoeologous exchange, has led to the recognition that such events are very frequent and segregating widely in the germplasm used by breeders of *B. napus* crops. The distribution of regions of the genome in which exchanges have occurred is not random, but predominantly in those showing greatest collinearity extending to telomeres. This is consistent with the hypothesis that HEs will occur where homoeologues are able to pair most efficiently and the observation that HEs mostly appear to extend to telomeres in both genomes (albeit often with alternating exchanges indicative of subsequent recombination events) is consistent with the hypothesis that subtelomeric regions are involved in partner recognition and selection (Corredor *et al.*, 2007). It is notable also

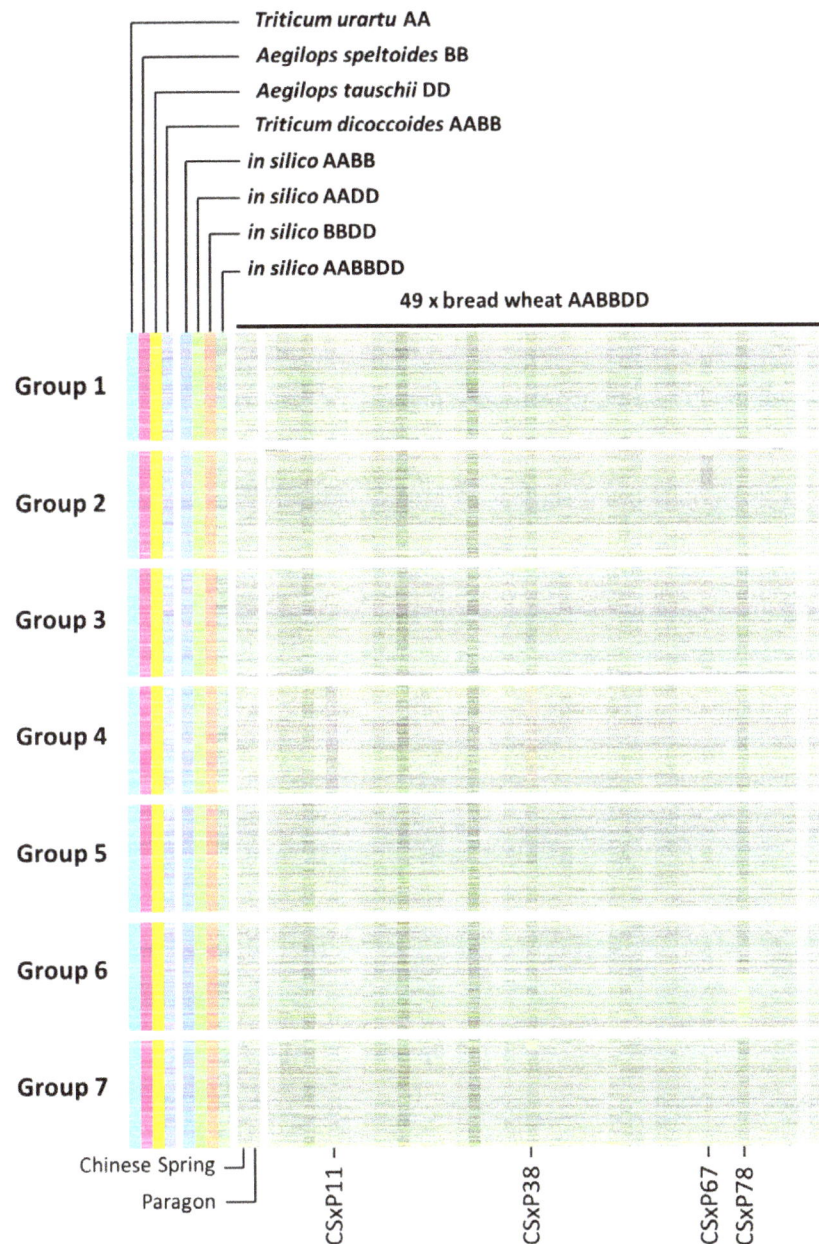

Figure 8 Visualization of homoeologous genome exchanges in wheat using Transcriptome Display Tile Plots based on mRNAseq data. The relative transcript abundance of A, B and D genome homoeologous gene triplets is represented in CMYK colour space, with cyan component representing transcript abundance of the wheat A genome copy, magenta component representing transcript abundance of the wheat B genome copy and yellow component representing transcript abundance of the wheat D genome copy. The triplets are plotted in genome order for the parents of a linkage mapping population (Chinese Spring × Paragon), 47 members of that population and controls comprising wheat parental species and *in silico* combinations to render resulting colours.

that HEs detected are conservative; that is, they involve substitution of genome segments. This may reflect selection imposed during breeding of individuals in which HEs have been nonconservative, *that is* where some of the exchanged region between the point of recombination and the telomere is nonhomoeologous, are impaired in growth or productivity. Although several of the regions frequently involved in HEs in *B. napus* show no bias, the majority mainly involves substitution of C genome sequences by A genome sequences, giving rise to the visual skewing towards cyan in the genetic diversity panel, as shown in Figure 2. Although the reasons for this bias are unclear, one possibility is that interspecific crosses have been conducted between *B. napus* and *B. rapa*, resulting in an increase in the genetic diversity of the A genome, but some of the HEs may have involved substitution of *B. napus* C genome by *B. rapa*-derived A genome sequences.

Genome visualization by TDTPs will have a broad impact on crop improvement due to the effects that HEs can have on traits

of importance. Two examples are illustrated in Figure 9. The first example is the HE that substitutes a region near the bottom of chromosome C2 with homoeologous A genome sequences. This exchange removes from varieties Cabriolet and Tapidor a functional orthologue of *HAG1*, which controls seed glucosinolate content, an important quality trait in rapeseed. The second example is the HE that substitutes a region towards the bottom of C1 with homoeologous A genome sequences. This exchange removes from variety Cabriolet a functional orthologue of *FAD2*, which controls the amount of polyunsaturated fatty acids synthesized and correspondingly the oleic acid content of seed oil, another important quality trait in rapeseed. Both events had originally been described as apparent deletions from the C genome as the increased copy number of the corresponding A genome sequences could not be identified by the analysis undertaken in either case (Harper *et al.*, 2012; Wells *et al.*, 2014). The ability to detect cost-effectively homoeologous

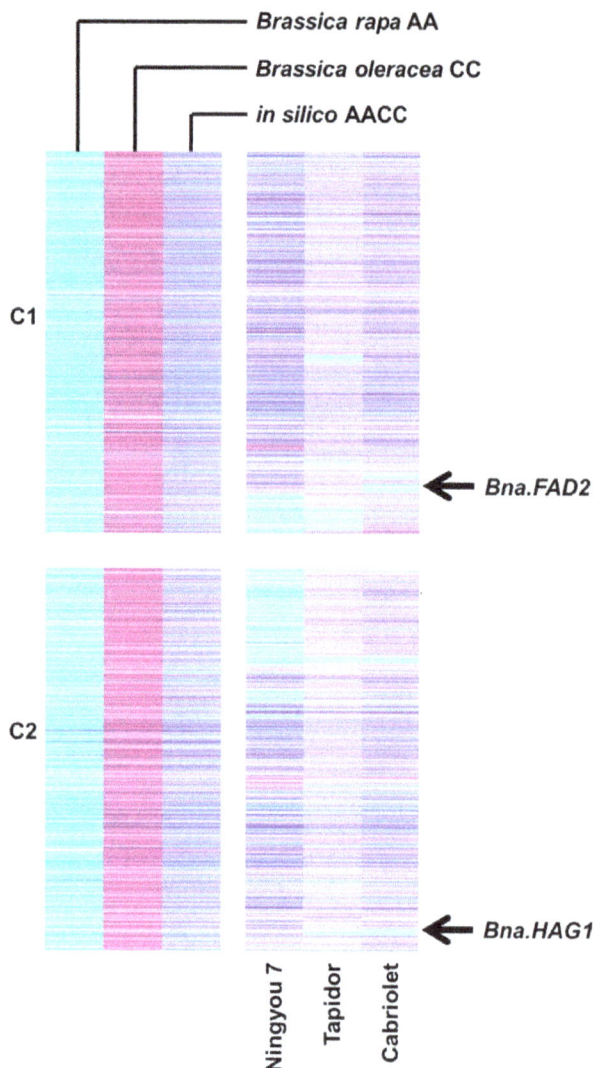

Figure 9 Visualization of homoeologous genome exchanges in *Brassica napus* causative of trait variation. The relative transcript abundance of A and C genome homoeologous gene pairs is represented in CMYK colour space, with cyan component representing transcript abundance of the *Brassica* A genome copy and magenta component representing transcript abundance of the *Brassica* C genome copy. The pairs are plotted in *Brassica* C genome order (only C1 and C2 shown) along with controls comprising parental species and their *in silico* combination. The positions of *B. napus* genes implicated in control of trait variation are marked.

genome exchanges across large panels of plants permits the systematic association of phenotypic variation with genome structural change, particularly the identification and selection of genome deletions or duplications encompassing genes with trait-controlling dosage effects.

Experimental procedures

Identification of homoeologous gene pairs and triplets

To assess genome exchange patterns in *Brassica* species, homoe-ologous gene pairs were identified via a two-way reciprocal BLASTn analysis (threshold E-value 1E-30). This analysis used the *Brassica* pan-transcriptome CDS models for *Brassica* A and C genomes (He *et al.*, 2015) and *Brassica* B genome CDS models

from *B. nigra* (I.A.P Parkin, unpublished). About 32 904 gene pairs were identified for *B. napus* (A and C genomes) and 25 167 gene pairs were identified for *B. juncea* (A and B genomes). For wheat, 15 527 homoeologous gene triplets had been identified previously (Harper *et al.*, 2016).

Growth and sampling of plants

Plants were sown on Levington professional F2 compost and grown in long day (16/8 h, 20 °C/14 °C) glasshouse conditions. Second true leaves from each of four plant replicates per accession were harvested when they reached ~3 cm in diameter, as close to the mid-point of the light period as possible. Leaves were harvested separately for processing as individual replicates or pooled when processed as a single sample per accession and immediately frozen in liquid nitrogen. Frozen leaf samples were stored at −80 °C.

RNA preparation

Pooled frozen leaf samples were ground in liquid nitrogen. RNA was extracted using the manufacturer's instruction for Omega Biotek EZNA Plant RNA Kit.

DNA preparation

DNA was extracted from individual replicate samples. The samples were homogenized in lysis buffer with 3-mm metal beads using Qiagen (Manchester, UK) TissueLyser II (30/s, 2 min). BioSprint 96 DNA Plant Kit and Qiagen BioSprint 96 Workstation system were used for the DNA extraction.

Transcriptome sequencing

Illumina sequencing, quality checking and processing were conducted as described previously (Higgins *et al.*, 2012) except that 100 base reads obtained from the HiSeq2500 platform were used. Maq was used for mapping with default parameters, meaning that reads with no more than two mismatches with summed $Q \geq 70$ were mapped.

Transcript quantification

Using methods and scripts described in Bancroft *et al.* (2011) and Higgins *et al.* (2012), expression of each of 116 098 genome-mapped *Brassica* CDS gene models was estimated for each of the accessions, using the recently developed ordered *Brassica* A and C pan-transcriptomes (He *et al.*, 2015) as reference sequences for *B. napus* panels of accessions. Expression of each of 87 630 genome-mapped *Brassica* CDS gene models representing the ordered *Brassica* A and B transcriptomes was estimated for each of the *B. juncea* accessions. Expression of each of 147 411 genome-mapped wheat unigenes was estimated for each of the wheat accessions, as described previously (Harper *et al.*, 2016). Transcript abundance was quantified and normalized as reads per kb per million aligned reads (RPKM) for each sample.

Analysis of differential gene expression

Differential gene expression was analysed in leaves for 27 accessions of *B. napus*, using four biological replicates (i.e. in leaves from four separate plants of each accession). Plants were grown, RNA extracted and purified, Illumina mRNA-seq data produced and transcript abundance quantified as described above, and genes showing low transcript abundance (mean RPKM across the respective panel <0.4) were removed. For each pair of homoeologous CDS models, a *t*-test analysis was performed between the expression profiles of a single accession (with four replicated samples) for one genome (A or C) against the mean of all samples of the 27 *B. napus* variety panel. The expression profiles of each homoeologous CDS model can be allocated into one of nine categories of genome dosage changes (Table 1). Differentially expressed CDS models are plotted with the same colour coding as the TDTP display. Homoeologue pairs showing significant ($P < 0.01$) differential expression were then placed in genome order for over-expression relative to the mean of all samples (Data S3), under-expression relative to the mean of all samples (Data S4) and over-expressed with the other genome under-expressed relative to the mean of all samples (Data S5).

Transcriptome display tile plots

A visualization approach involving in-house R script to display the relative transcript abundance of homoeologous gene pairs and triplets on a genome scale was used, as described previously (Harper *et al.*, 2016). Briefly, each member of the homoeologous pair/triplet has a value (normalized from 1 to 0 for the population so that darker colouring shows higher transcript abundance and lighter colouring shows lower transcript abundance) assigned in CMYK colour space where the contributions from each genome copy are coded to cyan, magenta or yellow channels, and the results displayed using tile plots. For *B. napus*, A genome copies are assigned to cyan and C genome copies are assigned to magenta, resulting in predominantly blue (cyan/magenta combination) where both genes are represented in an accession. For *B. juncea*, A genome copies are assigned to cyan and B genome copies are assigned to magenta, resulting in predominantly blue (cyan/magenta combination) where both genes are represented in an accession. For wheat, A genome copies are assigned to cyan, B genome copies are assigned to magenta, and D genome copies are assigned to yellow, resulting in predominantly grey (cyan/magenta/yellow) where all three genes are represented in a line.

Validation of homoeologous exchanges by low-pass genome resequencing

To differentiate clearly between structural genome changes and long-range gene silencing as an alternative explanation for apparent copy number reduction, we undertook low-pass genome sequencing on one individual of each of the same 27 varieties of *B. napus* studied by transcriptome analysis. Illumina sequencing, quality checking and processing were conducted as described previously (Higgins *et al.*, 2012) except that 100 base reads obtained from the HiSeq platform were used. BWA (Li and Durbin, 2009) sequence-alignment program (v0.7.12) was used for mapping with default parameters. The sequence reference being used shares the same gene id as the transcript quantification for direct comparison, but spliced introns are included for better mapping. The redundancy of representation of gene models was quantified and normalized as reads per kb per million aligned reads (RPKM). To visualize regions of the genome involved in homoeologous exchanges in a comparable way to the mRNAseq-based analysis, we used TDTPs to visualize the genome redundancy data, essentially as for visualization based on mRNAseq data. This involved assigning quantitative representation (as RPKM) for each member of the 19 083 A and C genome homoeologous gene pairs with representation >0.01 RPKM a value in CMYK colour space where the contributions from the A and C genome copies were coded to cyan and magenta channels, respectively, and displaying the results using tile plots.

Acknowledgements

Next-generation sequencing and library construction was delivered via the BBSRC National Capability in Genomics (BB/J010375/1) at The Genome Analysis Centre by members of the Platforms and Pipelines Group. We thank the High-Throughput Genomics Group at the Wellcome Trust Centre for Human Genetics (funded by Wellcome Trust grant reference 090532/Z/09/Z) for the generation of mRNAseq and genomic sequencing data. We would like to thank Carmel O'Neill of the John Innes Centre for her enthusiasm and contributions to preliminary work. This work was supported by UK Biotechnology and Biological Sciences Research Council (BB/L002124/1, BB/L011751/1, BB/H004351/1), including work carried out within the ERA-CAPS Research Program (BB/L027844/1). The authors declare no conflict of interest.

References

Adams, K.L. and Wendel, J.F. (2005) Polyploidy and genome evolution in plants. *Curr. Opin. Plant Biol.* **8**, 135–141.

Bancroft, I., Morgan, C., Fraser, F., Higgins, J., Wells, R., Clissold, L., Baker, D. *et al.* (2011) Dissecting the genome of the polyploid crop oilseed rape by transcriptome sequencing. *Nat. Biotechnol.* **29**, 762–766.

Bancroft, I., Fraser, F., Morgan, C. and Trick, M. (2015) Collinearity analysis of Brassica A and C genomes based on an updated inferred unigene order. *Data Brief*, **3**, 51–55.

Beckmann, J.S., Estivill, X. and Antonarakis, S.E. (2007) Copy number variants and genetic traits: closer to the resolution of phenotypic to genotypic variability. *Nat. Rev. Genet.* **8**, 639–646.

Bus, A., Körber, N., Snowdon, R.J. and Stich, B. (2011) Patterns of molecular variation in a species-wide germplasm set of *Brassica napus*. *Theor. Appl. Genet.* **123**, 1413–1423.

Chalhoub, B., Denoeud, F., Liu, S., Parkin, I.A.P., Tang, H., Wang, X., Chiquet, J. *et al.* (2014) Early allopolyploid evolution in the post-Neolithic *Brassica napus* oilseed genome. *Science*, **345**, 950–953.

Chantret, N., Salse, J., Sabot, F., Rahman, S., Bellec, A., Laubin, B., Dubois, I. *et al.* (2005) Molecular basis of evolutionary events that shaped the hardness locus in diploid and polyploid wheat species (Triticum and Aegilops). *Plant Cell*, **17**, 1033–1045.

Comai, L. (2005) The advantages and disadvantages of being polyploid. *Nat. Rev. Genet.* **6**, 836–846.

Corredor, E., Lukaszewski, A.J., Pachón, P., Allen, D.C. and Naranjo, T.S. (2007) Terminal regions of wheat chromosomes select their pairing partners in meiosis. *Genetics*, **177**, 699–706.

Cowan, C.R. and Cande, W.Z. (2002) Meiotic telomere clustering is inhibited by colchicine but does not require cytoplasmic microtubules. *J. Cell Sci.* **115**, 3747–3756.

Gregory, T.R. and Mable, B.K. (2005) Polyploidy in animals. *Evol. Genome*, **171**, 427–517.

Harper, A.L., Trick, M., Higgins, J., Fraser, F., Clissold, L., Wells, R., Hattori, C. *et al.* (2012) Associative transcriptomics of traits in the polyploid crop species *Brassica napus*. *Nat. Biotech.* **30**, 798–802.

Harper, A.L., Trick, M., He, Z., Clissold, L., Fellgett, A., Griffiths, S. and Bancroft, I. (2016) Genome distribution of differential homoeologue contributions to leaf gene expression in bread wheat. *Plant Biotechnol. J.* **14**, 1207–1214.

He, Z., Cheng, F., Li, Y., Wang, X., Parkin, I.A.P., Chalhoub, B., Liu, S. *et al.* (2015) Construction of Brassica A and C genome-based ordered pan-transcriptomes for use in rapeseed genomic research. *Data Brief*, **4**, 357–362.

Higgins, J., Magusin, A., Trick, M., Fraser, F. and Bancroft, I. (2012) Use of mRNA-seq to discriminate contributions to the transcriptome from the constituent genomes of the polyploid crop species *Brassica napus*. *BMC Genom.* **13**, 247.

Hillier, L.W., Miller, W., Birney, E., Warren, W., Hardison, R.C., Ponting, C.P., Bork, P. *et al.* (2004) Sequence and comparative analysis of the chicken genome provide unique perspectives on vertebrate evolution. *Nature*, **432**, 695–716.

Inaba, R. and Nishio, T. (2002) Phylogenetic analysis of Brassiceae based on the nucleotide sequences of the S-locus related gene, SLR1. *Theor. Appl. Genet.* **105**, 1159–1165.

Leitch, I.J. and Bennett, M.D. (1997) Polyploidy in angiosperms. *Trends Plant Sci.* **2**, 470–476.

Li, H. and Durbin, R. (2009) Fast and accurate short read alignment with Burrows-Wheeler transform. *Bioinformatics*, **25**, 1754–1760.

Marcussen, T., Sandve, S.R., Heier, L., Spannagl, M., Pfeifer, M., International Wheat Genome Sequencing, C., Jakobsen, K.S. *et al.* (2014) Ancient hybridizations among the ancestral genomes of bread wheat. *Science*, **345**, 1250092.

Möllers, C. and Iqbal, M. (2009) Doubled haploids in breeding winter oilseed rape. In *Advances in Haploid Production in Higher Plants* (Touraev, A., Forster, B.P. and Mohan Jain, S., eds), pp. 161–169. Netherlands: Springer.

Pradhan, A.K., Gupta, V., Mukhopadhyay, A., Arumugam, N., Sodhi, Y.S. and Pental, D. (2003) A high-density linkage map in *Brassica juncea* (Indian mustard) using AFLP and RFLP markers. *Theor. Appl. Genet.* **106**, 607–614.

The Arabidopsis Genome Initiative. (2000) Analysis of the genome sequence of the flowering plant *Arabidopsis thaliana*. *Nature*, **408**, 796–815.

Trick, M., Long, Y., Meng, J. and Bancroft, I. (2009) Single nucleotide polymorphism (SNP) discovery in the polyploid *Brassica napus* using Solexa transcriptome sequencing. *Plant Biotechnol. J.* **7**, 334–346.

U.N. (1935) Genome analysis in Brassica with special reference to the experimental formation of *B. napus* and peculiar mode of fertilization. *Jpn J. Bot.* **7**, 389–452.

Wang, X., Shi, X., Hao, B., Ge, S. and Luo, J. (2005) Duplication and DNA segmental loss in the rice genome: implications for diploidization. *New Phytol.* **165**, 937–946.

Wells, R., Trick, M., Soumpourou, E., Clissold, L., Morgan, C., Werner, P., Gibbard, C. *et al.* (2014) The control of seed oil polyunsaturate content in the polyploid crop species *Brassica napus*. *Mol. Breeding*, **33**, 349–362.

Wendel, J.F. (2000) Genome evolution in polyploids. *Plant Mol. Biol.* **42**, 225–249.

5

Stacking transgenic event DAS-Ø15Ø7-1 alters maize composition less than traditional breeding

Rod A. Herman[1,]*, Brandon J. Fast[1], Peter N. Scherer[1], Alyssa M. Brune[1], Denise T. de Cerqueira[2], Barry W. Schafer[1], Ricardo D. Ekmay[1], George G. Harrigan[1,†] and Greg A. Bradfisch[1]

[1]*Dow AgroSciences LLC, Indianapolis, IN, USA*
[2]*Dow AgroSciencies Sementes e Biotecnologia Brasil LTDA, Cravinhos, SP, Brazil*

*Correspondence
email raherman@dow.com
Present address: The Coca-Cola Company, 1 Coca Cola Plaza, Atlanta, GA 30313, USA.

Keywords: composition, breeding stacks, equivalence.

Summary

The impact of crossing ('stacking') genetically modified (GM) events on maize-grain biochemical composition was compared with the impact of generating nonGM hybrids. The compositional similarity of seven GM stacks containing event DAS-Ø15Ø7-1, and their matched nonGM near-isogenic hybrids (iso-hybrids) was compared with the compositional similarity of concurrently grown nonGM hybrids and these same iso-hybrids. Scatter plots were used to visualize comparisons among hybrids and a coefficient of identity (per cent of variation explained by line of identity) was calculated to quantify the relationships within analyte profiles. The composition of GM breeding stacks was more similar to the composition of iso-hybrids than was the composition of nonGM hybrids. NonGM breeding more strongly influenced crop composition than did transgenesis or stacking of GM events. These findings call into question the value of uniquely requiring composition studies for GM crops, especially for breeding stacks composed of GM events previously found to be compositionally normal.

Introduction

Compositional analysis of genetically modified (GM) crops is universally required by regulatory authorities that assess safety. Compositional analysis was proposed more than 20 years ago to account for uncertainties in how the types of compositional changes and the magnitude of these changes might differ between crops that have been engineered to contain transgenes compared with those developed through traditional breeding (OECD, 1993). The variability in biochemical composition associated with nonGM crop varieties is not typically considered a significant safety risk (Bradford *et al.*, 2005). Traditional breeding has a history of safe use, so the compositional variability among varieties developed through traditional breeding was the standard by which GM breeding techniques were to be assessed. After more than two decades of research, many published reports and hundreds of regulatory submissions, transgenesis has generally been found to have markedly less effect on crop composition compared with traditional breeding (Herman and Price, 2013). Advances in molecular biology have shown that the types of mutations that are possible during transgene insertion are similar to those associated with the intentional or unintentional random mutagenesis that occurs during traditional breeding, but that GM techniques typically have a smaller impact due to fewer genetic changes (Anderson *et al.*, 2016; Harrigan *et al.*, 2016; Schnell *et al.*, 2015; Venkatesh *et al.*, 2016).

While the potential for unintended compositional effects is now known to be markedly lower for GM crops compared with those developed using nonGM breeding techniques, government regulation and data requirements for GM crop composition have increased dramatically over the last 20 years, with a typical study now costing over one million US dollars (Herman and Price, 2013). Furthermore, when two or more previously approved GM events are crossed together (breeding stacks), many regulatory authorities require a new compositional study to be completed with equal complexity (and cost) to that required for novel transgenic events (EFSA Panel on Genetically Modified Organisms (GMO), 2011). Since transgenes are routinely incorporated into many new crop varieties after obtaining regulatory approvals, this requirement seems to be based on the premise that transgenes are more likely to interact with each other compared with interacting with the many different unique genotypes present in each new crop variety into which a GM event is introgressed.

Purpose and approach

To provide a science-based evaluation of the crop-compositional effects of stacking GM events and to provide context around the regulatory requirement for composition studies with GM breeding stacks, we examined the grain composition of seven breeding stacks containing DAS-Ø15Ø7-1 maize (Baktavachalam *et al.*, 2015) and other approved GM maize events. DAS-Ø15Ø7-1 expresses the Cry1F insecticidal protein and the phosphinothricin N-acetyltransferase (PAT) herbicide-tolerant enzyme. Event DAS-Ø15Ø7-1 has been crossed with other approved GM events expressing insecticidal and herbicide-tolerance traits. These traits have known modes of action that are not expected to modify the endogenous metabolism of the plant and thus are not expected to alter crop composition.

Since current regulations in the European Union require the inclusion of commercial nonGM reference lines (or hybrids) in crop composition studies (EFSA Panel on Genetically Modified Organisms (GMO), 2011), we were able to evaluate the compositional differences between the GM breeding stacks and matched near-isogenic nonGM hybrids (iso-hybrids), and contrast that with the differences between the iso-hybrids and nonGM reference hybrids. In this way, it was possible to directly compare the compositional effects of the multiple GM events in each breeding stack with compositional changes accompanying nonGM hybrid development.

Results

Purpose and context

The compositional profiles of nonGM near-isogenic maize hybrids (iso-hybrids) were compared with the same hybrids containing GM breeding stacks that included event DAS-Ø15Ø7-1 and were also compared with nonGM reference hybrids developed through traditional breeding. This approach allows the compositional effects of transgenesis, and the conventional crossing of GM events to be contrasted with non-GM hybrid development techniques that are generally accepted as safe. Results are relevant to the scientific evaluation of government regulation for GM breeding stacks because such results reflect relative risk.

Data analysis approach

While traditional statistical difference tests (e.g. ANOVA) have most commonly been used to compare crop composition between a GM variety and a nonGM near-isogenic line (isoline), statistical differences observed using such approaches most commonly reflect the statistical power of the study rather than the effects of transgenesis. This is because the GM line is derived from a single cell from a single plant and does not reflect the intravarietal genetic variability of the line into which the GM trait was introgressed and with which it is most commonly compared (Fasoula and Boerma, 2007; Harrigan et al., 2016; Tokatlidis et al., 2008; Venkatesh et al., 2016). With sufficient statistical power, one would expect every compositional analyte to be found statistically different between the GM line and its isoline due to genetic differences that are not related to transgenesis, but rather due to endogenous genetic differences that are expected between two different single cell- or plant-derived lines (intravarietal variation). Regulatory requirements for more and more complex and powerful experimental designs have led to greater and greater statistical detection of fleetingly small and biologically irrelevant compositional differences.

A complementary statistical approach has also been developed that is intended to represent the breadth of composition for nonGM varieties through construction of equivalence intervals based on a small number of reference lines included in the field trials, followed by determination of whether the composition for the GM line falls within these intervals (van der Voet et al., 2011; Ward et al., 2012). However, the conformation of the composition of the isoline (background genetics of GM line) to the average composition (or the compositional distribution) of the chosen reference lines may largely determine the results of such statistical tests. If the concentration of the measured compositional analyte in the isoline falls in the centre of the equivalence interval, then the GM line will likely also fall within the interval. The more distinct the isoline composition is from the average composition of the reference lines (or disparate from the compositional distribution of the reference lines), the more likely the composition of the GM line will be found to fall outside of the equivalence limits for the reference lines. That is, if the starting point for the isoline composition is in the centre of the equivalence interval for the reference lines, larger deviations from this composition will be

seen as normal for the crop. Conversely, if the starting point for the isoline composition is near or outside of the equivalence limits for the reference lines, smaller deviations will often result in a finding of nonequivalence to the reference lines.

In summary, current statistical approaches to evaluating compositional equivalency between a GM line and the isoline using difference tests, or evaluating the compositional normalcy of the GM line by constructing equivalence limits based on a small number of nonGM reference lines, do not properly take into consideration intravarietal variability or the potential disparity of the composition of the isoline from the reference lines, respectively. In addition, multiplicity is most often not addressed in the simultaneous statistical evaluation of the many compositional endpoints under consideration, resulting in a high probability of declaring statistically significant differences where none exist (van der Voet et al., 2011). For these reasons, we chose to compare the iso-hybrids in our analysis with both breeding stacks of DAS-Ø15Ø7-1 maize and with nonGM commercial hybrids using graphical methods intended to visualize the effects of transgenesis and GM breeding stacks on the profile of crop composition compared with non-GM breeding. We quantified these compositional differences using an I^2 statistic (coefficient of identity) that estimates the fraction of the variation in the data captured by the line of identity.

Focus is on transgenesis and stacking

We focused on GM entries that were treated with the same maintenance herbicides as the nonGM iso-hybrids and the nonGM commercial reference hybrids. This avoided confounding any possible effects of transgenesis or stacking GM events with any possible effects of the trait-associated herbicides, and because the use of herbicides on nonGM crop plants has a history of safe use. The application of herbicides to crops has never been associated with any adverse compositional change in a crop, and there is no reasonable hypothesis for why GM crops would respond differently to trait-related herbicides (Herman and Price, 2013; Herman et al., 2013).

Suitability of data for scatter plots

Previous analyses of soybean using six categories of analytes illustrated a reasonable spread of analyte concentrations along the regression lines and a subsequently adequate depiction of profiles within each analyte category (Fast et al., 2016). A similar situation was observed here for maize (Figures 1-6). This approach allows a more contextual interpretation of results by not isolating single analytes from related compositional components. This approach also weights more predominant analytes more heavily than less predominant analytes in calculating the I^2 statistic and does not inflate small absolute concentration changes for low-prevalence analytes by expressing them as a per cent of the isoline concentration. In the absence of a hypothesis for a GM trait causing a specific compositional change, this procedure seems reasonable for assessing unintended compositional changes.

To investigate the potential effects of lower prevalence compositional analytes on observed trends, the data were also transformed to the base-10 logarithm which weights smaller values more heavily than is the case in the natural scale. These data were plotted and used to generate I^2 statistics to supplement the analysis in the natural scale but are not discussed in detail (Table 1; Figures 1-6 insets). For some analyte profiles, data expressed as the base-10 logarithm are more evenly spread along the line of identity

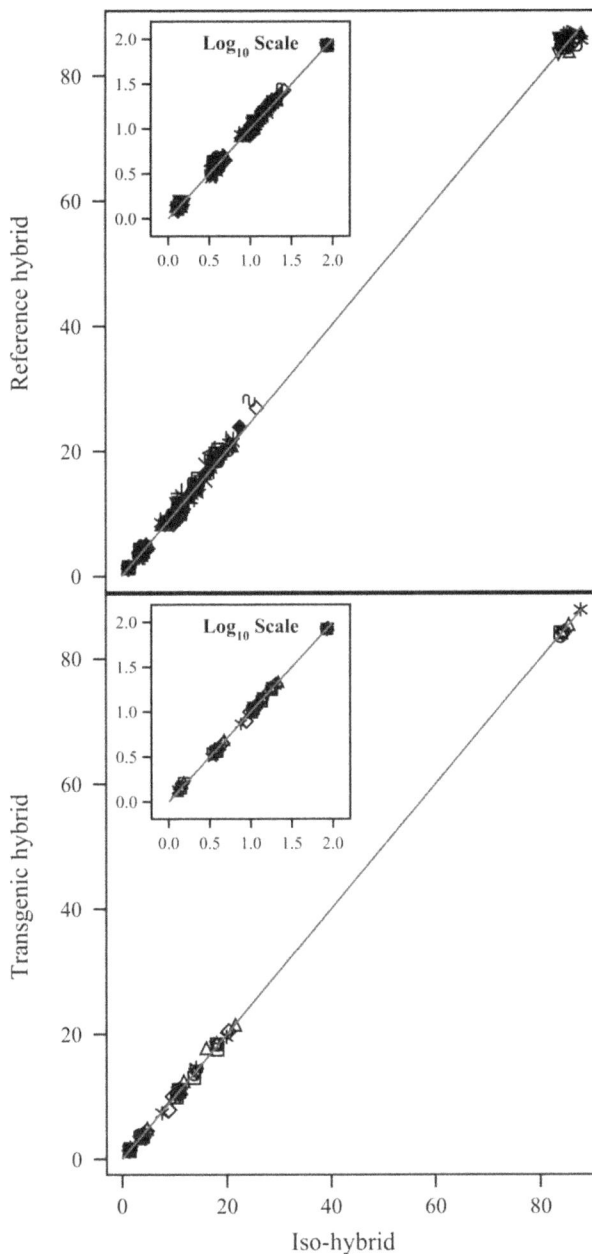

Figure 1 Proximates and fibre. Scatter plot of iso-hybrids composition vs. location-matched reference hybrids (upper panel) and DAS-Ø15Ø7-1 GM breeding stacks (lower panel). Analytes from left to right: ash, ADF, crude fat, NDF, crude protein, total dietary fibre, moisture and carbohydrates (moisture = %FW, all others = %DW). Line of identity ($y = x$) shown. The symbols representing each breeding stack and nonGM reference hybrid are listed in Table 1.

Figure 2 Amino acids. Scatter plot of iso-hybrids composition vs. location-matched reference hybrids (upper panel) and DAS-Ø15Ø7-1 GM breeding stacks (lower panel). Analytes from left to right: tryptophan, methionine, cystine, histidine, lysine, threonine, isoleucine, glycine, tyrosine, serine, valine, arginine, phenylalanine, aspartic acid, alanine, proline, leucine and glutamic acid (% of total amino acids). Line of identity ($y = x$) shown. The symbols representing each breeding stack and nonGM reference hybrid are listed in Table 1.

(e.g. proximates and fibre, Figure 1), but it is noteworthy that, unlike regression lines, the line of identity is fixed, and its position is not influenced by compositional deviations from the line.

NonGM reference line composition

As expected, the line of identity ($y = x$) approximated the relationship between the iso-hybrids and reference hybrids well with the exception of vitamins for a subset of the reference hybrids; vitamins in maize are known to vary widely (Figures 1-6, upper panels;

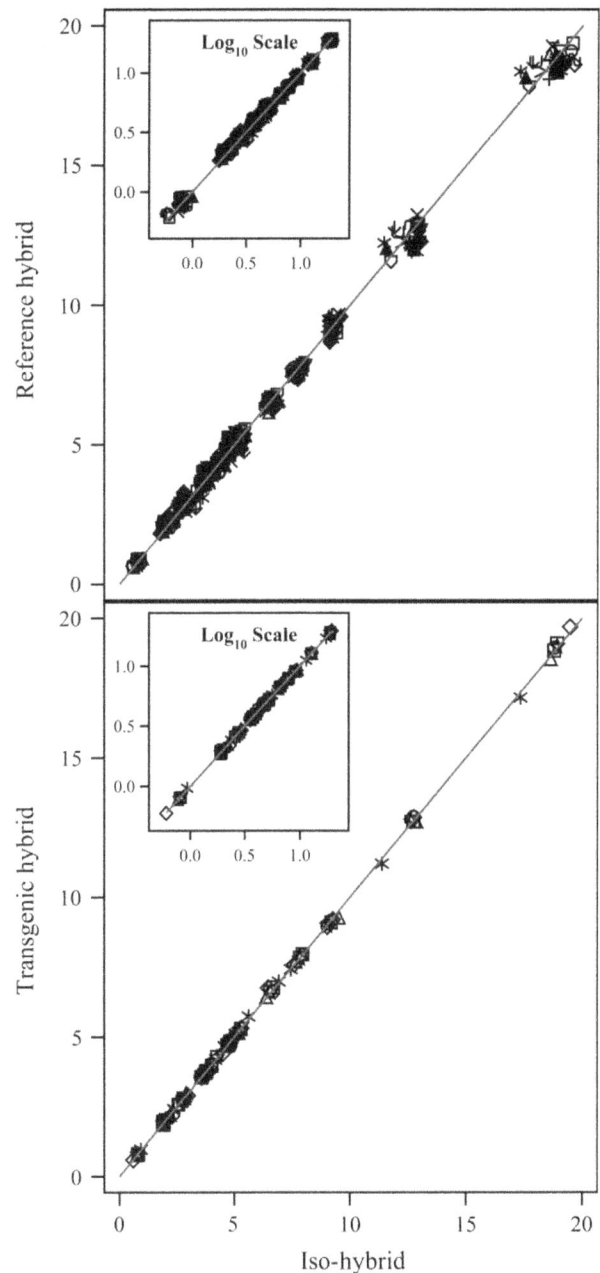

Table 1; Lundry *et al.*, 2013). Maize grain is typically sold as a generic commodity without price adjustments based on compositional variation (with the exception of high moisture which is docked to account for required drying to preserve quality). This practice is only practical if a reasonably consistent nutrient composition is observed. The r^2 values for the iso-hybrids predicting the twenty-five reference hybrids are ≥0.9964 for proximates and fibre, ≥0.9907 for amino acids, ≥0.8592 for fatty acids, ≥0.9685 for

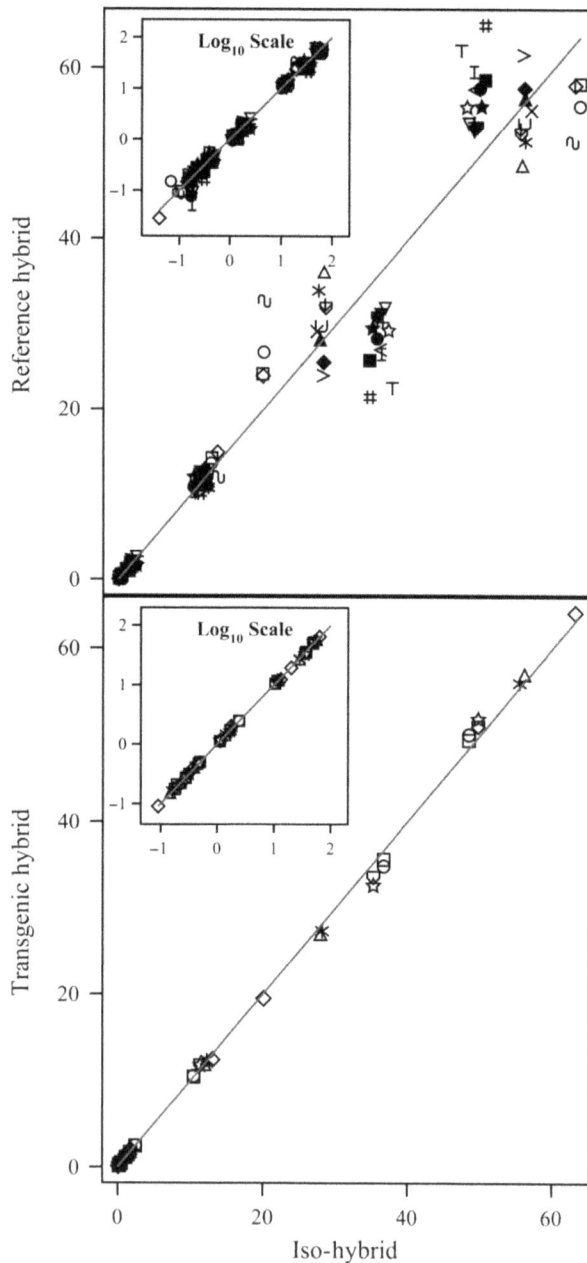

Figure 3 Fatty acids. Scatter plot of iso-hybrids composition vs. location-matched reference hybrids (upper panel) and DAS-Ø15Ø7-1 GM breeding stacks (lower panel). Analytes from left to right: 22 : 0 behenic, 20 : 1 eicosenoic, 20 : 0 arachidic, 18 : 3 linolenic, 18 : 0 stearic, 16 : 0 palmitic, 18 : 1 oleic and 18 : 2 linoleic (% of total fatty acids). Line of identity (y = x) shown. The symbols representing each breeding stack and nonGM reference hybrid are listed in Table 1.

Figure 4 Minerals. Scatter plot of iso-hybrids composition vs. location-matched reference hybrids (upper panel) and DAS-Ø15Ø7-1 GM breeding stacks (lower panel). Analytes from left to right: copper, manganese, zinc, iron, calcium, magnesium, phosphorus and potassium (mg/100 g DW). Line of identity (y = x) shown. The symbols representing each breeding stack and nonGM reference hybrid are listed in Table 1.

minerals, from 0.2255 to 0.9891 for vitamins and $\geq$0.8888 for secondary metabolites (Table 1).

DAS-Ø15Ø7-1 maize breeding stack composition

Also as expected, the line of identity approximated the relationship between the iso-hybrids and the DAS-Ø15Ø7-1 maize breeding stacks extremely well for all analyte profiles (Figures 1-6, lower panels; Table 1). The insertion of GM traits that are not expected to affect endogenous metabolic pathways (as is the case here), and their cross-breeding into stacks is not expected to alter the composition of

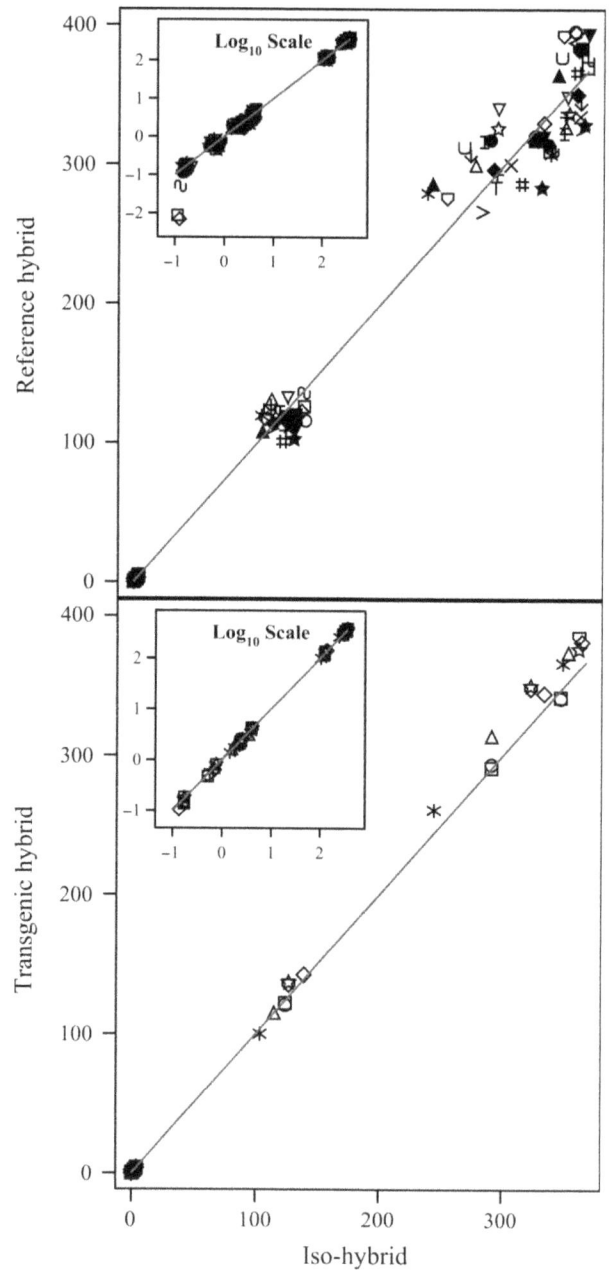

a crop as much as the recombination of many thousands of genes during traditional nonGM breeding. The r^2 values for the iso-hybrids predicting the seven matched DAS-Ø15Ø7-1 maize breeding stacks are $\geq$0.9993 for proximates and fibre, $\geq$0.9994 for amino acids, $\geq$0.9956 for fatty acids, $\geq$0.9937 for minerals, $\geq$0.9874 for vitamins and $\geq$0.9753 for secondary metabolites (Table 1).

GM breeding stack vs. traditional breeding

Our data analysis approach allows the direct comparison between traditional breeding and transgenic approaches

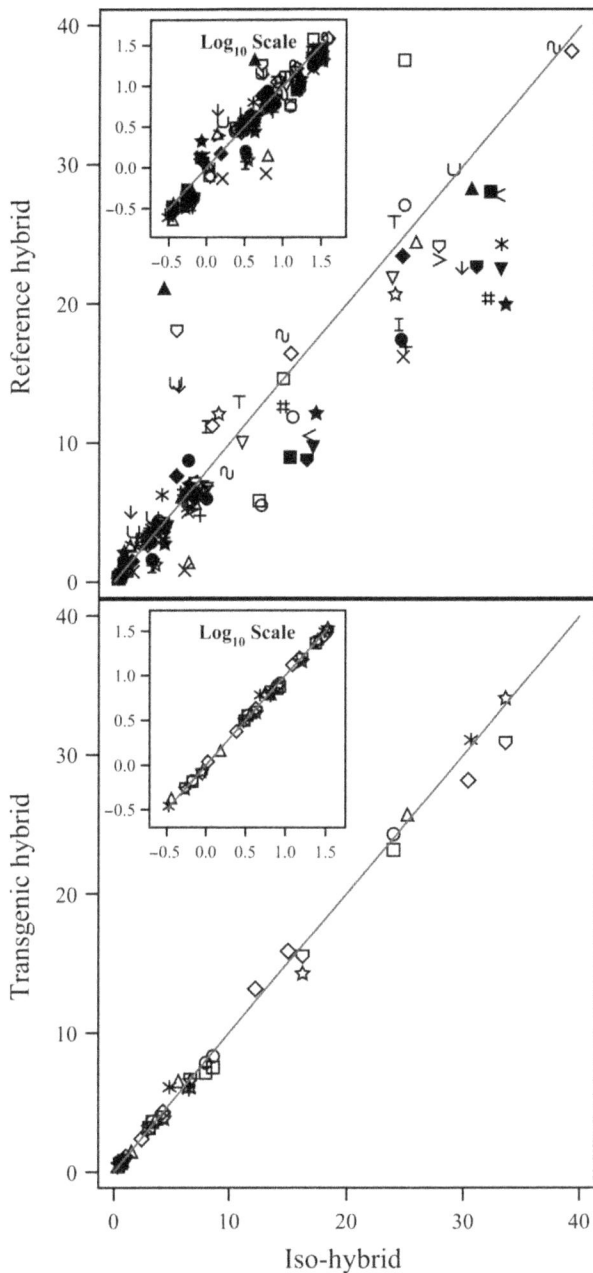

Figure 5 Vitamins. Scatter plot of iso-hybrids composition vs. location-matched reference hybrids (upper panel) and DAS-Ø15Ø7-1 GM breeding stacks (lower panel). Analytes from left to right: vitamin B_1 (thiamine HCl), β-carotene, vitamin B_6 (pyridoxine HCl), α-tocopherol and vitamin B_3 (niacin) (mg/kg DW). Line of identity ($y = x$) shown. The symbols representing each breeding stack and nonGM reference hybrid are listed in Table 1.

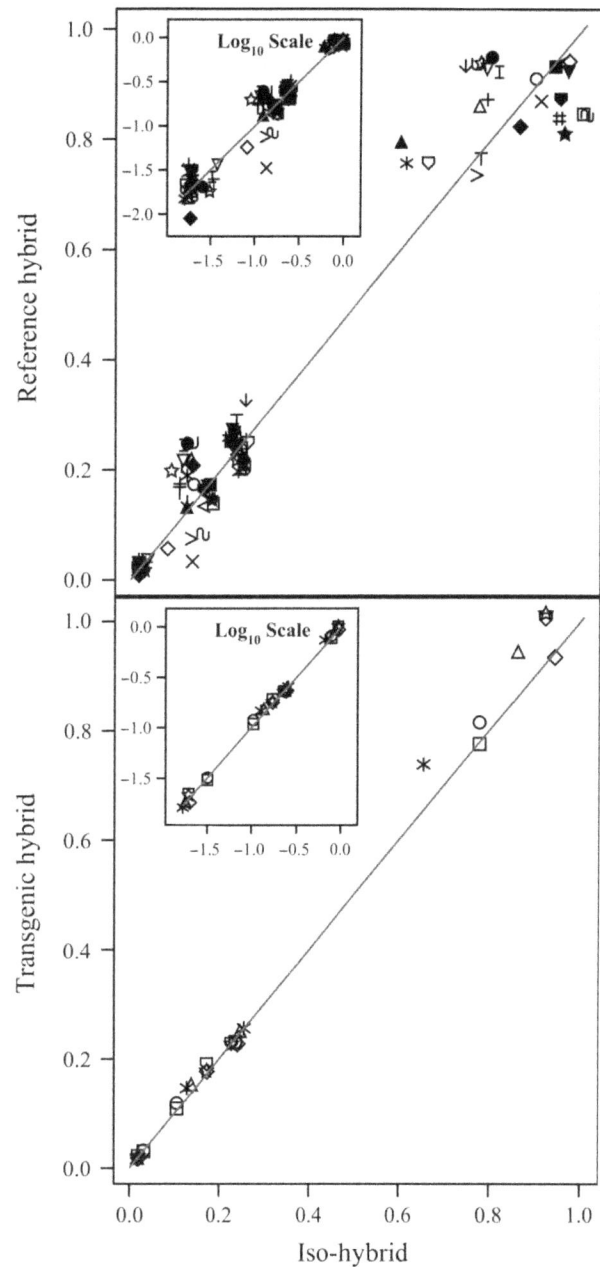

Figure 6 Secondary metabolites. Scatter plot of iso-hybrids composition vs. location-matched reference hybrids (upper panel) and DAS-Ø15Ø7-1 GM breeding stacks (lower panel). Analytes from left to right: p-coumaric acid, raffinose, ferulic acid and phytic acid (%DW). Line of identity ($y = x$) shown. The symbols representing each breeding stack and nonGM reference hybrid are listed in Table 1.

including stacking GM traits by traditional breeding (breeding stacks). We used the nonGM iso-hybrids as a calibrator against which these breeding techniques were evaluated. Several weaknesses of statistical difference tests and equivalence tests were avoided, including the isolation of one analyte from related analytes (avoided through examination of compositional profiles). Both graphically and through the statistical estimation of compositional identity (I^2), the greater influence of traditional breeding on composition, compared with transgenesis and

stacking GM traits by traditional breeding, is clearly evident (Table 1, Figures 1-6). Transformation of composition values to the base-10 logarithmic scale (weighting lower prevalence analytes more heavily than in the natural scale) does not change this conclusion (Table 1, Figures 1-6 insets). A comparison of the distributions of I^2 values for the iso-hybrids predicting the DAS-Ø15Ø7-1 maize breeding stacks vs. the nonGM commercial reference lines clearly indicates greater compositional changes associated with traditional breeding versus stacking GM events (Figure 7).

Table 1 Plot symbols and coefficients of identity (I^2) for iso-hybrids predicting indicated hybrid composition

Entry type	Entry name	Plot symbol	I^2 for data in natural scale (Log_{10} scale)					
			Proximates and fibre	Amino acids	Fatty acids	Minerals	Vitamins	Secondary metabolites
GM Breeding Stack	PowerCore™	□	0.9998 (0.9986)	0.9999 (0.9998)	0.9992 (1.0000)	0.9998 (0.9988)	0.9915 (0.9943)	0.9998 (0.9993)
	PowerCore™ Enlist™	○	1.0000 (0.9995)	0.9998 (0.9997)	0.9978 (0.9997)	0.9996 (0.9989)	0.9992 (0.9989)	0.9959 (0.9966)
	PowerCore™ Ultra	△	0.9993 (0.9985)	0.9995 (0.9995)	0.9994 (0.9998)	0.9943 (0.9994)	0.9972 (0.9944)	0.9862 (0.9973)
	PowerCore™ Ultra Enlist™	✳	0.9999 (0.9992)	0.9995 (0.9995)	0.9997 (0.9997)	0.9959 (0.9997)	0.9966 (0.9940)	0.9753 (0.9954)
	SmartStax®	▽	0.9999 (0.9995)	0.9998 (0.9996)	0.9983 (0.9997)	0.9937 (0.9996)	0.9874 (0.9972)	0.9879 (0.9956)
	SmartStax® Enlist™	☆	0.9999 (0.9993)	0.9999 (0.9999)	0.9956 (0.9996)	0.9944 (0.9996)	0.9943 (0.9968)	0.9857 (0.9974)
	SmartStax® PRO Enlist™	◇	0.9997 (0.9976)	0.9994 (0.9992)	0.9995 (0.9994)	0.9977 (0.9987)	0.9883 (0.9981)	0.9994 (0.9981)
Reference hybrid	AG Venture 7844	○	0.9996 (0.9974)	0.9989 (0.9970)	0.9597 (0.9774)	0.9888 (0.9958)	0.8511 (0.8744)	0.9966 (0.9897)
	Ayerza Semillas IMPERIO	△	0.9989 (0.9914)	0.9986 (0.9976)	0.9558 (0.9867)	0.9897 (0.9983)	0.9241 (0.7527)	0.9585 (0.9624)
	Ayerza Semillas OLYMPUS	✳	0.9985 (0.9910)	0.9918 (0.9926)	0.9754 (0.9926)	0.9760 (0.998)	0.7582 (0.9660)	0.9135 (0.9665)
	Beck's 5509	□	0.9989 (0.9967)	0.9989 (0.9986)	0.9827 (0.9964)	0.9948 (0.9276)	0.7848 (0.8969)	0.9375 (0.9861)
	Beck's 6158	◇	0.9991 (0.9970)	0.9933 (0.9912)	0.9854 (0.9952)	0.9916 (0.9090)	0.9891 (0.9715)	0.9973 (0.9808)
	Dekalb 6170	☆	0.9998 (0.9974)	0.9980 (0.9947)	0.9565 (0.9851)	0.9910 (0.9971)	0.9072 (0.8624)	0.9220 (0.8796)
	Don Atilio CARDENAL	▽	0.9994 (0.9947)	0.9988 (0.9957)	0.9918 (0.9904)	0.9844 (0.9964)	0.5821 (0.8740)	0.9499 (0.9546)
	Dow Semillas MILL 527	↓	0.9964 (0.9908)	0.9907 (0.9875)	0.9917 (0.9914)	0.9885 (0.9938)	0.5855 (0.7541)	0.9111 (0.8842)
	Golden Harvest 8920	▽	0.9998 (0.9969)	0.9979 (0.9955)	0.9812 (0.9849)	0.9875 (0.9948)	0.9644 (0.8703)	0.9354 (0.9174)
	KWS KM4321	>	0.9998 (0.9991)	0.9990 (0.9968)	0.9834 (0.9858)	0.9871 (0.9947)	0.9246 (0.9620)	0.9806 (0.8795)
	Merit Seed 5096	<	0.9980 (0.9934)	0.9937 (0.9920)	0.9472 (0.9935)	0.9940 (0.9944)	0.8696 (0.9630)	0.9959 (0.9908)
	Merit Seed 5145	#	0.9981 (0.9926)	0.9988 (0.9982)	0.8909 (0.9714)	0.9921 (0.9976)	0.5278 (0.9301)	0.9626 (0.9862)
	Middlekoop 5513	I	0.9997 (0.9988)	0.9990 (0.9990)	0.9326 (0.9543)	0.9877 (0.9970)	0.7925 (0.8472)	0.9386 (0.9036)
	Middlekoop 6614	+	0.9979 (0.9950)	0.9957 (0.9928)	0.9762 (0.9836)	0.9969 (0.9979)	0.5643 (0.8036)	0.9731 (0.9453)
	Middlekoop C110	⊤	0.9986 (0.9908)	0.9991 (0.9971)	0.8592 (0.9845)	0.9974 (0.9968)	0.9675 (0.8767)	0.9810 (0.9388)
	Pannar PAN 5E-203	∪	0.9981 (0.9908)	0.9956 (0.9937)	0.9964 (0.9953)	0.9810 (0.9959)	0.8480 (0.8598)	0.9086 (0.9376)
	Pfister 2874	℧	0.9971 (0.9939)	0.9973 (0.9974)	0.8887 (0.9891)	0.9948 (0.9847)	0.9856 (0.9878)	0.9224 (0.9413)
	Pioneer 31D06	✕	0.9996 (0.9976)	0.9974 (0.9959)	0.9975 (0.9953)	0.9948 (0.9984)	0.4327 (0.5546)	0.9739 (0.7990)
	Pioneer 33W82	●	0.9998 (0.9970)	0.9983 (0.9941)	0.9610 (0.9750)	0.9909 (0.9979)	0.6775 (0.8801)	0.9233 (0.9273)
	Pioneer P1979	▲	0.9994 (0.9965)	0.9959 (0.9938)	0.9996 (0.9896)	0.9842 (0.9983)	0.5105 (0.7848)	0.8888 (0.9822)
	Rupp XR1588	▼	0.9979 (0.9939)	0.9946 (0.9916)	0.9851 (0.9979)	0.9932 (0.9960)	0.4890 (0.9409)	0.9905 (0.9462)
	Rupp XR1612	■	0.9997 (0.9978)	0.9950 (0.9917)	0.9518 (0.9940)	0.9965 (0.9969)	0.8918 (0.9491)	0.9980 (0.9977)
	Rusticana NT 525	◆	0.9992 (0.9952)	0.9937 (0.9901)	0.9961 (0.9992)	0.9993 (0.9958)	0.9784 (0.9818)	0.9793 (0.9309)
	Viking 50-04N	★	0.9990 (0.9937)	0.9977 (0.9974)	0.9786 (0.9978)	0.9685 (0.9910)	0.2255 (0.8482)	0.9340 (0.9728)
	Viking 60-01N	◆	0.9985 (0.9964)	0.9936 (0.9896)	0.9854 (0.9977)	0.9956 (0.9951)	0.5981 (0.9359)	0.9814 (0.9578)

Discussion

Over 46 000 compositional data points are summarized here in a relatively small number of informative figures and tables. The results from this analysis are consistent with previous reports that transgenesis has a negligible influence on crop composition when traits are not expected to alter the endogenous metabolism of plants (Herman and Price, 2013). Specifically, there is extremely weak justification for requiring special compositional testing of GM breeding stacks where the component events have been

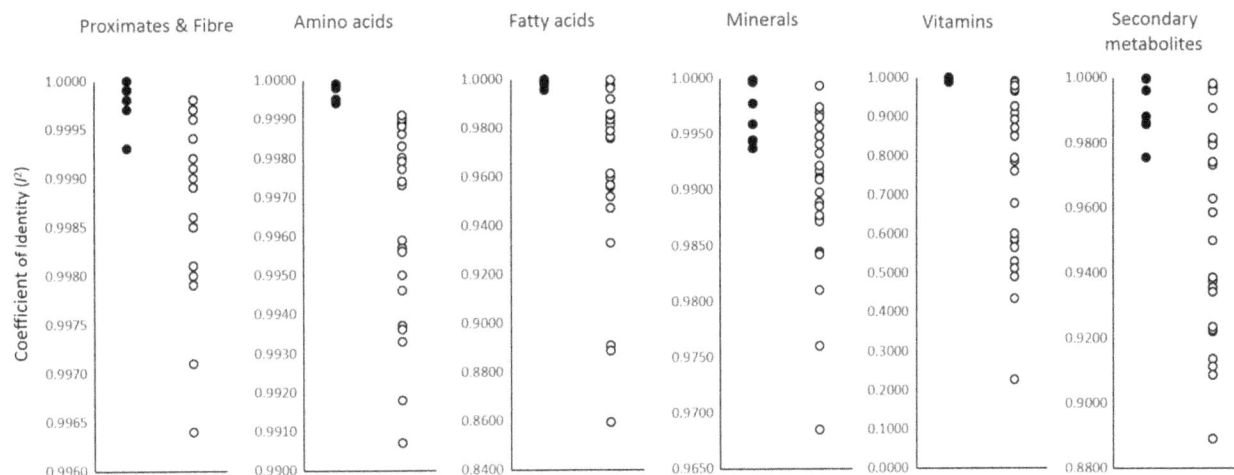

Figure 7 Coefficient of identity (I^2) for iso-hybrids predicting GM breeding stacks (solid circles) and nonGM commercial reference hybrids (open circles) for indicated compositional analyte profile.

found compositionally equivalent to the nonGM crop (Kok et al., 2014; Pilacinski et al., 2011). The use of difference and equivalence testing to analyse GM compositional equivalence does not adequately account for intravarietal variation, or the potential disparity of the composition of the isoline from the reference lines, respectively, nor is multiplicity typically addressed in analyses using these statistical approaches. The use of scatter plots and the estimation of compositional identity (I^2) for the isoline versus the GM lines and the reference lines allows the potential effects and risks of altering crop composition due to transgenesis to be put into context. Without putting the risks of transgenesis and stacking of GM traits into the context of traditional breeding and grower practices, regulatory authorities risk diverting limited resources to evaluating negligible risks while leaving appreciable risks uncharacterized (Buchholz et al., 2011). Scientific evidence strongly supports a proportionally low level of regulatory burden for GM breeding stacks based on their very low risk.

Experimental procedures

Field trials

Five field studies were conducted between 2010 and 2016 across two countries (USA and Argentina) with a total of seven DAS-Ø15Ø7-1 maize breeding stacks. The GM events in each breeding stack, the locations of the trials and the nonGM reference hybrids included at each location are listed in Table 2. The gene products expressed in the breeding stacks are listed in Table 3.

Plots were arranged in a randomized complete block design with four blocks at each location. Three commercial nonGM reference hybrids were planted at each location, and the reference hybrids (≥6 per study) were distributed randomly among locations. Plots at each location were 2-4 rows wide, and each plot was bordered by two rows of nonGM maize. Field sites were surrounded by a maize border of at least four rows. Blocks were separated by at least 1.5 m of bare soil. Seed was planted between 15 and 25 cm apart in the row. Plots were between 6 and 10 m long with row spacing ranging from 70 to 76 cm. Appropriate agronomic practices were implemented at each field site across all plots to produce a commercially acceptable crop. Grain samples were randomly collected at physiological maturity and consisted of

approximately 500 g (approximately 5 ears) that were stored frozen until analysis.

Grain analysis

Compositional analyses of maize grain samples focused on six key analyte categories (proximates and fibre, amino acids, fatty acids, minerals, vitamins and secondary metabolites; OECD, 2002). Methods of compositional analysis have been previously reported (Herman et al., 2010). Analytes within each category are listed in the figure captions (Figures 1-6).

Data analysis and interpretation

For each of the six categories of analytes, a scatter plot of each mean analyte level (across locations) for each DAS-Ø15Ø7-1 breeding stack was plotted against the corresponding mean iso-hybrid analyte level such that a finding of identical composition would result in points falling on the line of identity ($y = x$). Each GM breeding stack was represented by a unique symbol on these plots (Table 1). Plots were also generated where the mean analyte levels for each nonGM commercial reference hybrid were plotted against the mean levels of the iso-hybrid across the same sites as those where that reference hybrid was grown. Only reference hybrids grown at three or more sites were included in the analysis. For the four nonGM reference hybrids included in both Argentine studies (Table 2), means were calculated across both studies. The level of data scatter around the line of identity was used to compare the compositional effects of transgenesis and stacking with that of nonGM breeding.

To quantify the level of compositional identity between the iso-hybrid and each GM breeding stack or non-GM reference hybrid, the amount of data variation accounted for by the line of identity was calculated (defined as coefficient of identity or I^2). The I^2 statistic was calculated as for the coefficient of determination (R^2) except that the measured x value (x_i) was substituted for the predicted y value ($\hat{y}_i$) in the following equation.

$$R^2 = 1 - \frac{\sum_{i=1}^{N}(y_i - \hat{y})^2}{\sum_{i=1}^{N}(y_i - \bar{y})^2}$$

This statistic estimates the fraction of the variability in the data that is accounted for by the line of identity.

Table 2 Field trial information

GM event	Breeding stacks in indicated study				
	Study 101061	Study 101062	Study 141098	Study 150126	Study 151077
DAS-Ø15Ø7-1	X X	X X	X	X	X
DAS-59122-7	X X			X	
DAS-4Ø278-9	X	X		X	X
SYN-IR162-4			X		X
MON-89Ø34-3	X X	X X	X	X	X
MON-ØØ6Ø3-6		X X	X		X
MON-88Ø17	X X				
MON-87427-7				X	
MON-87411				X	
Commercial Name	SmartStax® & SmartStax® Enlist™	PowerCore™ & PowerCore™ Enlist™	PowerCore™ Ultra	SmartStax® PRO Enlist™	PowerCore™ Ultra Enlist™

Growing season	2010	2010	2014/2015	2015	2015/2016
Country	USA	USA	Argentina	USA	Argentina
Site 1	IA, Richland	IA, Richland	Buenos Aires, San Pedro	IA, Kimballton	Santa Fé, Amenábar
Site 2	IA, Jefferson	IA, Lime Springs	Buenos Aires, Berdier	IA, Cresco	Buenos Aires, Berdier
Site 3	IA, Lime Springs	IA, Atlantic	Buenos Aires, Carmen de Areco	IA, Richland	Buenos Aires, Blandengues
Site 4	IA, Atlantic	IL, Carlyle	Buenos Aires, Inés Indart	IL, Stewardson	Buenos Aires, Chacabuco
Site 5	IL, Wyoming	IL, Wyoming	Buenos Aires, El Crisol	IN, Pickard	Buenos Aires, Capitán Sarmiento
Site 6	MI, Deerfield	IN, Sheridan	Buenos Aires, Los Indos	MO, Kirkville	Córdoba, Mattaldi
Site 7	MN, Cherry Grove	MO, Fisk	Buenos Aires, San Patrico	NE, York	Buenos Aires, Tacuarí
Site 8	MN, Geneva	NE, Brunswick	Buenos Aires, Tacuarí	PA, Germansville	Santa Fé, Villa Cañás
Site 9	NE, Brunswick	NE, York	–	–	–
Site 10	PA, Germansville	PA, Germansville	–	–	–

	Reference hybrids[a] (planting sites)				
Reference 1	Viking 60-01N (1, 2, 4, 6, 10)	Golden Harvest 8920 (1, 4, 7, 9, 10)	Pannar PAN 5E-203 (1, 2, 5)	Pfister 2874 (1, 5, 6)	Pioneer P1979 (1, 6, 7)
Reference 2	Viking 50-04N (1, 3, 4, 5, 7, 8)	Middlekoop 6614 (1, 2, 3, 8, 9)	KWS KM4321 (2, 6, 7, 8)	Beck's 5509 (1, 2, 7)	Ayerza Semillas OLYMPUS (2, 4, 6, 8)
Reference 3	Rupp XR1588 (1, 5, 9, 10)	Middlekoop C110 (1, 3, 5, 6, 7)	Pioneer 31D06 (1, 2, 3, 4, 6)	Beck's 6158 (1, 3, 6)	Ayerza Semillas IMPERIO (5, 8)
Reference 4	Rupp XR1612 (2, 3, 7, 8, 10)	Middlekoop 5513 (2, 6, 7, 8, 10)	Dow Semillas MILL 527 (3, 4, 7)	*Becks 6175 (6, 8)*	Don Atilio CARDENAL (4, 5, 7)
Reference 5	Merit Seed 5096 (2, 5, 6, 8, 9)	Pioneer 33W82 (2, 3, 4, 5, 10)	Ayerza Semillas IMPERIO (1, 4, 5, 7, 8)	*Beck's 5828 (7)*	Dow Semillas MILL527 (1, 2, 5)
Reference 6	Merit Seed 5145 (3, 4, 6, 7, 9)	Dekalb 6170 (4, 5, 6, 8, 9)	Rusticana NT 525 (3, 5, 6, 8)	*Beck's 6272 (7)*	KWS KM4321 (2, 3, 7)
Reference 7	–	–	–	AG Venture 7844 (2, 3, 4)	*Rusticana NT 525 (6)*
Reference 8	–	–	–	*Master's Choice MC6060 (2, 8)*	*Rusticana NT 426 (3, 4)*
Reference 9	–	–	–	*Master's Choice MC6150 (4)*	Pannar PAN 5E-203 (1, 3, 8)
Reference 10	–	–	–	*Viking 52-11N (3, 5)*	–
Reference 11	–	–	–	*Mycogen 2H721 (4, 5)*	–
Reference 12	–	–	–	*Dairyland DS-1811 (8)*	–

[a]Reference hybrids in italics were represented at fewer than three field sites and were therefore excluded from analysis.

Table 3 Transgenic gene products

GM event	Insect protection	Herbicide tolerance
DAS-Ø15Ø7-1	Cry1F	PAT
DAS-59122-7	Cry34Ab1, Cry35Ab1	PAT
DAS-4Ø278-9	–	AAD-1
SYN-IR162-4	Vip3Aa20	–
MON-89Ø34-3	Cry1A.105, Cry2Ab2	–
MON-ØØ6Ø3-6	–	CP4 EPSPS
MON-88Ø17	Cry3Bb1	CP4 EPSPS
MON-87427-7	–	CP4 EPSPS
MON-87411	Cry3Bb1, DVsnf7	CP4 EPSPS

$$I^2 = 1 - \frac{\sum_{i=1}^{N}(y_i - x_i)^2}{\sum_{i=1}^{N}(y_i - \bar{y})^2}$$

Plots and I^2 statistics were generated for composition values in both the natural scale and for data transformed to the base-10 logarithmic scale.

Author contributions

All authors contributed to the writing of the text. RH drafted initial text and conceived of the data analysis approach. GB led the effort to publish composition results for GM breeding stacks. BF performed data analysis and production of figures. PS modified the coefficient of determination equation to calculate the coefficient of identity. BF, AB and DdC oversaw field trials under the direction of BS. RE and GH contributed to the text placing study results into the context of traditional breeding.

References

Anderson, J.E., Michno, J.-M., Kono, T.J., Stec, A.O., Campbell, B.W., Curtin, S.J. and Stupar, R.M. (2016) Genomic variation and DNA repair associated with soybean transgenesis: a comparison to cultivars and mutagenized plants. *BMC Biotechnol.* **16**, DOI: 10.1186/s12896-016-0271-z.

Baktavachalam, G.B., Delaney, B., Fisher, T.L., Ladics, G.S., Layton, R.J., Locke, M.E., Schmidt, J. *et al.* (2015) Transgenic maize event TC1507: global status of food, feed, and environmental safety. *GM Crops Food*, **6**, 80–102.

Bradford, K.J., van Deynze, A., Gutterson, N., Parrott, W. and Strauss, S.H. (2005) Regulating transgenic crops sensibly: lessons from plant breeding, biotechnology and genomics. *Nat. Biotechnol.* **23**, 439–444.

Buchholz, U., Bernard, H., Werber, D., Böhmer, M.M., Remschmidt, C., Wilking, H., Deleré, Y. *et al.* (2011) German outbreak of *Escherichia coli* O104: H4 associated with sprouts. *N. Engl. J. Med.* **365**, 1763–1770.

EFSA Panel ON Genetically Modified Organisms (GMO). (2011) Guidance for risk assessment of food and feed from genetically modified plants. *EFSA J.* **9**, 2150–2186.

Fasoula, V.A. and Boerma, H.R. (2007) Intra-cultivar variation for seed weight and other agronomic traits within three elite soybean cultivars. *Crop Sci.* **47**, 367–373.

Fast, B.J., Galan, M.P. and Schafer, A.C. (2016) Event DAS-444Ø6-6 soybean grown in Brazil is compositionally equivalent to non-transgenic soybean. *GM Crops Food*, **7**, 79–83.

Harrigan, G.G., Venkatesh, T.V., Leibman, M., Blankenship, J., Perez, T., Halls, S., Chassy, A.W. *et al.* (2016) Evaluation of metabolomics profiles of grain from maize hybrids derived from near-isogenic GM positive and negative segregant inbreds demonstrates that observed differences cannot be attributed unequivocally to the GM trait. *Metabolomics*, **12**, 1–14.

Herman, R.A. and Price, W.D. (2013) Unintended compositional changes in genetically modified (GM) crops: 20 years of research. *J. Agric. Food Chem.* **61**, 11695–11701.

Herman, R.A., Phillips, A.M., Lepping, M.D., Fast, B.J. and Sabbatini, J. (2010) Compositional safety of event DAS-40278-9 (AAD-1) herbicide-tolerant maize. *GM Crops Food Biotechnol. Agric.* **1**, 294–311.

Herman, R.A., Fast, B.J., Johnson, T.Y., Sabbatini, J. and Rudgers, G.W. (2013) Compositional safety of herbicide-tolerant DAS-81910-7 cotton. *J. Agric. Food Chem.* **61**, 11683–11692.

Kok, E.J., Pedersen, J., Onori, R., Sowa, S., Schauzu, M., de Schrijver, A. and Teeri, T.H. (2014) Plants with stacked genetically modified events: to assess or not to assess? *Trends Biotechnol.* **32**, 70–73.

Lundry, D.R., Burns, J.A., Nemeth, M.A. and Riordan, S.G. (2013) Composition of grain and forage from insect-protected and herbicide-tolerant corn, MON 89034× TC1507× MON 88017× DAS-59122-7 (SmartStax) is equivalent to that of conventional corn (*Zea mays* L.). *J. Agric. Food Chem.* **61** (8), pp 1991–1998.

OECD. (1993). *Safety evaluation of foods derived by modern biotechnology: concepts and principles.*

OECD. (2002). *Consensus document on compositional considerations for new varieties of maize (Zea Mays): key food and feed nutrients, anti-nutrients and secondary plant metabolites.*

Pilacinski, W., Crawford, A., Downey, R., Harvey, B., Huber, S., Hunst, P., Lahman, L.K. *et al.* (2011) Plants with genetically modified events combined by conventional breeding: an assessment of the need for additional regulatory data. *Food Chem. Toxicol.* **49**, 1–7.

Schnell, J., Steele, M., Bean, J., Neuspiel, M., Girard, C., Dormann, N., Pearson, C. *et al.* (2015) A comparative analysis of insertional effects in genetically engineered plants: considerations for pre-market assessments. *Transgenic Res.* **24**, 1–17.

Tokatlidis, I.S., Tsikrikoni, C., Tsialtas, J.T., Lithourgidis, A.S. and Bebeli, P.J. (2008) Variability within cotton cultivars for yield, fibre quality and physiological traits. *J. Agric. Sci.* **146**, 483–490.

Venkatesh, T.V., Bell, E., Bickel, A., Cook, K., Alsop, B., van de Mortel, M., Feng, P. *et al.* (2016) Maize hybrids derived from GM positive and negative segregant inbreds are compositionally equivalent: any observed differences are associated with conventional backcrossing practices. *Transgenic Res.* **25**, 83–96.

van der Voet, H., Perry, J., Amzal, B. and Paoletti, C. (2011) A statistical assessment of differences and equivalences between genetically modified and reference plant varieties. *BMC Biotechnol.* **11**, 15–20.

Ward, K., Nemeth, M., Brownie, C., Hong, B., Herman, R. and Oberdoerfer, R. (2012) Comments on the paper "A statistical assessment of differences and equivalences between genetically modified and reference plant varieties" by van der Voet et al. 2011. *BMC Biotechnol.* **12**, 13–20.

6

Barley *HvPAPhy_a* as transgene provides high and stable phytase activities in mature barley straw and in grains

Inger Bæksted Holme*,[†], Giuseppe Dionisio[†], Claus Krogh Madsen and Henrik Brinch-Pedersen

Department of Molecular Biology and Genetics, Faculty of Science and Technology, Research Centre Flakkebjerg, Aarhus University, Slagelse, Denmark

*Correspondence
email inger.holme@mbg.dk
[†]The two authors have contributed equally
to the experimental work.

Summary

The phytase purple acid phosphatase (HvPAPhy_a) expressed during barley seed development was evaluated as transgene for overexpression in barley. The phytase was expressed constitutively driven by the cauliflower mosaic virus 35S-promoter, and the phytase activity was measured in the mature grains, the green leaves and in the dry mature vegetative plant parts left after harvest of the grains. The T_2-generation of *HvPAPhy_a* transformed barley showed phytase activity increases up to 19-fold (29 000 phytase units (FTU) per kg in mature grains). Moreover, also in green leaves and mature dry straw, phytase activities were increased significantly by 110-fold (52 000 FTU/kg) and 57-fold (51 000 FTU/kg), respectively. The *HvPAPhy_a*-transformed barley plants with high phytase activities possess triple potential utilities for the improvement of phosphate bioavailability. First of all, the utilization of the mature grains as feed to increase the release of bio-available phosphate and minerals bound to the phytate of the grains; secondly, the utilization of the powdered straw either directly or phytase extracted hereof as a supplement to high phytate feed or food; and finally, the use of the stubble to be ploughed into the soil for mobilizing phytate-bound phosphate for plant growth.

Keywords: barley, genetic transformation, grain, HvPAPhy_a phytase, straw.

Introduction

Phytases (myo-inositol hexakisphosphate 3- and 6-phosphohydrolase; EC 3.1.3.8 and EC 3.1.3.26) are phosphatases that can hydrolyze phytic acid (InsP$_6$, myo-inositol-(1,2,3,4,5,6)-hexakisphosphate), the most important phosphorous (P) storage compound in plant seeds, representing 40%–80% of total seed P (Eeckhout and De Paepe, 1994; Lott, 1984). In its degradation of phytic acid, phytases play a fundamental biological role, as it ensures bioavailable phosphate needed for regular progression of germination. However, also from applied perspectives, phytase in plants has a range of potentials. First of all, high phytase activity in feed and food is most wanted because the digestive tracts of nonruminants possess negligible phytase activity. Moreover, the activity provided by food and feedstuffs is often nonexisting or insufficient for efficient hydrolysis of the InsP6, which is therefore excreted via the manure along with chelated essential minerals such as zinc, calcium and iron (Brinch-Pedersen et al., 2014; Rostami and Giri, 2013).

Increment of the phytase activities in seeds through genetic transformation has been accomplished in several crops including wheat, rice, maize, soybean and canola (Brinch-Pedersen et al., 2000, 2003, 2006; Chen et al., 2013; Denbow et al., 1998; Gao et al., 2007; Hong et al., 2004; Lucca et al., 2001; Peng et al., 2006). Focus has not been on using the plants own phytases but on using microbial enzymes belonging to the group of histidine acid phosphatases. Basically, the phytases used are the same as the ones used in industrial production of feed enzymes. In addition to increasing seed phytase activity, microbial phytase has also been expressed in green vegetative tissues of tobacco, alfalfa and potato. The purpose of this was to use green leaves for extraction of phytase (Ullah et al., 1999, 2002, 2003). Accumulation of functional phytase in mature dead vegetative plant parts left after seeds harvest has so far not been reported but constitutes a potential valuable alternative for phytase production. With a yearly estimated production of mature barley straw, on around 3 tons per hectare or 55 kg straw per 100 kg grain (Statistic Denmark; http://www.dst.dk/en) valorization of this tissue is highly attractive.

Scientific initiatives in recent years have led to a substantially increased knowledge base on the complement of cereal phytases and paved the way for using the plants own enzymes for improving phytase activity in seeds and vegetative tissues. Cereal grains contain HAP phytases (the multiple inositol polyphosphate phosphatase (MINPP) phytase) but the bulk of activity can be attributed to phytases belonging to a different group of phosphatases, the purple acid phosphatase (Dionisio et al., 2007, 2011). In Triticeae tribe cereals, purple acid phosphatase phytase (PAPhy) genes generally consist of a set of paralogues, PAPhy_a and PAPhy_b. The promoters share a conserved core, but the PAPhy_a promoter have acquired a novel cis-acting regulatory element for expression during grain filling while the PAPhy_b promoter has maintained the archaic function and drives expression during germination. PAPhy_a accumulates in the aleurone layer and scutellum during grain development and contributes to the mature grain phytase activity (MGPA) of Triticeae cereals that is barley, wheat and rye (Madsen et al., 2013). Non-Triticeae cereals for example maize and rice only has one PAPhy gene which is predominantly expressed during germination (Dionisio et al., 2011). In agreement with this, their mature grains have very low phytase activity. The cereal PAPhy pH optimum is 5.5 ± 0.14, and the temperature profile is broad with optimum at $55\,°C \pm 1.8\,°C$ (both values for TaPAPhy_a1) (Dionisio et al., 2011). These biochemical parameters are in comparable range to the commercial phytase products the *Aspergillus niger*-based phytase, Natuphos and the *Perniphora lycii* phytase, Ronozyme NP (Menezes-Blackburn et al., 2015).

The purpose of this study is to evaluate the transgenic potential of the barley grain phytase gene *HvPAPhy_a* constitutively expressed by the 35S-promoter for (I) increased mature grain phytase activity in barley, (II) for accumulation of functional enzyme in green leaves and (III) for its accumulation in mature plant tissues as potential side product from barley production. A series of transgenic barley plants constitutively expressing the HvPAPhy_a gene were analysed, and highly elevated phytase activities were achieved in mature grains, green leaves and also in mature straw. HvPAPhy_a accumulation in vegetative tissues were monitored by nanoLC-MS, and high levels of easy extractable phytase activities were recorded in a series of mature vegetative tissues of HvPAPhy_a transformed barley.

Results

MGPA in primary transformants

We infected 200 embryos with Agrobacterium strain AGL0 containing the construct 35S:PAPhy_a (Figure 1). This leads to 20 primary transformants (T_0). PCR analysis for the hygromycin resistance gene was utilized to verify that the plants contained the T-DNA of the vectors (data not shown). Seventeen of the 35S:PAPhy_a transformants were fertile.

As T_0 grains are segregating for the transgene, mature grain phytase activity (MGPA) measurements were performed on pools of 20–25 grains from each T_0-plant. With exception of transformant 2 which were not significantly different from the control MGPA (1863 phytase units (FTU) per kg flour), all other 35S:PAPhy_a-transformed plant showed significantly higher MGPAs, ranging from 3672 to 42 700 FTU per kg flour (Figure 2).

Progeny from primary transformants

Three T_0-plants in the lower MGPA range (≤8000 FTU/kg) and three plants in the higher MGPA range (>12 000 FTU/kg) were selected for further investigations. PCR analysis was utilized to identify progeny positive for the transgene. To distinguish from the endogenous PAPhy_a, primer combinations were used where forward priming in the 35S promoter was combined with reverse priming in the PAPhy_a coding area (see Figure 1 for indication of primer sites). Inserts were detected in 11 of 13 progenies (Figure S1).

With the exception of plant 16.1 (with an MGPA of 2697 FTU), progeny containing the 35S:PAPhy_a transgene, all had significantly increased MGPA as compared with the nontransformed control (Figure 3). The two progenies with no 35S:PAPhy_a

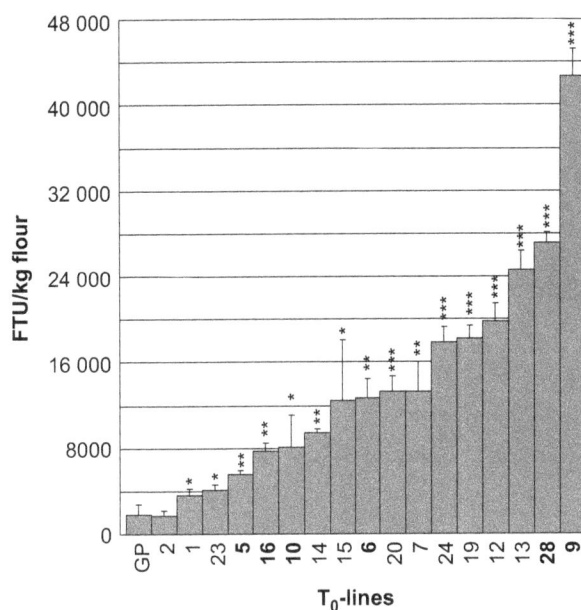

Figure 2 Mature grain phytase activity of T_0-plants transformed with 35S:PAPhy_a. GP: nontransformed control. Bars represent standard deviations (SD). Asterisks indicate significance levels: * significant different from nontransformed control (GP) at the 5% level, **significant different from nontransformed control (GP) at the 1% level, ***significant different from nontransformed control (GP) at the 1‰ level. The plant numbers that are further investigated in the T_1-generation are in bold.

inserts had nonsignificant MGPAs as their activities were within the range of the nontransformed control.

The differences in MGPA between the 35S:PAPhy_a plants could be influenced by integration number effects of the transgenes. Therefore, the number of integration sites in progeny of plants with varying MGPAs was investigated by Southern blotting (Figure S2). For the 35S:PAPhy_a plants 5.1 (6280 FTU/kg), 6.1 (13 680 FTU/kg), 28.4 (11 913 FTU/kg) and 9.4 (21 551 FTU/kg), one, two, two and three PAPhy_a integrations, respectively, were revealed by the Southern blot. This supports a relationship between integration site number and the MGPA. However, a direct linear comparison between integration site number and phytase activity is difficult as T_1-plants can be either heterozygous for the insert with the insert in only one of the two homologous chromosomes or homozygous for the insert with an insert in both homologous chromosomes. We were not able to distinguish these in the Southern blots.

PAPhy_a protein in green leaves

The presence of the PAPhy_a protein in green flag leaves of T_1-plant 35S:PAPhy_a-28.4 and a nontransformed control were analysed by nanoLC-MS/MS. Flag leaves were collected just before the appearance of the awns. Peptide mapping of PAPhy_a is shown in Figure S3. The nanoLC-MS/MS analysis showed that PAPhy_a was present at 1225 ng/g fresh weight (FW) among the leaf soluble proteins of 35S:PAPhy_a-28.4 whereas none of them were present among the leaf soluble proteins of the nontransformed control (Table 1). As a control, three common enzymes present in the flag leaves were also quantified using the same procedure and were found to be present in similar amounts in both plants investigated (Table 1). Furthermore, the quantitative values of all the proteins investigated were very similar over the

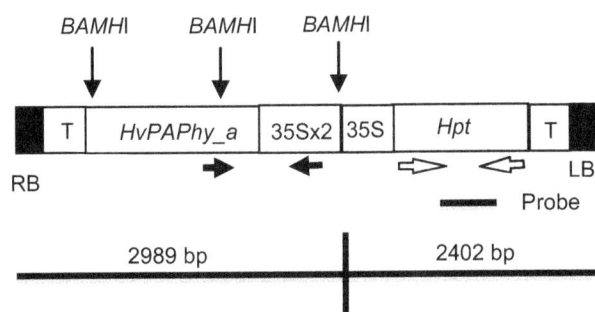

Figure 1 Diagram of the T-DNA of the 35S:PAPhy_a vector. The BamHI restriction sites are indicated. The primers utilized for PCR analysis are shown with arrows, and the position of the Hpt probe employed for Southern blotting is indicated. RB: right border; LB: left border; T: 35S-terminator.

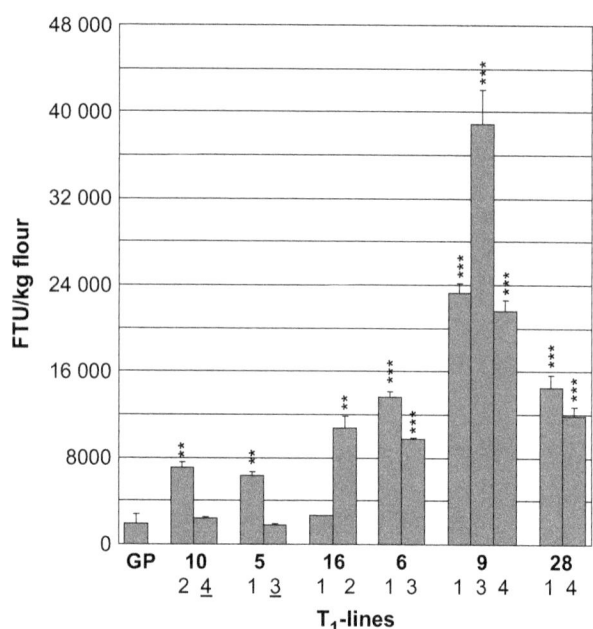

Figure 3 Mature grain phytase activity of T_1-plants transformed with *35S:PAPhy_a*. GP: nontransformed control. The numbers of the T_1 progeny correspond to numbers on Figure 2 and Figure S1. Bars represent SD. Asterisks indicate significance levels: for explanation see legend for Figure 2. The numbers underlined indicate the T_1-plants where no transgene was detected in the PCR analysis (Figure S1).

triplicate biological harvests and triplicate LC-MS runs and identifications with standard deviation between 4.5% and 6% between triplicates.

Phytase activity in green and dry mature vegetative plant parts of *35S:PAPhy_a* transgenic plants

The phytase activity was measured in green flag leaves excised just before the appearance of the awns of two progenies from T_1-plant *35S:PAPhy_a*-28.4 and in two progenies of T_1-plant *35S:PAPhy_a*-6.3 (Figure 4). The phytase activities in the leaves were very high ranging from 26 405 FTU/kg for plant *35S:PAPhy_a*-28.4.2 to 52 330 FTU/kg for plant *35S:PAPhy_a*-6.3.1.

Just after harvest of the mature plants, the level of preserved functional phytases was also assayed in the mature dry leaves, stems, rachis, chaff and awns in three of these T_2-plants (*35S:*

PAPhy_a-28.4.1, *35S:PAPhy_a*-28.4.2, *35S:PAPhy_a*-6.3.1). Initial experiments measuring free phosphate in the dry vegetative material protein extracts by the molybdovanadate method before incubation revealed that the extracts from nontransformed plants had a free phosphate content. Thus, in the subsequent activity assays, the free phosphate in the dry vegetative material was subtracted from the assay result. As shown in Figure 5, the phytase activities in the leaves, stem, rachis and chaff and awns of the nontransformed controls are low (320–890 FTU/kg). In contrast, the phytase activities in *35S:PAPhy_a*-28.4.2 leaves, stem, rachis and chaff and awns accounted for 7621, 11 181, 8357 and 22 816 FTU/kg dry tissue, respectively. Even higher phytase activities were measured in leaves, stem, rachis and chaff and awns of *35S:PAPhy_a*-28.4.1 with 44 922, 26 384, 16 794 and 22 816 FTU/kg dry tissue, respectively and in *35S:PAPhy_a*-6.3.3.1 with 50 850, 18 707, 21 465 and 31 751 FTU/kg dry tissue, respectively (Figure 5). Overall, these results show that high levels of PAPhy can be synthesized in green vegetative barley tissue in a fully functional form and that most of this remains intact during senescence and can be found in freshly harvested mature dry tissue.

Phytase activity in 3-year-old stored dry vegetative plant parts of *35S:PAPhy_a* transgenic plants

The level of preserved functional phytase was assayed in mature rachis, chaff and awns from two T_2-progenies from *35S:PAPhy_a*-28.4 and one T_2-progeny plant from *35S:PAPhy_a*-6.3 that had been stored for 3 years.

As shown in Figure 6, the phytase activities in *35S:PAPhy_a*-28.4.3 rachis and chaff and awns both accounted for 4800 FTU/kg dry tissue, respectively. Even higher phytase activities were measured in rachis and chaff and awns of *35S:PAPhy_a*-28.4.5 with 6100 and 14 800 FTU/kg dry tissue, respectively and in *35S:PAPhy_a*-6.3.2 with 10 000 and 9000 FTU/kg dry tissue, respectively. Overall, these results show that a high percentage of the PAPhy_a initially synthesized in the vegetative barley tissue can be stored in a fully functional form in mature dry tissue for minimum 3 years.

Discussion

In the present study, we have evaluated *in planta* the HvPAPhy_a phytase as transgene in barley. The results clearly demonstrate that 35S:HvPAPhy_a-transformed plants reached highly elevated phytase activities in both grain and green leaves. Yet, the most

Table 1 Relative and absolute quantification of HvPAPhy_a (in bold), HvMINPPII_a and some representative barley leaf soluble enzymes in nontransformed barley (control) and 35S: PAPhy_a (T_1-plant 28.4). Data are presented as a single LC-MS run identification. However, standard deviations over triplicate experiments were below 6% for all the data reported

Acces. no.	Description	MW Da	Control			35S:PAPhy_a (28.4)		
			Score	fmol	ng/g FW	Score	fmol	ng/g FW
P00924	Eno1_yeast	4664	164	150	–	150	150	–
C4PKL2	**PAPhy_a**	**60 297**	**0**	**0**	**0**	**24**	**24**	**1225**
A0FHA8	MINPPII_a	58 145	0	0	0	0	0	0
P26517	GAPDH1	36 513	60	55	1345	53	53	1285
F2CR16	Aldolase A	37 896	59	54	1359	38	38	952
K7X0F7	GS1	38 774	9.9	9.1	234	9.7	9.7	251

Eno1 yeast: yeast enolase 1, added and utilized as relative quantification control; GAPDH1: Glyceraldehyde-3-phosphate dehydrogenase (cytosolic); Aldolase A: Fructose-bisphosphate aldolase; GS1: Glutamine synthetase 1.

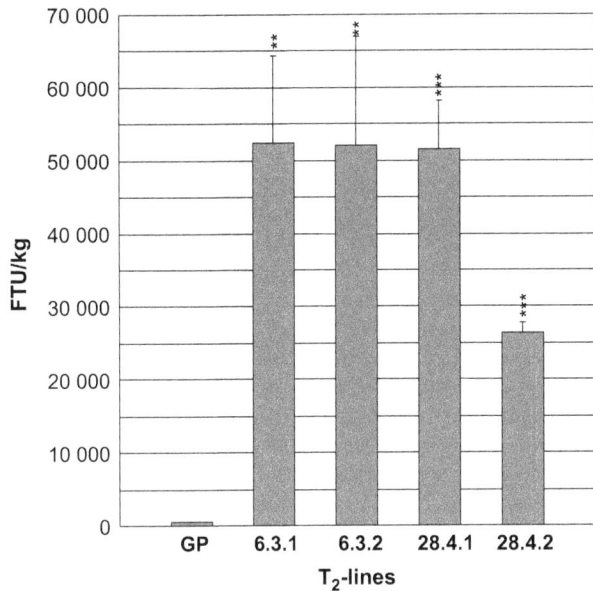

Figure 4 Phytase activity in green leaves of T_2-35S:PAPhy_a plants. Bars represent SD. Asterisks indicate significance levels: for explanation see legend for Figure 2.

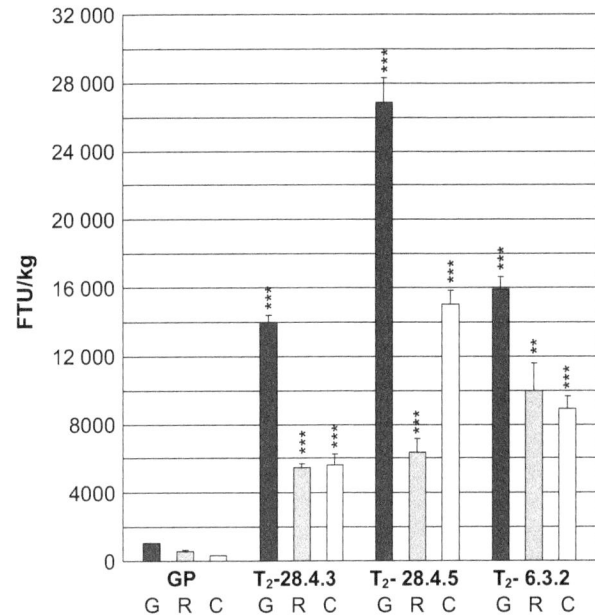

Figure 6 Phytase activity in 3-year-old stored dry grains (G), rachis (R) and chaff and awns (C) of a nontransformed control (GP) and T_2-35S: PAPhy_a plants. Bars represent SD. Asterisks indicate significance levels: for explanation see legend for Figure 2.

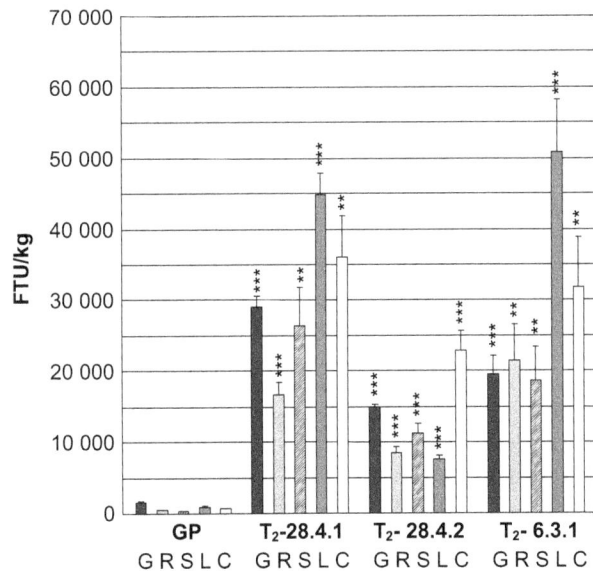

Figure 5 Phytase activity in freshly harvested dry mature grains (G), rachis (R), stems (S), leaves (L) and chaff and awns (C) of a nontransformed control (GP) and T_2-35S:PAPhy_a plants. Bars represent SD. Asterisks indicate significance levels: for explanation see legend for Figure 2.

important finding of this study is that we also observed high phytase activities in mature dead vegetative tissue not only just after harvest but also in the corresponding tissue stored for 3 years.

The increases in mature grain phytase activity (MGPA) were stably inherited to the T_1- and T_2-transgenic plants that were selected for further investigations. The MGPAs found in some of the barley plants constitutively expressing the HvPAPhy_a are very high. However, similar or higher MGPAs have been reported in other transgenic crops expressing microbial-derived phytases

(reviewed by Gontia et al., 2012). Also, large differences in MPGA were observed between the barley plants constitutively expressing the HvPAPhy_a gene. This is most likely caused by position effects of the transgenes and/or differences in the number of transgene integrations. Southern blots of four selected T_1-plants might indicate a possible correlation between the number of 35S:PAPhy_a transgene integrations and MGPA. We have previously found a positive correlation between phytase activity and plants being heterozygous or homozygous for the HvPAPhy_a cisgene (the genomic clone of HvPAPhy_a including the native promoter and terminator of the gene) (Holme et al., 2012). Here, plants heterozygous (one insert) for the HvPAPhy_a cisgene resulted in a twofold increase while plants homozygous (two inserts) for the cisgene resulted in a 2.7-fold increase in MGPA. Although we cannot distinguish between heterozygous and homozygous inserts in the Southern blots of this study, the present study shows that one insert of the 35S:PAPhy_a transgene can lead to a higher increase (3.5-fold) in MGPA than one or two inserts of the cisgene. The results therefore indicate that utilization of the constitutive promoter causes higher increases in MGPA than the use of the native promoter which controls expression in only the aleurone layer and scutellum and not in the entire endosperm.

Likewise, expression of the 35S:PAPhy_a transgene in vegetative tissue was expected due to the constitutive promoter. This was confirmed by the nanoLC-MS/MS investigation of young flag leaves from the T_1-plants 35S:PAPhy_a-28.4. Here, the PAPhy_a enzyme was detected at a concentration of 1225 ng/g FW while the PAPhy_a phytase could not be detected in leaves of the nontransformed control. The concentration of 35S:PAPhy_a-28.4 in leaves was almost similar to the glycolytic enzymes glycealdehyde-3-phosphate dehydrogenase and fructose-bisphosphate aldolase which are some of the most abundant enzymes in cells (Table 1).

We subsequently measured the phytase activities in the green leaves of progeny from the *35S:PAPhy_a*-28.4 plant. In these green leaves, we found very high increases in phytase activities of up to 110-fold. High increases in phytase activity of green vegetative tissue ranging from fivefold to 273-fold have previously been reported in different plants expressing microbial phytases including tobacco, alfalfa, potato and rice (Hamada *et al.*, 2005; Ullah *et al.*, 1999, 2002, 2003). In these studies, the purpose was to purify the phytase or use the green tissue directly for feed before the start of senescence.

However, in the present study, we also focused on phytase activity in mature dry dead vegetative tissue left after harvest of the grains. The high phytase activities that we found in milled mature dry vegetative tissue of *35S:PAPhy_a* plants just after harvest of the grains clearly show that preformed HvPAPhy_a phytase could be stored in the straw, leaves, rachis, chaff and awns with little degradation and remobilized during senescence. As both microbial and plant grain phytases are functional in highly proteolytically active environments, special properties of the enzymes may allow the enzyme also to be very resistant to proteolysis during senescence in leaves and stems. This is supported by a previous study where a heterologous *A. niger* phytase accumulating in tobacco leaves was found to be present at much higher levels than other proteins in senescent leaves (Verwoerd *et al.*, 1995). In that study, they compared the degradation of two different heterologous proteins that is the *Aspergillus* phytase and the human serum albumin and found that only the phytase was present in high amounts in senescent leaves. Therefore, the authors suggested that the phytase protein is more resistant to degradation during senescence due to the heavy glycosylation and folding of the protein in a way that makes it resistant to degradation (Verwoerd *et al.*, 1995). Similarly, a heterologous yeast phytase transgene with the phytase derived from *Schwanniomyces occidentalis* showed high activity levels in young leaves of transformed rice of up to 10 600 FTU/kg FW. Most importantly, the leaf extracts utilized for phytase assays could be stored for at least 4 weeks with minimal loss of phytase activity whereas the concentration of other soluble proteins in these extracts showed strong degradation after only 1 week of storage (Hamada *et al.*, 2005). These results indicate that phytase enzymes in general possess a native resistance towards proteolytic degradation. Phytase activities in mature dry vegetative transgenic tissue have, however, not been previously reported in any crop. Additionally, our results show that the HvPAPhy_a enzyme is not only protected from degradation during senescence but also retains much of the phytase activity in the mature vegetative parts during storage at room temperature for 3 years.

The presence of such high preformed phytase levels in the mature dry vegetative material of the *35S:PAPhy_a*-plants adds further potentials to the utilization of these transgenic plants than using the high phytase grains for feed and food (Figure 7). The straw from *35S:PAPhy_a* plants can be collected during harvest of the grains, milled to a very fine powder and added to feed or food as an extra supplement of phytase activity either directly or as extracts hereof as the phytase can very easily be water extracted from the tissue. This could be a great substitute for commercially microbial phytase that is often added to the feed of monogastric animals. Taken into account that supplement of microbial phytase to the feed at a concentration of 1500 FTU/kg makes about 60% of the phosphate bound in phytate bioavailable (Kerr *et al.*, 2010), then addition of extracts from only 50 g of nonstored

powder from plant *35S:PAPhy_a* 6.3.1 to one kilo of the feed would have the same effect on phosphate bioavailability. Similarly extracts from the fine powder could be supplemented to high phytate human food to increase the nutritional value of the food not only with respect to phosphate but also the other essential minerals bound to phytate that is zinc, calcium and iron (Brinch-Pedersen *et al.*, 2007). As importantly, the phytase activity in the stubble left after harvest of fields with these plants may, if the stubble is ploughed down into the soil, contribute to the soil fertility by a faster release of phosphate and other minerals bound to phytate present in the soil.

As previously mentioned, PAPhy_a phytases are only present in Triticeae cereals. Among these, we chose to overexpress the *HvPAPhy_a* gene in barley. Barley is considered a model species for the Triticeae cereals as it is a well-characterized diploid cereal with the genome almost fully sequenced (The International Barley Genome Sequencing Consortium 2012). Barley also has a number of molecular and genetic techniques are available, including a relatively efficient *Agrobacterium*-mediated transformation system (Mrizova *et al.*, 2014). Particularly, barley has the advantage in biosafety aspects of being a self-pollinating species with for many cultivars; a closed type of flowering and outcrossings with other barley plants or related species are extremely rare (Ritala *et al.*, 2002). Despite the reduced environmental risk of transgenic barley as compared to many other crops, chances of acceptance for deliberate release of any transgenic crop in EU are very low. Also, the costs and timely procedures for obtaining approval of GM crops for cultivation are a major impediment (Holme *et al.*, 2013). Still, the *35S:PAPhy_a* plants possess the advantage that the HvPAPhy_a enzyme is a barley endogenous enzyme, for which it is already known that there are no allergenic or toxic risks. Hence, the magnitude of risk assessment data requirements for food and feed safety of the *35S:PAPhy_a* plants can be reduced. As regulations in USA is less stringent, approval of the *35S:PAPhy_a* plants would probably be somewhat faster and less expensive than in the EU. But as a consequence of the high prices of GM approval worldwide, increases in bioavailable phosphate from feed are still more economically achieved by a simple supplement of microbial phytases to the feed. However, the great potentials of phytase production in the straw can only be accomplished by a transgenic approach.

In conclusion, the present study demonstrates for the first time the transgenic potentials of the *HvPAPhy_a* gene in barley. Transgenic expression of the gene led to highly increased phytase levels in grains, in green leaves and as a novel discovery also in dead mature straw. High phytase levels were detected in mature vegetative tissues after up to 3 years of storage. The current results demonstrate that *HvPAPhy_a* is a potent phytase in transgenic barley and that it has multiple applications with potential implications for nutrient bioavailability and environment.

Experimental procedures

Vector designs

The *PAPhy_a* cDNA was cloned and characterized by Dionisio *et al.*, 2011;. *PAPhy_a* was PCR amplified from the cDNA clone using USER primers (Hebelstrup *et al.*, 2010; Nour-Eldin *et al.*, 2006). The primers were Hv_PAPHy_a_user_fw GGCTTAAUATGCCAAGCAA-CAACATCAA and Hv_PAPhy_a_user_rv GGTTTAAUTTACG-GACCGTGTGCGGGC. The PCR reaction was carried out using PfuTurbo Cx Hotstart DNA polymerase (Stratagene) according to the manufacturer's instructions. Subsequently, each amplicon was

Barley containing the transgene *35S:PAPhy_a*

Utilization of plant parts after harvest

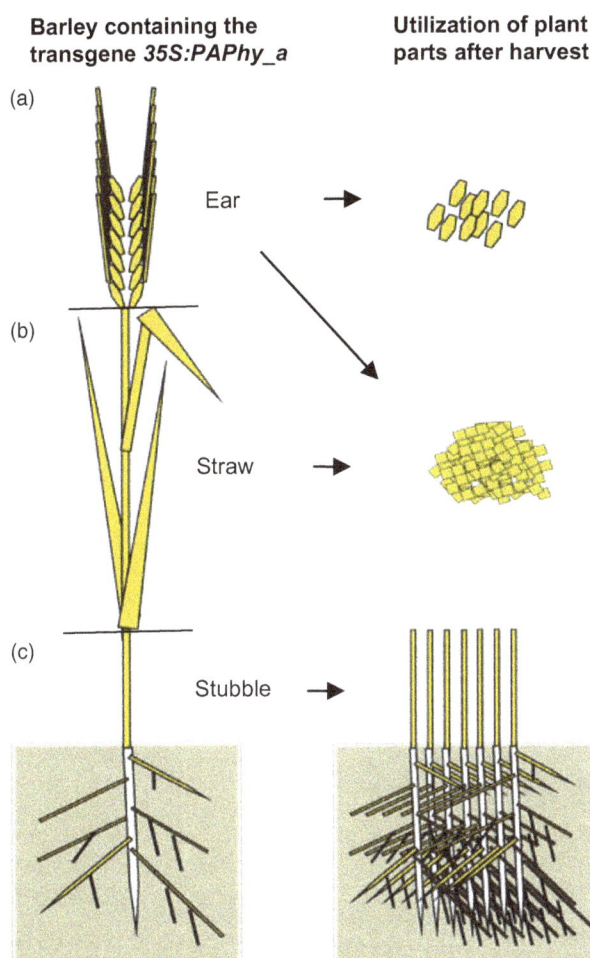

Figure 7 Illustration of the possible potentials of growing barley containing the transgene *35S:PAPhy_a*. (a) Grains with high phytase activities can be utilized directly for food and feed to increase the release of bioavailable phosphate and minerals bound to the phytate of the grains. (b) Powdered straw, rachis, chaff and awns with high phytase activities can be used either directly or the phytase can be extracted and used as supplements to high phytate feed and food. (c) The stubble can be ploughed into the soil to mobilize phytate-bound phosphate for plant growth.

cloned using USER™ cloning into the pCambia130035Su vector (Nour-Eldin *et al.*, 2006). In the resulting vector, *35S:PAPhy_a*, the *PAPhy_a* open reading frames are under control of the cauliflower mosaic virus 35-S promoter and terminator (Figure 1). The vector was transformed into *Agrobacterium* strain AGL0 using the freeze/thaw method and selected on medium with 50 mg/L kanamycin and 25 mg/L rifampicin.

Transformation

The spring barley cultivar Golden Promise was grown in growth cabinets with a 15/10 °C day/night temperature regime and 16 h light period at a light intensity of 350 µE/m^2/s. Immature embryos isolated 12–14 days after pollination were used for *Agrobacterium* transformation following the procedure of Bartlett *et al.* (2008) with the exception that 500 µM acetosyringone was included in the cocultivation medium (Hensel *et al.*, 2008) and 30 mg/L hygromycin was used instead of 50 mg/L in the transition, regeneration and rooting medium and that 0.1 mg/L

benzylaminopurine was added to the regeneration medium. Regenerated T_0-plants and their progenies were grown in a greenhouse.

PCR analysis of T_0-plants, T_1-progeny and T_2-progeny

The transgene in the primary transformants (T_0) was verified by PCR analysis using primers for the hygromycin resistance gene. The forward primer 5'-ACTCACCGCGACGTCTGTCG-3' and the reverse primer 5'GCGCGTCTGCTGCTCCATA'3 were used to amplify a 727-bp fragment of the *hpt* gene (Vain *et al.*, 2003). The PCR conditions were 95 °C for 60 s, 39 cycles of 95 °C for 60 s, 63 °C for 40 s, 72 °C for 60 s. After the last cycle, the reaction was subjected to 72 °C for 6 min.

For the analysis of T_1-progeny, forward primers annealed in the 35S promoter and reverse primers were located in the coding region of *PAPhy_a* (Figure 1a). The following primers were used to amplify a 882 bp fragment: forward: 5'CTGACGTAAGGGATGACGCA'3, reverse 5'GCTGGTAGGTCTCGTGGATG'3. PCR conditions were 95 °C for 2 min, then 34 cycles of 95 °C for 30 s, 63 °C for 40 s, 72 °C for 40 s and then 72 °C for 2 min.

The transgene in T_2-progenies was verified by PCR analysis utilizing the primers for the hygromycin resistance gene described above.

Genomic DNA gel blot analysis of T_1-progeny

Genomic DNA (10.0 µg) was digested overnight with *Bam*HI and separated in a 1% gel. After transfer to Hybond N+, the membrane was hybridized with probes labelled with [^{32}P]. The probe was a 420-bp fragment of the hygromycin resistance gene (*Hpt*) released by the digestion of pVec8GFP (Murray *et al.*, 2004) with *Bam*HI and *Pst*I. The probe was labelled by Ready-To-Go DNA-labelling beads (Amersham Biosciences). Transfer and hybridization was performed as described by Holme *et al.* (2006). *Bam*HI cleaves the vectors upstream the *Hpt* gene, releasing the *Hpt* intact with a minimum size of 2150 bp to be recognized by the probe (Figure 1).

Analysis of phytase activity

The phytase activities in grains of T_0-, T_1- and T_2-plants were analysed in milled flour from randomly selected samples of 20–25 grains from each transformant. The phytase activities in green flag leaves of T_2-plants were analysed in grinded leaves. Mature dry leaves, stems, rachis, chaff and awns utilized for phytase assays were freshly harvested or stored for 3 years at room temperature. Prior to analysis, the tissues were milled to fine powder in an IKA TUBE Mill. Protein extraction, assaying and incubation for 1 h were performed as described already (Brinch-Pedersen *et al.*, 2000; Engelen *et al.*, 1994). Soluble inorganic phosphate (Pi) present in the extract already before phytase assaying was quantified by adding molybdovanadate reagent to aliquots of the soluble protein extracts before incubation. The Pi content of this extract was subtracted from the Pi quantified after assaying the phytase activity. The final value appearing after this subtraction represents only the Pi released by the phytase during incubation. The phytase activity of all material was determined in four repetitions.

Samples preparation for nanoLC-MS/MS

Flag leaves from nontransformed Golden Promise and *35S:PAPhy_a*-28.4 were collected just before the appearance of the awns and frozen in liquid nitrogen. One hundred milligram of leaf material was homogenized in 1 mL of extraction buffer containing 25 mM Tris-HCl buffer, pH 8.0, 200 mM Mannitol, 10 mM ascorbic acid, 2 mM EDTA and 0.4% (v/v) protease and phosphatase inhibitor cocktail (Sigma P8849 and P0044) by the help of liquid nitrogen in a mortar. The homogenate was centrifuged for 5 min at 20 000 **g**, and the supernatant was used for proteomics. Sample preparation for proteomics analysis was performed as described in Christensen *et al.* (2014).

NanoLC-MS/MS

A nanoflow UHPLC instrument system (Easy nLCDionex Ultimate 3000 Rapid separation (RSLC) nano, Proxeon Biosystems) was coupled online to a Q-Exactive mass spectrometer (Thermo Fisher Scientific) by a Z-spray nanoelectrospray ion source. Chromatography column was packed in-house with ReproSil-Pur C18-AQ 3-μm resin (Dr. Maisch GmbH) in buffer A (0.5% acetic acid). The peptide mixture (0.5 μg) was loaded (3 μL) onto a C18-reversed phase column (50 cm long, 75 μm inner diameter) and separated with a linear gradient of 3%–40% buffer B (100% acetonitrile and 0.1% formic acid) at a flow rate of 250 nL/min controlled by IntelliFlow technology over 210 min. MS data were acquired using a data-dependent acquisition (DDA) Top10 method dynamically choosing the most abundant precursor ions from the survey scan (400–2000 m/z) for HCD fragmentation. Survey scans were acquired at a resolution of 70 000 at m/z 200, and resolution for HCD spectra was set to 17 500 at m/z 200. Normalized collision energy was 30 eV, while the underfill ratio was 0.1%. Data analysis was performed by Proteome Discoverer 1.4 (ThermoScientific). The proteome discoverer search workflow included a filter for raw file mass deisotope and charge deconvolution, a trypsin/chymotrypsin search into the integrated Sequest algorithm using 10 ppm peptide tolerance and 0.2 Da fragment tollerance, carbamidomethylation C (fixed) and as variable carbamylation N-term, oxidation MWHPKDWFRY, dioxidation WFR, formyl (N-term), ethanolamine (N-term), acetyl (N-term), deamidation NQ, methylation, proline oxidation to pyrrolidinone and phosphorylation STY. Automatic annotation was downloaded from the Proxeon web server and validation of phosphorylation sites with PhosphoRS algorithm. *Hordeum vulgare* proteins have been retrieved as Swissprot/TrEMBL (www.uniprot.org) fasta file and used as protein sequence search database. Protein annotation has been performed by Blast2GO tool (Conesa *et al.*, 2008) and further manually blasted in GenBank for extra annotation/curation of the selected proteins. For absolute relative quantification, 50 fmoles/μL of yeast enolase 1 (Uniprot accession number P00924) tryptic digest was spiked to each sample. Relative quantification to ENO1 was performed by Proteome Discoverer protein score and taking into account unique peptides related to different isoforms of the proteins. Peptide mapping was performed by Peptide Finder ver 2.0 (Thermo Fisher Scientific). All determinations were performed in triplicates on tree biological samples.

Statistics

The Welch's *t*-test for unequal variances was employed in all phytase activity experiments to test significant differences in phytase activity between the nontransformed control plants and the transformants.

Acknowledgements

The authors thank Lis Holte and Ole Bråd Hansen for skilful technical assistance. We are very grateful to Willy Bjørklund (Thermo Fisher Scientific, Denmark), Gary Woffendin and Jenny Ho (Thermo Fisher Scientific, Hemel Hempstead, Hertfordshire, United Kingdom) for allowing us to run our samples in their nanoUPLC/Q-exactive MS instrument. The research was funded by a grant from the Danish Ministry of Food, Agriculture and Fisheries.

References

Bartlett, J.G., Alves, S.C., Smedley, M., Snape, J.W. and Harwood, W.A. (2008) High-throughput *Agrobacterium*-mediated barley transformation. *Plant Methods*, **4**, 1–12.

Brinch-Pedersen, H., Olesen, A., Rasmussen, S.K. and Holm, P.B. (2000) Generation of transgenic wheat (Triticum aestivum L.) for constitutive accumulation of an Aspergillus phytase. *Mol. Breed.* **6**, 195–206.

Brinch-Pedersen, H., Hatzack, F., Sørensen, L.D. and Holm, P.B. (2003) Concerted action of endogenous and heterologous phytase on phytic acid degradation in seed of transgenic wheat (*Triticum aestivum* L.). *Transgenic Res.* **12**, 649–659.

Brinch-Pedersen, H., Hatzack, F., Stöger, E., Arcalis, E., Pontopidan, K. and Holm, P.B. (2006) Heat-stable phytases in transgenic wheat (*Triticum aestivum* L.): Disposition pattern, thermostability, and phytate hydrolysis. *J. Agric. Food Chem.* **54**, 4624–4632.

Brinch-Pedersen, H., Borg, S., Tauris, B. and Holm, P.B. (2007) Molecular genetic approaches to increasing mineral availability and vitamin content of cereals. *J. Cereal Sci.* **46**, 308–326.

Brinch-Pedersen, H., Madsen, C.K., Holme, I.B. and Dionisio, G. (2014) Increased understanding of the cereal phytase complement for better mineral bio-availability and resource management. *J. Cereal Sci.* **59**, 373–381.

Chen, R., Zhang, C., Yao, B., Xue, G., Yang, W., Zhou, X., Zhang, J. *et al.* (2013) Corn seeds as bioreactors for the production of phytase in the feed industry. *J. Biotechnol.* **165**, 120–126.

Christensen, J.B., Dionisio, G., Poulsen, H.D. and Brinch-Pedersen, H. (2014) Effect of pH and recombinant barley (*Hordeum vulgare* L.) endoprotease B2 on degradation of proteins in soaked barley. *J. Agric. Food Chem.* **62**, 8562–8570.

Conesa, A., Bro, R., Garcia-Garcia, F., Prats, J.M., Götz, S., Kjeldahl, K., Montaner, D. *et al.* (2008) Direct functional assessment of the composite phenotype through multivariate projection strategies. *Genomics*, **92**, 273–383.

Denbow, D.M., Grabau, E.A., Lacy, G.H., Kornegay, E.T., Russell, D.R. and Umbeck, P.F. (1998) Soybeans transformed with a fungal phytase gene improve phosphorus availability in broilers. *Poultry Sci.* **77**, 878–881.

Dionisio, G., Holm, P.B. and Brinch-Pedersen, H. (2007) Wheat (*Triticum aestivum* L.) and barley (*Hordeum vulgare* L.) multiple inositol polyphosphate phosphatases (MINPPs) are phytases expressed during grain filling and germination. *Plant Biotechnol. J.* **5**, 325–338.

Dionisio, G., Madsen, C.K., Holm, P.B., Welinder, K.G., Jørgensen, M., Stöger, E., Arcalis, E. *et al.* (2011) Cloning and characterization of purple acid phosphatases from wheat, barley, maize and rice. *Plant Physiol.* **156**, 1087–1100.

Eeckhout, W. and De Paepe, M. (1994) Total phosphorus, phytate-phosphorus and phytase activity in plant feedstuffs. *Anim. Feed Sci. Technol.* **47**, 19–29.

Engelen, A.J., van der Heeft, F.C., Randsdorp, P.H.G. and Smit, E.L.C. (1994) Simple and rapid-determination of phytase activity. *J. AOAC Int.*, **77**, 760–764.

Gao, X.R., Wang, G.K., Su, Q., Wang, Y. and An, L.J. (2007) Phytase expression in transgenic soybeans: stable transformation with a vector-less construct. *Biotechnol. Lett.* **29**, 1781–1787.

Gontia, I., Tantwai, K., Rajput, L.P.S. and Tiwari, S. (2012) Transgenic plants expressing phytase gene of microbial origin and their prospective application as feed. *Food Technol. Biotechnol.* **50**, 3–10.

Hamada, A., Yamaguchi, K., Ohnishi, N., Harada, M., Nikumaru, S. and Honda, H. (2005) High-level production of yeast (*Schwanniomyces occidentalis*) phytase in transgenic rice plants by a combination of signal sequence and codon modification of the phytase gene. *Plant Biotechnol. J.* **3**, 43–55.

Hebelstrup, K.H., Christiansen, M.W., Carciofi, M., Tauris, B., Brinch-Pedersen, H. and Holm, P.B. (2010) UCE: A uracil excision (User™)-based toolbox for transformation of cereals. *Plant Methods*, **6**, 15.

Hensel, G., Valkov, V., Middlefell-Willians, J. and Kumlehn, J. (2008) Efficient generation of transgenic barley: the way forward to module plant-microbe interactions. *J. Plant Physiol.* **165**, 71–82.

Holme, I.B., Brinch-Pedersen, H., Lange, M. and Holm, P.B. (2006) Transformation of barley (*Hordeum vulgare* L.) by *Agrobacterium tumefaciens* infection of in vitro cultured ovules. *Plant Cell Rep.* **25**, 1325–1335.

Holme, I.B., Dionisio, G., Brinch-Pedersen, H., Wendt, T., Madsen, C.K., Vincze, E. and Holm, P.B. (2012) Cisgenic barley with improved phytase activity. *Plant Biotechnol. J.* **10**, 237–247.

Holme, I.B., Wendt, T. and Holm, P.B. (2013) Intragenesis and cisgenesis as alternatives to transgenic crop development. *Plant Biotechnol. J.* **11**, 395–407.

Hong, C., Cheng, K., Tseng, T., Wang, C., Liu, L. and Yu, S.M. (2004) Production of two highly bacterial phytases with broad pH optima in germinated transgenic rice seeds. *Transgenic Res.* **13**, 29–39.

Kerr, B.J., Weber, T.E., Miller, P.S. and Southern, L.L. (2010) Effect of phytase on apparent total tract digestibility of phosphorus in corn-soybean meal diets fed to finishing pigs. *J. Anim. Sci.* **88**, 238–247.

Lott, J.N.A. (1984) Accumulation of seed reserves of phosphorus and other minerals. In *Seed Physiology* (Murray, D.R., ed), pp. 139–166. New York, USA: Academic Press.

Lucca, P., Hurrell, R. and Potrykus, I. (2001) Genetic engineering approaches to improve the bioavailability and the level of iron in rice grains. *Theor. Appl. Genet.* **102**, 392–397.

Madsen, C.K., Dionisio, G., Holme, I.B., Holm, P.B. and Brinch-Pedersen, H. (2013) High mature grain phytase activity in the Triticeae has evolved by duplication followed by neofunctionalization of the purple acid phosphatase phytase (*PAPhy*) gene. *J. Exp. Bot.* **64**, 3111–3123.

Menezes-Blackburn, D., Gabler, S. and Greiner, R. (2015) Performance of seven commercial phytases in an in vitro simulation of poultry digestive tract. *J. Agric. Food Chem.* **63**, 6142–6149.

Mrizova, K., Holaskova, E., Oz, M.T., Jiskrova, E., Frebort, I. and Galuszka, P. (2014) Transgenic barley: A prospective tool for biotechnology and agriculture. *Biotechnol. Adv.* **32**, 137–157.

Murray, F., Brettell, R., Matthews, P., Bishop, D. and Jacobsen, J. (2004) Comparison of *Agrobacterium* mediated transformation of four barley cultivars using GFP and GUS reporter genes. *Plant Cell Rep.*, **22**, 397–402.

Nour-Eldin, H.H., Hansen, B.G., Nørholm, M.H.H., Jensen, J.K. and Halkier, B.A. (2006) Advancing uracil-excision based cloning towards an ideal technique for cloning PCR fragments. *Nucleic Acids Res.* **34**, e122.

Peng, R., Yao, Q., Xiong, A., Cheng, Z. and Li, Y. (2006) Codon- modifications and an endoplasmic reticulum-targeting sequence additively enhance expression of an *Aspergillus* phytase gene in transgenic canola. *Plant Cell Rep.* **25**, 124–132.

Ritala, A., Nuutila, A.M., Aikasalo, R., Kauppinen, V. and Tammisola, J. (2002) Measuring gene flow in the cultivation of transgenic barley. *Crop Sci.* **42**, 278–285.

Rostami, H. and Giri, A. (2013) An overview on microbial phytase and its biotechnological applications. *Intl. J. Adv. Biotech. Res.* **4**, 62–71.

The International Barley Genome Sequencing Consortium. (2012) A physical, genetic and functional sequence assembly of the barley genome. *Nature*, **491**, 711–716.

Ullah, A.H.J., Sethumadhavan, K., Mullaney, E.J., Ziegelhoffer, T. and Austin-Phillips, S. (1999) Characterization of recombinant fungal phytase (*phyA*) expressed in tobacco leaves. *Biochem. Biophys. Res. Commun.* **264**, 201–206.

Ullah, A.H.J., Sethumadhavan, K., Mullaney, E.J., Ziegelhoffer, T. and Austin-Phillips, S. (2002) Cloned and expressed fungal *phyA* gene in alfalfa produce a stable phytase. *Biochem. Biophys. Res. Commun.* **290**, 1343–1348.

Ullah, A.H.J., Sethumadhavan, K., Mullaney, E.J., Ziegelhoffer, T. and Austin-Phillips, S. (2003) Fungal *phyA* gene expressed in potato leaves produces active and stable phytase. *Biochem. Biophys. Res. Commun.* **306**, 603–609.

Vain, P., Afolabi, A.S., Worland, B. and Snape, J.W. (2003) Transgene behavior in populations of rice plants transformed using a new dual binary vector system: pGreen/pSoup. *Theor. Appl. Genet.* **107**, 210–217.

Verwoerd, T.C., van Paridon, P.A., van Ooyen, A.J.J., van Lent, J.W.M., Hoekema, A. and Pen, J. (1995) Stable accumulation of *Aspergillus niger* phytase in transgenic tobacco leaves. *Plant Physiol.* **109**, 1199–1205.

7

Overexpression of the leucine-rich receptor-like kinase gene *LRK2* increases drought tolerance and tiller number in rice

Junfang Kang[†], Jianmin Li[†], Shuang Gao, Chao Tian and Xiaojun Zha*

College of Chemistry and Life Sciences, Zhejiang Normal University, Jinhua, China

*Correspondence

email zhaxj@zjnu.cn
[†]These authors contributed equally to this work.

Keywords: rice, *LRK2*, drought stress, tiller.

Summary

Drought represents a key limiting factor of global crop distribution. Receptor-like kinases play major roles in plant development and defence responses against stresses such as drought. In this study, *LRK2*, which encodes a leucine-rich receptor-like kinase, was cloned and characterized and found to be localized on the plasma membrane in rice. Promoter–GUS analysis revealed strong expression in tiller buds, roots, nodes and anthers. Transgenic plants overexpressing *LRK2* exhibited enhanced tolerance to drought stress due to an increased number of lateral roots compared with the wild type at the vegetative stage. Moreover, ectopic expression of *LRK2* seedlings resulted in increased tiller development. Yeast two-hybrid screening and bimolecular fluorescence complementation (BiFC) indicated a possible interaction between LRK2 and elongation factor 1 alpha (OsEF1A) *in vitro*. These results suggest that *LRK2* functions as a positive regulator of the drought stress response and tiller development via increased branch development in rice. These findings will aid our understanding of branch regulation in other grasses and support improvements in rice genetics.

Introduction

Rice (*Oryza sativa*) is one of the most important crops, feeding more than one-third of the world's population. This major staple requires large amounts of water during growth and is therefore susceptible to drought stress. At each stage of growth, drought hinders crop development and decreases yield. Molecular genetic tools aimed at improving rice drought tolerance and maintaining production, while expanding development into regions with limited water resources is therefore important (Fernie *et al.*, 2006; Pennisi, 2008).

Leucine-rich repeat receptor-like kinase (LRK), which belongs to the largest subfamily of kinases in plants, contains a leucine-rich extracellular domain, a transmembrane domain and a C-terminal intracellular kinase domain (Shiu *et al.*, 2004; Walker, 1993). At least 223 *LRKs* are known in *Arabidopsis* and more than 300 in rice (Shiu *et al.*, 2001). *LRKs* function in a number of developmental processes and defence responses such as meristem maintenance (Clark *et al.*, 1997), cellular proliferation (Matsubayashi *et al.*, 2002), brassinosteroid signalling (Li and Chory, 1997), floral organ abscission (Jinn *et al.*, 2000; Taylor *et al.*, 2016) and defence bacterial flagellin (Zipfel *et al.*, 2006). However, known functions in rice remain limited to, for example, *ERECTA* (Shen *et al.*, 2015), *Xa21* (Jiang *et al.*, 2013; Song *et al.*, 1997), *OsSIK1* (Ouyang *et al.*, 2010) and *FON1* (Feng *et al.*, 2014), with the majority of genes yet to be elucidated.

Eukaryotic elongation factor (eEF) proteins can be divided into eEF1 and eEF2. eEF1 proteins, which can further be divided into eEF1A, eEF1Bα, eEF1Bβ and eEF1Bγ subunits, are highly conserved in a number of species (Browning, 1996). eEF1-mediated ammonia acyl-tRNA has also been shown to bind to ribosomes (Riis *et al.*, 1990). Moreover, recent studies have shown that *eEF1A* is not only important for translation, but is also an important multifunctional protein (Ejiri, 2002; Sasikumar *et al.*, 2012), playing a role in processes such as cell proliferation (Pecorari *et al.*, 2009; Sanders *et al.*, 1992), cell apoptosis (Byun *et al.*, 2009; Shepherd *et al.*, 1989; Zhang *et al.*, 2015), cell morphogenesis (Gross and Kinzy, 2005) and signal transduction (Numata *et al.*, 2000). However, most of these studies have focused on animals and humans, with few findings in plants. In *Arabidopsis*, *eEF1B* is associated with cell wall biosynthesis and plant development (Hossain *et al.*, 2012). Moreover, the *AtEF2* gene has been shown to be involved in low temperature signalling (Guo *et al.*, 2002). Since 1998, four *EF1A* genes have also been cloned in rice, but their functions have yet to be reported (Kidou and Ejiri, 1998).

Previously, an eight leucine-rich LRK gene (*LRK1-LRK8*) cluster, which has been shown to increase grain yield, was cloned from the rice quantitative trait locus (QTL) *qGY2-1* (Li *et al.*, 2002a). Haplotype divergence of the *LRK* locus was subsequently found to be associated with the origin and differentiation of cultivated rice. Moreover, *LRK2* was found to be highly expressed in *Oryza sativa* L. ssp. *indica* var. 9311, but was undetectable by RT-PCR in *Oryza sativa* L. ssp. *japonica* cv. Nipponbare (He *et al.*, 2006). In the present study, we cloned and characterized *LRK2* and revealed an increase in branch number and drought tolerance in *LRK2*-overexpressing transgenic lines. *pLRK2::GUS* was largely expressed in tiller buds, nodes, roots and anthers. Furthermore, yeast two-hybrid screening and bimolecular fluorescence complementation (BiFC) analysis suggested an interaction between *LRK2* and *OsEF1A*. The molecular mechanisms underlying the response of rice to drought was also discussed with the aim of improving crop growth under potentially adverse conditions.

(a)

SP   LRR motifs   TM   KD

(b)

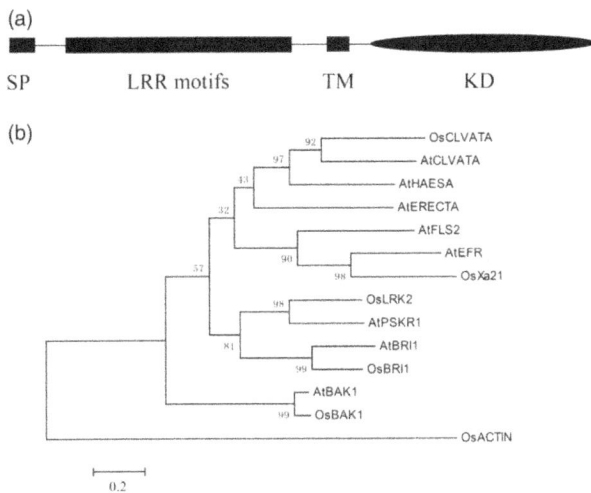

Figure 1 LRK2 encodes a leucine-rich receptor-like kinase. (a) Schematic diagram of the LRK2 protein. SP, signal peptide; LRR motif, leucine-rich repeat region; TM, transmembrane domain; KD, intercellular kinase domain. (b) Phylogenetic analysis of deduced amino acid sequences of LRK2 compared with other homologous sequences.

Results

LRK2 encodes a leucine-rich receptor-like kinase

LRK2 contains extracellular LRR motifs, a transmembrane domain (TM) and a cytoplasmic kinase domain (Figure 1a). To investigate the relationships between LRK2 (GenBank accession no. AY756174.4 GI:54306232) and LRK members from other plant species, phylogenetic analysis of LRKs from rice and *Arabidopsis thaliana* was performed using Clustalx1.83 and MEGA 6. Analysis involved 1000 bootstrap replicates with the following sequences: OsCLVATA: EAY84170; AtCLVATA: AAB58929; AtHAESA: XP_002869498; AtERECTA: XP_002880777; AtFLS2: AAO41929.1; AtEFR: AAL77697.1; OsX21: NC_008395.2; LRK2: AY756174.4; AtBRI1: XP_002866847.1; OsBRI1: AAK52544.1; AtBAK1: NP_567920.1; OsBAK1: EEC82980.1; and AtPSKR1:At2g02220. Based on a comparison of homologous amino acid sequences, LRK2 was found to share a close genetic relationship with phytosulfokine receptor 1 (PSKR1) from *A. thaliana* (Hartmann *et al.*, 2014; Matsubayashi *et al.*, 2002, 2006; Figure 1b). In addition, the function of LRK2 as a novel leucine-rich repeat receptor-like kinase (LRR-RLK) was reported for the first time.

Features of LRK2 in rice

To confirm the subcellular localization of *LRK2*, *LRK2* with an enhanced *GFP* was constructed under control of the cauliflower mosaic virus 35S promoter (*35S::LRK2:eGFP*), then infiltrated into tobacco (*Nicotiana benthamiana*) leaves by *Agrobacterium*-mediated transient transformation. The resulting construct, *35S::LRK2:eGFP*, exhibited *eGFP* expression in the plasma membrane compared with the control, *35S::eGFP*, which showed expression throughout the cell (Figure 2A). In addition, to determine expression patterns, the 2-kb promoter region of the *LRK2* gene was cloned into the expression vector *pBIN121* with the β-glucuronidase (*GUS*) reporter gene (Figure S1A). The construct *pLRK2:GUS* was subsequently transformed into rice (Figure S1B). A *GUS* staining assay of T2 transgenic lines revealed expression in the tiller buds, nodes, roots and anthers (Figure 2B).

To examine whether *LRK2* expression was regulated by drought stress, 7-day-old rice seedlings were subjected to 20% PEG6000 then harvested for RNA extraction at different time points and the transcripts of whole plants were quantified by real-time PCR. As a result, transcript levels were found to be rapidly and strongly induced by drought (Figure 2C).

Expression of LRK2 in transgenic rice lines

To obtain further insight into *LRK2*, *2X35S::LRK2* and *2X35S::antiLRK2* (Figure 3a), plasmids containing the entire *LRK2* gene were introduced into Nipponbare, and transformants selected on medium containing hygromycin. The transgenic plants were simultaneously examined by PCR using genomic DNA as a template with specific primers. Eleven and eight independent transformants (T_0) were regenerated from hygromycin-resistant calli of the *2X35S::LRK2* and *2X35S::antiLRK2* lines, respectively (Figure S2). Four T_2 transgenic lines plus the wild type were subsequently selected for expression analysis (M2 and M6 to represent *2X35S::LRK2*, and AM5 and AM8 for *2X35S::antiLRK2*). Expression levels during the three-leaf stage were determined by semiquantitative and RT-PCR. *LRK2* was strongly expressed in M2 and M6, but reduced in AM5 and AM8 (Figure 3b,c), suggesting that the *2X35S::LRK2* and *2X35S::antiLRK2* plasmids were genetically transmitted to the next generation.

Overexpression of LRK2 increases drought tolerance in rice

LRKs reportedly regulate a number of stress responses, including drought, salinity and low temperature (Ouyang *et al.*, 2010; Shen *et al.*, 2015; Yang *et al.*, 2014). The performance of *LRK2*-overexpressing, *LRK2*-antisense and wild-type plants under drought stress conditions was therefore examined (Figure 4). Under normal conditions, all plants grew well. After 12 days, plants were cultured in 20% PEG6000 solution and then 5 days later, phenotypic changes in wild-type and transgenic plants were observed. Leaf rolling and wilting was delayed in the M2 and M6 transgenic lines compared with the wild type, while in AM5 and AM8 symptoms of drought stress were severe. After 8-day treatment, the M2 and M6 lines showed increased tolerance to drought stress compared with the wild type, while AM5 and AM8 remained sensitive.

Following drought treatment, the plants were watered to induce recovery, and then, the growth status was examined. M2 and M6 recovered and grew more vigorously than the wild type, with survival rates of 73.5% and 65.5%, respectively, compared with 49.2%. In contrast, only 12.2% of the AM5 and 10.2% of the AM8 plants recovered, compared with 28.6% of the wild type. All values were significantly different ($P < 0.05$, t-test; Figure 5A), suggesting that *LRK2* positively regulates the drought stress response in rice.

Large root systems are known to be more conducive to extraction of water from deep soil layers compared with small root systems. The root structures of transgenic and wild-type plants after drought treatment were therefore examined under a stereoscopic microscope. The transgenic plants had significantly more lateral roots and a larger overall root system than the wild type (Figure 5B). Root activity in rice at different stages directly affects plant growth. In this study, root activity appeared higher in *LRK2*-overexpressing lines compared with the wild type (Figure 5C). These results suggest that increased root activity in the *LRK2*-overexpressing lines may be one reason for the increase in lateral root development.

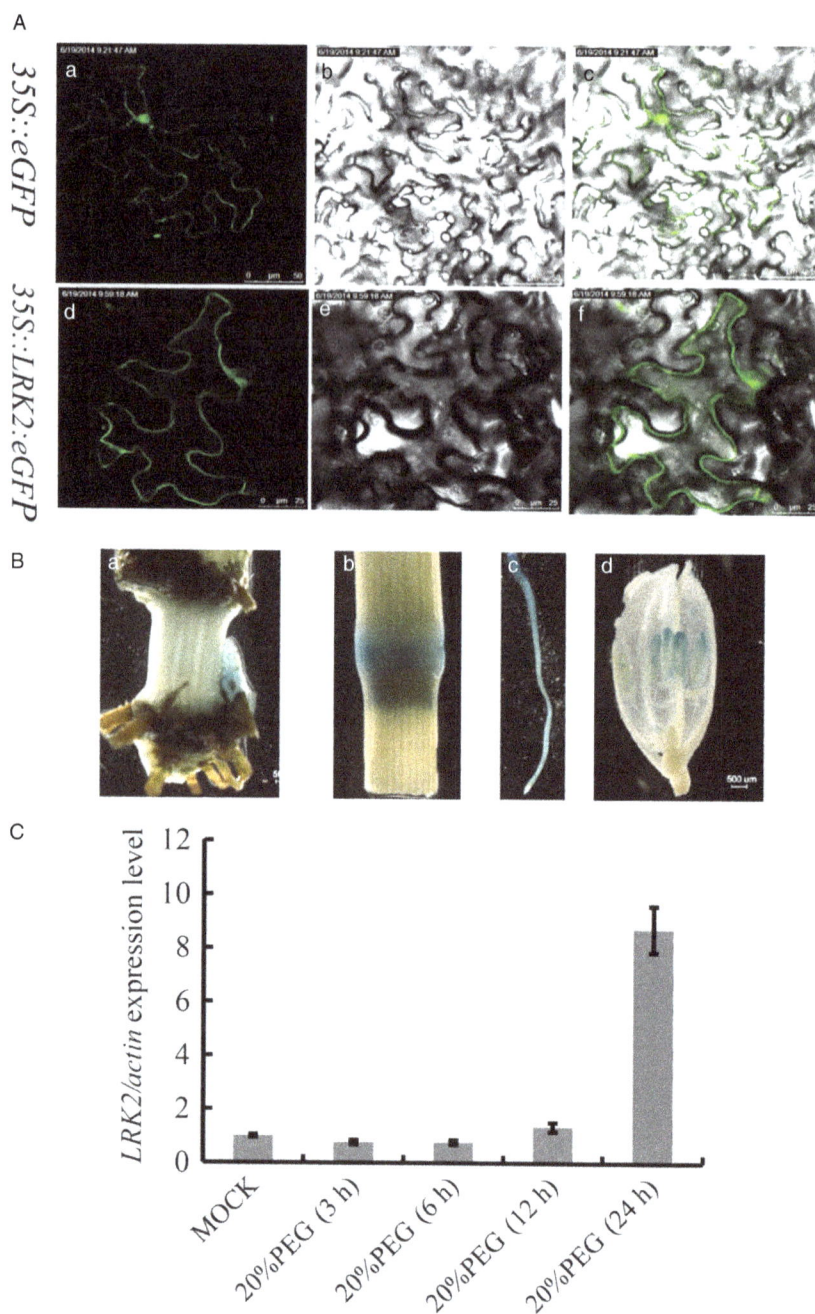

Figure 2 Expression patterns and subcellular localization of *LRK2* in rice. (A) Subcellular localization of *LRK2*. a,d: under the laser (430 nm); b,e: bright field; c,f: merged image; a, b,c: sections for *35S::eGFP*; d,e,f: sections for *LRK2* construction. Bars = 25 µm. (B) GUS staining of different organs in the overexpressing line: a, tiller bud; b, node; c, root; d, anther. (C) Expression analysis of *LRK2* in Nipponbare rice seedlings under drought treatment using real-time PCR. Each column represents an average of three replicates and bars indicate the SD.

To analyse the expression of genes involved in drought tolerance, expression levels of 9-cis-epoxycarotenoid dioxygenase 1 (*NCED1*), *NCED2*, plasma membrane intrinsic protein 2;3 (*PIP2;3*) and ABA-responsive element binding factor (*ABF*) were detected using real-time PCR in wild-type and transgenic seedlings under drought (20% PEG6000) for 12 h (Hatmi *et al.*, 2015; Redillas *et al.*, 2012; Shi *et al.*, 2015; Wu *et al.*, 2015; Yoshida *et al.*, 2010). The data showed that *PIP2;3* was strongly expressed in *LRK2* overexpression lines (Figure S3).

Effect of *LRK2* on rice tiller development

To investigate the effect of *LRK2* on yield traits, 14 transgenic lines were analysed (M1 to M6, AM1 to AM8; 20 individual plants per line). After cultivating transgenic lines and Nipponbare wild-type plants under identical conditions, the following yield components were examined: numbers of tillers per plant, grains

on the main panicle, grains per panicle and grains per plant (Table 1). The tiller is a specialized grain-bearing branch that forms on unelongated basal internodes. At the tillering stage, the *LRK2*-overexpressing lines exhibited increased tiller development (Figure S4A). At maturity, M2 and M6 produced 36% and 32% more panicles than the wild type, respectively, while AM5 and AM8 showed a decrease in panicles of 29.7% and 44.9% compared with the wild type, respectively (Table 1, Figures 6, S4B). The number of grains on the main panicle increased in overexpressing lines compared with the wild type (M2: 8.57%; M6: 5.46%), while the number of grains per panicle showed a slight decrease (M2: 10.13%; M6: 18.79%). In contrast, the number of grains per plant increased in the overexpressing lines (M2: 29.9%; M6; 25.3%) compared with the wild type (Table 1, Figure S4C). The number of grains per panicle is affected by the number of primary, secondary and sometimes higher-order

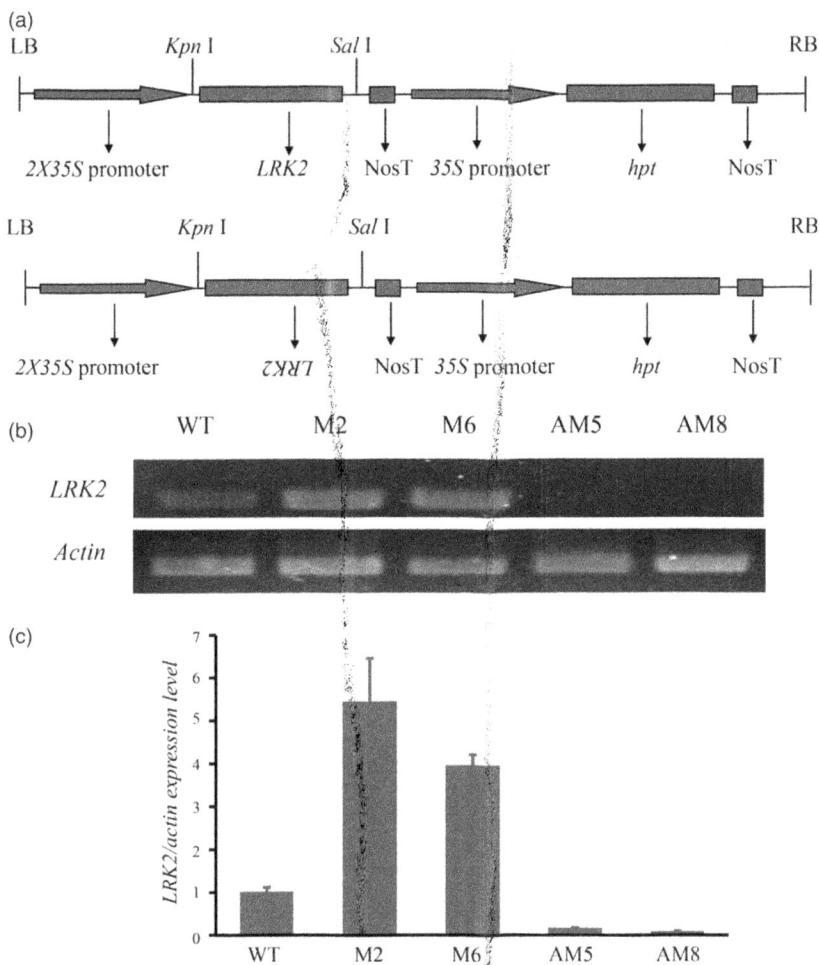

Figure 3 Expression of *LRK2* in transgenic rice. (a) Schematic diagram of the plant expression vectors *pCAMBIA1300-2X35S::LRK2* and *pCAMBIA1300-2X35S::antiLRK2* used for *LRK2* overexpression and decreased expression in transgenic plants. (b, c) Expression analysis of *LRK2* in transgenic plants using semiquantitative PCR and real-time PCR with gene-specific primers.

panicle branches. The number of primary and secondary panicle branches was therefore counted. The overexpressing lines showed a slight increase in the number of secondary branches on the main panicle compared with the wild type (M2: 7.86%; M6: 6.46%); however, the number of primary branches did not significantly differ (Table 2, Figure S4C). These results suggest that *LRK2* improves rice tiller number and the number of grains per plant.

Interaction between LRK2 and eukaryotic elongation factor 1A

To identify interactors of LRK2, LRK2 kinase domain (*LRK2D*) was used as bait in yeast two-hybrid analysis. As a result, six potential interacting molecules were identified. Nucleotide sequence analysis further revealed one cDNA fragment encoding the eukaryotic elongation factor *OsEF1A* (GenBank Accession no. GQ848073.1), which encodes a protein involved in protein synthesis and cell proliferation. To confirm the direct interaction between LRK2 and OsEF1A, a yeast two-hybrid assay was performed via cotransformation of the *pGBKT7-LRK2D* bait construct and full-length *pGADT7-OsEF1A*. As expected, only yeast cells harbouring both *LRK2D* and *OsEF1A* grew vigorously on both SD/Leu-Trp- and SD/Leu-Trp-His-/ 50 mm 3-amino-1,2,4-triazole (3-AT) plus X-gal media. In contrast, the wild type grew well on SD/Leu-Trp- but not SD/ Leu-Trp-His-/50 mm 3-AT/ X-gal medium (Figure 7A). *pGBKT7-LRK2D* and a truncated version of *OsEF1A* were also

cotransformed with *pGADT7*, revealing that the C-terminal region alone is sufficient, and indeed necessary, for binding with the LRK2 kinase domain (Figure 7B). To further verify the interaction between LRK2D and OsEF1A, BiFC analysis of *Nicotiana benthamiana* leaves was performed (Figure 7C). Under 513 nm illumination, cells harbouring *LRK2D:YEP^{CE}* and *OsEF1A:YEP^{NE}* emitted yellow fluorescence, confirming the interaction between OsEF1A and LRK2.

Discussion

It is predicted that global climate change will increase temperatures, alter geographical patterns of rainfall and increase the frequency of extreme climatic events (Harrison *et al.*, 2014). Drought is a major constraint of crop development and production worldwide. Accordingly, a number of studies have suggested that overexpression of stress-related genes may help improve drought tolerance in cereal crops (Cheng *et al.*, 2015; Uga *et al.*, 2013). In this study, a new rice leucine-rich repeat receptor-like kinase, *LRK2*, was found to have the ability to increase drought stress by promoting root growth and significantly increasing tiller number, while reducing plant height (Figures 6 and S4; Table S1). The data indicated that the *LRK2* gene encodes a protein localized to the plasma membrane, and expressed in tiller buds, nodes, roots and anthers (Figure 2). These results suggest that the *LRK2* gene is essential for stable and adequate crop production in drought-prone areas.

| WT | M2 | M6 | WT | AM5 | AM8 |

0 day

5 days

8 days

Recovery 20 days

1 day

3 days

11 days

Recovery 8 days

Figure 4 Performance of *LRK2* transgenic plants under drought stress. 0 day: Wild-type, M2 and M6 seedlings grown for 12 days under normal conditions. 5, 8 days: Performance of wild-type, M2 and M6 seedlings treated with 20% PEG6000 for 5 and 8 days, respectively. 1, 3, 11 days: Performance of wild-type, AM5 and AM8 seedlings grown under drought stress for 1, 3 and 11 days, respectively. Recovery 20 days: Recovery of the treated wild-type, M2 and M6 seedlings for 20 days. Recovery 8 days: Recovery of the treated wild-type, AM5 and AM8 seedlings for 8 days.

The mature rice fibrous root system is composed of adventitious and lateral roots. Adventitious root branching results in both large and small lateral roots (Coudert *et al.*, 2010), which are essential for water and nutrient uptake and critical for increased yield under stress (Atkinson *et al.*, 2014; Coudert *et al.*, 2010). In rice, *DRO1* was previously found to be negatively regulated by auxin and involved in cell elongation, resulting in an enlarged root system and increased drought avoidance (Uga *et al.*, 2013). Moreover, rice *OsAHP1* and *OsAHP2* knockdown plants were previously found to exhibit phenotypes representative of a deficiency in cytokinin signalling, including enhanced lateral root growth and resistant to osmotic stress compared with wild-type plants (Sun *et al.*, 2014). Moreover, ectopic *OCI* expression was found to increase the lateral root density and drought tolerance in *Arabidopsis* and soya bean (Quain *et al.*, 2014). In the present study, *LRK2* overexpression resulted in an improved root system with an increased number of both large and small lateral roots compared with the wild type. This is one reason for the increased drought tolerance in *LRK2*-overexpressing lines. The 2,3,5-triphenyltetrazolium chloride (TTC) test was used here to determine root activity in the *LRK2*-overexpressing plants (Hu *et al.*, 2016; Steponkus and Lanphear, 1967). Roots of the transgenic

plants were significantly more active than those of the wild-type plants (Figure 5C), further contributing to increased drought resistance in the *LRK2*-expressing lines. PIP2;3, one of the aquaporins, plays a crucial role in response to drought stress (Yu *et al.*, 2006). SIRK1, a member of the LRK family in *Arabidopsis*, was shown to interact with and activate PIP2;3 by phosphorylation (Wu *et al.*, 2013). Experiments should be undertaken to evaluate whether LRK2 can specifically phosphorylate PIP2;3.

Plant architecture, such as the structure of the roots, shoots and inflorescences, is affected by branching. Shoot branches in rice are referred to as tillers (Tanaka *et al.*, 2015). Root branches are located underground, while tiller and inflorescence structures in rice undergo lateral branching above ground during the vegetative and reproductive stages, respectively. Tiller number is generally regarded as the determining factor of yield as tillers are specialized panicle-bearing branches (Grillo *et al.*, 2009). Several genes related to tiller development have been characterized; for example, *MOC1* is important in both tiller bud formation and outgrowth, while *OsTB1* negatively regulates axillary bud outgrowth (Li *et al.*, 2003; Takeda *et al.*, 2003). Moreover, both *MOC1* and *OsTB1* expression was found in the axillary meristem and tiller buds. *MOC1* was also

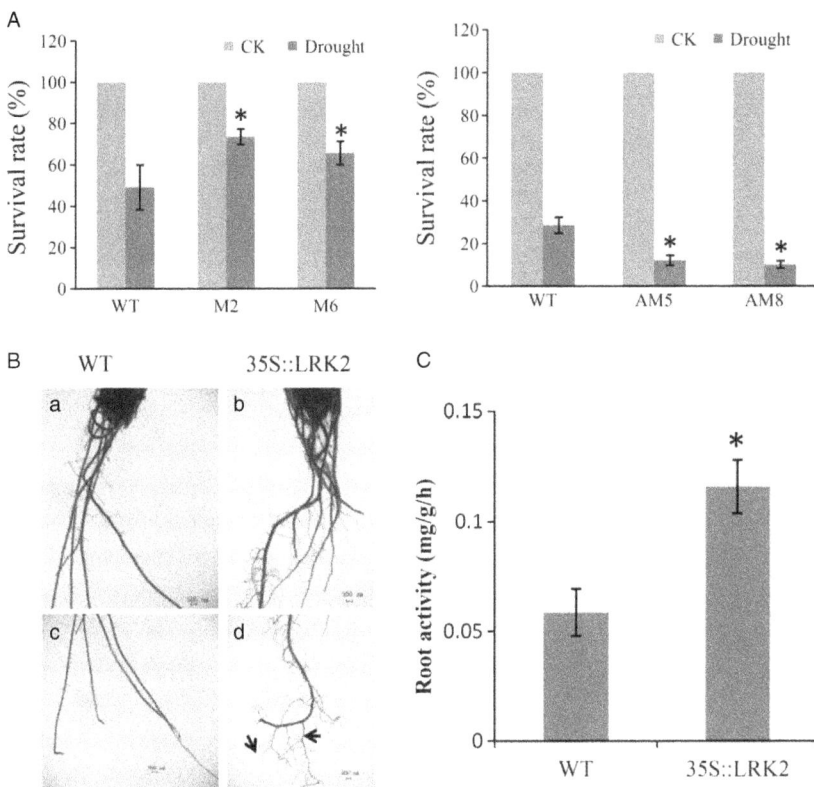

Figure 5 Performance of *LRK2* transgenic plants under drought stress. (A) Survival rates of the wild-type and transgenic plants after recovery. (B) (a–d) Root characters of the plants under drought stress. Arrows indicate the large and small lateral roots. Scale bars: 500 μm. (C) Root activity of the wild-type and *LRK2*-overexpressing lines (LRK2). *, $P < 0.05$, *t*-test.

Table 1 Yield components of wild-type (WT) and transgenic plants M2, M6, AM5 and AM8

Plant line	Number of tillers per plant	Number of grains on the main panicle	Number of grains per panicle	Number of grains per plant
WT	16.38 ± 3.57	85.91 ± 4.9	76.89 ± 6.37	1036.63 ± 115.01
M2	22.28 ± 4.79	93.27 ± 5.4	69.10 ± 4.58	1347.5 ± 156.81
M6	21.34 ± 5.73	90.6 ± 5.64	62.44 ± 8.26	1298.53 ± 104.28
AM5	11.5 ± 2.56	77.56 ± 3.3	52.56 ± 8.3	752.23 ± 82.02
AM8	9.02 ± 1.41	75.13 ± 4.5	50.13 ± 7.5	612.58 ± 98.3
P value	$P < 0.05$	$P < 0.05$	$P < 0.05$	$P < 0.05$

Data were obtained from random samples at maturity and represent the mean ± SD of 20 individuals.

detected at the leaf axils. The *LRK2* gene studied here was expressed in tiller buds and nodes, suggesting a role in tiller formation and outgrowth. In line with this, *LRK1* was previously found to increase both tiller and grain number (Zha *et al.*, 2009). In the current study, overexpression of *LRK2* resulted in an increase in tiller number, but no significant change in the number of grains per panicle. Similarly, both *LRK1*- and *LRK2*-overexpressing plants exhibited an increase in tiller number. These results suggest that *LRK2* and *LRK1* are expressed in different tissues, despite belonging to the same gene cluster. However, it is also possible that the diverse phenotypes of the transgenic lines were partly due to the different genetic backgrounds of the rice cultivars used for transformation (i.e. 9311 and Nipponbare).

Plants have evolved a number of morphophysiological and biochemical strategies at both the cellular and molecular levels to allow them to adapt to biotic and abiotic stresses. When plants encounter abiotic stresses, membrane-localized receptors rapidly sense environmental signals and transmit them downstream, thereby activating stress-related responses. During these processes, *LRKs* play important roles as both sensors and transducers

(Lease *et al.*, 1998; Shiu *et al.*, 2004; Torii, 2004). Meanwhile, *RPK1* (receptor-like protein kinase 1) is required for embryonic pattern formation and enhances both water and oxidative stress tolerance in *Arabidopsis* (Mandel *et al.*, 2014; Masle *et al.*, 2005; Meng *et al.*, 2012; Shen *et al.*, 2015; Shpak *et al.*, 2004; Van Zanten *et al.*, 2009). Similarly, *BAK1* (*BRI1*-associated receptor kinase 1) was found to play a role in a number of diverse processes, including brassinosteroid signalling, the phytosulfokine (PSK) signal pathway, light responses, cell death and plant innate immunity (Chinchilla *et al.*, 2007, 2009; Ingram, 2007; Ladwig *et al.*, 2015; Li *et al.*, 2002b; Nam and Li, 2002; Sun *et al.*, 2013). In this study, the novel *LRK* gene *LRK2* was found to function in drought tolerance and tiller development. Identification and further elucidation of the molecular mechanisms of such genes would be valuable in rice production management and genetic improvement studies.

EF1A was first shown to function in protein synthesis, and since then has been implicated in a number of biochemical processes such as interactions with the cytoskeleton (Gross *et al.*, 2005), apoptosis (Byun *et al.*, 2009; Zhang *et al.*, 2015) and cell

(a)

(b)

Figure 6 Effects of *LRK2* in transgenic plants. (a) Comparison of wild-type and transgenic lines at the flowering stage. (b) Number of panicles per plant in the wild-type and transgenic plants. *, $P < 0.05$, *t*-test.

Table 2 Number of branches on the main panicle of wild-type (WT) and transgenic lines M2 and M6

Plant lines	Primary branches	Secondary branches
WT	7.45 ± 0.82	15.64 ± 3.44
M2	7.93 ± 1.1	16.87 ± 2.33*
M6	7.54 ± 0.9	16.65 ± 2.28*

Data were obtained from random samples at maturity and represent the mean ±SD of 20 individuals.

*$P < 0.05$, *t*-test.

proliferation (Sanders *et al.*, 1992). *EF1A* also interacts with phospho-Akt in breast cancer cells and regulates their proliferation, survival and motility (Pecorari *et al.*, 2009), and is expressed ubiquitously in humans. Although a highly abundant cellular protein associated with the cytoskeleton, the function of *EF1A* in rice remains unknown. Phylogenetic analysis revealed a close genetic relationship between LRK2 and PSKR1 from *A. thaliana*. PSKR1 is a PSK receptor that stimulates plant growth and differentiation (Igarashi *et al.*, 2012). In this study, the LRK2 intracellular domain was also found to directly interact with OsEF1A in double molecule fluorescence analysis and a yeast two-hybrid assay. The findings further suggest that *LRK2* interacts with *OsEF1A* to regulate plant developmental processes such as cell proliferation, thereby increasing plant branching.

Based on the results of this study, a model was proposed to describe the functions of rice *LRK2* in regulating tiller size and the drought stress response (Figure 8). By interacting with other molecules such as the eukaryotic translation elongation factor, *LRK2* is thought to regulate cellular proliferation, promote branch

development and subsequently increase tiller number. The larger root system subsequently contributes to an increase in drought tolerance. As rice is an important food crop and drought one of the main factors affecting crop growth and yield, methods aimed at increasing yield and creating drought-resistant varieties are crucial. The findings of this study demonstrate the potential of *LRK2* as a useful tool for crop improvement, particularly with regard to drought tolerance, helping enhance agronomically useful traits such as tiller number, yield and the number of grains per plant. Future research will facilitate further improvements in abiotic stress tolerance in crops through genetic manipulation, thereby paving the way for a new green revolution.

Experimental procedures

Phylogenetic analysis

Phylogenetic analysis of LRK2 and other LRK proteins was carried out using MEGA 6.0 and a phylogenetic tree constructed using ClustalX 1.83 and UltraEdit21 software.

Generation of transgenic rice

Full-length cDNA of *LRK2* was amplified from the rice cultivar Nipponbare, and the confirmed sense and antisense sequences inserted into the *pCAMBIA 1300-2 × 35S* vector under control of the cauliflower mosaic virus 35S promoter to produce *LRK2*-overexpressing and knockdown lines. The primers 5′-GTC GGTACCATGCAGCCACCTCATTCTTCATGCAAC-3′ and 5′-CAG GTCGAC TCAGTCGGAGCCTACACTGTCCAG-3′ were used to construct *2 × 35S::LRK2*, and primers 5′-GTCGGTACCTTATA TCTTTATTTCAGTGCCTATACTGTC-3′ and 5′-CAGGTCGACATGC AGCTACTTCATTACAAGAAACACAG-3′ to construct *2 × 35S:: antiLRK2*. The constructs were transformed into Nipponbare mediated by Agrobacterium-mediated transformation as described previously (Attia *et al.*, 2005). The primers 5′-CAAGA CCTGCCTGAAACCGAACTG-3′ and 5′-GCGCGTCTGCTGCTCCA TACA-3′ were used to confirm the transgenic plants.

Subcellular localization of rice *LRK2*

To investigate the subcellular localization of *LRK2*, the *LRK2* coding sequence was amplified with the cDNA clone as the template, using primers 5′-CGGGGTACCATG CAG CCACCTCAT TCTTCATGCA-3′ and 5′-TCCCCCGGGGTCGGA GCCTACACT GTCCAGGCAG-3′. The PCR product was used to construct a *35S::LRK2-eGFP* fusion plasmid, which was transformed into tobacco leaves by *Agrobacterium*-mediated transformation, as described previously (Yang *et al.*, 2000). The transformed plants were cultured at 22 °C under 16-h light for 24–48 h. The GFP image was subsequently obtained using a Leica TCS SP5 AOBS confocal laser microscope.

Promoter–GUS analysis

The *LRK2* promoter, an approximately 2000-bp DNA fragment upstream of the translation start site, was amplified using primers 5′-TTGAAGCTTCCTCCCACCTCCAAGTGTTCAAC-3′ and 5′-CCA GGATCCGGTTTTCTGGTGATACTAGCATGGAAG-3′ from 9311. The DNA fragment was then cloned into the *pBI121* expression vector between the *Hin*dIII and *Bam*HI sites using the above transformation method. The primers 5′-CTGGATCCGTAGATCTG AGGAACCGACGA-3′ and 5′-GAGGACGTCTCACACGTGGTGGT GGTGGT-3′ were used to confirm the transgenic plants. A GUS assay was performed at various developmental stages as described previously (Jefferson *et al.*, 1987).

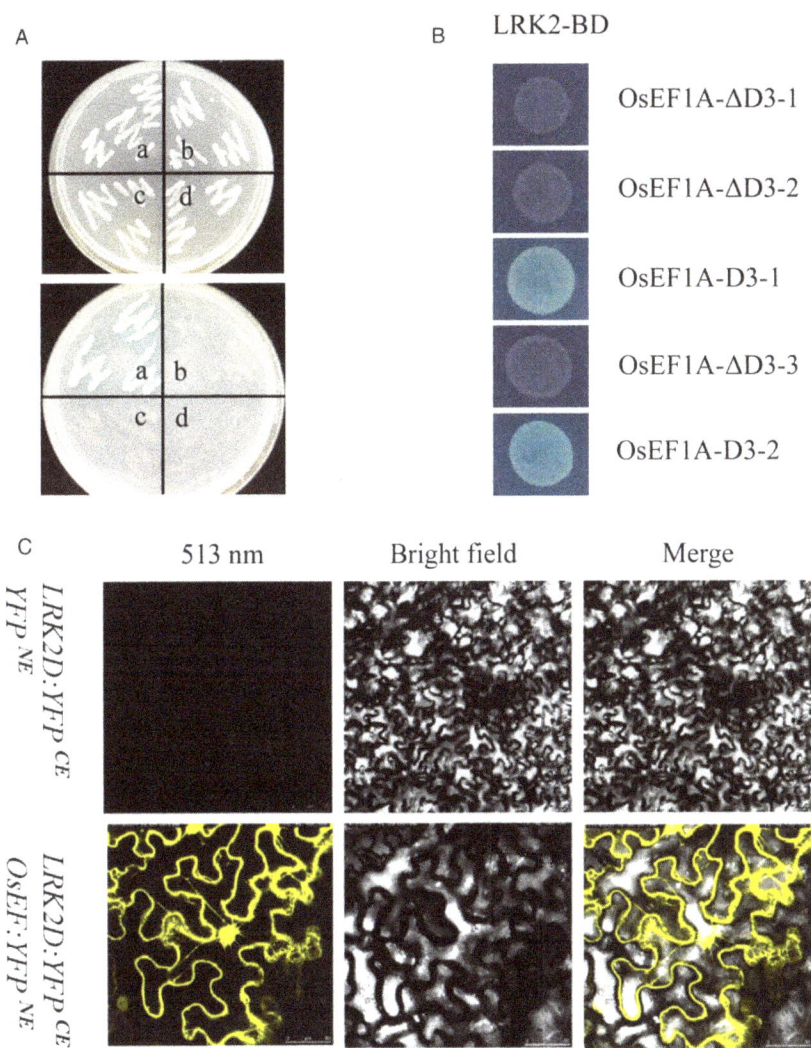

Figure 7 Interaction between LRK2D and OsEF1A *in vitro*. (A) Interaction between OsEF1A and the LRK2 kinase domain as bait in a yeast two-hybrid system. Photograph shows the growth behaviour of transformants on SD/Leu-Trp medium (upper) and SD/Leu-Trp-His-/50 mM3-AT medium (lower). The yeast cells harboured various pairs of plasmids: a. *pGBKT7-LRK2D* with *pGADT7-OsEF1A*; b. *pGBKT7* with *pGADT7-OsEF1A*; c. *pGBKT7-LRK2D* with *pGADT7*; d. *pGBKT7* with *pGADT7*. (B) The carboxyl domain of OsEF1A was found sufficient, and necessary, for interaction with the LRK2 kinase domain in yeast. Residues of OsEF1A present in the constructs were 1 to 230aa (OsEF1A-ΔD3-1), 231 to 320aa (OsEF1A-ΔD3-2), 321 to 447aa (OsEF1A-D3-1), 1 to 320aa (OsEF1A-ΔD3-3), 231 to 447aa (OsEF1A-D3-2). (C) Interaction between LRK2D and OsEF1A as examined by BiFC. The indicated constructs were transformed into leaves of *Nicotiana benthamiana*. Scale bars in upper images: 20 mm. Scale bars in lower images: 50 mm.

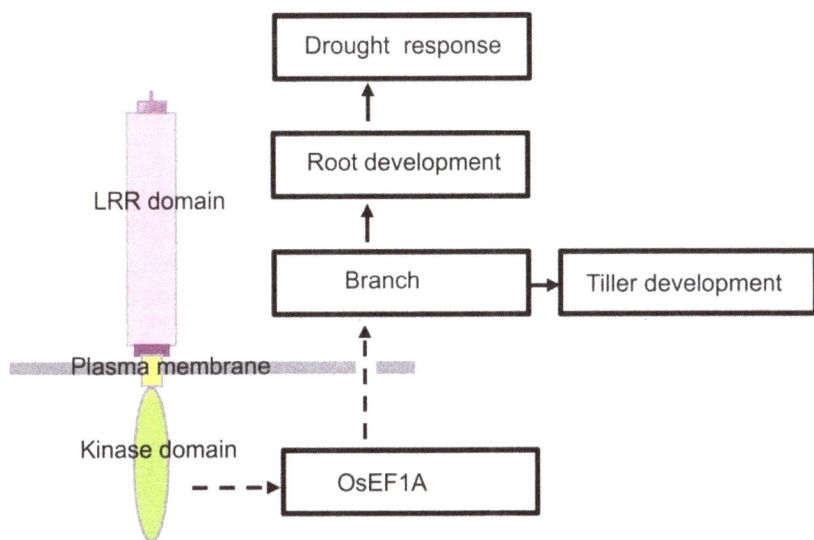

Figure 8 Proposed model of the role of the *LRK2* gene in drought tolerance and tiller development in rice. LRR: leucine-rich repeat.

Total RNA isolation, RT-PCR and real-time quantitative PCR analysis (qRT-PCR)

Total RNA extraction was performed using an RNeasy Plant Mini Kit following the manufacturer's instructions (Qiagen, Germany). Translation of RNA into cDNA was performed using a ReverTra Ace qPCR-RT Kit following the manufacturer's instructions (TOYOBO, Japan). Amplification of genes from the cDNA template was performed using specific primers with high-fidelity primeSTAR HS DNA Polymerase according the user's manual

(TaKaRa, Japan). Diluted reaction products were used as templates for RT-PCR and real-time quantitative PCR analysis. The following *LRK2*-specific primers were used for RT-PCR: 5'-GTCGGTACCATGCAGCCACCTCATTCTTCATGCAAC-3' and 5'-CAGGTCGACTCAGTCGGAGCCTACACTGTCCAG-3', and the following specific primers for qRT-PCR analysis: 5'-TCAGCATC-CAAAAAACAGTTGAAC-3' and 5'-CTCTGGATCAGAGGTGAAC-GAAC-3'. Each data point represents three replicates, and each experiment was repeated twice.

Growth conditions

Rice seeds (*O. sativa* ssp. japonica cv. Nipponbare) and transgenic plants were sown in pots after 3 day of germination at 37 °C. All plants were grown under 28 °C/16-h light and 25 °C/8-h dark conditions at 75% relative humidity in a glasshouse. For gene expression analysis, the roots of Nipponbare rice seedlings at the three-leaf stage were immersed in PEG6000 (20%) or water for 3, 6, 12 and 24 h. After treatment, the seedlings were harvested and used for total RNA isolation.

Drought stress treatment

For drought stress treatment, 12-day-old *2 × 35S::LRK2*, *2 × 35S::antiLRK2* and wild-type plants were grown in pots under the indicated conditions then treated with 20% PEG6000, respectively, until the leaves of the wild type were rolled as a result of drought stress. Plants were then rewatered. The phenotypes of the plants were subsequently observed and photographed at various time points. After recovery, survival rates were calculated by counting plants with green healthy young leaves.

Measurements of root activity

Root activity in terms of TTC reduction was measured as described previously (Hu *et al.*, 2016; Steponkus and Lanphear, 1967). Fresh roots from wild-type and transgenic plants were placed in six test tubes, each containing 2 mg sodium thiosulfate and 5 mL of various concentrations of TTC or distilled water, followed by the addition of 5 mL phosphate buffer (pH 7.5). The roots were incubated at 37 °C for 2 h after which 2 mL 1 M sulphuric acid was added to terminate the reaction. The roots were removed and ground in a mortar containing 3 mL ethyl acetate to extract the triphenylformazan (TTCH). TTCH was measured based on the absorbance of the supernatant at 485 nm. A standard curve was constructed with TTCH on the *x*-axis and OD on the *y*-axis. The root activity was determined by TTCH concentration for each fresh root as follows: root activity (mg g^{-1} h^{-1}) = TTCH reduction (TTCH mg)/fresh root weight (FW g)/time (h). For each root activity measurement, data points represent the average of three replicates.

Yeast two-hybridization analysis

Yeast two-hybrid library construction and screening was performed using BD Matchmaker library construction and screening kits (Clontech). The *LRK2D* coding region was fused in-frame with the GAL4 DNA binding domain in the *pGBKT7* vector to generate the bait vector. Primers included *LRK2D-F* (5'-CTG CATATG CTTTTCTCGCTCAGGGATGC-3') and *LRK2D-R* (5'-CGC GTCGAC GTCGGAGCCTACACTGTCCAG-3'). Rice ds-cDNA, the vector *pGADT7-Rec2* and bait construct *pGBKT7-LRK2D* were then cotransformed into yeast strain AH109, which was subsequently plated directly onto SD/ Leu-Trp-His-/50 mM 3-AT medium followed by incubation at 30 °C for 4 days. Positive clones were screened according to the manufacturer's protocol. The *OsEF1A* coding region was cloned into pGADT7 and yeast two-hybrid analysis performed using primers *OsEF1A-F* (5'-CCAGAATTC ATGGGTAAGGAGAAGACGCACATCA-3'), *OsEF1A-R* (5'-TTT GG ATCCTTATTTCTTCTTGGCGGCAGCCTTG-3'), *OsEF1A D1-AD-EcoRI-F* (5'-CCAGAATTCATTGTGGTCATTGGCCACG-3'), *OsEF1A D1-AD-BamHI-R* (5'-TTTGGATCCGGGCTCGTTGATCTGGTCAAG-3'), *OsEF1A D2-AD-EcoRI-F* (5'-CCAGAATTCGACAAGCCCCTACGT CTTCCC-3'), *OsEF1A D2-AD-BamHI-R* (5'-TTTGGATCCGTCATCC TTGGAGTTGGAGGC-3'), *OsEF1A D3-AD-EcoRI-F* (5'-CCAGAAT TCGAGGCTGCCAGCTTCACCTC-3') and *OsEF1AD3-AD-BamHI-R* (5'-TTTGGATCCGATGACGCCAACAGCCACC-3'). Yeast cells harbouring *pGBKT7-LRK2D+ pGADT7-OsEF1A*, *pGBKT7+ pGA DT7-OsEF1A*, *pGBKT7-LRK2D+ pGADT7*, *pGBKT7+ pGBKT7*, *pGBKT7-LRK2D+ OsEF1A-ΔD3-1*, *pGBKT7-LRK2D+ OsEF1A-ΔD3-2*, *pGBKT7-LRK2D+ OsEF1A-D3-1*, *pGBKT7-LRK2D+ OsE F1A-ΔD3-3*, *pGBKT7-LRK2D+ OsEF1A-D3-2* were selected on SD/ Leu-Trp- and SD/Leu-Trp-His-/50 mM 3-AT with X-gal plates.

Bimolecular fluorescence complementation assay

Bimolecular fluorescence complementation in tobacco was carried out as described previously (Schweiger and Schwenkert, 2014). *LRK2D* and *OsEF1A* cDNA fragments were amplified with the following primers *OsEF1A-pSPYNE-F* (5'-CACACTAGTAT GGGTAAGGAGAAGACGCACAT-3'), *OsEF1A-pSPYNE-R* (5'-CGA CCCGGGTT TCTTCTTGGCGGCAGCC-3'), *LRK2D-pSPYCE-F* (5'-CACTCTCGAATGCT TTTCTCGCTCAGGGATG-3') and *LRK2D-pSPYCE-R* (5'-CGAGTCGACGTC GGAGCCTACACTGTCC-3'). They were then cloned into the following split-YFP vectors with nonoverlapping coding regions: *35S::SPYNE* and *35S::SPYCE* (Walter *et al.*, 2004). The constructs were verified by sequencing and transformed, respectively, into *Agrobacterium*. *Agrobacterium* were grown at 28 °C to a final OD$_{600}$ of 0.8 for agroinfiltration. Next, we mixed equal volumes of *Agrobacterium* culture carrying the constructs *OsEF1A-pSPYNE* and *LRK2D-pSPYCE*, selected the leaves of 3-week-old tobacco plants, and infiltrated the *Agrobacterium* suspension carefully into the tobacco leaves by pressing a syringe without a needle. Plants were watered and cultured at 22 °C under 16-h light for 2 days. Images were collected using a Leica TCS SP5 AOBS confocal laser microscope.

Acknowledgements

This work was supported by the National Natural Science Foundation of China (Grant Nos. 31000741, 31671650), the Natural Science Foundation of Zhejiang Province, China (Grant No. LY13C130009), the Zhejiang Province University Students' Science and Technology Innovation Program (Grant No. 2015R404005) and the Open Funds for Key Modern Agricultural Biotechnology and Crop Disease Prevention and Control Program of Zhejiang Province, China (Grant No. 2012KFJJ0014).

References

Atkinson, J.A., Rasmussen, A., Traini, R., Voss, U., Sturrock, C., Mooney, S.J., Wells, D.M. *et al.* (2014) Branching out in roots: uncovering form, function, and regulation. *Plant Physiol.* **166**, 538–550.

Attia, K., Li, K.G., Wei, C., He, G.M., Su, W. and Yang, J.S. (2005) Transformation and functional expression of the rFCA-RRM2 gene in rice. *J. Integr. Plant Biol.* **47**, 823–830.

Browning, K.S. (1996) The plant translational apparatus. *Plant Mol. Biol.* **32**, 107–144.

Byun, H.O., Han, N.K., Lee, H.J., Kim, K.B., Ko, Y.G., Yoon, G., Lee, Y.S. *et al.* (2009) Cathepsin D and eukaryotic translation elongation factor 1 as promising markers of cellular senescence. *Cancer Res.* **69**, 4638–4647.

Cheng, S., Zhou, D.X. and Zhao, Y. (2016) WUSCHEL-related homeobox gene WOX11 increases rice drought resistance by controlling root hair formation and root system development. *Plant Signal. Behav.* **11**, e1130198.

Chinchilla, D., Zipfel, C., Robatzek, S., Kemmerling, B., Nurnberger, T., Jones, J.D., Felix, G. *et al.* (2007) A flagellin-induced complex of the receptor FLS2 and BAK1 initiates plant defence. *Nature*, **448**, 497–500.

Chinchilla, D., Shan, L., He, P., de Vries, S. and Kemmerling, B. (2009) One for all: the receptor-associated kinase BAK1. *Trends Plant Sci.* **14**, 535–541.

Clark, S.E., Williams, R.W. and Meyerowitz, E.M. (1997) The CLAVATA1 gene encodes a putative receptor kinase that controls shoot and floral meristem size in Arabidopsis. *Cell*, **89**, 575–585.

Coudert, Y., Perin, C., Courtois, B., Khong, N.G. and Gantet, P. (2010) Genetic control of root development in rice, the model cereal. *Trends Plant Sci.* **15**, 219–226.

Ejiri, S. (2002) Moonlighting functions of polypeptide elongation factor 1: from actin bundling to zinc finger protein R1-associated nuclear localization. *Biosci. Biotechnol. Biochem.* **66**, 1–21.

Feng, L., Gao, Z., Xiao, G., Huang, R. and Zhang, H. (2014) Leucine-rich repeat receptor-like kinase FON1 regulates drought stress and seed germination by activating the expression of ABA-responsive genes in rice. *Plant Mol. Biol. Rep.* **32**, 1158–1168.

Fernie, A.R., Tadmor, Y. and Zamir, D. (2006) Natural genetic variation for improving crop quality. *Curr. Opin. Plant Biol.* **9**, 196–202.

Grillo, M.A., Li, C., Fowlkes, A.M., Briggeman, T.M., Zhou, A., Schemske, D.W. and Sang, T. (2009) Genetic architecture for the adaptive origin of annual wild rice, oryza nivara. *Evolution*, **63**, 870–883.

Gross, S.R. and Kinzy, T.G. (2005) Translation elongation factor 1A is essential for regulation of the actin cytoskeleton and cell morphology. *Nat. Struct. Mol. Biol.* **12**, 772–778.

Guo, Y., Xiong, L., Ishitani, M. and Zhu, J.K. (2002) An Arabidopsis mutation in translation elongation factor 2 causes superinduction of CBF/DREB1 transcription factor genes but blocks the induction of their downstream targets under low temperatures. *Proc. Natl Acad. Sci. USA*, **99**, 7786–7791.

Harrison, M.T., Tardieu, F., Dong, Z., Messina, C.D. and Hammer, G.L. (2014) Characterizing drought stress and trait influence on maize yield under current and future conditions. *Glob. Chang. Biol.* **20**, 867–878.

Hartmann, J., Fischer, C., Dietrich, P. and Sauter, M. (2014) Kinase activity and calmodulin binding are essential for growth signaling by the phytosulfokine receptor PSKR1. *Plant J.* **78**, 192–202.

Hatmi, S., Gruau, C., Trotel-Aziz, P., Villaume, S., Rabenoelina, F., Baillieul, F., Eullaffroy, P. *et al.* (2015) Drought stress tolerance in grapevine involves activation of polyamine oxidation contributing to improved immune response and low susceptibility to Botrytis cinerea. *J. Exp. Bot.* **66**, 775–787.

He, G., Luo, X., Tian, F., Li, K., Zhu, Z., Su, W., Qian, X. *et al.* (2006) Haplotype variation in structure and expression of a gene cluster associated with a quantitative trait locus for improved yield in rice. *Genome Res.* **16**, 618–626.

Hossain, Z., Amyot, L., McGarvey, B., Gruber, M., Jung, J. and Hannoufa, A. (2012) The translation elongation factor eEF-1Bbeta1 is involved in cell wall biosynthesis and plant development in *Arabidopsis thaliana*. *PLoS ONE*, **7**, e30425.

Hu, Y., Xia, S., Su, Y., Wang, H., Luo, W., Su, S. and Xiao, L. (2016) Brassinolide increases potato root growth in vitro in a dose-dependent way and alleviates salinity stress. *Biomed Res. Int.* **3**, 1–11.

Igarashi, D., Tsuda, K. and Katagiri, F. (2012) The peptide growth factor, phytosulfokine, attenuates pattern-triggered immunity. *Plant J.* **71**, 194–204.

Ingram, G.C. (2007) Cell signalling: the merry lives of BAK1. *Curr. Biol.* **17**, R603–R605.

Jefferson, R.A., Kavanagh, T.A. and Bevan, M.W. (1987) GUS fusions: beta-glucuronidase as a sensitive and versatile gene fusion marker in higher plants. *EMBO J.* **6**, 3901–3907.

Jiang, Y., Chen, X., Ding, X., Wang, Y., Chen, Q. and Song, W.Y. (2013) The XA21 binding protein XB25 is required for maintaining XA21-mediated disease resistance. *Plant J.* **73**, 814–823.

Jinn, T.L., Stone, J.M. and Walker, J.C. (2000) HAESA, an Arabidopsis leucine-rich repeat receptor kinase, controls floral organ abscission. *Genes Dev.* **14**, 108–117.

Kidou, S. and Ejiri, S. (1998) Isolation, characterization and mRNA expression of four cDNAs encoding translation elongation factor 1A from rice (*Oryza sativa* L.). *Plant Mol. Biol.* **36**, 137–148.

Ladwig, F., Dahlke, R.I., Stuhrwohldt, N., Hartmann, J., Harter, K. and Sauter, M. (2015) Phytosulfokine regulates growth in Arabidopsis through a response module at the plasma membrane that includes CYCLIC NUCLEOTIDE-GATED CHANNEL17, H+-ATPase, and BAK1. *Plant Cell*, **27**, 1718–1729.

Lease, K., Ingham, E. and Walker, J.C. (1998) Challenges in understanding RLK function. *Curr. Opin. Plant Biol.* **1**, 388–392.

Li, J. and Chory, J. (1997) A putative leucine-rich repeat receptor kinase involved in brassinosteroid signal transduction. *Cell*, **90**, 929–938.

Li, D.J., Sun, C.Q., Fu, Y.C., Li, C., Zhu, Z.F., Chen, L., Cai, H.W. *et al.* (2002a) Identification and mapping of genes for improving yield from Chinese common wild rice (*O. rufipogon* Griff.) using advanced backcross QTL analysis. *Chin. Sci. Bull.* **47**, 1533–1537.

Li, J., Wen, J., Lease, K.A., Doke, J.T., Tax, F.E. and Walker, J.C. (2002b) BAK1, an Arabidopsis LRR receptor-like protein kinase, interacts with BRI1 and modulates brassinosteroid signaling. *Cell*, **110**, 213–222.

Li, X.Y., Qian, Q., Fu, Z.M., Wang, Y.H., Xiong, G.S., Zeng, D.L., Wang, X.Q. *et al.* (2003) Control of tillering in rice. *Nature*, **422**, 618–621.

Mandel, T., Moreau, F., Kutsher, Y., Fletcher, J.C., Carles, C.C. and Eshed Williams, L. (2014) The ERECTA receptor kinase regulates Arabidopsis shoot apical meristem size, phyllotaxy and floral meristem identity. *Development*, **141**, 830–841.

Masle, J., Gilmore, S.R. and Farquhar, G.D. (2005) The ERECTA gene regulates plant transpiration efficiency in Arabidopsis. *Nature*, **436**, 866–870.

Matsubayashi, Y., Ogawa, M., Morita, A. and Sakagami, Y. (2002) An LRR receptor kinase involved in perception of a peptide plant hormone, phytosulfokine. *Science*, **296**, 1470–1472.

Matsubayashi, Y., Ogawa, M., Kihara, H., Niwa, M. and Sakagami, Y. (2006) Disruption and overexpression of Arabidopsis phytosulfokine receptor gene affects cellular longevity and potential for growth. *Plant Physiol.* **142**, 45–53.

Meng, X., Wang, H., He, Y., Liu, Y., Walker, J.C., Torii, K.U. and Zhang, S. (2012) A MAPK cascade downstream of ERECTA receptor-like protein kinase regulates Arabidopsis inflorescence architecture by promoting localized cell proliferation. *Plant Cell*, **24**, 4948–4960.

Nam, K.H. and Li, J.M. (2002) BRI1/BAK1, a receptor kinase pair mediating brassinosteroid signaling. *Cell*, **110**, 203–212.

Numata, O., Kurasawa, Y., Gonda, K. and Watanabe, Y. (2000) Tetrahymena elongation factor-1 alpha is localized with calmodulin in the division furrow. *J. Biochem.* **127**, 51–56.

Ouyang, S.Q., Liu, Y.F., Liu, P., Lei, G., He, S.J., Ma, B., Zhang, W.K. *et al.* (2010) Receptor-like kinase OsSIK1 improves drought and salt stress tolerance in rice (*Oryza sativa*) plants. *Plant J.* **62**, 316–329.

Pecorari, L., Marin, O., Silvestri, C., Candini, O., Rossi, E., Guerzoni, C., Cattelani, S. *et al.* (2009) Elongation Factor 1 alpha interacts with phospho-Akt in breast cancer cells and regulates their proliferation, survival and motility. *Mol. Cancer.*, **8**, 58.

Pennisi, E. (2008) Plant genetics. The blue revolution, drop by drop, gene by gene. *Science*, **320**, 171–173.

Quain, M.D., Makgopa, M.E., Marquez-Garcia, B., Comadira, G., Fernandez-Garcia, N., Olmos, E., Schnaubelt, D. *et al.* (2014) Ectopic phytocystatin expression leads to enhanced drought stress tolerance in soybean (*Glycine max*) and *Arabidopsis thaliana* through effects on strigolactone pathways and can also result in improved seed traits. *Plant Biotechnol. J.* **12**, 903–913.

Redillas, M.C.F.R., Jeong, J.S., Kim, Y.S., Jung, H., Bang, S.W., Choi, Y.D., Ha, S.-H. *et al.* (2012) The overexpression of OsNAC9 alters the root architecture of rice plants enhancing drought resistance and grain yield under field conditions. *Plant Biotechnol. J.* **10**, 792–805.

Riis, B., Rattan, S.I., Clark, B.F. and Merrick, W.C. (1990) Eukaryotic protein elongation factors. *Trends Biochem. Sci.* **15**, 420–424.

Sanders, J., Maassen, J.A. and Moller, W. (1992) Elongation factor-1 messenger-RNA levels in cultured cells are high compared to tissue and are not drastically affected further by oncogenic transformation. *Nucleic Acids Res.* **20**, 5907–5910.

Sasikumar, A.N., Perez, W.B. and Kinzy, T.G. (2012) The many roles of the eukaryotic elongation factor 1 complex. *Wiley Interdiscip. Rev. RNA*, **3**, 543–555.

Schweiger, R. and Schwenkert, S. (2014) Protein-protein interactions visualized by bimolecular fluorescence complementation in tobacco protoplasts and leaves. *J. Vis. Exp.* **85**, e51327.

Shen, H., Zhong, X., Zhao, F., Wang, Y., Yan, B., Li, Q., Chen, G. *et al.* (2015) Overexpression of receptor-like kinase ERECTA improves thermotolerance in rice and tomato. *Nat. Biotechnol.* **33**, 996–1003.

Shepherd, J.C., Walldorf, U., Hug, P. and Gehring, W.J. (1989) Fruit flies with additional expression of the elongation factor EF-1 alpha live longer. *Proc. Natl Acad. Sci. USA*, **86**, 7520–7521.

Shi, L., Guo, M., Ye, N., Liu, Y., Liu, R., Xia, Y., Cui, S. *et al.* (2015) Reduced ABA accumulation in the root system is caused by ABA exudation in upland rice (*Oryza sativa* L. var. Gaoshan1) and this enhanced drought adaptation. *Plant Cell Physiol.* **56**, 951–964.

Shiu, S.H. and Bleecker, A.B. (2001) Receptor-like kinases from Arabidopsis form a monophyletic gene family related to animal receptor kinases. *Proceedings of the National Academy of Sciences of the United States of America*, **98**, 10763–10768.

Shiu, S.H., Karlowski, W.M., Pan, R.S., Tzeng, Y.H., Mayer, K.F.X. and Li, W.H. (2004) Comparative analysis of the receptor-like kinase family in Arabidopsis and rice. *Plant Cell*, **16**, 1220–1234.

Shpak, E.D., Berthiaume, C.T., Hill, E.J. and Torii, K.U. (2004) Synergistic interaction of three ERECTA-family receptor-like kinases controls Arabidopsis organ growth and flower development by promoting cell proliferation. *Development*, **131**, 1491–1501.

Song, W.-Y., Pi, L.-Y., Wang, G.-L., Gardner, J., Holsten, T. and Ronald, P.C. (1997) Evolution of the rice Xa2I disease resistance gene family. *Plant Cell* **9**, 1279–1287.

Steponkus, P.L. and Lanphear, F.O. (1967) Refinement of the triphenyl tetrazolium chloride method of determining cold injury. *Plant Physiol.* **42**, 1423–1426.

Sun, Y., Li, L., Macho, A.P., Han, Z., Hu, Z., Zipfel, C., Zhou, J.M. *et al.* (2013) Structural basis for flg22-induced activation of the Arabidopsis FLS2-BAK1 immune complex. *Science*, **342**, 624–628.

Sun, L., Zhang, Q., Wu, J., Zhang, L., Jiao, X., Zhang, S., Zhang, Z. *et al.* (2014) Two rice authentic histidine phosphotransfer proteins, OsAHP1 and OsAHP2, mediate cytokinin signaling and stress responses in rice. *Plant Physiol.* **165**, 335–345.

Takeda, T., Suwa, Y., Suzuki, M., Kitano, H., Ueguchi-Tanaka, M., Ashikari, M., Matsuoka, M. *et al.* (2003) The OsTB1 gene negatively regulates lateral branching in rice. *Plant J.* **33**, 513–520.

Tanaka, W., Ohmori, Y., Ushijima, T., Matsusaka, H., Matsushita, T., Kumamaru, T., Kawano, S. *et al.* (2015) Axillary Meristem formation in rice requires the WUSCHEL Ortholog TILLERS ABSENT1. *Plant Cell*, **27**, 1173–1184.

Taylor, I., Wang, Y., Seitz, K., Baer, J., Bennewitz, S., Mooney, B.P. and Walker, J.C. (2016) Analysis of phosphorylation of the receptor-like protein kinase HAESA during arabidopsis floral abscission. *PLoS ONE*, **11**, e0147203.

Torii, K.U. (2004) Leucine-rich repeat receptor kinases in plants: structure, function, and signal transduction pathways. *Int. Rev. Cytol.* **234**, 1–46.

Uga, Y., Sugimoto, K., Ogawa, S., Rane, J., Ishitani, M., Hara, N., Kitomi, Y. *et al.* (2013) Control of root system architecture by DEEPER ROOTING 1 increases rice yield under drought conditions. *Nat. Genet.* **45**, 1097–1102.

Van Zanten, M., Snoek, L.B., Proveniers, M.C. and Peeters, A.J. (2009) The many functions of ERECTA. *Trends Plant Sci.* **14**, 214–218.

Walker, J.C. (1993) Receptor-like protein kinase genes of *Arabidopsis thaliana*. *Plant J.* **3**, 451–456.

Walter, M., Chaban, C., Schutze, K., Batistic, O., Weckermann, K., Nake, C., Blazevic, D. *et al.* (2004) Visualization of protein interactions in living plant cells using bimolecular fluorescence complementation. *Plant J.* **40**, 428–438.

Wu, X.N., Sanchez Rodriguez, C., Pertl-Obermeyer, H., Obermeyer, G. and Schulze, W.X. (2013) Sucrose-induced receptor kinase SIRK1 regulates a plasma membrane aquaporin in Arabidopsis. *Mol. Cell Proteomics*, **12**, 2856–2873.

Wu, F., Sheng, P., Tan, J., Chen, X., Lu, G., Ma, W., Heng, Y. *et al.* (2015) Plasma membrane receptor-like kinase leaf panicle 2 acts downstream of the DROUGHT AND SALT TOLERANCE transcription factor to regulate drought sensitivity in rice. *J. Exp. Bot.* **66**, 271–281.

Yang, Y., Li, R. and Qi, M. (2000) In vivo analysis of plant promoters and transcription factors by agroinfiltration of tobacco leaves. *Plant J.* **22**, 543–551.

Yang, L., Wu, K., Gao, P., Liu, X., Li, G. and Wu, Z. (2014) GsLRPK, a novel cold-activated leucine-rich repeat receptor-like protein kinase from Glycine soja, is a positive regulator to cold stress tolerance. *Plant Sci.* **215–216**, 19–28.

Yoshida, T., Fujita, Y., Sayama, H., Kidokoro, S., Maruyama, K., Mizoi, J., Shinozaki, K. *et al.* (2010) AREB1, AREB2, and ABF3 are master transcription factors that cooperatively regulate ABRE-dependent ABA signaling involved in drought stress tolerance and require ABA for full activation. *Plant J.* **61**, 672–685.

Yu, X., Peng, Y.H., Zhang, M.H., Shao, Y.J., Su, W.A. and Tang, Z.C. (2006) Water relations and an expression analysis of plasma membrane intrinsic proteins in sensitive and tolerant rice during chilling and recovery. *Cell Res.* **16**, 599–608.

Zha, X., Luo, X., Qian, X., He, G., Yang, M., Li, Y. and Yang, J. (2009) Over-expression of the rice LRK1 gene improves quantitative yield components. *Plant Biotechnol. J.* **7**, 611–620.

Zhang, Z., Lin, W., Li, X., Cao, H., Wang, Y. and Zheng, S.J. (2015) Critical role of eukaryotic elongation factor 1 alpha 1 (EEF1A1) in avian reovirus sigma-C-induced apoptosis and inhibition of viral growth. *Arch. Virol.* **160**, 1449–1461.

Zipfel, C., Kunze, G., Chinchilla, D., Caniard, A., Jones, J.D., Boller, T. and Felix, G. (2006) Perception of the bacterial PAMP EF-Tu by the receptor EFR restricts Agrobacterium-mediated transformation. *Cell*, **125**, 749–760.

8

The rice OsNAC6 transcription factor orchestrates multiple molecular mechanisms involving root structural adaptions and nicotianamine biosynthesis for drought tolerance

Dong-Keun Lee[1,†], Pil Joong Chung[1,†], Jin Seo Jeong[1,†], Geupil Jang[2], Seung Woon Bang[1], Harin Jung[1], Youn Shic Kim[1], Sun-Hwa Ha[3], Yang Do Choi[2] and Ju-Kon Kim[1,*]

[1]Graduate School of International Agricultural Technology and Crop Biotechnology Institute/GreenBio Science and Technology, Seoul National University, Pyeongchang, Korea

[2]Department of Agricultural Biotechnology, Seoul National University, Seoul, Korea

[3]Department of Genetic Engineering and Graduate School of Biotechnology, Kyung Hee University, Yongin, Korea

*Correspondence
email jukon@snu.ac.kr
[†]These authors contributed equally to this work.

Keywords: biotechnology, drought, NAC transcription factor, nicotianamine, rice, root.

Summary

Drought has a serious impact on agriculture worldwide. A plant's ability to adapt to rhizosphere drought stress requires reprogramming of root growth and development. Although physiological studies have documented the root adaption for tolerance to the drought stress, underlying molecular mechanisms is still incomplete, which is essential for crop engineering. Here, we identified *OsNAC6*-mediated root structural adaptations, including increased root number and root diameter, which enhanced drought tolerance. Multiyear drought field tests demonstrated that the grain yield of *OsNAC6* root-specific overexpressing transgenic rice lines was less affected by drought stress than were nontransgenic controls. Genome-wide analyses of loss- and gain-of-function mutants revealed that OsNAC6 up-regulates the expression of direct target genes involved in membrane modification, nicotianamine (NA) biosynthesis, glutathione relocation, 3′-phophoadenosine 5′-phosphosulphate accumulation and glycosylation, which represent multiple drought tolerance pathways. Moreover, overexpression of *NICOTIANAMINE SYNTHASE* genes, direct targets of OsNAC6, promoted the accumulation of the metal chelator NA and, consequently, drought tolerance. Collectively, OsNAC6 orchestrates novel molecular drought tolerance mechanisms and has potential for the biotechnological development of high-yielding crops under water-limiting conditions.

Introduction

Drought is a major environmental factor contributing to loss of crop yield worldwide, and in rice (*Oryza sativa*), this is due to drought-induced phenomena such as delayed flowering time, a reduction in the number of spikelets and poor grain filling rate (Ekanayake *et al.*, 1989; O'Toole and Namuco, 1983). Moreover, the proportion of agriculturally important areas with an inadequate water supply has increased substantially as a consequence of global warming and an explosive increase in human population (Mittler, 2006). Thus, the identification of plant drought tolerance mechanisms for deployment in crops is an important objective. To avoid and cope with drought stress, plants have evolved molecular mechanisms that coordinate the expression of suites of genes that protect them from drought-induced damage, minimize loss of water and modulate their growth and development in arid environments (Shinozaki and Yamaguchi-Shinozaki, 2007). Most drought-inducible genes are regulated by drought-responsive transcription factors (TFs), such as members of the AP2/ERF, MYB, bZIP and NAC families, which directly, or indirectly, regulate drought stress tolerance mechanisms (Abe *et al.*, 2003; Fujita *et al.*, 2004; Kang *et al.*, 2002; Oh *et al.*, 2009; Tran *et al.*, 2004).

The NAC (NAM, ATAF and CUC) superfamily constitutes one of the largest plant-specific TF families: 117 in *Arabidopsis thaliana*, 151 in rice, 163 in poplar (*Populus trichocarpa*) and 152 in both soybean (*Glycine max*) and tobacco (*Nicotiana tabacum*; Puranik *et al.*, 2012). NAC TFs are involved in a wide range of abiotic and biotic stress responses. For example, *A. thaliana*, *AtNAC72* (*RD29*), *AtNAC109* and *AtNAC55* contribute to drought tolerance by promoting the detoxification of aldehydes in the glyoxalase pathway (Fujita *et al.*, 2004; Tran *et al.*, 2004), while *AtNAC2* is involved in responses to salt stress through ethylene and auxin signalling pathways (He *et al.*, 2005). In rice, overexpression of *OsNAC9*, *OsNAC45*, *OsNAC52* and *OsNAC63* enhances tolerance to multiple abiotic stresses via the up-regulation of genes involved in osmolyte production, detoxification activities, redox homeostasis and the protection of macromolecules (Hu *et al.*, 2006; Redillas *et al.*, 2012).

One key adaptation to drought stress involves changes in root growth and development in response to water-deficit conditions (Sharp *et al.*, 2004). Roots detect insufficient water availability in soils and release uncharacterized signals to induce resistance and/or adapt their architecture for optimal growth (Sieburth and Lee, 2010). Previous studies showed that rice inbred lines (IR20 × MGL-2) with long and thick roots exhibit enhanced drought tolerance (Ekanayake *et al.*, 1985). Moreover, overexpression of *TaNAC2* and *HRD* (*HARDY*) in *A. thaliana* promotes primary and lateral root growth and thus increasing root numbers (Karaba et al., 2007; Mao et al., 2012), while overexpression of OsNAC5, OsNAC9 and OsNAC10 in rice roots activates radial root growth (Jeong et al., 2010, 2013; Redillas et al., 2012), all of which result in enhanced drought tolerance. Recently, mechanisms involving the phytohormone auxin, regulated by DEEPER ROOTING 1, were shown to confer drought tolerance to rice by altering root growth angle (Uga et al., 2013). Thus, modification of root architecture is closely associated with drought tolerance; however, the underlying molecular mechanisms that confer root-mediated drought tolerance are not fully understood.

OsNAC6 is previously identified as a key regulator for rice stress responses (Nakashima et al., 2007; Ohnishi et al., 2005). Overexpression rice plants of OsNAC6 show various stress tolerances to drought, high salinity and blast disease. The OsNAC6 acts as a transcriptional activator and up-regulates stress-inducible genes including lipoxygenase and peroxidase for stress tolerance (Nakashima et al., 2007), indicating that the OsNAC6 is sufficient to confer stress tolerance in rice plant. Interestingly, the OsNAC6 controls root growth at early vegetative stage through chromatin modification (Chung et al., 2009). It suggests a possible connection between the root structure modifications by OsNAC6 and OsNAC6-mediated drought tolerance.

In this study, we investigated the molecular mechanisms of OsNAC6-mediated drought tolerance. Transgenic rice lines overexpressing OsNAC6 under the control of either the root-specific or the constitutive promoters showed improved drought tolerance, whereas nac6 mutant exhibited drought susceptibility. In addition, multiyear field drought tests confirmed that root-

specific overexpression of OsNAC6 significantly enhanced drought tolerance. We further characterized OsNAC6-mediated root phenotypes related to drought tolerance. RNA-seq and ChIP-seq analyses led to the identification of the direct target genes of OsNAC6, which together constitute the OsNAC6-mediated drought tolerance pathways.

Results

OsNAC6 overexpression in roots is sufficient to confer drought tolerance

OsNAC6 is a drought-responsive TF that is also regulated by the abscisic acid as well as by low temperature and salinity stresses (Figure S1; Jeong et al., 2010; Nakashima et al., 2007). To investigate its biological roles, we designed two different constructs for OsNAC6 overexpression in rice (Nipponbare): root-specific RCc3::OsNAC6 and constitutive GOS2::OsNAC6. To eliminate somaclonal variation, successive field selection of T_{1-4} plants was performed to identify elite lines that grew normally, without stunting. Six independent homozygous lines (#7, 24 and 38 for RCc3::OsNAC6 and #18, 53 and 62 for GOS2::OsNAC6) were selected for further analysis.

To assess drought resistance, 4-week-old OsNAC6 overexpressors (T_5 generation) and nontransgenic (NT, Nipponbare) plants were subjected to progressive drought stress by withholding water for 5 days under greenhouse conditions. NT plants showed drought-associated visual symptoms, such as leaf rolling and wilting earlier than the transgenic plants (Figure 1a). Moreover, after re-watering, both types of OsNAC6

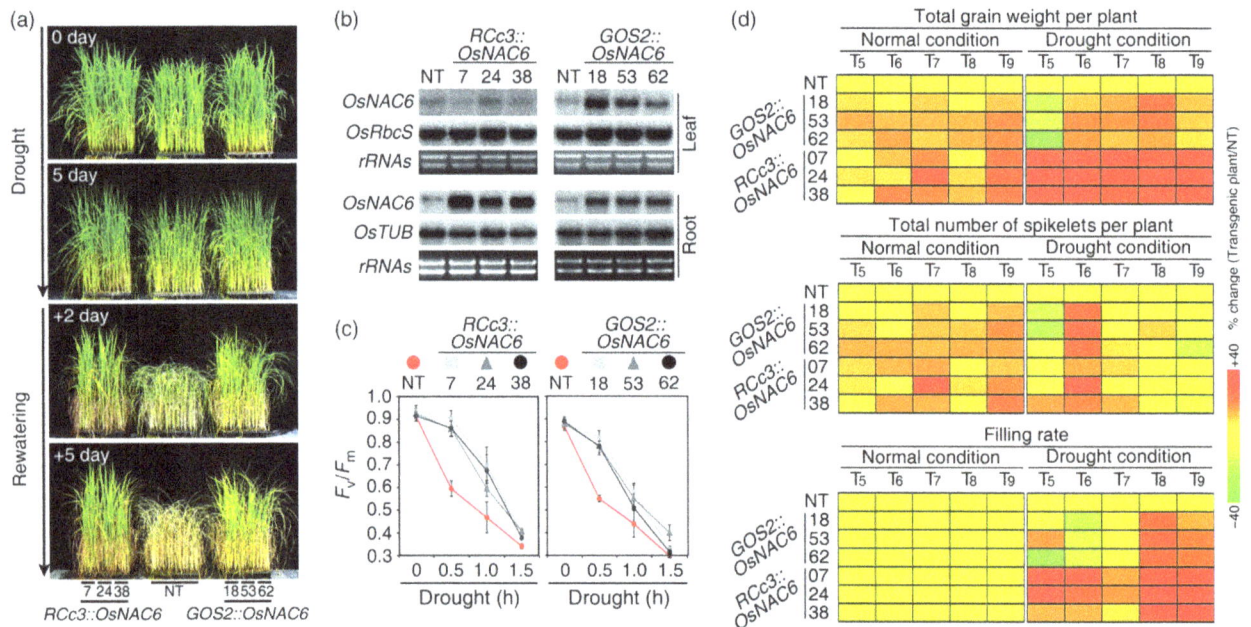

Figure 1 Drought tolerance of RCc3::OsNAC6 and GOS2::OsNAC6 transgenic plants. (a) Drought tolerance of three independent RCc3::OsNAC6 and GOS2::OsNAC6 lines (T_5 generation) at a vegetative development stage. Four-week-old plants were exposed to drought for 5 days, followed by re-watering. The number of days on the images indicates the duration of the drought and re-watering. (b) RNA gel blot analysis using total RNA from leaves and roots of 4-week-old RCc3::OsNAC6, GOS2::OsNAC6 and NT plants, grown under normal growth condition. OsRbcS and OsTUB were used as internal controls. (c) Photochemical efficiency test, measuring the leaf chlorophyll fluorescence (F_v/F_m). Values for each time point represent the mean ± SE of three-replicate experiments (n = 30 for each genotype). (d) Heatmap of agronomic traits of three independent homozygous lines (from T_5 to T_9 generation) grown in the field under normal and drought conditions. Mean values for each category (n = 30 for each condition of each line) are listed in Table S1. Change (%) of total grain weight, or total number of spikelets per line in each year, was calculated as (mean value of the agronomic trait per each line/ mean value of agronomic trait of NT) × 100. Filling rate (%) was (total filled grain/[total filled grain + total unfilled grain]) × 100.

overexpressors recovered better from the drought stress than the NT plants, which continued to wilt and finally died (Figure 1a). The *RCc3::OsNAC6* lines showed high levels of *OsNAC6* expression only in roots, while the *GOS2::OsNAC6* lines showed high levels of *OsNAC6* expression in both leaves and roots (Figure 1b). To independently confirm the conferred drought tolerance, we carried out a leaf chlorophyll fluorescence assay, measuring F_v/F_m (F_v: variable fluorescence and F_m: maximum fluorescence), an indicator of photochemical efficiency of photosystem II (PSII), which can be reduced by drought stress. NT leaves exhibited a rapid decrease in F_v/F_m values as early as 0.5 h after the onset of the drought treatment, while the transgenic leaves showed a delayed decrease in F_v/F_m values that were ~1.5-fold higher than those of the NT (Figure 1c), indicating that PS II of the *OsNAC6* overexpressors was less affected by drought stress. Notably, *OsNAC6* overexpression in roots alone was sufficient to confer drought tolerance during the vegetative stage of growth.

Multiyear field tests of the *OsNAC6* overexpressors

As reproductive development is highly vulnerable to drought stress (Ekanayake *et al.*, 1989; O'Toole and Namuco, 1983), and field tests represent a more informative approach to evaluate effective crop traits under agronomically relevant conditions (Nuccio *et al.*, 2015), we performed drought studies of the *OsNAC6* overexpressors in a rice paddy field and focused on the reproductive development over the course of 5 years (T_{5-9} generation). *OsNAC6* overexpressors, along with NT plants, were transplanted in a paddy field in Gunwi, Korea, and grown to maturity. Yield parameters, such as total grain weight, the total number of spikelets and grain filling rate, were scored for 30 plants per transgenic event and for the NT control. Under normal growth conditions, total grain weight increased by 3%–25% in *RCc3::OsNAC6* plants and by 3%–18% in *GOS2::OsNAC6* compared with NT control plants (Figure 1d; Table S1). This increase in grain weight was mostly caused by an increase in the number of spikelets, rather than an increased filling rate (Figure 1d; Table S1), indicating that overexpression of *OsNAC6* in roots affects reproductive development, especially grain yield, under normal growth conditions.

Under drought conditions (plants exposed to intermittent drought stress at the transition stage from vegetative to reproductive development), the total grain weight of the *RCc3::OsNAC6* plants was 26%–74% greater than that of the NT controls, whereas *GOS2::OsNAC6* plants showed similar values, or only a slight increase (−32% to 22%), compared with NT plants (Figure 1d; Table S1). Given the similar levels of drought tolerance shown by *RCc3::OsNAC6* and *GOS2::OsNAC6* plants at the vegetative stage, this difference in their total grain weight under drought conditions was unexpectedly large. We determined that this was mainly due to a higher grain filling rate in the *RCc3::OsNAC6* plants than in either the NT or the *GOS2::OsNAC6* plants under drought conditions (Figure 1d; Table S1). Taken together, these results indicate that root-specific overexpression of *OsNAC6* increases drought tolerance at the reproductive stage of growth under field drought conditions.

OsNAC6 expression in roots controls tiller development

To further investigate the higher spikelet number of the *OsNAC6* overexpressors under normal growth conditions, we first grew *RCc3::OsNAC6* and *GOS2::OsNAC6* plants, together with NT control, until the panicle developmental stage (approximately

3-month-old plants) in rice paddy fields, and counted the number of tillers. The *OsNAC6* overexpressors showed a slight increase ($P > 0.05$) in tiller number compared with NT plants (Figure 2a, c). However, as one tiller typically produces approximately 90 spikelets, the small difference in tiller number accounts for the substantially higher spikelet number in the *OsNAC6* overexpressing lines. We also evaluated *nac6*, a null mutant (Hwayoung) that has a T-DNA insertion in *OsNAC6* (Chung *et al.*, 2009), with NT (Hwayoung) plants, grown until the panicle stage in rice paddy fields. *nac6* showed reduced grain productivity, mainly due to a reduced number of spikelets and poor grain filling rate (Table S2). In addition, *nac6* produced significantly fewer ($P < 0.05$) tillers than NT plants (Figure 2b; Table S2), with averages over 2 years of 5 and 9, respectively (Figure 2c). This phenotype was rescued in complementation lines (*nac6*^COM), in which an *OsNAC6* genomic region was inserted into the *nac6* mutant (Figure S2a; Table S3). To further verify the role of *OsNAC6* in tiller development, we grew *OsNAC6* overexpressors and *nac6* together with NT controls (Nipponbare and Hwayoung) under long-day growth conditions in a greenhouse, under which rice plants produce more tillers (Figure 2i). NT (Nipponbare) plants had ~40 tillers at the panicle stage, whereas *GOS2::OsNAC6* and *RCc3::OsNAC6* transgenic lines had ~54 and ~49 tillers, respectively. In addition, NT (Hwayoung) plants produced ~43 tillers, whereas *nac6* had ~25. The opposite tiller number phenotype in the *OsNAC6* overexpressors compared with *nac6* indicates that *OsNAC6* regulates tiller development, and the root-specific overexpression of *OsNAC6* is sufficient to promote tiller development.

Root development is regulated by *OsNAC6*

We next compared the roots of the *OsNAC6* overexpressors and *nac6* with those of NT controls grown at the panicle stage under long-day conditions. Root length was similar among all the genotypes (Figure 2j; Figure S2b). However, root number and diameter were significantly different between *OsNAC6* overexpressors and NT plants: NT plants had ~2,100 crown roots and an average root diameter of 0.8 mm, whereas the *OsNAC6* overexpressors had ~2,600 crown roots and a 1.1 mm root diameter (Figure 2k, l). Conversely, *nac6* had ~1,300 crown roots and an average root diameter of 0.8 mm (Figure 2k, l). We noted that large aerenchyma cells in roots of *OsNAC6* overexpressors mainly contributed to their wider root diameter (Figure 2d–f) and that cell layers in the cortex regions of *nac6* roots were substantially reduced in the elongation zones compared with NT plants (Figure 2g, h). These opposite root phenotypes were caused by expression level of the *OsNAC6* in roots of *nac6* mutants and *OsNAC6* overexpressors (Figure 2m). As root number positively correlates with tiller number (Hockett, 1986), we evaluated the average root number per tiller, which was found to be similar among all the genotypes (Figure 2n). We therefore concluded that the root number variation was caused by *OsNAC6* overexpression and the *nac6* mutation leads to abnormal tiller development.

OsNAC6 is necessary for rice drought tolerance

To test whether the *OsNAC6* is necessary for drought response, 4-week-old *nac6*, *nac6*^COM and WT (Hwayoung) plants were subjected to progressive drought stress and we monitored drought-induced visual symptoms (Figure 3a). Leaf rolling and wilting were detected in *nac6* earlier than *nac6*^COM and WT plants (Figure 3a). Moreover, after re-watering, both *nac6*^COM

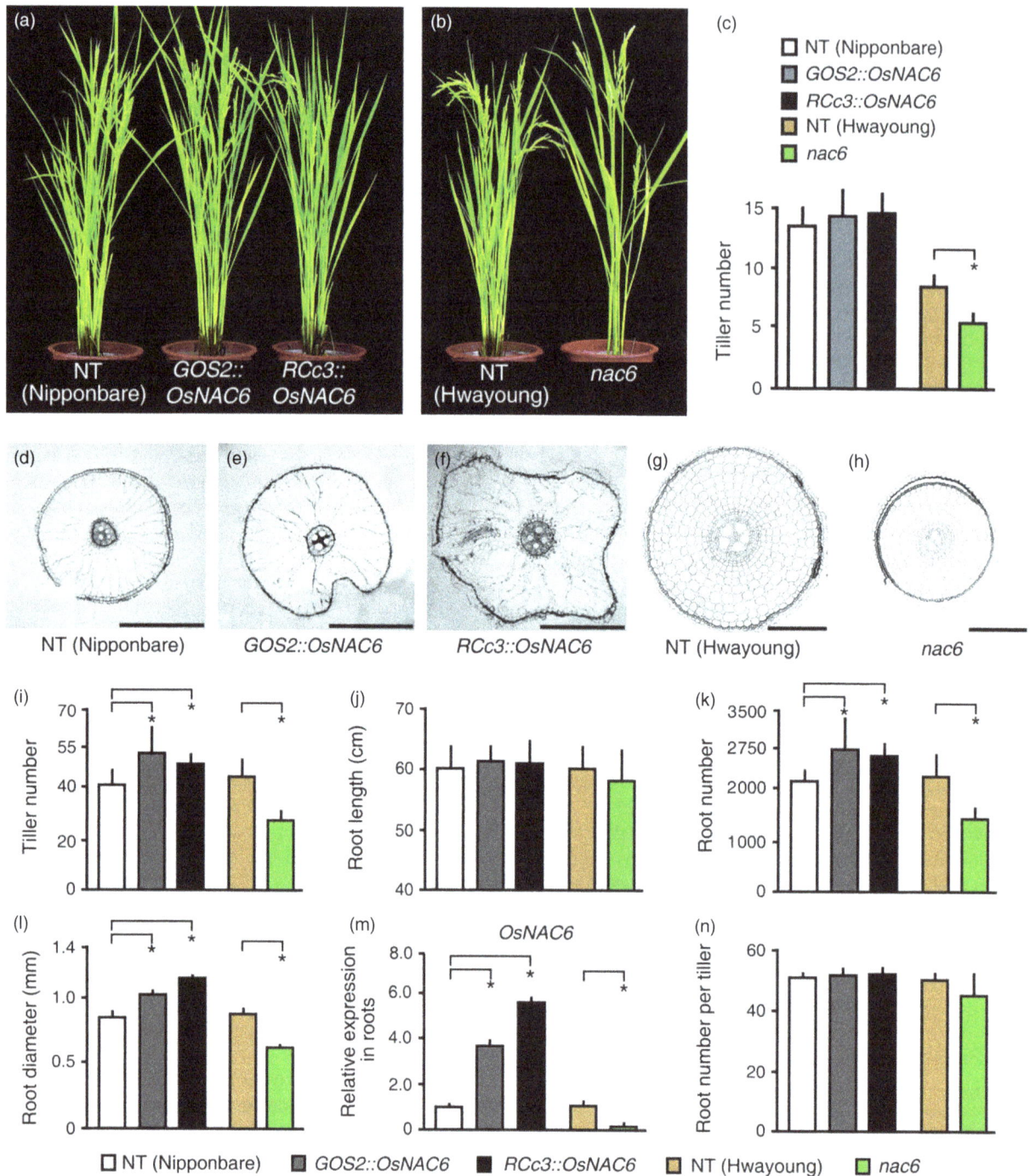

Figure 2 Phenotypes of *RCc3::OsNAC6*, *GOS2::OsNAC6* and *nac6* plants. (a–c) Phenotypes of 3-month-old *OsNAC6* overexpressors and *nac6* knockout mutants grown in a rice paddy field. Representative plants were transferred to pots for photographing. (a) *OsNAC6* overexpressors. (b) *nac6* mutants. (c) Tiller number of 3-month-old plants. Values shown are the mean + SD (n = 20 for each genotype). Asterisks indicate significant differences compared to NT control plants ($P < 0.05$, Student's *t*-test). (d–m) Phenotypes of 3-month-old *OsNAC6* overexpressors and *nac6* knockout mutants grown under long-day conditions in the greenhouse. (d–f) Cross sections of root maturation zones (~30 cm from the root apex). (g, h) Cross sections of roots in the elongation zone (~3 mm from the root apex). (i–m) Quantitative analysis of phenotypes. Values shown are the mean + SD (n = 5 plants for each genotype). (i) Tiller number. (j) Root length. (k) Crown root number. (l) Root diameter (~30 cm from the root apex; 50 roots from five plants for each genotype). (m) qRT-PCR of *OsNAC6* in 2-week-old roots. *UBIQUITIN 1* expression was used as an internal control. Values shown as the mean + SD of two biological replicates, each of which had two technical replicates. (n) Crown root number per tiller. Asterisks indicate significant differences compared to NT control plants ($P < 0.05$, Student's *t*-test). Scale bars, 500 μm in d–f and 100 μm in g and h.

and NT plants recovered better from the drought stress than the *nac6* mutants, which continued to wilt and finally died (Figure 3a). To independently confirm the drought susceptibility of *nac6* mutants, we carried out a leaf chlorophyll fluorescence assay (F_v/F_m). *nac6* leaves exhibited a rapid decrease in F_v/F_m values at 1 h after the drought treatment, while the *nac6*[COM] and

NT leaves showed a delayed decrease in F_v/F_m values (Figure 3b), indicating that PS II of the *nac6* mutants was more damaged by drought stress and *nac6*[COM] plants showed normal drought response compared to WT plants. Collectively, *OsNAC6* is necessary to confer rice drought tolerance.

OsNAC6 is predominantly expressed in the root endodermis, pericycle and phloem

To determine the expression patterns of *OsNAC6* in roots, we generated transgenic rice plants (*OsNAC6::GUS*) harbouring an *OsNAC6* promoter region driving expression of the β-*GLUCUR-ONIDASE* (*GUS*) gene. Four-week-old *OsNAC6::GUS* plants showed GUS activity in the root apical meristem that diminished in the root elongation zone (Figure 4a). In addition, *in situ* hybridization analysis in the elongation zone of crown roots revealed the expression of *OsNAC6* in the endodermis and pericycle, as well as the vasculature, where it was predominantly expressed in the phloem (Figure 4b, c). The phloem-dominant *OsNAC6* expression was also observed in the base of shoots (Figure 4d–f). Expression in the endodermis and pericycle is consistent with *OsNAC6* influencing root radial growth, while expression in the vasculature suggests an association with long-distance regulation.

Identification of *OsNAC6*-regulated downstream genes

To identify molecular pathways by which *OsNAC6* regulates root development and confers drought tolerance, we performed RNA-seq analyses of five different roots from 2-week-old *RCc3::OsNAC6*, *GOS2::OsNAC6*, *nac6* and two NT control plants (Nipponbare and Hwayoung). As OsNAC6 functions as a transcriptional activator (Nakashima *et al.*, 2007), candidate target genes regulated by OsNAC6 were identified using the following cut-off criteria: genes with ≥2-fold higher expression in *RCc3::OsNAC6* or *GOS2::OsNAC6* (log2 ratio ≥1.0) than in NT and genes with ≥2-fold lower expression in *nac6* (log2 ratio ≤−1.0) than in NT. Accordingly, a total of 1,825 and 1,294 genes were up-regulated in *RCc3::OsNAC6* and *GOS2::OsNAC6* lines, respectively, of which 479 genes were present in both sets (Figure 5a; Figure S3; Table S4). We expanded the analysis by comparing these genes with the 1,715 genes that were down-regulated in *nac6*. Based on these RNA-seq profiles, 51 genes were identified as potential key genes in *OsNAC6*-mediated drought tolerance when up-regulated in roots.

Among the 51 genes, putative direct targets were identified by considering the results of a chromatin immunoprecipitation (ChIP-

seq) analysis of roots from transgenic rice plants (*RCc3::6xmyc-OsNAC6*; Figure S4). After removing background peaks, a total of 11,969 peaks were found to be associated with rice gene models. Among the 51 up-regulated genes through the RNA-seq analyses, 13 were found to be present in the ChIP-seq profiles (Figure 5b; Table S4). The potential direct target genes included *NICOTIANAMINE SYNTHASE 1* and *NICOTIANAMINE SYNTHASE 2* (*OsNAS1* and *OsNAS2*), *DEHYDRASE-ENLOASE PHOSPHATASE* (*OsDEP*), *GLUTATHIONE TRANSPORTER* (*GSHT*), *PEPTIDE TRANS-PORTER*, *SULFOTRANSFERASE* (*SOT*), *GLUCOSYLTRANSFERASE* (*GT*), *GLYCEROL ACYLTRANSFERASE* (*GAT*) and *PROTEIN KINASE* (Figure 5c). *OsNAS1*, *OsNAS2* and *OsDEP* are key regulators of the biosynthesis of nicotianamine (NA), which acts as an Fe chelator and a potential antioxidant (Inoue *et al.*, 2003; Itai *et al.*, 2013; Lee *et al.*, 2009, 2012). *GSHT* and *SOT* contribute to antioxidant activity through glutathione relocation and 3′-phosphoadenosine 5′-phosphate (PAP) biosynthesis, respectively (Klein and Papenbrock, 2004; Noctor *et al.*, 2010). The accumulation of these antioxidants and PAP has been shown to alleviate drought-induced oxidative damage (Cheng *et al.*, 2015; Wilson *et al.*, 2009). GAT is involved in membrane modification through lipid metabolism that affects abiotic stress tolerance (Li *et al.*, 2007; Murata *et al.*, 1992), while GT genes control the glycosylation of a structurally diverse range of substrates, including auxin, in association with water stress tolerance (Tognetti *et al.*, 2010; Vogt and Jones, 2000). The expression of the target genes was verified by qRT-PCR using independently isolated total RNAs from roots of the *OsNAC6* overexpressors and *nac6* (Figure 5d).

OsNAC6 up-regulates NA biosynthesis in roots, thereby conferring drought stress tolerance

As *OsNAS1*, *OsNAS2* and *OsDEP* were found to be direct targets of OsNAC6, we expanded our analysis of the RNA-seq data to include genes in the whole NA biosynthesis pathway, including the methionine (Met) cycle (Figure 6a; Table S4). This revealed that the expression of *OsNAS3*, which encodes another enzyme in NA biosynthesis, was also up-regulated in roots of the *OsNAC6* overexpressors and down-regulated in *nac6* roots. Similarly, the expression of *METHYLTHIORIBOSE KINASE* (*OsMTK1* and *OsMTK2*), *OsDEP*, *METHYLTHIORIBUROSE-1-PHOSPHATE ISO-MERASE* (*OsIDI2*), *ACIREDUCTONE DIOXYGENASE* (*OsIDI1*) and *AROMATIC AMINOTRANSFERASE* (*OsIDI4*) genes, which encode key enzymes that generate *S*-adenosyl-Met, a NA precursor in Met cycle (Itai *et al.*, 2013), were all up-regulated in the *OsNAC6* overexpressors and down-regulated in *nac6* (Figure 6a; Table S4).

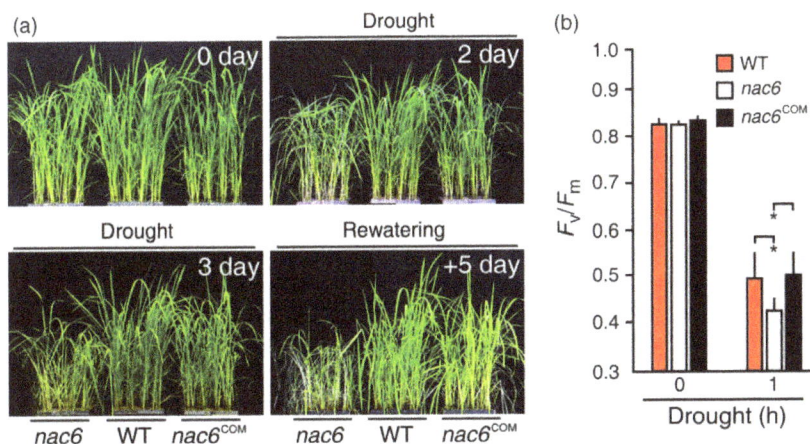

Figure 3 Drought susceptibility of *nac6* mutant. (a) Drought response of *nac6* and *nac6*[COM]. Four-week-old plants were exposed to drought for 5 days, followed by re-watering. The number of days on the images indicates the duration of the drought and re-watering. (b) Photochemical efficiency (F_v/F_m) of 4-week-old *nac6* and *nac6*[COM] plants. Data are shown as the mean + SD of two replicates experiments (n = 20 for each genotype). Asterisks indicate significant differences compared to WT control plants ($P < 0.05$ by Student's *t*-test).

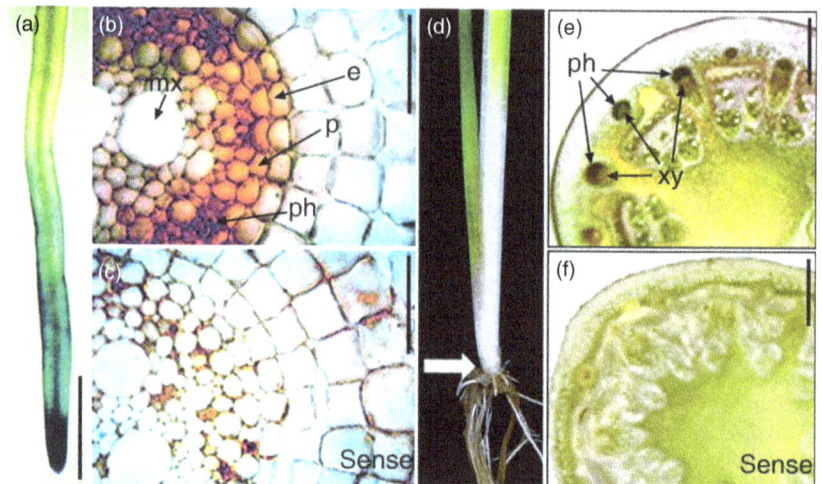

Figure 4 Expression patterns of *OsNAC6*. (a) Expression pattern in 4-week-old *OsNAC6::GUS* roots. (b, c) *In situ* hybridization analysis of *OsNAC6* in the elongation zone of 3-month-old roots. (b) Antisense probe. Signals were detected in the endodermis, pericycle and phloem. (c) Sense probe. (d–f) *In situ* hybridization analysis of *OsNAC6* at the base of 3-month-old shoots. (d) Position of a shoot cross section (arrow). (e) Antisense probe. Signals were detected in the phloem. (f) Sense probe. e, endodermis; mx, metaxylem; p, pericycle; ph, phloem; xy, xylem. Scale bars, 2 mm in a, e and f, and 100 μm in b and c.

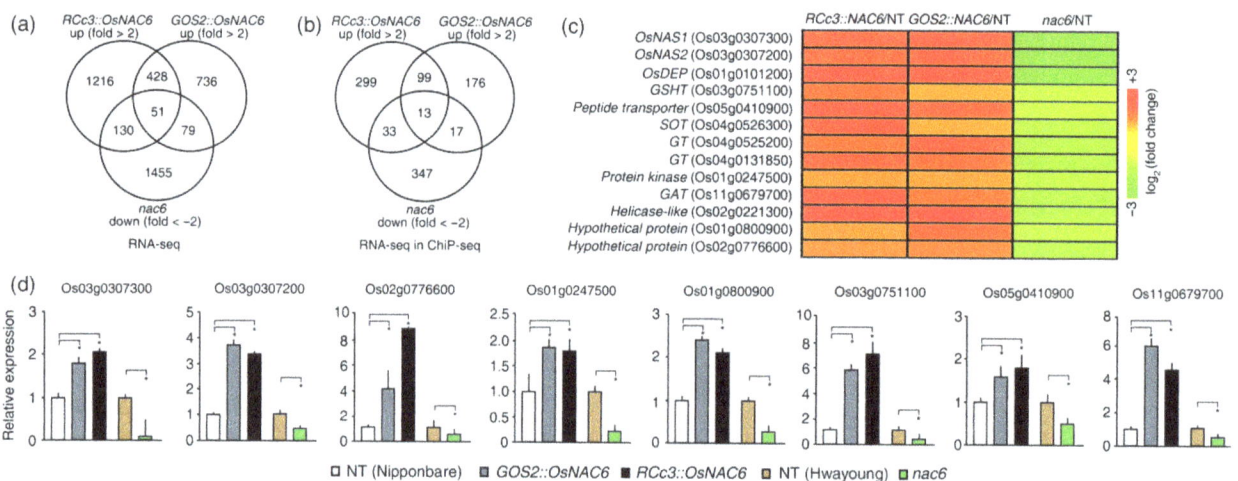

Figure 5 Transcriptomic analysis of *RCc3::OsNAC6*, *GOS2::OsNAC6* and *nac6* roots and a comparison between RNA-seq and ChIP-seq data. (a) Venn diagram of up-regulated genes in roots of 2-week-old *RCc3::OsNAC6* and *GOS2::OsNAC6* relative to NT plants (cut-off, ≥ 2.0-fold) and down-regulated genes in roots of *nac6* relative to NT plants (cut-off, ≤−2.0-fold), using RNA-seq. (b) Venn diagram showing the number of potential genes directly up-regulated by OsNAC6 by comparing RNA-seq with ChIP-seq data. (c) Thirteen high-confidence target genes up-regulated by OsNAC6 and a heatmap of their expression levels based on RNA-seq data. (d) qRT-PCR verification of genes up-regulated by *OsNAC6*. *UBIQUITIN 1* expression was used as an internal control. Values shown as the mean + SD of two biological replicates, each of which had two technical replicates. Asterisks indicate significant differences compared to NT control plants ($P < 0.05$, Student's *t*-test).

From this, we inferred that high NA accumulation in the roots of *OsNAC6* overexpressors might promote root development, leading to drought tolerance.

To determine whether increased NA accumulation in rice plants affects their responses to drought, we exposed two activation-tagged rice lines, *OsNAS2-D1* and *OsNAS3-D1*, which have high levels of NA (Lee *et al.*, 2009, 2012), together with nullizygotes (Null) from *OsNAS2-D1* and NT (Dongjin) plants to drought conditions during vegetative development. NT and Null plants showed drought-associated visual symptoms, such as leaf rolling and wilting, earlier than *OsNAS2-D1* and *OsNAS3-D1* plants (Figure 6b). After re-watering, *OsNAS2-D1* and *OsNAS3-D1* recovered, whereas the NT and Null plants continued to wilt and finally died (Figure 6b). To further confirm the drought tolerance, we measured F_v/F_m of leaves after 2 days of drought stress treatment. NT and Null leaves exhibited a sharp decrease in F_v/F_m values after 2 days of drought treatment, whereas there was no significant difference in the F_v/F_m values of leaves from *OsNAS2-D1* and *OsNAS3-D1* plants that were untreated or drought-treated for 2 days (Figure 6c). These results suggested that *OsNAC6*-mediated NA accumulation in roots is sufficient to confer drought tolerance.

Discussion

We demonstrated here that the rice transcription factor *OsNAC6* is a regulator of drought tolerance pathways and represents a potentially valuable candidate for genetic engineering of drought-tolerant high-yielding crops. Root-specific (*RCc3::OsNAC6*) and whole-body (*GOS2::OsNAC6*) rice overexpression lines showed enhanced drought tolerance at the vegetative stage. Notably, only *RCc3::OsNAC6* lines showed improved drought tolerance at the reproductive stage under field drought conditions. Similar observations were made when *OsNAC5*, *OsNAC9* and *OsNAC10*

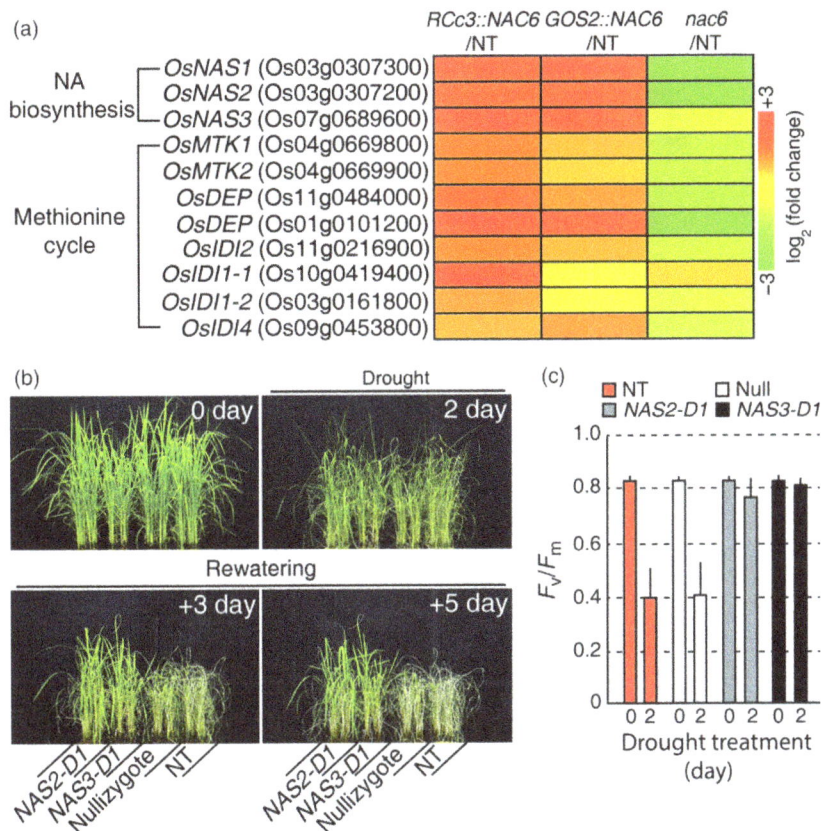

Figure 6 Drought tolerance of *OsNAS2-D1* and *OsNAS3-D1*. (a) Expression levels of genes related to NA biosynthesis and the methionine cycle regulated by *OsNAC6*, based on RNA-seq data. (b) Drought tolerance of *OsNAS2-D1* and *OsNAS3-D1* plants at a vegetative developmental stage. Four-week-old plants were exposed to drought for 5 days, followed by re-watering. The numbers of days on the images indicate the duration of the drought (2 day means plants that were exposed to drought stress for 2 days) and re-watering (+3 and +5 day mean plants that were re-watered for three and 5 days, respectively). (c) Photochemical efficiency (F_v/F_m) of *OsNAS2-D1* and *OsNAS3-D1* rice plants, grown in a greenhouse at 28–30 °C for 4 weeks and drought treated for 2 days. Each value represents the mean + SE (n = 30 for each genotype).

were overexpressed under the control of the *RCc3* and *GOS2* promoters (Jeong *et al.*, 2010, 2013; Redillas *et al.*, 2012). It is possible that whole-body overexpression of *OsNAC6* and its target genes may perturb reproductive development under drought conditions, resulting in a trade-off in grain yield. This idea is supported by the observations that both *UBIQUITIN* promoter-driven *OsNAC6* plants and the stress-inducible *LIP9* promoter-driven *OsNAC6* plants exhibit low reproductive yields (Nakashima *et al.*, 2007). Furthermore, these strong promoter-driven *OsNAC6* plants exhibit growth retardation at 14 days after germination, whereas the overexpression plants show no growth retardation at reproductive stage (Nakashima *et al.*, 2007), indicating that the strong *OsNAC6* expression causes vegetative growth retardation. Collectively, these data suggest that root-specific overexpression of *OsNAC6* represents a more effective approach than whole-body overexpression.

Root structural adaptations to drought stresses were observed in *RCc3::OsNAC6* and *GOS2::OsNAC6* plants, which both showed higher numbers and a thicker diameter of roots, while the opposite phenotypes were observed in *nac6* mutants. Root structural modification for enhanced drought tolerance is associated with root elongation, high number of root and increased radial root growth. Overexpression of *TaNAC2* and *HRD* in *A. thaliana* or rice activates primary and lateral root elongation, which may promote the acquisition of water in deep soils (Karaba *et al.*, 2007; Mao *et al.*, 2012). Guidance of the roots to water sources in deep soils is regulated by *DEEPER ROOTING 1*, which modifies root growth angle (Uga *et al.*, 2013). Overexpression of *HRD* also results in an increase in root number, which may enhance water uptake by increasing the total root surface area in contact with drying soils. Radial root growth, including the formation of larger aerenchyma, also enhances drought tolerance

(Jeong *et al.*, 2010, 2013; Redillas *et al.*, 2012). In maize, the formation of root cortical aerenchyma promotes drought tolerance as it reduces the metabolic cost of soil exploration under water stress, permitting greater root growth and water acquisition from drying soil (Zhu *et al.*, 2010). Moreover, the radial root growth maintains plant water potential under drought conditions (Karaba *et al.*, 2007; Price *et al.*, 1997). Taken together, modulation of root architecture by controlling root development can provide an effective strategy to combat to water-deficit conditions, and our results suggest that such mechanisms occur in rice.

Our studies identified indirect and direct gene targets of OsNAC6. The RNA-seq analyses revealed 51 up-regulated genes by OsNAC6. After a comparison of these genes with ChIP-seq data, we finally identified 13 up-regulated genes as direct targets of OsNAC6, including *OsNAS1*, *OsNAS2*, *OsDEP*, *GSHT*, *SOT*, *GT* and *GAT*. These genes are involved in membrane modification, NA biosynthesis, glutathione relocation, PAP accumulation and glycosylation (Figure S5). In a previous study, OsNAC6 was found to directly regulate a peroxidase (AK104277) and a hypothetical protein (AK110725) gene, when analysed using DEX-treated *UBIQUITIN::OsNAC6-GR* plants (Nakashima *et al.*, 2007). However, these genes were not identified through our root-based RNA-seq and ChIP-seq analyses. This discrepancy may be due to differences in the tissues and promoters used in the two studies.

Drought inhibits photosynthesis due to stomatal closure, resulting in the increased production of reactive oxygen species (ROS) and the induction of drought-mediated oxidative damage (Mittler, 2002). When ROS levels exceed the capacity of a plant to scavenge them, membrane damage can occur, due to the susceptibility of the unsaturated fatty acid components to the effects of ROS (Sharma *et al.*, 2012). This was reported to be alleviated through the overexpression of *GAT* in *Nicotiana tabacum* by increasing the unsaturated fatty acid content of membranes, resulting in a stress-tolerant phenotype (Murata *et al.*, 1992). Similar phenotypes were observed when the *A. thaliana* glycerol-3-phosphate acyltransferase genes *GPAT4* and *GPAT8* were overexpressed, which altered the accumulation of cutin and suberin (Li *et al.*, 2007). Consistent with these, *OsNAC6* overexpression was observed to mediate the *GAT* up-regulation and enhance drought tolerance (Figure S5).

To protect themselves from drought-mediated oxidative stress, plants produce antioxidants or osmoprotectants (Mittler, 2002; Sharma *et al.*, 2012). For example, the antioxidant glutathione scavenges ROS and plays a protective role in abiotic stresses (Cheng *et al.*, 2015). *OsNAC6* was found to directly up-regulate *GSHT* expression in roots, suggesting that *OsNAC6* overexpressors activate glutathione relocation via regulation of *GSHT*, which may have alleviated drought-mediated oxidative stress (Figure S5). In addition, sulphur metabolites play important roles in drought tolerance, and among them, PAP produced by SOT is known to accumulate during drought in *A. thaliana*, thereby contributing to drought tolerance (Chan *et al.*, 2012; Wilson *et al.*, 2009). Expression of *SOT* was up-regulated in the *OsNAC6* overexpressors, suggesting that *OsNAC6* overexpressors might accumulate PAP, giving rise to drought-tolerant phenotypes (Figure S5).

Drought can also cause an increase in Fe concentration in plants that are producing ROS, resulting in drought-mediated oxidative damage (Moran *et al.*, 1994; Price and Hendry, 1991). Fe homeostasis is influenced by the Fe chelator nicotianamine (NA), which is produced by the enzymes OsNAS1 and OsNAS2 (Inoue *et al.*, 2003; Lee *et al.*, 2009, 2012). In roots of plants fed with Fe, *OsNAS1* and *OsNAS2* were reported to be specifically expressed in companion and pericycle cells (Inoue *et al.*, 2003). This expression pattern was similar to that of *OsNAC6*, suggesting a role for *OsNAC6* in a long-distance Fe transport and homeostasis. Fe is an essential element for plant growth and development, especially through meristem-specific callose deposition, which regulates cell-to-cell communication for root radial growth and normal root stem cell maintenance (Muller *et al.*, 2015). However, excessive amount of Fe is highly toxic (Moran *et al.*, 1994; Price and Hendry, 1991) and can perturb primary root elongation and lateral root formation via interaction with the auxin and ethylene signalling pathways (Giehl *et al.*, 2012; Li *et al.*, 2015; Ward *et al.*, 2008). Despite being identified as one of the up-regulated genes during drought conditions in *A. thaliana*, rice and wheat (Ergen *et al.*, 2009; Shaik and Ramakrishna, 2013), the roles of NAS in drought tolerance mechanisms have not been reported to date. NA overproduction by OsNAC6-mediated up-regulation of NAS genes may promote the binding of excess Fe, thereby preventing the production of hydroxyl radicals, which consequently confers drought tolerance (Figure S5), indicating a role for NA in drought tolerance. In conclusion, overexpression of *OsNAC6* not only improves drought tolerance but also increases grain yield, further indicating its potential importance for crop improvement.

Experimental procedures

Plasmid construction and rice transformation

Total RNA was extracted from 2-week-old japonica rice roots (*Oryza sativa* cv Nipponbare), grown in a greenhouse (16-h light/8-h dark cycle), and used to generate total cDNAs, from which the *OsNAC6* (Os01g0884300) cDNA was amplified by PCR, using PrimeSTAR HS DNA Polymerase (Takara, Kusatsu, Japan) and the Reverse Transcription System (Promega, Madison, WI), according

to the manufacturer's instructions. The primers were forward 5'-CACCATGAGCGGCGGTCAGGACC-3' and reverse 5'-CTAGA ATGGCTTGCCCCAG-3'. The PCR product was cloned into the entry vector, pENTR/SD (Invitrogen, Carlsbad, CA), and then ligated downstream of 2.2 kb of the *GOS2* (Os07g0529800) promoter in the rice transformation vector, *p700-GOS2* (Jeong *et al.*, 2010) for constitutive expression, or 1.3 kb of the *RCc3* (Os02g0662000) promoter in the rice transformation vector, *p700-RCc3* (Jeong *et al.*, 2010) for root-specific expression, using the Gateway System (Invitrogen, Carlsbad, CA). The resulting vectors were named *GOS2::OsNAC6* and *RCc3::OsNAC6*, respectively. To generate *OsNAC6::GUS* transgenic plants, a 2-kb promoter region (upstream region of the ATG start codon) of *OsNAC6* was amplified using PrimeSTAR HS DNA Polymerase (5'-CTGCAGTGTGCAAACTTTCAATG TTGAC-3' and 5'-GAATTCCTC TCTCCCCCTTCTCCGGT-3') and ligated upstream of the β-*glucuronidase* (*GUS*) reporter gene in the rice transformation vector pCAMBIA1391Z using the *Eco*R1 and *Pst*1 restriction sites. Transgenic plants were obtained by *Agrobacterium tumefaciens* (LBA4404)-mediated embryogenic callus (Nipponbare) transformation.

For complementation of the *nac6* mutant, 5,081 bp of a *OsNAC6* genomic fragment, including 2,240 bp of promoter (upstream region of the start ATG), 1,869 bp exons and introns (from ATG to stop codon) and 972 bp 3' of then untranslated region, were amplified from genomic DNA of 2-week-old Nipponbare, using PrimeSTAR HS DNA Polymerase. The amplified genomic fragment was cloned into the rice transformation vector, pSB11 (Komori *et al.*, 2007). Complementation lines were obtained by *A. tumefaciens* (LBA4404)-mediated transformation of *nac6* (Chung *et al.*, 2009) embryogenic callus.

Stress treatments for RNA gel blot analysis

Rice (Nipponbare) seeds were germinated in soil and grown for 14 days in a greenhouse at 28–30 °C. For the drought treatment, seedlings were air-dried under continuous light (~1000 μmol/m^2/s), and for high-salinity and ABA treatments, seedlings were transferred to a nutrient solution (Inoue *et al.*, 2003) including 400 mM NaCl or 100 μM ABA, respectively. For low-temperature treatments, seedlings were placed in a 4 °C cold chamber under continuous light (150 μmol/m^2/s). Total RNA was extracted from these samples using TRIzol® reagent (Invitrogen, Carlsbad, CA), and 10 μg from each sample was fractionated on a 1.2% denatured agarose gel and blotted onto a Hybond N+ nylon membrane (Amersham Bioscience, Piscataway, NJ). A radiolabeled *OsNAC6* cDNA fragment was used to probe the membrane, corresponding to a 376-bp fragment of the 3' untranslated region, which was generated by PCR amplification (5'-CCTCCTCCAGGA-CATCCTCA-3' and 5'-CGAATCAATCACCATGTACT-3'). *OsDip1* and *OsRbcS* probes were used as markers for the stress treatments (Jeong *et al.*, 2013). To determine the *OsNAC6* expression patterns and levels in the *OsNAC6* overexpressors, total RNA was extracted from the roots and leaves of three homozygous T_5 lines of *RCc3::OsNAC6* and *GOS2::OsNAC6* plants. Ten micrograms of each total RNA sample was used for RNA gel blot analysis, as above.

Drought stress treatment during vegetative development

Transgenic and NT control plants were germinated on Murashige and Skoog (MS) media (Duchefa, Haarlem, Netherlands) at 28 °C for 4 days, and eighteen seedlings of each transgenic lines and NT were transplanted into soil pots (4 × 4 × 6 cm; four plants per pot) and grown for 4 weeks in a greenhouse (16-h light/8-h dark cycle) at 28–30 °C. Each pot had the same size of holes in the bottom, and they were all placed in a single tray to synchronize watering. Drought stress was simultaneously applied to all the rice plants by first adding no water to the soil pots for 5 days and then re-watering. Drought-induced symptoms were monitored by imaging transgenic and NT plants at the indicated time points using a NEX-5N camera (Sony, Tokyo, Japan).

Measurement of chlorophyll fluorescence

RCc3::OsNAC6, *GOS2::OsNAC6* and NT plants were grown in greenhouse at 28–30 °C for 2 weeks. *nac6*, *nac6*[COM], *OsNAS2-D1*, *OsNAS3-D1* and NT plants were grown in greenhouse at 28–30 °C for 4 weeks. Thirty leaves from ten seedlings were collected before each stress treatment. Samples were adapted in dark conditions for 10 min. To simulate drought, the leaf discs were air-dried for the indicated time points at 28 °C. To measure F_v/F_m values, representing the activity of PSII, a PAM test was carried out with a pulse modulation fluorometer, Mini-PAM (Walz, Effeltrich, Germany) as described previously (Redillas *et al.*, 2012). The dark-treated leaf was given a measuring light of 0.15 μmol photon m^{-2} s^{-1} for a minimal level of fluorescence and then a 0.8 s actinic light of 10 000 μmol photon m^{-2} s^{-1} for a maximal level of fluorescence.

Phenotypic and anatomical analysis of rice roots grown under long-day conditions

RCc3::OsNAC6, *GOS2::OsNAC6*, *nac6* and NT plants were transplanted to PVC tubes (1.2 m in length and 0.2 m in diameter) that were filled with natural paddy soil and placed into container filled with water. At the panicle development stage, roots were quantified and used for sectioning for anatomical analysis. The diameter of 50 individual roots from *RCc3::OsNAC6*, *GOS2::OsNAC6*, *nac6* and NT plants (five plants for each genotype) was measured. Internal root anatomy was examined using a Technovit 7100 system, as previously described (Jang *et al.*, 2011) with minor modifications. Technovit saturation was carried out for 3 days. Sections (3 μm) were generated using an ultramicrotome (MTX, RMC, USA), and images were captured with an Olympus DP70 camera mounted on Olympus BX 500 light microscope.

Agronomic trait analysis in rice paddy fields over a 5-year period

The field experiments, including the use of fertilizers, drought treatments and analysis of agronomic traits, were as described previously (Oh *et al.*, 2009). Briefly, to evaluate yield components of the transgenic plants under normal growth conditions, three independent homozygous lines from T$_5$ (2009) to T$_9$ (2013) for the *RCc3::OsNAC6* and *GOS2::OsNAC6* lines, together with NT, were planted in a rice paddy field at Gunwi (36°06′48.0″N, 128°38′38.0″E), Kyungpook National University, Korea, and grown to maturity. A randomized design was employed for three replicates using three different 10 m^2 plots. Yield parameters were scored for 30 plants per line, collected from three different plots. To evaluate the yield components of the transgenic plants under drought field conditions, we built rain-off shelters to cover rice plants and made a semi-field condition before drought treatment. Intermittent drought stress was applied twice during the panicle development by draining othe water from the bottom of the container. When full leaf rolling was observed in the NT plants after the first drought treatment, they were irrigated

overnight and subjected to a second round of drought stress until complete leaf rolling occurred again. After two drought stress treatments, the plants were irrigated until harvesting. Yield parameters were scored for 30 plants per line collected from three different plots corresponding to the drought field conditions. The results from three independent lines were compared with those of the NT controls, using one-way ANOVA analysis.

Quantitative real-time PCR analysis

For quantitative real-time PCR (qRT-PCR) experiments, a Super-Script™ III Platinum® One-Step qRT-PCR System (Invitrogen, Carlsbad, CA) was used to generate first-strand cDNAs. qRT-PCR was carried out using a Platinum® SYBR® Green qPCR SuperMix-UDG (Invitrogen) and a Mx3000p Real-Time PCR machine (Stratagene, La Jolla, CA). To validate the RNA-seq data, total RNA was extracted from the roots of 2-week-old *OsNAC6* overexpressors, *nac6* and NT rice seedlings grown under normal growth conditions. Rice *UBIQUITIN 1* (Os06g0681400) was used as an internal control, and two biological replicates, each with two technical replicates, were analysed. Gene-specific primers used for qRT-PCR are listed in Table S5.

In situ hybridization

Three-month-old crown roots were used for *in situ* hybridization using Technovit resin previously described (Jang *et al.*, 2011), with minor modifications. Briefly, roots were incubated in FAA fixing solution (50% [v/v] ethanol, 5% [v/v] acetic acid and 3.7% [v/v] formaldehyde) for 3 h, then dehydrated and finally embedded with Technovit resin. The sections (3 μm) were made, and *OsNAC6* probes were generated by *in vitro* transcription with a DIG RNA labeling Kit (Roche, Mannheim, Germany) targeting the 376-bp 3′ untranslated region of the *OsNAC6* mRNA, as described above.

RNA-seq analysis

Total RNA was extracted from the roots of 2-week-old *OsNAC6* overexpressors, *nac6* and NT rice seedlings grown under normal growth conditions, using an RNeasy Plant Mini Kit (Qiagen, Hilden, Germany), according to the manufacturer's instructions. A modified TruSeq method was used to construct a strand-specific RNA-seq library with different index primers, and libraries were sequenced using an Illumina HiSeq 2000 system in the National Instrumentation Center of Environmental Management College of Agriculture and Life Science, Seoul National University. Genes were defined as being differentially expressed if their transcript abundance was ≥2-fold higher in *RCc3::OsNAC6* or *GOS2::OsNAC6* compared to NT (Nipponbare) or ≥2-fold lower in *nac6* compared to NT (Hwayoung). These data can be found at http://www.ncbi.nlm.nih.gov/geo/ (Accession number: GSE81069).

ChIP-seq analysis

To produce *RCc3::6 × myc-OsNAC6* plants, the coding sequence of *OsNAC6* was amplified using PrimeSTAR DNA polymerase (Takara) with the *OsNAC6-B* F-primer (GGATCCATGAGCGGC GGTCAGGACC) and the *OsNAC6-N* R-primer (GCGGCCGCG CTAGAATGGCTTGCCCCAG). After digestion of the PCR products with *Bam*HI and *Not*I, the coding sequence was ligated into the multiple cloning site of the pE3n vector (Dubin *et al.*, 2008), which is flanked with a 6 × *myc* tag coding sequence. Finally, the 6 × *myc-OsNAC6* sequence from the *pE3n-OsNAC6* was sub-cloned into the p700-RCc3 vector (Jeong *et al.*, 2010) carrying a

1.3-kb *RCc3* promoter sequence, using the Gateway system. Chromatin immunoprecipitation (ChIP) was performed with roots of 2-week-old rice seedlings, as described previously (Chung *et al.*, 2009). These data can be found at http://www.ncbi.nlm. nih.gov/geo/ (Accession number: GSE80986).

Acknowledgements

We thank the Kyungpook National University for providing rice paddy fields and G. An (Kyung Hee University) for providing the *OsNAS2-D1* and *OsNAS3-D1* mutants. This research was supported by the Rural Development Administration under the Next-Generation BioGreen 21 Program (PJ011829012016) and by the Basic Science Research Program through the National Research Foundation of Korea (NRF-2014R1A2A1A11051690, NRF-2014R1A6A3A04053795 and NRF-2013R1A6A3A04060627).

References

Abe, H., Urao, T., Ito, T., Seki, M., Shinozaki, K. and Yamaguchi-Shinozaki, K. (2003) Arabidopsis AtMYC2 (bHLH) and AtMYB2 (MYB) function as transcriptional activators in abscisic acid signaling. *Plant Cell* **15**, 63–78.

Chan, K.X., Wirtz, M., Phua, S.Y., Estavillo, G.M. and Pogson, B.J. (2012) Balancing metabolites in drought: the sulfur assimilation conundrum. *Trends Plant Sci.* **18**, 18–29.

Cheng, M.C., Ko, K., Chang, W.L., Kuo, W.C., Chen, G.H. and Lin, T.P. (2015) Increased glutathione contributes to stress tolerance and global translational changes in Arabidopsis. *Plant J.* **83**, 926–939.

Chung, P.J., Kim, Y.S., Jeong, J.S., Park, S.H., Nahm, B.H. and Kim, J.K. (2009) The histone deacetylase OsHDAC1 epigenetically regulates the *OsNAC6* gene that control seedling root growth in rice. *Plant J.* **59**, 764–776.

Dubin, M.J., Bowler, C. and Benvenuto, G. (2008) A modified gateway cloning strategy for overexpression tagged proteins in plants. *Plant Methods* **4**, 3.

Ekanayake, I.J., O'Toole, J.C., Garrity, D.P. and Masajo, T.M. (1985) Inheritance of root characters and their relations to drought resistance in rice. *Crop Sci.* **25**, 927–933.

Ekanayake, I.J., De Datta, S.K. and Steponkus, P.L. (1989) Spikelet sterility and flowering response of rice to water stress at anthesis. *Ann. Bot.* **63**, 257–264.

Ergen, N.Z., Thimmapuram, J., Bohnert, H.J. and Budak, H. (2009) Transcriptome pathways unique to dehydration tolerant relatives of modern wheat. *Funct. Integr. Genomics* **9**, 377–396.

Fujita, M., Fujita, Y., Maruyama, K., Seki, M., Hiratsu, K., Ohme-Takagi, M., Tran, L.S. *et al.* (2004) A dehydration-induced NAC protein, RD26, is involved in a novel ABA-dependent stress-signaling pathway. *Plant J.* **39**, 863–876.

Giehl, R.F.H., Lima, J.E. and von Wiren, N. (2012) Localized iron supply triggers lateral root elongation in Arabidopsis by altering the AUX1-mediated auxin distribution. *Plant Cell* **24**, 33–49.

He, X.J., Mu, R.L., Cao, W.H., Zhang, Z.G., Zhang, J.S. and Chen, S.Y. (2005) AtNAC2, a transcription factor downstream of ethylene and auxin signaling pathways, is involved in salt stress response and lateral root development. *Plant J.* **44**, 903–916.

Hockett, E.A. (1986) Relationship of adventitious roots and agronomic characteristics in barley. *Can. J. Plant Sci.* **66**, 257–280.

Hu, H., Dai, M., Yao, J., Xiao, B., Li, X., Zhang, Q. and Xiong, L. (2006) Overexpressing a NAM, ATAF and CUC (NAC) transcription factor enhances drought resistance and salt tolerance in rice. *Proc. Natl Acad. Sci. USA* **103**, 12987–12992.

Inoue, H., Higuchi, K., Takahashi, M., Nakanishi, H., Mori, S. and Nishizawa, H.K. (2003) Three rice nicotianamine synthase genes, *OsNAS1*, *OsNAS2*, and *OsNAS3* are expressed in cells involved in long-distance transport of iron and differentially regulated by iron. *Plant J.* **36**, 366–381.

Itai, R.N., Ogo, Y., Kobayashi, T., Nakanishi, H. and Nishizawa, N.K. (2013) Rice genes involved in phytosiderophore biosynthesis are synchronously regulated during the early stages of iron deficiency in roots. *Rice* **6**, 16.

Jang, G., Yi, K., Pires, N.D., Menand, B. and Dolan, L. (2011) RSL genes are sufficient for rhizoid system development in early diverging land plants. *Development* **138**, 2273–2281.

Jeong, J.S., Kim, Y.S., Baek, K.H., Jung, H., Ha, S.H., Do, C.Y., Kim, M. *et al.* (2010) Root-specific expression of *OsNAC10* improves drought tolerance and grain yield in rice under field drought conditions. *Plant Physiol.* **153**, 185–197.

Jeong, J.S., Kim, Y.S., Redillas, M.C., Jang, G., Jung, H., Bang, S.W., Choi, Y.D. *et al.* (2013) *OsNAC5* overexpression enlarges root diameter in rice plants leading to enhanced drought tolerance and increased grain yield in the field. *Plant Biotechnol. J.* **11**, 101–114.

Kang, J.Y., Choi, H.I., Im, M.Y. and Kim, S.Y. (2002) Arabidopsis basic leucine zipper proteins that mediate stress-responsive abscisic acid signaling. *Plant Cell* **14**, 343–357.

Karaba, A., Dixit, S., Greco, R., Aharoni, A., Trijatmiko, K.R., Marsch-Martinez, N., Krishnan, K.N. *et al.* (2007) Improvement of water use efficiency in rice by expression of *HARDY*, an Arabidopsis drought and salt tolerance gene. *Proc. Natl Acad. Sci. USA* **104**, 15270–15275.

Klein, M. and Papenbrock, J. (2004) The multi-protein family of Arabidopsis sulphotransferases and their relatives in other plant species. *J. Exp. Bot.* **55**, 1809–1820.

Komori, T., Imayama, T., Kato, N., Ishida, Y., Ueki, J. and Komari, T. (2007) Current status of binary vectors and superbinary vectors. *Plant Physiol.* **145**, 1155–1160.

Lee, S., Jeon, U.S., Lee, S.J., Kim, Y.K., Persson, D.P., Husted, S., Schjorring, J.K. *et al.* (2009) Iron fortification of rice seeds through activation of the *nicotianamine synthase* gene. *Proc. Natl Acad. Sci. USA* **106**, 22014–22019.

Lee, S., Kim, Y.S., Jeon, U.S., Kim, Y.K., Schjorring, J.K. and An, G. (2012) Activation of rice *nicotianamine synthase 2* (*OsNAS2*) enhances iron availability for biofortification. *Mol. Cells* **33**, 269–275.

Li, Y., Beisson, F., Koo, A.J., Molina, I., Pollard, M. and Ohlrogge, J. (2007) Identification of acyltransferases required for cutin biosynthesis and production of cutin with suberin-like monomers. *Proc. Natl Acad. Sci. USA* **104**, 18339–18344.

Li, G., Xu, W., Kronzucker, H.J. and Shi, W. (2015) Ethylene is critical to the maintenance of primary root growth and Fe homeostasis under Fe stress in Arabidopsis. *J. Exp. Bot.* **66**, 2041–2054.

Mao, X., Zhang, H., Qian, X., Li, A., Zhao, G. and Jing, R. (2012) TaNAC2, a NAC-type wheat transcription factor conferring enhanced multiple abiotic stress tolerances in Arabidopsis. *J. Exp. Bot.* **63**, 2933–2946.

Mittler, R. (2002) Oxidative stress, antioxidants and stress tolerance. *Trends Plant Sci.* **7**, 405–410.

Mittler, R. (2006) Abiotic stress, the field environment and stress combination. *Trends Plant Sci.* **11**, 15–19.

Moran, J.F., Becana, M., Iturbe-Ormaetxe, I., Frechilla, S., Klucas, R.V. and Aparicio-Tejo, P. (1994) Drought induces oxidative stress in pea plants. *Planta* **194**, 346–352.

Muller, J., Toev, T., Heisters, M., Teller, J., Moore, K.L., Hause, G., Dinesh, D.C. *et al.* (2015) Iron-dependent callose deposition adjusts root meristem maintenance to phosphate availability. *Dev. Cell* **33**, 216–230.

Murata, N., Ishizaki-Nishizawa, O., Higashi, S., Hayashi, H., Tasaka, Y. and Nishida, I. (1992) Genetically engineered alteration in the chilling sensitivity of plants. *Nature* **356**, 710–713.

Nakashima, K., Tran, L.S., Van Nguyen, D., Fujita, M., Maruyama, K., Todaka, D., Ito, Y. *et al.* (2007) Functional analysis of a NAC-type transcription factor OsNAC6 involved in abiotic and biotic stress-responsive gene expression in rice. *Plant J.* **51**, 617–630.

Noctor, G., Queval, G., Mhamdi, A., Chaouch, S. and Foyer, C.H. (2010) Glutathione. *Arabidopsis book* **9**, e0142.

Nuccio, M.L., Wu, J., Mowers, R., Zhou, H.P., Meghji, M., Primavesi, L.F., Paul, M.J. *et al.* (2015) Expression of trehalose-6-phosphate phosphatase in maize ears improves yield in well-watered and drought conditions. *Nat. Biotechnol.* **33**, 862–869.

Oh, S.J., Kim, Y.S., Kwon, C.W., Park, H.K., Jeong, J.S. and Kim, J.K. (2009) Overexpression of the transcription factor AP37 in rice improves grain yield under drought conditions. *Plant Physiol.* **150**, 1368–1379.

Ohnishi, T., Sugahara, S., Yamada, T., Kikuchi, K., Yoshiba, Y., Hirano, H.Y. and Tsutsumi, N. (2005) *OsNAC6*, a member of the NAC gene family, is induced by various stresses in rice. *Genes Genet. Syst.* **80**, 135–139.

O'Toole, J.C. and Namuco, O.S. (1983) Role of panicle exsertion in water stress induced sterility. *Crop Sci.* **23**, 1093–1097.

Price, A.H. and Hendry, A.F. (1991) Iron-catalysed oxygen radical formation and its possible contribution to drought damage in nine native grasses and three cereals. *Plant Cell Environ.* **14**, 477–484.

Price, A.H., Tomos, A.D. and Virk, D.S. (1997) Genetic dissection of root growth in rice (*Oryza sativa L.*) I: a hydrophonic screen. *Theor. Appl. Genet.* **95**, 132–142.

Puranik, S., Sahu, P.P., Srivastava, P.S. and Prasad, M. (2012) NAC proteins: regulation and role in stress tolerance. *Trends Plant Sci.* **17**, 369–381.

Redillas, M.C., Jeong, J.S., Kim, Y.S., Jung, H., Bang, S.W., Choi, Y.D., Ha, S.H. *et al.* (2012) The overexpression of *OsNAC9* alters the root architecture of rice plants enhancing drought resistance and grain yield under field conditions. *Plant Biotechnol. J.* **10**, 792–805.

Shaik, R. and Ramakrishna, W. (2013) Genes and co-expression modules common to drought and bacterial stress responses in Arabidopsis and rice. *PLoS ONE* **8**, e77261.

Sharma, P., Jha, A.B., Dubey, R.S. and Pessarakli, M. (2012) Reactive oxygen species, oxidative damage, and antioxidative defense mechanism in plants under stressful conditions. *J. Bot* **2012**, 1–26.

Sharp, R.E., Poroyko, V., Hejlek, J.G., Spollen, W.G., Springer, G.K., Bohnert, H.J. and Nguyen, H.T. (2004) Root growth maintenance during water deficits: physiology to functional genomics. *J. Exp. Bot.* **55**, 2343–2351.

Shinozaki, K. and Yamaguchi-Shinozaki, K. (2007) Gene networks involved in drought stress response and tolerance. *J. Exp. Bot.* **58**, 221–227.

Sieburth, L.E. and Lee, D.K. (2010) BYPASS1: how a tiny mutant tell a big story about root-to-shoot signaling. *J. Integr. Plant Biol.* **52**, 77–85.

Tognetti, V.B., Van Aken, O., Morreel, K., Vandenbroucke, K., van de Cotte, B., De Clercg, I., Chiwocha, S. *et al.* (2010) Perturbation of indole-3-butryric acid homeostasis by the UDP-glucosyltransferase *UGT74E2* modulates Arabidopsis architecture and water stress tolerance. *Plant Cell* **22**, 2660–2679.

Tran, L.S., Nakashima, K., Sakuma, Y., Simpson, S.D., Fujita, Y., Maruyama, K., Fujita, M. *et al.* (2004) Isolation and functional analysis of Arabidopsis stress-inducible NAC transcription factors that bind to a drought-responsive cis–element in the early responsive to *dehydration stress 1* promoter. *Plant Cell* **16**, 2481–2498.

Uga, Y., Sugimoto, K., Ogawa, S., Rane, J., Ishitani, M., Hara, N., Kitomi, Y. *et al.* (2013) Control of root system architecture by *DEEPER ROOTING 1* increases rice yield under drought conditions. *Nat. Genet.* **45**, 1097–1102.

Vogt, T. and Jones, P. (2000) Glycosyltransferases in plant natural product synthesis: characterization of a supergene family. *Trends Plant Sci.* **5**, 380–386.

Ward, J.T., Lahner, B., Yakubova, E., Salt, D.E. and Raghothama, K.G. (2008) The effect of iron on the primary root elongation of Arabidopsis during phosphate deficiency. *Plant Physiol.* **147**, 1181–1191.

Wilson, P.B., Estavillo, G.M., Field, K.J., Pornsiriwong, W., Carroll, A.J., Howell, K.A., Woo, N.S. *et al.* (2009) The nucleotidase/phosphatase SAL1 is a negative regulator of drought tolerance in Arabidopsis. *Plant J.* **58**, 299–317.

Zhu, J., Brown, K.M. and Lynch, J.P. (2010) Root cortical aerenchyma improves the drought tolerance of maize (*Zea mays L.*). *Plant Cell Environ.* **33**, 740–749.

9

Bt rice in China — focusing the nontarget risk assessment

Yunhe Li[1,*], Qingling Zhang[1,2], Qingsong Liu[1], Michael Meissle[3], Yan Yang[1], Yanan Wang[1], Hongxia Hua[2], Xiuping Chen[1], Yufa Peng[1,*] and Jörg Romeis[1,3,*]

[1]State Key Laboratory for Plant Diseases and Insect Pests, Institute of Plant Protection, Chinese Academy of Agricultural Sciences, Beijing, China

[2]College of Plant Science & Technology, Huazhong Agricultural University, Wuhan, China

[3]Agroscope, Biosafety Research Group, Zurich, Switzerland

*Correspondence
email liyunhe@caas.cn (Y.L.) or email
yfpeng@ippcaas.cn (Y.P.) or email
joerg.romeis@agroscope.admin.ch (J.R.)

Summary

Bt rice can control yield losses caused by lepidopteran pests but may also harm nontarget species and reduce important ecosystem services. A comprehensive data set on herbivores, natural enemies, and their interactions in Chinese rice fields was compiled. This together with an analysis of the Cry protein content in arthropods collected from *Bt* rice in China indicated which nontarget species are most exposed to the insecticidal protein and should be the focus of regulatory risk assessment.

Keywords: *Bt* rice, ecosystem services, environmental risk assessment, nontarget effects, surrogate species.

Introduction

Rice (*Oryza sativa*) suffers massive yield losses from attacks by a complex of lepidopteran pests (Chen *et al.*, 2011). To control these pests, researchers have developed genetically engineered (GE) rice lines that produce insecticidal Cry proteins derived from *Bacillus thuringiensis* (*Bt*) (Cohen *et al.*, 2008; Li *et al.*, 2016; Liu *et al.*, 2016). Before a *Bt* rice line can be cultivated, the risks to the environment must be assessed. This includes the evaluation of potential adverse effects on valued nontarget arthropods (NTAs) and the ecosystem services they provide (Devos *et al.*, 2015; Romeis *et al.*, 2008). This premarket NTA risk assessment requires information about which species live in the receiving environment and which species are most likely exposed to the Cry proteins (Romeis *et al.*, 2013, 2014; Todd *et al.*, 2008). We have therefore conducted a comprehensive literature search to identify (i) the taxonomic groups and species of aboveground arthropods present in the rice-growing regions of Central and Southern China, and (ii) the known food web interactions of those species. Furthermore, we have (iii) collected arthropods in a field experiment with *Bt* rice to assess the level of plant-produced Cry2A-protein to which the NTA species are exposed. Based on this information surrogate species for laboratory toxicity studies to support the regulatory risk assessment of *Bt* rice are suggested.

Results and discussion

Aboveground rice arthropods

The literature search identified 201 publications that contained relevant information on aboveground arthropods present in rice fields in Central and Southern China. From those publications, a total of 3266 records were retrieved for 930 arthropod species belonging to 14 taxonomic orders and four functional groups (Figure 1; Table S1). Of the 930 species, 23.7% were represented by herbivores (26.3% of records), 49.0% by predators (45.9% of records), 26.5% by parasitoids (27.5% of records) and 0.9% by pollinators (0.3% of records). The group with most species and records was spiders (Araneae), which represented 34.1% of the species (and 34.0% of the records); all spiders are predators. The second most species-rich and abundant group was the Hymenoptera (25.6% of species, 26.6% of records), which contains parasitoids, predators and pollinators, followed by the Hemiptera and Coleoptera, which constituted 13.4% and 9.2% of the total number of arthropod species, and 14.8% and 8.4% of records, respectively. The species belonging to those two orders are either herbivores or predators. Orthoptera (5.5% of species) and Lepidoptera (4.8%) are herbivores. Diptera (4.4%) are mainly herbivores, while seven predator and 21 parasitoid species were identified. The remaining seven orders (Odonata, Thysanoptera, Mantodea, Neuroptera, Strepsiptera, Trichoptera, and Megaloptera) contained only 27 species in total, representing 2.9% of the total number of species (Figure 1). In general, the number of species is linked to the number of records of a particular group. Exception is the Lepidoptera that contained the most important rice pests for which a relatively high number of published records is available.

Rice arthropod food web

Published information on the trophic interactions of the most abundant rice arthropods (abundance was roughly estimated based on the number of published records, see Table S1) was used to construct a simplified food web (Figure 2; Table S2). Interactions with natural enemies have been reported for 26 herbivorous species from six orders.

Most reports concern the three most important Lepidoptera pests, that is, *Chilo suppressalis*, *Scirpophaga incertulas*, and

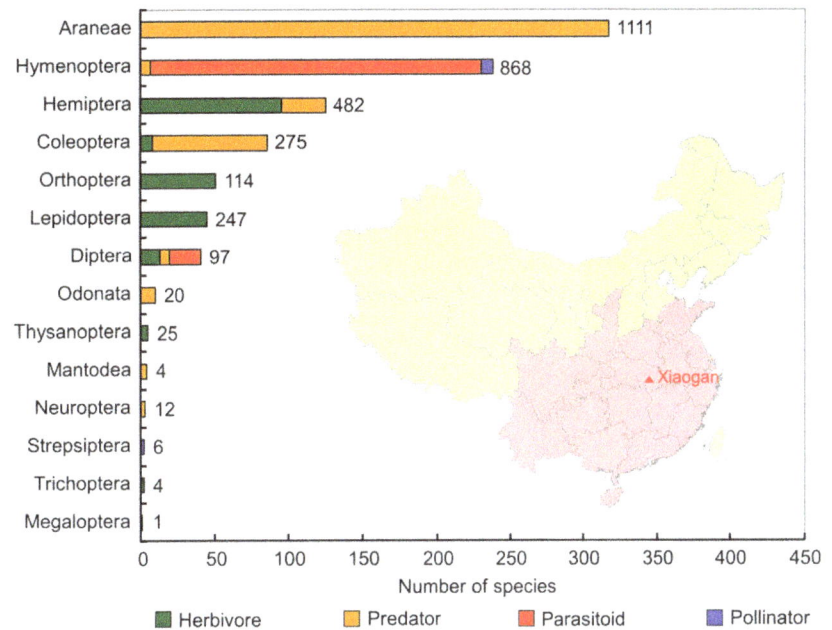

Figure 1 Numbers of arthropod species (sorted by order) recorded in the rice-planting regions of Central and Southern China (area indicated in red). Total numbers of records available in the published literature are indicated beside the bars. The red triangle represents Xiaogan where arthropods were collected in a Cry2A-transgenic rice field. Map has been adapted from http://d-maps.com/carte.php?num_car=17501&lang=en.

Cnaphalocrocis medinalis (all Crambidae), which are the targets of *Bt* rice, and the two most important Hemiptera pests, that is, *Sogatella furcifera* and *Nilaparvata lugens* (both Delphacidae).

According to the literature, the major predators of the lepidopteran rice pests belong to ten families of Araneae, and there are other predatory species from the Coleoptera, Hemiptera, and Neuroptera (Figure 2; Table S2). The parasitoids of lepidopteran rice pests are mainly from six families of Hymenoptera with a few species from two families of Diptera (i.e., Tachinidae and Sarcophagidae) (Figure 2; Table S2). In addition to the three major lepidopteran insect pest species, the Lepidoptera *Naranga aenescens* (Noctuidae), *Parnara guttata* (Hesperiidae), *Mycalesis gotama* (Nymphalidae), and *Pseudaletia separata* (Noctuidae) are also recorded as rice pests. Because they do not cause substantial rice losses, however, they are rarely investigated, and little information is available regarding their natural enemies (Table S2).

The natural enemies of the two major hemipteran pests, *S. furcifera* and *N. lugens*, have been extensively studied. Most predator species reported to attack hemipteran herbivores are the same species that attack lepidopteran herbivores (Figure 2; Table S2), because they are mainly generalists. The parasitoids of hemipteran pests belonging to the Delphacidae are mainly from the hymenopteran families Trichogrammatidae, Mymaridae, and Dryinidae (Figure 2). Similarly, hymenopteran parasitoids are known for the plant bug *Nezara viridula* (Pentatomidae) (Table S2). The thrips *Stenchaetothrips biformis* (Thysanoptera: Thripidae) is also a common rice pest in Southern China, and its predators are mainly from the Coleoptera and Hemiptera, while no parasitoid has been recorded (Figure 2; Table S2). Orthopteran species such as *Oxya chinensis* (Acrididae) are also commonly found in rice fields, and their predators mainly include species belonging to the Araneae, Coleoptera, and Mantodea (Table S2). *Oulema oryzae* (Chrysomelidae), an important Coleoptera pest in China, is attacked by coleopteran predators and hymenopteran parasitoids (Figure 2; Table S2). The major natural enemies of dipteran pests are hymenopteran parasitoids (Table S2).

Exposure of arthropods to Cry2A produced by *Bt* rice

To assess the level at which arthropods are exposed to Cry proteins in *Bt* rice fields, a replicated field experiment was conducted near Xiaogan (Hubei Province, China) in the years 2011 and 2012.

The concentrations of Cry2A detected in rice tissues collected in 2011 and 2012 were similar (Table S3). Rice leaves contained the highest concentrations of Cry2A (from 54 to 115 μg/g DW), followed by rice pollen (from 33 to 46 μg/g DW). The stems contained the lowest concentrations (from 22 to 32 μg/g DW).

Different sampling techniques (including suction sampling, beating sheet and visual searching) were used to collect the 29 most frequently encountered plant-dwelling arthropod species in the *Bt* and control rice plots during and after anthesis in 2011 and before, during and after anthesis in 2012. The highest measured concentrations of Cry2A in the collected arthropods at any of the sampling dates are indicated in Figure 2.

A total of 13 nontarget herbivores from 11 families belonging to the Hemiptera, Orthoptera, Diptera, and Thysanoptera were collected and analysed (Figure 2; Table S3). In the order Hemiptera adults of *S. furcifera* and nymphs and adults of *N. lugens* contained trace amounts of Cry2A (<0.06 μg/g DW) while the protein was not detected in other species. In contrast, larger amounts of Cry2A (from 0.15 to 50.7 μg/g DW) were detected in all but one sample of the Diptera, Thysanoptera, and Orthoptera. The thrips *S. biformis* contained the highest concentrations of Cry2A of all collected arthropods, which were close to the concentrations in the rice tissues. During anthesis, *S. biformis* contained Cry2A at 51 μg/g DW, which was higher than the concentration in specimens collected before anthesis (35 μg/g DW). Similarly, the protein level in *Agromyza* sp. (Diptera: Agromyzidae) was >2 times higher in samples collected during rice anthesis than before or after anthesis (Table S3). In contrast, the level in *Euconocephalus thunbergii* (Orthoptera: Tettigoniidae) was almost 2.5 times higher in samples collected after anthesis than during anthesis.

Target herbivores

Lepidoptera
Crambidae: *Chilo suppressalis*[1]; *Scirpophaga incertulas*[1]; *Cnaphalocrocis medinalis*[1]
Noctuidae: *Sesamia inferens*[1], *Naranga aenescens*[1]
Hesperiidae: *Parnara guttata*[1]
Nymphalidae: *Mycalesis gotama*[1]

Nontarget herbivores

Hemiptera
Delphacidae: *Sogatella furcifera*[1] ●; *Nilaparvata lugens*[1] ●; *Laodelphax striatellus*[1]
Cicadellidae: *Nephotettix cincticeps*[1] ●; *Thaia subrufa*[1]; *Recilia dorsalis*[1]
Aphididae: *Rhopalosiphum padi*
Pentatomidae: *Nezara viridula*[1]
Coreidae: *Cletus punctiger*

Orthoptera
Acrididae: *Oxya chinensis*[1]
Tetrigidae: *Eucriotettix oculatus* ●
Tettigoniidae: *Euconocephalus thunbergii* ●

Thysanoptera
Thripidae: *Stenchaetothrips biformis*[1] ●

Diptera
Ephydridae: *Hydrellia griseola* ●
Cecidomyiidae: *Orseolia oryzae*[1]
Chloropidae: *Chlorops* sp. ●
Agromyzidae: *Agromyza* sp. ●

Coleoptera
Chrysomelidae: *Oulema oryzae*[1]
Erirhinidae: *Lissorhoptrus oryzophilus*[1]

Predators

Araneae
Lycosidae[1] ● ●
Linyphiidae[1] ●
Araneidae[1]
Theridiidae[1] ●
Oxyopidae[1]
Tetragnathidae[1]
Salticidae[1] ● ●
Thomisidae[1]
Clubionidae[1]

Coleoptera
Staphylinidae[1] ●
Carabidae[1]
Coccinellidae[1] ●

Hemiptera
Miridae[1] ●
Reduviidae ●
Veliidae[1]

Neuroptera
Chrysopidae ●

Odonata
Coenagrionidae ●

Parasitoids

Hymenoptera
Braconidae[1]
Chalcididae[1]
Trichogrammatidae[1]
Scelionidae
Ichneumonidae[1]
Eulophidae[1]
Mymaridae[1]
Dryinidae[1]

Diptera
Tachinidae

Bt rice plant
Leaves, stems, pollen ●

Mean Cry2A content (μg/g dry weight)
○ <LOD
● <0.1
● 0.1 – 1
● 1 – 10
● 10 – 100

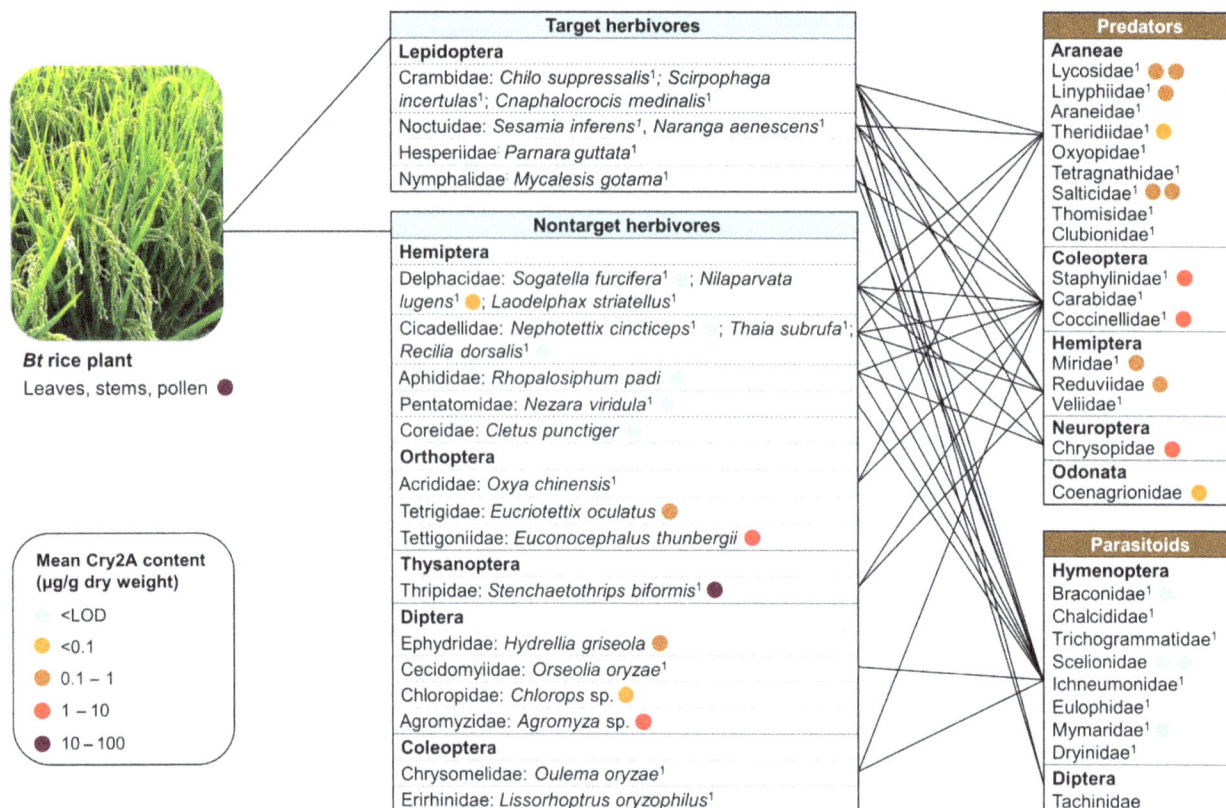

Figure 2 Simplified arthropod food web of rice in Central and Southern China and the content of plant-derived Cry2A protein in each arthropod species (other than target herbivores). Species are grouped into orders and families. Lines indicate a verified interaction between different families reported in the literature (details are provided in Table S2). For herbivores, only species are listed for which ≥10 records were available ([1]) or that were collected in the field experiment; the latter were analysed by ELISA (coloured circles). For natural enemies, families are listed that contain species with ≥10 records ([1]) or that were represented by species that were collected in the field experiment; the latter were analysed by ELISA (coloured circles). The coloured circles following arthropod species or families indicate the highest Cry2A content measured at any of the sampling dates (one circle represents one species). <LOD = below the limit of detection. Taxa not followed by a coloured circle are commonly reported in the literature but have not been collected in the field experiment.

Cry2A was estimated in a total of 12 predatory arthropod species belonging to 10 families in five orders (Table S3). Araneae represent the most abundant aboveground predators in the rice ecosystem, and six species from four families were analysed. While spiders collected before or after anthesis generally did not contain measurable amounts of Cry2A, the protein was detected in spiders collected during anthesis at concentrations ranging from 0.02 to 0.57 μg/g DW; the concentrations in spiders were two to three orders of magnitude lower than those in plant tissues (Figure 2; Table S3). All samples of hemipteran species contained Cry2A protein at comparable levels (concentrations ranged from 0.05 to 0.34 μg/g DW). No Cry2A protein was detected in samples of predatory beetles (Coccinellidae and Staphylinidae) collected before rice anthesis. However, the beetles and lacewings (Neuroptera) contained significant amounts during anthesis ranging from 0.20 to 2.60 μg/g DW. These concentrations were one to two orders of magnitude lower than those in plant tissues (Figure 2; Table S3). Adults of four species of parasitoids were collected from three families of Hymenoptera, but Cry2A levels were below the limit of detection (LOD) for all of them (Figure 2; Table S3).

Overall our data show a reduction in Cry protein concentrations from lower to higher trophic levels. This is in accordance with field studies from other *Bt*-transgenic crops producing different Cry proteins, including maize (Cry1Ab: Harwood *et al.*, 2005; Obrist *et al.*, 2006; Cry3Bb1: Meissle and Romeis, 2009), cotton (Cry1Ac: Torres *et al.*, 2006) and soybean (Cry1Ac: Yu *et al.*, 2014).

Implications for nontarget risk assessment

Risk assessments of *Bt* rice should focus on taxonomic and functional groups that are both common and highly exposed to the produced Cry proteins (Romeis *et al.*, 2013). These groups include the common predators in the Araneae, Coleoptera, Hemiptera, and Neuroptera, all of which contained significant amounts of Cry protein in our field experiment. Predators might be exposed to the Cry protein when they consume prey but also when they consume rice pollen or plant sap as a supplemental food source. Elevated Cry2A concentrations during anthesis indicate that several predatory species also consume rice pollen (Table S3). The following predators are abundant in Chinese rice fields, are exposed to plant-produced Cry protein, and are available and amenable for testing under controlled laboratory conditions: *Cyrtorhinus lividipennis* (Hemiptera: Miridae), *Chrysoperla nipponensis* (Neuroptera: Chrysopidae), *Propylea japonica* (Coleoptera: Coccinellidae), *Paederus fuscipes* (Coleoptera: Staphylinidae), and the two spiders *Pirata subpiraticus* (Araneae: Lycosidae) and *Ummeliata insecticeps* (Araneae: Linyphiidae) (Table 1). These thus represent a suitable set of test species for initial, early-tier risk assessment studies.

Table 1 Predatory arthropod species recommended as surrogate test species to support the environmental risk assessment of insecticidal GM rice

Species	Order: Family	Food and feeding mode	Studies demonstrating testability of the species
Cyrtorhinus lividipennis	Hemiptera: Miridae	Larvae and adults are predators on arthropod herbivores (mainly planthoppers); adults also suck sap of plant leaf or stem when prey is lacking; plant-dwelling	Han *et al.* (2014)
Chrysoperla nipponensis (syn.: *C. sinica*)	Neuroptera: Chrysopidae	Larvae feed on arthropods (mainly aphids) (piercing-sucking). Adults feed on pollen and nectar. Both stages are plant-dwelling	Li *et al.* (2014a,b)
Propylea japonica	Coleoptera: Coccinellidae	Both larvae and adults feed on arthropods (mainly aphids/planthoppers), and consume pollen during plant anthesis; plant-dwelling	Zhang *et al.* (2014); Li *et al.* (2015)
Paederus fuscipes	Coleoptera: Staphylinidae	Larvae and adults feed on arthropods; soil- and plant-dwelling	Cheng *et al.* (2014)
Pirata subpiraticus	Araneae: Lycosidae	Larvae and adults feed on arthropods; soil-dwelling	Chen *et al.* (2009)
Ummeliata insecticeps	Aranea: Linyphiidae	Larvae and adults feed on arthropods; web building	Tian *et al.* (2010)

Materials and methods

Identification of the aboveground arthropods in rice

Bt rice could be planted in the rice-growing regions of Central and Southern China. To identify the taxonomic groups and species of aboveground arthropods in those regions, literature searches were conducted. The Web-of-Science Core Collection (Thomson Reuters, New York, NY) and the China National Knowledge Infrastructure (CNKI) (Tsinghua Tongfang Knowledge Network Technology Co., Ltd., Beijing, China) were used to cover both international and Chinese peer-reviewed literature. The following broad search terms were used for retrieving the references from both sources: *rice* and (*arthropod* or *invertebrate* or *predator* or *parasitoid* or *insect*). The collected references were manually screened, and references that fulfilled one of the following criteria were excluded from further analyses: (i) the reference did not contain original data; (ii) the data presented were not from a field survey in rice; (iii) the data were collected in regions other than Central and Southern China (area shaded in red in Figure 1); (iv) the reference contained no data at the species level, or (v) the reference was published in a noncore Chinese academic journal of uncertain quality (high potential for erroneous species identification). The remaining references were used to compile the list of rice arthropods. The total number of records of each species was used as a rough indicator for its abundance in rice fields (Meissle *et al.*, 2012; Romeis *et al.*, 2014). For each arthropod species, the scientific name and the taxonomic classification (family, order) were confirmed using the Catalogue of Life (http://www.catalogueof life.org/). For species not included in the Catalogue of Life, databases including the Global Biodiversity Information facility (http://www.gbif.org/), Insektoid.info (http://insektoid.info), or other specialised sources were consulted (Table S1). Species belonging to two functional groups, for example, Syrphidae with predatory larvae and pollinating adults, were assigned to the function rated as most important (e.g., Syrphidae were considered predators).

Establishment of the arthropod food web

NTAs may be exposed to the plant-produced insecticidal proteins through various routes, but mainly via consumption of GE plant tissues or by predation or parasitism of herbivores of the GE crop (Raybould *et al.*, 2007; Romeis *et al.*, 2009). To illustrate the routes by which natural enemies may be exposed to insecticidal proteins in GE rice fields in China, a simplified arthropod food web was constructed. This food web included the most abundant species (species for which ≥10 records were available; Table S1) based on reported trophic links between herbivores and natural enemies from the retrieved literature (Figure 2; Table S2).

Exposure of rice arthropods to Cry2A produced by *Bt* rice

Experimental *Bt* rice field

In 2011 and 2012, a *Bt* rice line (T2A-1) expressing a modified *cry2A* gene under the control of the maize ubiquitin promoter (Chen *et al.*, 2005) and the corresponding nontransformed near-isoline Minghui 63 (hereafter referred to as 'control rice') were grown in an experimental field in a suburb of Xiaogan City (29.25°N, 108.21°E) in Hubei Province, China (Figure 1). This *Bt* rice line was selected for the field experiment as it contains high concentrations of Cry protein when compared to other *Bt* rice lines (Liu *et al.*, 2016). The field was divided into eight plots, and each plot was approximately 109 m^2. Plots were separated by a nonplanted buffer of 1 m, and the whole field was surrounded by a 1-m buffer of conventional rice. Both *Bt* rice and control rice were planted in four plots arranged in a randomised block design. The rice seeds were sown in a seeding bed on 25 May 2011 and on 17 May 2012, and the seedlings were transplanted to the experimental plots at the four-leaf stage (23 June 2011 and 17 June 2012). The plots were cultivated according to the common local agricultural practices, but no insecticide was applied.

Collection of arthropods

For determination of the Cry2A concentrations in rice arthropods, the most frequent aboveground arthropod species were collected in the four *Bt* and four control rice plots during (August) and after (September) anthesis in 2011 and before (June), during (August) and after (September) anthesis in 2012. Arthropods were collected by suction sampling, by hand (visual collection), or by using a beating sheet. Plant tissue was also collected on each sampling date; leaf tissue and pollen were collected in 2011, and leaf tissue, pollen and rice stems were collected in 2012. The collected arthropods and plant tissues were immediately placed in a portable refrigerator and transported to the laboratory. After taxonomic identification, individuals (1 to more than 100 depending on size and availability) of the different arthropod species collected on the same date in the same plot were pooled as one replicate and placed in a 5-mL centrifuge tube. All samples were stored at −80°C before ELISA measurements.

ELISA analyses

The concentration of Cry2A was quantified by double-antibody sandwich enzyme-linked immunosorbent assays (DAS-ELISA) using QuantiPlate Kits from EnviroLogix (Portland, ME). Arthropods collected in the field were washed with deionised water to minimise contamination by debris and pollen. All samples of arthropods, rice leaf tissue, stem tissue and pollen were lyophilised before being weighed on an electronic balance (CPA224S, Sartorius, Göttingen, Germany; accuracy = 0.1 ± 0.1 mg). Phosphate-buffered saline with Tween (PBST) at a ratio of 1:30 to 1:100 mg DW/mL buffer was added to the samples. If sample DW was <3 mg, 300 µL buffer was used (Meissle and Romeis, 2009). For maceration, two 3-mm tungsten carbide balls were added to each sample, and the samples were shaken for 3 min at 30 Hz in a MM400 mixer mill (Retsch, Haan, Germany) fitted with 24-tube adapters for microreaction tubes. After centrifugation at 15 800 *g*, the supernatants were diluted with PBST according to the expected Cry2A concentration. Antibody-coated plates were loaded with enzyme conjugate and Cry2A standards provided with the kit, negative controls (buffer only) and the diluted sample extracts. After the plates were incubated for 1 h under ambient conditions, they were washed four times with PBST, and the provided substrate solution was added. After 30 min of incubation at ambient temperature, 100 µL of stop solution (1.0 N hydrochloric acid) was added per well. After 15 min of incubation, the optical density (OD) was measured at a light wavelength of 450 nm with a microplate spectrophotometer (PowerWave XS2; BioTek, Winooski, VT, USA). Cry2A concentrations (µg/g DW) were calculated using regression analysis. For the clear separation of positive readings from controls, the LOD of the test was determined based on the standard deviation of the OD values of buffer-only controls multiplied by three (Meissle and Romeis, 2009). Subsequently, the detection limit of each sample was calculated from the dilution, sample weight and amount of added buffer.

Because arthropod and plant tissue samples from the control rice treatment in the field did not show OD values systematically different from those of buffer-only controls, no cross-reaction of arthropod proteins with ELISA was apparent.

Acknowledgements

We thank Dr. Hongbo Qiao (Henan Agricultural University) for help in making the map of China in Figure 1. The study was supported by the National GMO New Variety Breeding Program of PRC (2014ZX08011-02B and 2016ZX08011-001).

References

Chen, H., Tang, W., Xu, C., Li, X., Lin, Y. and Zhang, Q. (2005) Transgenic indica rice plants harboring a synthetic *cry2A** gene of *Bacillus thuringiensis* exhibit enhanced resistance against lepidopteran rice pests. *Theor. Appl. Genet.* **111**, 1330–1337.

Chen, M., Ye, G.-Y., Liu, Z.-C., Fang, Q., Hu, C., Peng, Y.-F. and Shelton, A.M. (2009) Analysis of Cry1Ab toxin bioaccumulation in a food chain of Bt rice, an herbivore and a predator. *Ecotoxicology* **18**, 230–238.

Chen, M., Shelton, A. and Ye, G.Y. (2011) Insect-resistant genetically modified rice in China: from research to commercialization. *Annu. Rev. Entomol.* **56**, 81–101.

Cheng, Z.X., Huang, J.H., Liang, Y.Y., Cheng, S.D., Xiong, H.B., Chen, N.P. and Hu, S.X. (2014) Effect of transgenic Bt rice on the survival rate and predation of *Paederus fuscipes* Curtis adults. *Chinese J. Appl. Entomol.* **51**, 1184–1189.

Cohen, M.B., Chen, M., Bentur, J.S., Heong, K.L. and Ye, G. (2008) *Bt* rice in Asia: potential benefits, impact, and sustainability. In *Integration of Insect-Resistant Genetically Modified Crops within IPM Programs* (Romeis, J., Shelton, A.M. and Kennedy, G.G., eds), pp. 223–248. Dordrecht: Springer Science+Business Media B.V.

Devos, Y., Romeis, J., Luttik, R., Maggiore, A., Perry, J.N., Schoonjans, R., Streissl, F. *et al.* (2015) Optimising environmental risk assessments. Accounting for ecosystem services helps to translate broad policy protection goals into specific operational ones for environmental risk assessments. *EMBO Rep.* **16**, 1060–1063.

Han, Y., Meng, J., Chen, J., Cai, W., Wang, Y., Zhao, J., He, Y. *et al.* (2014) *Bt* rice expressing Cry2Aa does not harm *Cyrtorhinus lividipennis*, a main predator of the nontarget herbivore *Nilapavarta lugens*. *PLoS One*, **9**, e112315.

Harwood, J.D., Wallin, W.G. and Obrycki, J.J. (2005) Uptake of Bt endotoxins by nontarget herbivores and higher order arthropod predators: molecular evidence from a transgenic corn agroecosystem. *Mol. Ecol.* **14**, 2815–2823.

Li, Y., Hu, L., Romeis, J., Wang, Y., Han, L., Chen, X. and Peng, Y. (2014a) Use of an artificial diet system to study the toxicity of gut-active insecticidal compounds on larvae of the green lacewing *Chrysoperla sinica. Biol. Control*, **69**, 45–51.

Li, Y., Hu, L., Romeis, J., Chen, X. and Peng, Y. (2014b) Bt rice producing Cry1C protein does not have direct detrimental effects on the green lacewing *Chrysoperla sinica* (Tjeder). *Environ. Toxicol. Chem.* **33**, 1391–1397.

Li, Y., Zhang, X., Chen, X., Romeis, J., Yin, X. and Peng, Y. (2015) Consumption of Bt rice pollen containing Cry1C or Cry2A does not pose a risk to *Propylea japonica* (Thunberg) (Coleoptera: Coccinellidae). *Sci. Rep.* **5**, 7679.

Li, Y., Hallerman, E.M., Liu, Q., Wu, K. and Peng, Y. (2016) The development and status of *Bt* rice in China. *Plant Biotech. J.* **10**, 839–848.

Liu, Q., Hallerman, E., Peng, Y. and Li, Y. (2016) Development of *Bt* rice and *Bt* maize in China and their efficacy in target pest control. *Int. J. Mol. Sci.* **17**, 1561.

Meissle, M. and Romeis, J. (2009) The web-building spider *Theridion impressum* (Araneae: Theridiidae) is not adversely affected by *Bt* maize resistant to corn rootworms. *Plant Biotech. J.* **70**, 645–656.

Meissle, M., Álvarez-Alfageme, F., Malone, L.A. and Romeis, J. (2012) *Establishing a database of bio-ecological information on non-target arthropod species to support the environmental risk assessment of genetically modified crops in the EU*. Supporting Publications 2012: EN-334. Parma: European Food Safety Authority (EFSA). http://www.efsa.europa.eu/en/supporting/pub/334e.htm

Obrist, L.B., Dutton, A., Albajes, R. and Bigler, F. (2006) Exposure of arthropod predators to Cry1Ab toxin in Bt maize fields. *Ecol. Entomol.* **31**, 143–154.

Raybould, A., Stacey, D., Vlachos, D., Graser, G., Li, X. and Joseph, R. (2007) Non-target organisms risk assessment of MIR604 maize expressing mCry3A for control of corn rootworms. *J. Appl. Entomol.* **131**, 391–399.

Romeis, J., Bartsch, D., Bigler, F., Candolfi, M.P., Gielkens, M.M.C., Hartley, S.E., Hellmich, R.L. *et al.* (2008) Assessment of risk of insect-resistant transgenic crops to non-target arthropods. *Nat. Biotech.* **26**, 203–208.

Romeis, J., Meissle, M., Raybould, A. and Hellmich, R.L. (2009) Impact of insect-resistant transgenic crops on above-ground non-target arthropods. In *Environmental Impact of Genetically Modified Crops* (Ferry, N. and Gatehouse, A.M.R., eds), pp. 165–198. Wallingford: CABI Publishing.

Romeis, J., Raybould, A., Bigler, F., Candolfi, M.P., Hellmich, R.L., Huesing, J.E. and Shelton, A.M. (2013) Deriving criteria to select arthropod species for laboratory tests to assess the ecological risks from cultivating arthropod-resistant transgenic crops. *Chemosphere*, **90**, 901–909.

Romeis, J., Meissle, M., Álvarez-Alfageme, F., Bigler, F., Bohan, D.A., Devos, Y., Malone, L.A. et al. (2014) Potential use of an arthropod database to support the non-target risk assessment and monitoring of transgenic plants. Transgenic Res. 23, 995–1013.

Tian, J.C., Liu, Z.C., Chen, M., Chen, Y., Chen, X.X., Peng, Y.F., Hu, C. et al. (2010) Laboratory and field assessments of prey-mediated effects of transgenic Bt rice on Ummeliata insecticeps (Araneida: Linyphiidae). Environ. Entomol. 39, 1369–1377.

Todd, J.H., Ramankutty, P., Barraclough, E.I. and Malone, L.A. (2008) A screening method for prioritizing non-target invertebrates for improved biosafety testing of transgenic crops. Environ. Biosafety Res. 7, 35–56.

Torres, J.B., Ruberson, J.R. and Adang, M.J. (2006) Expression of Bacillus thuringiensis Cry1Ac protein in cotton plants, acquisition by pests and predators: a tritrophic analysis. Agric. Forest Entomol. 8, 191–202.

Yu, H., Romeis, J., Li, Y., Li, X. and Wu, K. (2014) Acquisition of Cry1Ac protein by non-target arthropods in Bt soybean fields. PLoS One, 9, e103973.

Zhang, X., Li, Y., Romeis, J., Yin, X., Wu, K. and Peng, Y. (2014) Use of a pollen-based diet to expose the ladybird beetle Propylea japonica to insecticidal proteins. PLoS One 9, e85395.

10

OsCESA9 conserved-site mutation leads to largely enhanced plant lodging resistance and biomass enzymatic saccharification by reducing cellulose DP and crystallinity in rice

Fengcheng Li[1,2,3,4,†], Guosheng Xie[1,2,3,†], Jiangfeng Huang[1,2,3], Ran Zhang[1,2,3], Yu Li[1,2,3], Miaomiao Zhang[1,2,5], Yanting Wang[1,2,3], Ao Li[1,2,3], Xukai Li[1,2,3], Tao Xia[1,2,5], Chengcheng Qu[6], Fan Hu[7,8], Arthur J. Ragauskas[7,8] and Liangcai Peng[1,2,3,*]

[1]Biomass and Bioenergy Research Centre, Huazhong Agricultural University, Wuhan, China

[2]National Key Laboratory of Crop Genetic Improvement, Huazhong Agricultural University, Wuhan, China

[3]College of Plant Science and Technology, Huazhong Agricultural University, Wuhan, China

[4]Key Laboratory of Crop Physiology, Ecology, Genetics and Breeding, Ministry of Agriculture, Rice Research Institute, Shenyang Agricultural University, Shenyang, China

[5]College of Life Science and Technology, Huazhong Agricultural University, Wuhan, China

[6]State Key Laboratory of Agricultural Microbiology, Huazhong Agricultural University, Wuhan, China

[7]Department of Chemical and Biomolecular Engineering, The University of Tennessee- Knoxville, Knoxville, TN, USA

[8]Department of Forestry, The University of Tennessee-Knoxville, Knoxville, TN, USA

*Correspondence

emails lpeng@mail.hzau.edu.cn;
pengliangcai2007@sina.com
[†]These authors contributed equally.

Keywords: biomass saccharification, cellulose, CESA, lodging resistance, rice.

Summary

Genetic modification of plant cell walls has been posed to reduce lignocellulose recalcitrance for enhancing biomass saccharification. Since cellulose synthase (CESA) gene was first identified, several dozen *CESA* mutants have been reported, but almost all mutants exhibit the defective phenotypes in plant growth and development. In this study, the rice (*Oryza sativa*) *Osfc16* mutant with substitutions (W481C, P482S) at P-CR conserved site in CESA9 shows a slightly affected plant growth and higher biomass yield by 25%–41% compared with wild type (Nipponbare, a *japonica* variety). Chemical and ultrastructural analyses indicate that *Osfc16* has a significantly reduced cellulose crystallinity (CrI) and thinner secondary cell walls compared with wild type. CESA co-IP detection, together with implementations of a proteasome inhibitor (MG132) and two distinct cellulose inhibitors (Calcofluor, CGA), shows that CESA9 mutation could affect integrity of CESA4/7/9 complexes, which may lead to rapid CESA proteasome degradation for low-DP cellulose biosynthesis. These may reduce cellulose CrI, which improves plant lodging resistance, a major and integrated agronomic trait on plant growth and grain production, and enhances biomass enzymatic saccharification by up to 2.3-fold and ethanol productivity by 34%–42%. This study has for the first time reported a direct modification for the low-DP cellulose production that has broad applications in biomass industries.

Introduction

Cellulose is the most abundant biomass convertible for biofuels and chemical products. As a principal component of plant cell walls, cellulose plays a central role in plant mechanical strength and morphogenesis (Somerville, 2006), but its features determine lignocellulose recalcitrance, leading to a costly biomass process (Himmel *et al.*, 2007; Pauly and Keegstra, 2008). To reduce recalcitrance, genetic modifications of wall polymers (hemicelluloses and lignin) have been applied to enhance biomass saccharification (Bonawitz *et al.*, 2014; Chen and Dixon, 2007; Chiniquy *et al.*, 2012; Ding *et al.*, 2012; Li *et al.*, 2015; Wilkerson *et al.*, 2014), but little has been reported about a direct alteration of cellulose in plants (Burton and Fincher, 2014).

Cellulose consists of β-1,4-linked glucan chains that form microfibrils by intra- and intermolecular hydrogen bonds. The formed hydrogen bonds significantly determine cellulose crystallinity, which is reportedly a key parameter negatively affecting biomass digestibility (Harris *et al.*, 2012; Li *et al.*, 2013; Zhang

et al., 2013). The crystallinity index (CrI) has been broadly used to account for cellulose crystallinity and could be detected by X-ray diffraction (XRD) patterns (Segal *et al.*, 1959). Besides cellulose crystallinity, the degree of polymerization (DP) of crystalline cellulose is also regarded as an important cellulose feature (Zhang *et al.*, 2013). Recent reports have indicated that cellulose CrI is positively correlated with its DP in *Miscanthus* samples (Zhang *et al.*, 2013), and both cellulose features (CrI, DP) are the main factors that could negatively affect either plant lodging resistance or biomass enzymatic saccharification in plants (Li *et al.*, 2015; Zhang *et al.*, 2013). However, it remains largely unknown how cellulose biosynthesis process determines the cellulose features in plants.

In higher plants, cellulose is synthesized at the plasma membrane by cellulose synthase (CESA) enzymes that are organized into cellulose synthase complexes (CSCs) (Taylor *et al.*, 2003). Since the first higher plant cellulose synthase gene was cloned from cotton in 1996 (Pear *et al.*, 1996), the CESA superfamily has been characterized with eight transmembrane domains and a central cytoplasmic domain with D,D,D,QXXRW motif. The central cytoplasmic domain contains the plant-conserved region (P-CR) and class-specific region (CSR), which may play a role in CESA protein association and assembly (Olek et al., 2014; Sethaphong et al., 2013). To dissect CESA biological functions, more than fifty distinct CESA mutants have been identified in different plant species through multiple genetic approaches (Table S1). Nevertheless, almost all mutants exhibit markedly reduced cellulose and defective growth phenotypes, and several mutants are examined with low cellulose crystallinity for high biomass enzymatic digestibility (Table S1). To our knowledge, however, little is yet reported about cellulose DP alteration from the mutants. Furthermore, the homologous and heterologous overexpression of CESA genes could not enhance cellulose products but did affect plant growth in transgenic plants (Table S1). Exceptionally, the recent rice bc13 mutant with one amino acid alteration in CESA9 showed normal plant growth and cadmium tolerance, despite a reduction in cellulose (Song et al., 2013).

Rice is a major food crop over the world with enormous biomass residues for biofuels and chemical products. In this study, we identified a novel rice CESA9 allele Osfc16 that showed a normal plant growth and high biomass production. Mutation of the CESA9 protein reduced two cellulose features (CrI, DP), leading to improved plant lodging resistance and enhanced biomass enzymatic saccharification. Further analysis revealed that the P-CR region mutation of CESA9 protein could affect stability

of secondary wall CSCs, which may early terminate the CSC track in the plasma membrane resulting in low-DP cellulose synthesis.

Results

CESA9 conserved-site mutation and improved agronomic traits in Osfc16

Using map-based cloning approach, the rice Osfc16 mutant was identified as a single recessive gene, which encodes the CESA9 protein with two amino acid substitutions ($W^{481}P^{482}GN \rightarrow C^{481}S^{482}GN$) in the site of P-CR region (Figure 1a). In particular, the substituted amino acids (Trp and Pro) are fully conserved in all CESA family proteins of the eight plant species examined (Figure S1). Although several dozens of CESA mutants and overexpressed transgenic plants have been previously identified with remarkably defective phenotypes in different plant species (Table S1), the Osfc16 mutant exhibited a normal plant growth as observed in wild type (Nipponbare (NPB), a japonica variety) (Figure 1b). In 3-year (2012–2014) independent field experiments, the Osfc16 mutant maintained grain yields (dry spike) similar to wild type (Figure 1c and Table S2). Notably, despite the relatively short height (Figure 1d), the Osfc16 mutant had significantly improved plant lodging resistance (lodging index reduced by 18%–24%) and enhanced biomass production (dry straw increased by 25%–41%), compared with wild type (Figure 1e,f and Table S2). In particular, tillers numbers (tillers/

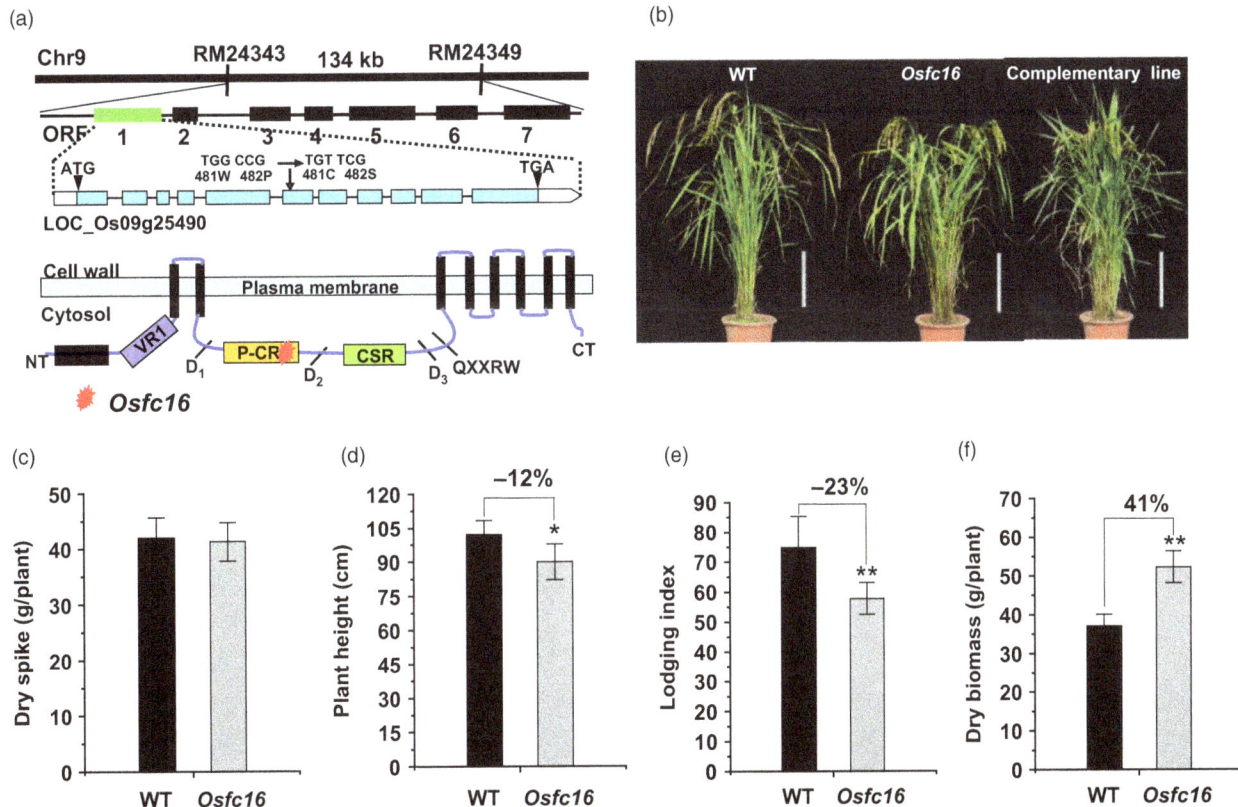

Figure 1 Osfc16 mutant identification and agronomic trait observation. (a) Location of Osfc16 mutation with substitutions of Trp and Pro residues with Cys and Ser at the 481 and 482 position of the CESA9 protein. (b) Plant growth in wild-type (WT), Osfc16 mutant and complementary line (scale bar = 20 cm). (c) Dry spike. (d) Plant height. (e) Lodging index. (f) Dry biomass. * and ** indicate significant differences between WT and Osfc16 mutant by t-test at P < 0.05 and 0.01, respectively, with the increased or decreased percentage (%) calculated by subtraction of the values between mutant and WT divided by WT. The error bar indicates SD values (n = 3).

plant) were much increased in the *Osfc16* mutant by 59%–68%, attributing for its higher biomass production (Table S2).

To verify the *Osfc16* mutation as the single recessive gene, the full-length cDNA of *CESA9* gene was expressed in the *Osfc16* mutant. As a result, the *Osfc16* mutant phenotype was fully complemented (Figure 1b), and the related major agronomic traits (lodging index and dry straw) were restored in three independent complementary transgenic lines at significant levels (Table S3).

Enhanced biomass saccharification and ethanol production in *Osfc16*

Using mature stem materials, we detected biomass enzymatic digestibility (saccharification) in the *Osfc16* mutant by calculating the hexose yields released from enzymatic hydrolysis of pretreated biomass (Figure 2a). The *Osfc16* mutant exhibited higher yields of hexoses by up to 2.3-fold than that of wild type, under pretreatments with three concentrations of alkali (0.5%, 1% and 4% NaOH) and acid (0.5%, 1% and 2% H_2SO_4) or upon enzymatic hydrolysis with three dosages of cellulase (3.5, 7 and 14 FPU/g cellulose) (Figure 2b,c; Figure S2; Table S4). Such large enhancements were confirmed by visualizations of more violent destruction of stem tissue *in situ* (Figure 2e) and of rougher biomass residue surfaces *in vitro* (Figure 2f) in the *Osfc16* mutant from 1% NaOH and 1% H_2SO_4 pretreatments and sequential enzymatic hydrolyses. Furthermore, the *Osfc16* mutant, compared with wild type, exhibited higher ethanol yields by 34%–42% obtained by yeast fermentation of the sugars released from biomass enzymatic hydrolysis of rice straw upon the mild chemical (7.5% CaO, 1% H_2SO_4) pretreatments (Figure 2d;

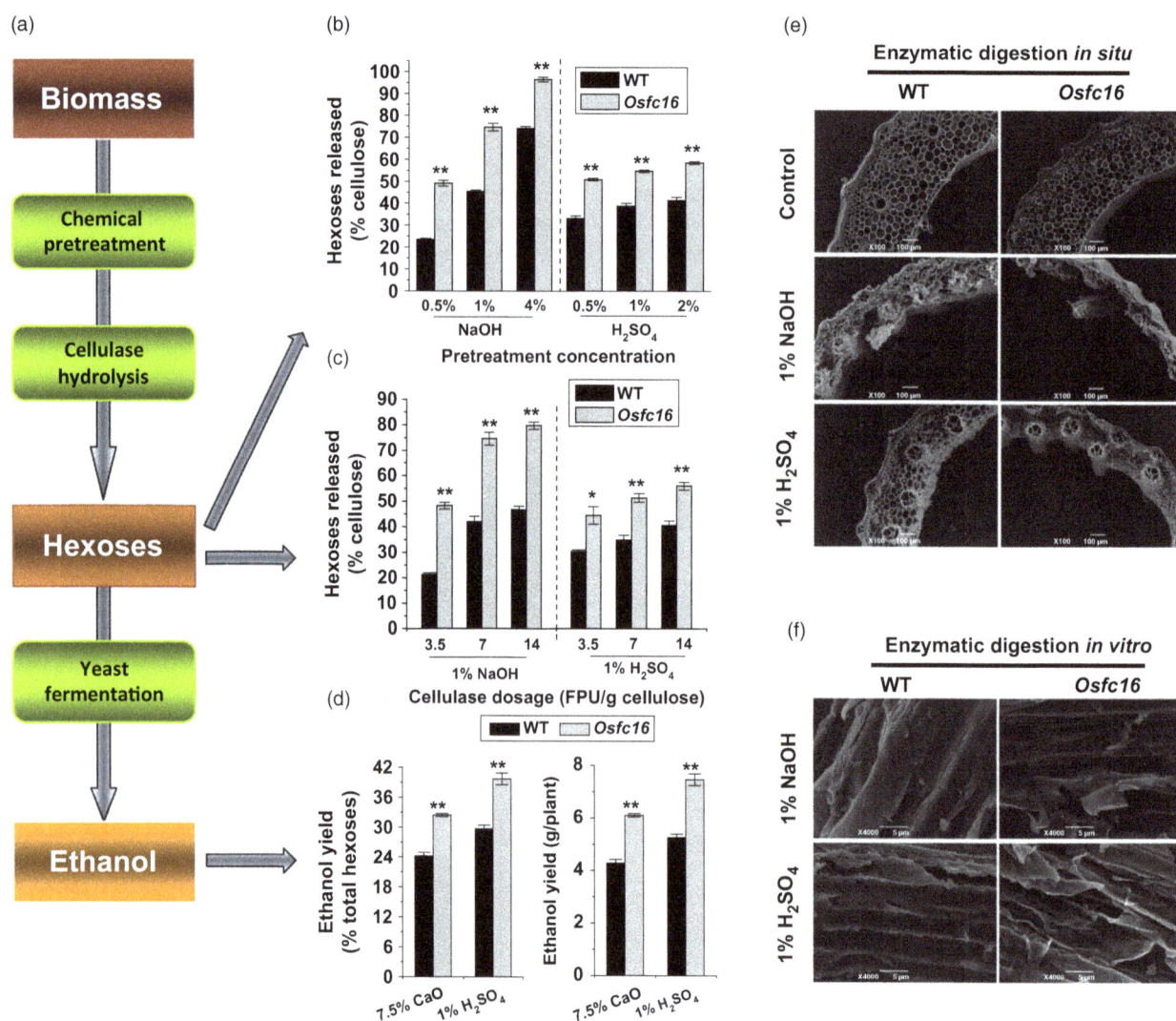

Figure 2 Biomass enzymatic saccharification and ethanol production. (a) Scheme for biomass enzymatic saccharification and ethanol yield. (b) Hexose yields released from enzymatic (mixed-cellulase) hydrolysis after pretreatment with NaOH and H_2SO_4 at three concentrations. (c) Hexose yields released from three dosages of mixed-cellulase hydrolysis after pretreatment with 1% NaOH and 1% H_2SO_4. (d) Ethanol yield obtained by yeast fermentation of the sugars from biomass enzymatic hydrolysis of the mature stems after pretreatment with 7.5% CaO or 1% H_2SO_4. Ethanol yield was expressed as either percentage of total hexoses in the biomass residues or ethanol yield per plant. (e) SEM images of *in situ* enzymatic digestion of stems at heading stage after 1% NaOH or 1% H_2SO_4 pretreatment and sequential enzymatic hydrolysis. (f) SEM images of *in vitro* enzymatic digestion of biomass residues released from enzymatic hydrolysis after 1% NaOH or 1% H_2SO_4 pretreatment. * and ** indicate significant differences between WT and *Osfc16* mutant by *t*-test at $P < 0.05$ and 0.01, respectively, and the error bar indicates SD values ($n = 3$).

Table S5). This study demonstrated that the CESA9 site mutation could lead to largely enhanced biomass saccharification and ethanol productivity in the *Osfc16* mutant.

Altered cell wall composition and structure in *Osfc16*

To understand the improved agronomic traits and enhanced biomass digestibility in *Osfc16* mutant, we examined its cell wall composition and structure. Besides relatively smaller-diameter stems (Figure 3a), the *Osfc16* mutant showed thinner secondary cell walls than wild type (Figure 3b). Chemical analysis indicated that the *Osfc16* mutant had reduced cellulose levels by 18% and increased hemicellulose levels by 16% with lignin level similar to wild type in the mature stems (Figure 3c). Furthermore, the *Osfc16* mutant did not show much difference from wild type in monosaccharide composition of hemicelluloses and three monomer constituents (G, S and H) of lignin (data not shown). In addition, the cell wall composition of *Osfc16* mutant could be fully restored in three independent complementary transgenic lines.

Reduced cellulose crystallinity in *Osfc16*

As a major cellulose feature, cellulose crystallinity has been characterized by determining crystalline index (CrI) of biomass samples (Li *et al.*, 2015; Xu *et al.*, 2012; Zhang *et al.*, 2013). Using four internodes of stems at heading stage of rice (Figure 4a), a standard development from primary to secondary cell walls (Xie *et al.*, 2013), the *Osfc16* mutant exhibited a significant reduction of the cellulose CrI in the second, third and fourth internodes by 3.9%, 7.8% and 23.4%, respectively, compared with wild type (Figure 4b). Notably, the *Osfc16* mutant had much lower CrI value than wild type by 36% in the mature stem that is rich in secondary cell walls (Figure 4b,c). Because

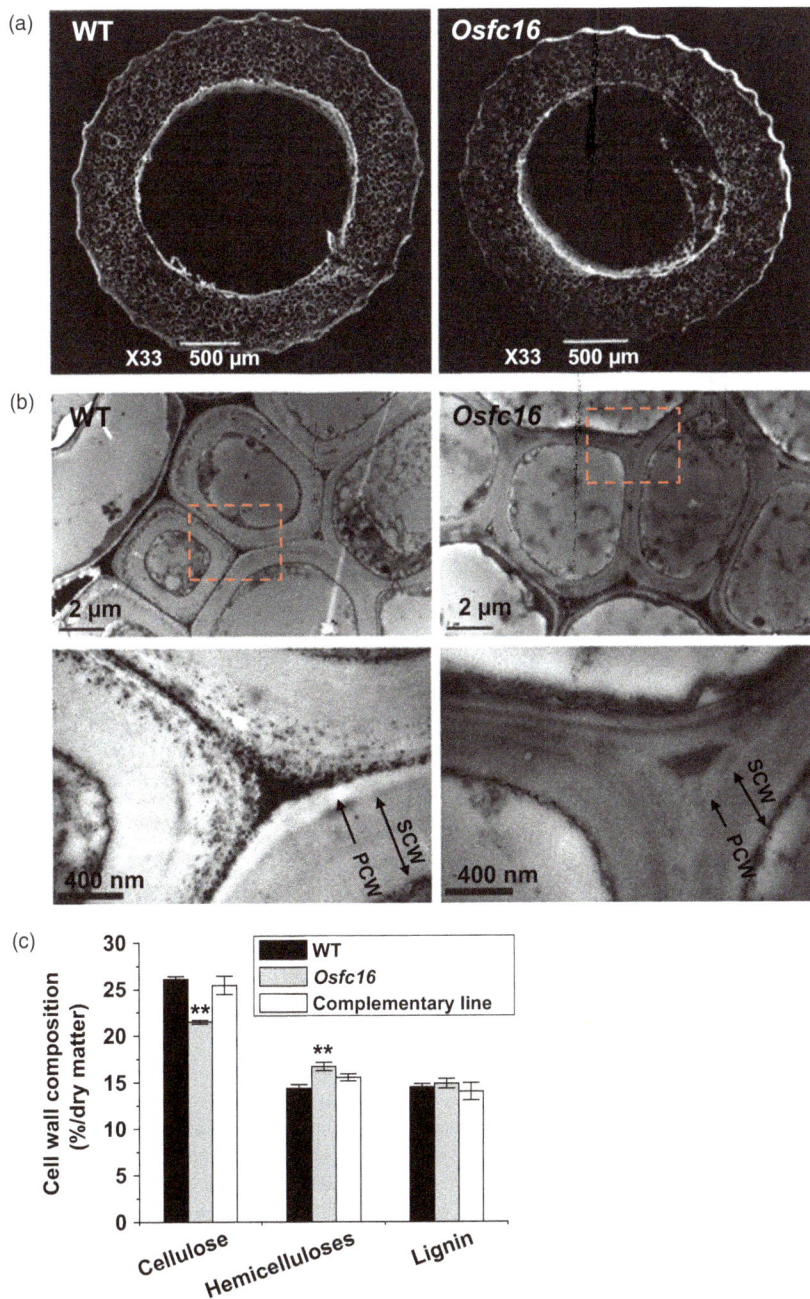

Figure 3 Observations of stem tissues and cell wall structures. (a) SEM images of the second-internode stem at the heading stage of rice. (b) TEM images of the sclerenchyma cell walls. PCW: primary cell wall. SCW: secondary cell wall. (c) Cell wall composition of mature stems. ** indicates significant differences between WT and *Osfc16* or complementary line by *t*-test at $P < 0.01$, and the error bar indicates SD values ($n = 3$).

slight different CrI values were detected between *Osfc16* and wild type in the second-internode stems that are predominately composed of primary cell walls, the data thus indicated that a major reduction of cellulose CrI occurred in the secondary cell walls of *Osfc16* mutant, consistent with its thinner secondary cell walls.

To further confirm the reduction of cellulose CrI in *Osfc16* mutant, we applied two distinct cellulose inhibitors (Calcofluor, CGA325′615-CGA) to treat with rice seedlings. While the germinated rice seeds were incubated with Calcofluor, an inhibitor of cellulose crystallization (Haigler *et al.*, 1980), the *Osfc16* mutant showed less retarded root growth than did the

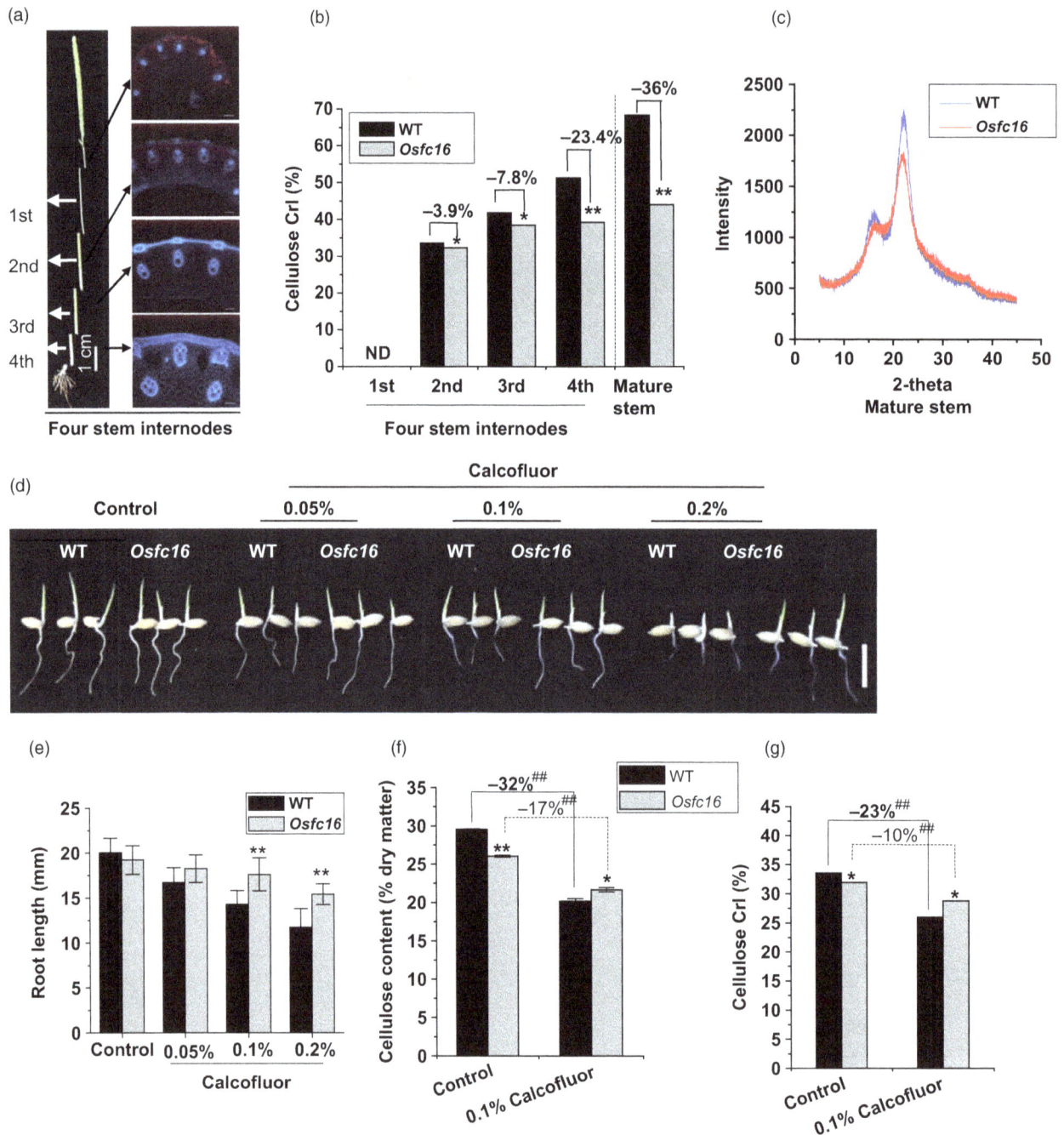

Figure 4 Detection of cellulose crystallinity. (a) Four-internode stems at heading stage used for CrI and DP detection. (b) Cellulose CrI of the four internodes and mature stems using the X-ray diffraction (XRD) method. (c) The XRD scanning patters applied for CrI calculation. (d, e) Root lengths of the germinated seedlings treated with Calcofluor for 48 h. (f, g) Cellulose content and CrI in roots of the seedlings treated with 0.1% Calcofluor for 48 h. * and ** indicate significant differences between WT and *Osfc16* by *t*-test at $P < 0.05$ and 0.01, respectively, with the increased or decreased percentage (%) calculated by subtraction of the values between mutant and WT divided by WT. The error bar indicates SD values. ## indicates significant differences between the Calcofluor treatment and control by *t*-test at $P < 0.01$, with the increased or decreased percentage (%) calculated by subtraction of the values between the Calcofluor and control divided by control.

wild type (Figure 4d,e and Table S6). As Calcofluor influences microfibril crystallization by competing for hydrogen binding sites that form the crystalline lattice (Haigler et al., 1980), the Osfc16 mutant, which is rich in low-CrI cellulose, should have less binding capability with Calcofluor, ultimately leading to less inhibited plant growth and relatively higher cellulose level and CrI value, compared with wild type (Figure 4f,g). Furthermore, while treated with CGA, the Osfc16 mutant also showed much less retarded root growth and reduced cellulose level, compared with wild type (Figure 5a–c and Table S6). Notably, the Osfc16 mutant treated with CGA had a significantly higher CrI value than wild type (Figure 5d), a similar phenomenon observed in the Calcofluor treatment (Figure 4g). Because CGA is presumed to affect CESA complex association on the plasma membrane (Crowell et al., 2009; Kurek et al., 2002; Peng et al., 2001, 2002), the wild type may be much more affected by CGA to produce low-CrI cellulose (Figure 5d), whereas the Osfc16 mutant was less sensitive to CGA, probably due to its unstable CESA complexes as described below. Hence, in terms of its low sensitivity to two distinct cellulose inhibitors, the Osfc16 mutant had much less reduction of cellulose CrI by 10% and 7% relative to the control, whereas the wild type showed the reduced CrI by 23% and 33% (Table S6), which on the contrary confirmed that the Osfc16 mutant had a significantly reduced cellulose crystallinity.

Reduced cellulose DP in Osfc16

As cellulose CrI is positively correlated with its DP (Zhang et al., 2013), it remains essential to examine cellulose DP in the Osfc16 mutant. In this study, we focused on detecting cellulose DP of stem and hull tissues in both Osfc16 mutant and wild type (Figure 6a), because both tissues are of predominately secondary cell walls containing extremely high cellulose and lignin for biomass application (Table S7). However, to distinguish cellulose DP in primary and secondary cell walls, we established a novel approach to extract intact cellulose samples by fully removing hemicelluloses and lignin with 4 M KOH and 8% $NaClO_2$ under mild conditions and consequently gradated the purified cellulose into relatively low- and high-DP cellulose fractions using ionic liquid (1-butyl-3-methylimidazolium acetate) and DMSO chemicals (Figure 6b). Using the viscometry method, a classic assay for cellulose DP (Kumar et al., 2009; Li et al., 2014; Zhang et al., 2013), we examined that the Osfc16 mutant in the high-DP cellulose fractions exhibited much lower cellulose DP values by 28%–30% than did the wild type in hull and stem tissues from two independent biological replicate experiments (Figure 6c and Table S8). By contrast, much different DP values were not determined between wild type and mutant in the low-DP fractions (Figure 6d). These findings were confirmed by atomic force microscopy (AFM) observations in which the Osfc16 mutant exhibited much smaller cellulose particles by 44%–57% than did wild types in the high-DP fraction (Figure 6e,g). Because the high-DP fractions cover 10%–40% of total cellulose in the hull and stem tissues (Table S8), their cellulose is thus derived from the secondary cell walls, whereas the low-DP fractions should contain the cellulose from primary cell walls and partial secondary cell walls. Hence, the results indicated that the Osfc16 mutant could partially synthesize the low-DP cellulose in the secondary cell walls of hull and stem tissues, compared with wild type. In addition, although relatively small particles were observed in the low-DP fractions, the Osfc16 mutant had significantly smaller particles by 28% than did the wild type in the hull and in the stem (Figure 6f, h). It should explain that the hull contained much more cellulose

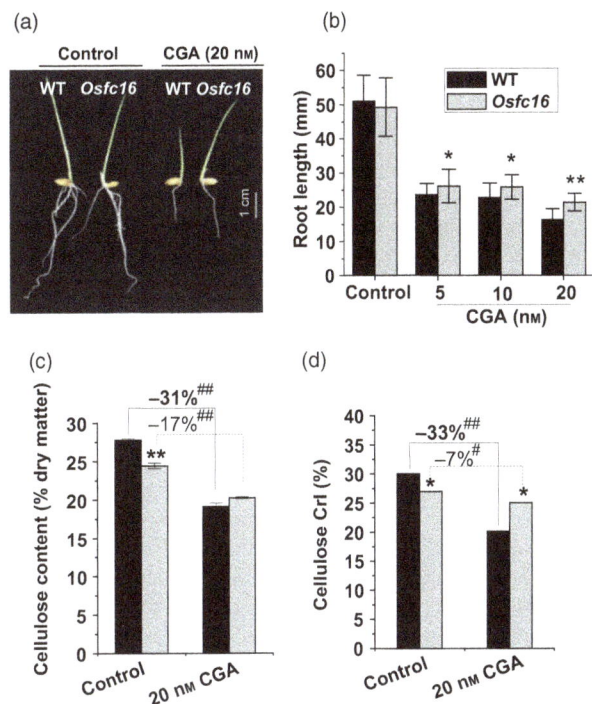

Figure 5 CGA effects on plant growth and cellulose crystallinity. (a, b) Root lengths of the germinated seeds treated with CGA for 72 h. (c, d) Cellulose content and CrI in roots of the seedlings treated with 20 nM CGA for 72 h. * and ** indicate significant differences between WT and Osfc16 mutant by t-test at $P < 0.05$ and 0.01, respectively, and the error bar indicates SD ($n = 3$). # and ## indicate significant differences between the CGA treatment and control by t-test at $P < 0.05$ and 0.01, respectively, with the increased or decreased percentage (%) calculated by subtraction of the values between CGA and control divided by control.

from secondary cell walls than did the stem as described above (Table S7). Taken all together, the results demonstrated that the Osfc16 mutant could synthesize low-DP cellulose in the secondary cell walls, which should lead to thinner secondary cell wall and reduced cellulose level and CrI relative to wild type.

Affected CESA4/7/9 complex association in Osfc16

Because the CESA4/7/9 are required to form a functional cellulose synthase complexes for secondary cell wall synthesis in rice (Huang et al., 2015; Liu et al., 2013; Tanaka et al., 2003; Wang et al., 2010), the three CESA proteins were detected by Western blot analysis of microsomal membrane extracts. Compared with wild type, the Osfc16 mutant showed much lower CESA9 protein levels by 71% as well as reduced CESA4 and CESA7 protein levels by 34% and 22%, respectively (Figure 7a). To sort out the CESA9 protein reduction in the mutant, we used MG132, a proteasome inhibitor (Smalle and Vierstra, 2004), to treat rice plants at tillering stage (Figure 7b). When treated with MG132, both Osfc16 and wild-type plants exhibited higher CESA9 protein levels than did those treated only with DMSO (control), indicating that CESA9 is degraded in a proteasome-dependent manner in plant cells. Notably, the Osfc16 mutant treated with MG132 had increased CESA9 protein biosynthesis rates by onefold compared with the control, whereas wild type only showed biosynthesis rate that increased by 15%, suggesting a rapid and massive proteasome degradation of the CESA9 protein in the Osfc16 mutant. Furthermore, we detected the levels of CESA9 in the CESA4/7/9 complexes pulled down by anti-CESA4 and anti-CESA7, respectively (Figure 7c,d). Although the levels of CESA4 and 7 proteins were reduced by 22% and 4%, the *Osfc16* mutant showed much lower CESA9 protein levels by 49% and 29% than did the wild type, indicating that the *Osfc16* mutant had reduced CESA9 in proportion to the CESA4/7/9 complexes. Therefore, the CESA9 conserved-site mutation affects its association with the CESA complexes, leading to a rapid proteasome degradation. On the other hand, because CGA could affect CESA complex association, this result may also explain why the *Osfc16* mutant was less sensitive to CGA treatment than was the wild type as described above.

Discussion

It has been defined that genetic modification of plant cell walls should not only enhance biomass enzymatic saccharification, but also have little effect on plant growth and development (Abramson *et al.*, 2010). Although previous *CESA* mutation alleles exhibited enhanced biomass digestibility by reducing cellulose crystallinity, various defective plant growth phenotypes had been observed in almost all *CESA* mutants and *CESA*-overexpressed transgenic plants (Table S1). Therefore, this study indicates a new genetic strategy on a direct cellulose modification by CESA mutation at plant fully conserved sites. As recent CRISPR/Cas9 technology is well developed (Doudna and Charpentier, 2014), it could be applied to generate a bunch of mutants from other conserved-site mutations in three CESA4/7/9 isoforms, which may lead to finding out optimal mutants in rice and beyond. In addition, characterization of those generated mutants should further interpret why the *CESA* mutants with CESA9 conserved-site mutations could maintain a normal plant growth and grain production in plants.

Notably, the *Osfc16* mutant has exhibited much higher biomass production and plant lodging resistance than did the wild type. As the plant height is negatively correlated with tiller number in rice (Li *et al.*, 2003; Zhao *et al.*, 2014), the relatively thin stems and short height of *Osfc16* mutant may cause its increased tiller number per plant for high biomass production. Plant lodging resistance is a major and integrated agronomic trait on plant growth and grain production (Li *et al.*, 2015). In particular, rice lodging resistance is negatively affected with plant height and fresh weight (Crook and Ennos, 1994; Islam *et al.*, 2007). Importantly, cellulose crystallinity has been recently demonstrated as the main factor negatively determining plant lodging resistance in rice (Li *et al.*, 2015). Therefore, the *Osfc16* mutant showing much higher lodging resistance should be due to reductions of related factors, such as shorter height, less fresh weight per tiller and lower cellulose CrI. In addition, it remains interesting to test whether the CESA conserved-site mutation could enhance lodging resistance in other plants.

Cellulose CrI reflects the relative amount of crystalline material in cellulose, and highly crystalline cellulose is less accessible to cellulase attack than amorphous cellulose on biomass hydrolysis (Himmel *et al.*, 2007). However, cellulose DP is another important factor on biomass digestibility, because decreasing cellulose DP could increase both number of β-1,4-glucan chain-reducing ends and proportion of amorphous cellulose. In this study, it has been demonstrated that the OsCESA9 site mutation could much reduce cellulose DP and CrI for largely enhanced biomass enzymatic saccharification in the *Osfc16* mutant, which is distinct from the lignin and hemicellulose modifications that increase biomass digestion by improving enzyme accessibility to the cellulose surface (Bonawitz *et al.*, 2014; Chen and Dixon, 2007; Chiniquy *et al.*, 2012; Ding *et al.*, 2012; Li *et al.*, 2015; Wilkerson *et al.*, 2014). In addition, because hemicelluloses negatively affect

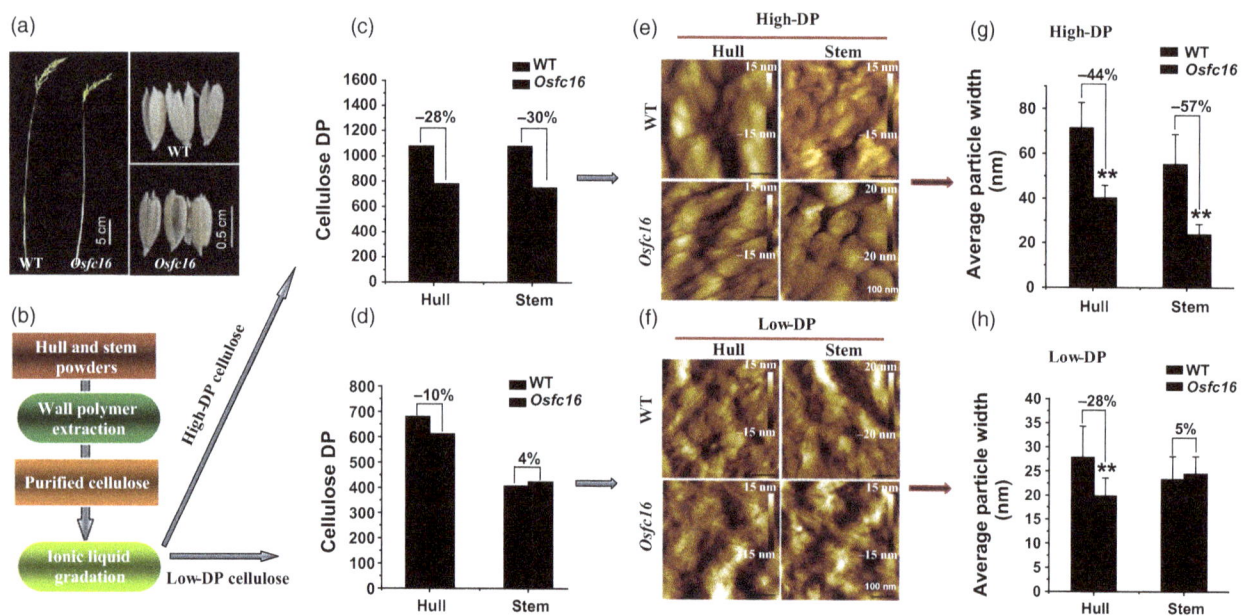

Figure 6 Measurements of cellulose DP in the gradated cellulose fractions of stem and hull tissues in wild type and *Osfc16* mutant. (a, b) Mature stem and hull tissues collected for cellulose extraction and gradation into high- and low-DP cellulose fractions. (c, d) Detection of cellulose DP in the high- and low-DP fractions from one independent biological experiment (Table S8). (e, f) AFM observation of cellulose surfaces in the high- and low-DP fractions. (g, h) Quantitative analysis of AFM imagine by randomly selecting ten dots in the high- and low-DP factions. ** indicates significant differences between the WT and *Osfc16* mutant by *t*-test at $P < 0.01$, with the increased or decreased percentage (%) calculated by subtraction of the DP values between WT and mutant divided by WT.

cellulose crystallinity (Li *et al.*, 2013; Xu *et al.*, 2012), the relatively high level of hemicelluloses in the *Osfc16* mutant (Figure 3c) should be an additional contributor to its biomass enzymatic saccharification.

Plant cellulose biosynthesis process principally involves in three major steps: β-1,4-glucan chain initiation, elongation and termination (Peng *et al.*, 2002). Although CESA complexes are presumed to synthesize the β-1,4-glucan chains, little is yet known about the chain termination that determines cellulose DP. Hence, this study proposed a hypothetic model that the low-DP cellulose synthesis in *Osfc16* mutant should be due to the CESA9 site mutation that may reduce lifetime of CESA4/7/9 complexes towards a relatively early β-1,4-glucan chain termination (Figure 8). Here are four evidences: (i) CESA9 site mutation occurs in the P-CR region that has been proposed to function in CESA protein association and assembly (Olek *et al.*, 2014; Sethaphong *et al.*, 2013); (ii) all three CESA4/7/9 proteins are reduced in the *Osfc16* mutant from co-immunoprecipitation assays; (iii) *Osfc16* mutation mimics the CGA inhibition mode that could disassociate CESA complexes in plants; and (iv) *Osfc16* mutation leads to a rapid proteasome degradation of CESA proteins. On the other hand, as the Cys and Ser substitution with Trp and Pro in the *Osfc16* mutant may play a role in protein interaction and modification, it remains interesting to test whether both amino acids could affect CESA complex association by generating new mutants in the future.

In conclusion, the CESA9 conserved-site mutation could affect its association with the CESA complexes towards a rapid proteasome degradation and cause the low-DP cellulose synthesis for a reduced lignocellulose crystallinity, which largely enhances plant lodging resistance and biomass enzymatic saccharification in *Osfc16* mutant. In addition, this study provides the perspective to find out the optimal mutants from other conserved-site mutations in CESA4/7/9 using CRISPR/Cas9 technique. It also suggests a potential genetic manipulation on the genes that could lead to defective phenotypes from overexpression and knockout in plants.

Experimental procedures

Plant sample collections and physical character measurements

The homozygous *Osfc16* mutant and wild-type plants (*japonica* cultivar Nipponbare (NPB)) were respectively grown in the experimental fields of Huazhong Agricultural University, Wuhan, China, in 2012, 2013 and 2014. The collected mature stem tissues were dried at 55 °C, cut into small pieces, ground through 40-mesh screen (0.425 × 0.425 mm) and stored in the dry container until use.

Rice dry spike and dry biomass were respectively weighed after the samples were dried in the oven at 60 °C. Plant lodging index was detected at six independent biological duplicates using the stem tissues at 30 days after flowering. The breaking resistance of the third internode was detected using a prostrate tester (DIK 7401, Japan), with the distance between fulcra of the tester at 5 cm. Fresh weight (W) of the upper portion of the plant was measured including panicle and the three internodes, leaf and leaf sheath. Bending moment (BM) and lodging index (LI) were calculated using the following formulae: BM = Length from the third internode to the top of panicle × W; and LI = BM/breaking resistance.

Genetic identification of *Osfc16* mutant

The *Osfc16* mutant was selected in 2008 from *japonica* variety Nipponbare T-DNA mutagenesis pools. To identify the *Osfc16* mutant, a F_2 mapping population was generated from the crossing between *Osfc16* mutant and SH838, an *indica* fertility-restoring line in China. The segregation ratio in F_2 population showed that the normal plants and brittle culm plants segregated as 3 : 1. Map-based cloning approach was then used for gene

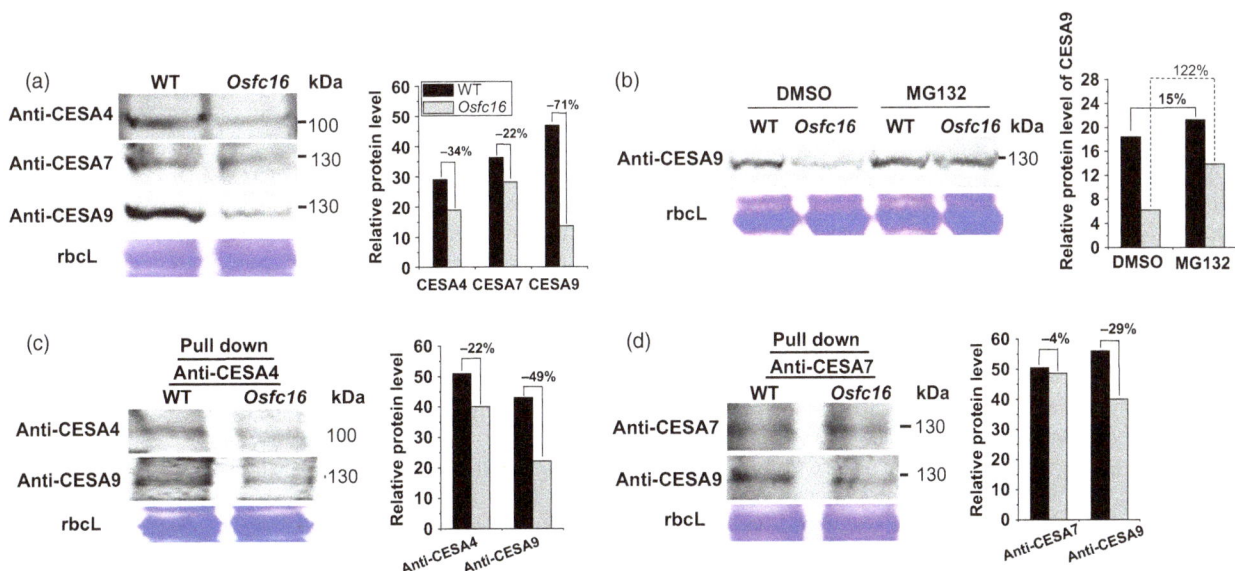

Figure 7 Western blot analysis of CESA proteins. (a) Detection of total CESA4, CESA7 and CESA9 proteins using microsomal membrane extracts of stems at heading stage. (b) CESA9 proteins in the stems treated with 150 μM MG132 and an equivalent dilution of DMSO (control) for 4 h. (c, d) CESA4, CESA7 and CESA9 proteins using co-immunoprecipitation with anti-CESA4 and anti-CESA7. The decreased percentage (%) was calculated by subtraction of the relative protein levels between mutant and WT divided by WT.

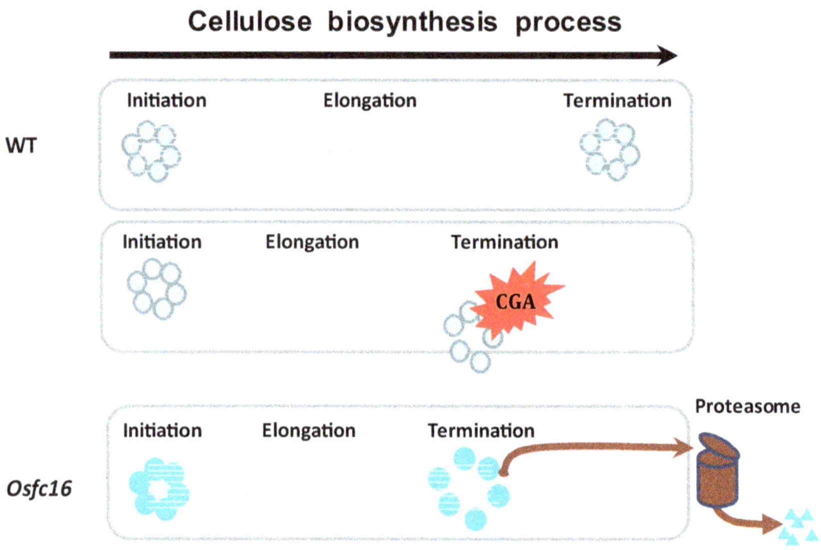

Cellulose biosynthesis process

Figure 8 A hypothesis model on cellulose biosynthesis process involved in initiation, elongation and termination of β-1,4-glucan chains synthesized by CESA complexes on plasma membrane, which highlights that the CESA complexes in the *Osfc16* mutant or in the WT treated with CGA may have a reduced lifetime, leading to relatively early termination of β-1,4-glucan chains for low-DP cellulose biosynthesis. [Colour figure can be viewed at wileyonlinelibrary.com]

identification of the *Osfc16*, based on ~5000 F$_2$ mutant plants with SSR molecular markers. The *Osfc16* gene was localized between RM24343 (forward primer: 5′-AACTGCCACTGCCAAT CATCG-3′; and reverse primer: 5′-CTCCAGCTCTCTCCAC GACTCC-3′) and RM24349 (forward primer: 5′-GT ACTA CTAGCTCGGCTGCTCTGC-3′; and reverse primer: 5′-GTAGTG GAGAGC GTGGACAGC-3′) on chromosome 9 within a 134 kb in the rice genome (Figure 1a). The 134-kb genomic region containing the *CESA9* gene was amplified from the mutants and their corresponding wild-type plants by PCR with KOD-Plus (TOYOBO, Japan) and sequenced with a 3730 sequencer (ABI, Massachusetts, USA).

For genetic complementation of *Osfc16* mutant, a 3401-bp cDNA fragment containing the entire *CESA9* coding region driven by the *ubiquitin* promoter was cloned into the binary vector pCAMBIA 3300 to generate the binary plasmid with forward primer 5′-CTTCTA-GACTCCTCTCCTCCTTCCTGCGTC-3′ and reverse primer 5′-TTC CTGCAGGGCCATCTGTCCATTCCCTCTTC-3′. This binary plasmid was introduced into *Agrobacterium tumefaciens* strain *EHA105* and transformed into the *Osfc16* mutant. The complementary transgenic plant lines of *Osfc16* mutant that expressed full-length cDNA of wild-type *CESA9* gene were characterized as shown in Figures 1b and 3c and Table S3. To examine the T-DNA insertion in the *Osfc16* mutant, TAIL-PCR was performed to analyse the genetic cosegregation of the mutant phenotype with T-DNA insertion based on the known T-DNA sequences. As a result, the T-DNA insertion occurred in the nonfunctional gene and did not exhibit association with the brittle culm phenotypes.

Plant cell wall fractionations

Plant cell wall fractionations were performed as described previously (Peng *et al.*, 2000), with minor modifications as follows: the dry biomass powder (40 mesh) samples (0.1–1.0 g) were washed twice with 5.0 mL buffer and twice with 5.0 mL distilled water. The remaining pellet was stirred with 5.0 mL chloroform–methanol (1 : 1, v/v) for 1 h at 40 °C and washed twice with 5.0 mL methanol, followed by 5.0 mL acetone. The pellet was washed once with 5.0 mL distilled water. The remaining pellet was added with 5.0-mL aliquot of DMSO–water (9 : 1, v/v), rocked gently on a shaker overnight. After centrifugation, the pellet was washed twice with 5.0 mL DMSO–water and then with 5.0 mL distilled water three times. The remaining

pellet was defined as crude cell wall. The remaining crude cell wall was suspended in 0.5% (w/v) ammonium oxalate (5.0 mL) and heated for 1 h in a boiling water bath. During this step, the sample was stirred vigorously every 10 min to prevent the accumulation of materials at the tube surface. After centrifugation and washing the pellet once with 5.0 mL ammonium oxalate and twice with 5.0 mL distilled water, the pellet was suspended in 4 M KOH containing 1.0 mg/mL sodium borohydride (5.0 mL) and incubated for 1 h at 25 °C. During this step, the sample was stirred vigorously every 10 min. After centrifugation, the pellet was washed once with 5.0 mL 4 M KOH and twice with 5.0 mL distilled water. The remaining pellet was defined as crude cellulose. Meanwhile, the remaining pellet from KOH extraction (crude cellulose) was also suspended in 5.0 mL acetic acid–nitric acid–water (8 : 1 : 2, v/v/v) and heated for 1 h in a boiling water bath with stirring every 10 min. After centrifugation, the pellet was washed twice with 5.0 mL water and the remaining pellet was defined as crystalline cellulose sample.

Cellulose extraction and gradation

The dry biomass powders (0.2–1 g) of hull and stem samples were treated with 4 M KOH containing 1.0 mg/mL sodium borohydride (10 mL) at 25 °C for 1 h and then centrifuged (2810 **g**) for 5 min. The pellet was retreated with 4 M KOH for one more time and washed with distilled water five times until pH at 7.0. The remaining pellet was further added with 8% NaClO$_2$ (10 mL) at 25 °C for 72 h (NaClO$_2$ change every 12 h). After centrifugation, the pellet samples were washed with distilled water for five times until pH 7.0 and then further treated with 50 U xylanase (Lot 91101c; Megazyme, Ireland) at 60 °C for 24 h. The remaining pellet was retreated with 8% NaClO$_2$ for one more time and dried as purified cellulose sample, which were verified with nondetectable lignin and less than 1%–2% (of dry matter) pentoses.

The purified cellulose samples (40 mg) were further treated with 3 mL 1-butyl-3-methylimidazolium acetate at 70 °C for 25 min (stem) and 40 min (hull), respectively. The samples were added with 3 mL DMSO and then fully suspended. After centrifugation (2810 **g**) for 5 min, the supernatant was collected as low-DP cellulose sample for cellulose level assay and atomic force microscopy (AFM) observation as described below. The remaining pellet was retreated with 2 mL 1-butyl-3-methylimidazolium acetate at 90 °C until fully dissolved, and then well mixed with 2 mL DMSO as collection of high-DP cellulose sample for cellulose level assay and AFM observation. The high- and low-DP cellulose samples were respectively mixed with distilled water (1 : 1, v/v) at 50 °C and centrifuged (2810 **g**) for 5 min. The precipitated residues were then collected as high- and low-DP cellulose samples for DP detection by the viscometry method described below.

Cell wall composition determinations

Cellulose level was determined using the anthrone/H_2SO_4 method (Fry, 1988), and total hemicellulose contents were calculated subjective to total hexoses and pentoses in the hemicellulose fraction. Total pentoses were detected using the orcinol/HCl method (Dische, 1962). To eliminate the interference of pentoses on hexoses reading at 620 nm, a deduction from pentoses reading at 660 nm was carried out for final hexoses calculation. A standard curve referred for the deduction was drawn using a series of xylose concentrations, which was confirmed by GC-MS analysis. Total lignin content was determined by the two-step acid hydrolysis method according to the Laboratory Analytical Procedure of the National Renewable Energy Laboratory, USA (Sluiter et al., 2008). All experiments were conducted in the biological triplicates.

Cellulose CrI and DP detections

The X-ray diffraction (XRD) method was applied for detection of the lignocellulose crystallinity index (CrI) in the crude cell wall materials using Rigaku-D/MAX instrument (Ultima III; Japan) as described by Zhang et al. (2013). The XRD method was detected with SD at ±0.05–0.15 using five representative samples in triplicate. The relative DP of cellulose was independently measured by the viscometry method as described by Zhang et al. (2013).

Microscopic observations

Scanning electron microscopy (SEM; JSM-6390/LV, Hitachi, Tokyo, Japan) was applied for observations of biomass residues and plant tissues obtained from pretreatments and sequential enzymatic hydrolysis as described by Li et al. (2015). For plant tissue in situ enzymatic digestion, the second-stem transverse sections at heading stages were pretreated with 1% NaOH or 1% H_2SO_4 as described below, washed with distilled water until pH 7.0 and incubated with 1 g/L mixed cellulase for 2 h at 50 °C. After enzymatic hydrolysis, the tissue samples were sputter-coated with gold and observed for 5–10 times with the photography of representative images. The mixed cellulase containing β-glucanase (≥6 × 10^4 U), cellulase (≥600 U) and xylanase (≥1.0 × 10^5 U) was commercially available from Imperial Jade Bio-technology Co., Ltd (Ningxia, 750002, China).

Transmission electron microscopy (TEM) was used to observe cell wall structures in the third leaf veins of three-leave-old seedlings. The samples were post-fixed in 2% (w/v) OsO4 for 1 h after extensively washing in the PBS buffer and embedded with Super Kit (Sigma). Sample sections were cut with an Ultracut E ultramicrotome (Leica) and picked up on formvar-coated copper grids. After poststaining with uranyl acetate and lead citrate, the specimens were viewed under a Hitachi H7500 transmission electron microscope.

AFM was applied to observe cellulose particles. The cellulose samples obtained as previously described in the 'Cellulose extraction and gradation' section were suspended in ultrahigh-purity water and placed on mica using a pipette. The mica was glued onto

a metal disc (15 mm diameter) after removal of extra water under nitrogen and then placed on the piezo scanner of AFM (MultiMode VIII; Bruker, Santa Barbara, CA). AFM imaging was carried out in ScanAsyst-Air mode using Bruker ScanAsyst-Air probes (tip radius, 2 nm; and silicon nitride cantilever; spring constant, 0.4 N/m) with a slow scan rate of 1 Hz. All AFM images were third-flattened and analysed quantitatively using NanoScope Analysis software (Bruker). Ten dots of each AFM image were randomly selected, and the width (nm) of each dot was measured by NanoScope Analysis software (Bruker). The average particle width of each image was calculated from the selected ten particles.

Biomass pretreatment and enzymatic hydrolysis

The chemical (H_2SO_4, NaOH) pretreatment and sequential enzymatic hydrolysis were performed as described by Xu et al. (2012). The CaO pretreatment was performed as follows: the well-mixed biomass powder samples were treated with CaO (7.5% w/w) and shaken at 150 rpm for 36 h at 50 °C. SEM observation was described above using the biomass residues obtained from pretreatment and enzymatic digestion.

Yeast fermentation and ethanol measurement

Saccharomyces cerevisiae (Angel yeast Co., Ltd, Yichang, China) was used in all the fermentation reactions, and the yeast powder was dissolved in 0.2 M phosphate buffer (pH 4.8) for 30 min for activation prior to use. The well-mixed biomass powders were pretreated with CaO (7.5% w/w) and 1% H_2SO_4 as described above. After pretreatments, the biomass residues and supernatants were neutralized to pH 4.8 using appropriate amounts of CaO or H_2SO_4 and were autoclaved for 20 min. Then, mixed cellulases were loaded into each solution with the final enzyme concentration at 3.2 g/L (64 mg/g dry matter) and incubated at 50 °C under 150 rpm for 48 h. After that, the activated yeast was inoculated into the mixture of enzymatic hydrolysates and residues, and to the initial cell mass concentration at 0.5 g/L. The fermentation experiments were performed at 37 °C for 48 h, and the tube cover was loosened a bit to remove the generated CO_2. The fermentation solution was distilled after 48 h for determination of ethanol content. All samples were carried out in the biological triplicates.

Ethanol content was measured using the dichromate oxidation method (Fletcher and van Staden, 2003) with minor modifications (Li et al., 2014).

Calcofluor, CGA325'615 and MG132 treatments in the plant growth

The germinated seeds of Osfc16 mutant and wild type were transferred onto the MS media supplied with Calcofluor White dye (Sigma-Aldrich Co. LLC, California, USA) at different concentrations. After 24-h incubation, the root tissues were measured every 24 h and harvested after 48 h for cellulose content, DP and CrI assays. For CGA325'615 (CGA) treatment, the germinated seeds were incubated with 20 nM CGA (kindly provided by Syngenta Com., Switzerland) in the MS media for 72 h. The root tissues were then measured and harvested for cellulose content, DP and CrI assays. All experiments were performed in the biological triplicates.

For the MG132 treatment, 6-week-old seedlings were incubated with 150 mM MG132 (dissolved in 1% DMSO; purchased from Alabiochem Tech. Co., Ltd, China) for 4 h. The seedlings were also treated with 1.5% DMSO as control. After treatments, total proteins of the seedlings were extracted in the extraction buffer (50 mM MOPS/NaOH buffer, pH 7.5, 0.25 M sucrose, 1.0 mM PMSF, 1.0 μM pepstatin A and 1.0 μM leupeptin), transferred to 15-mL tubes and centrifuged at 2000 *g* for 10 min at 4 °C. The supernatant was incubated with 100 mM MG132 or the solvent DMSO for 1 h in room temperature. The protein concentration was determined using the BCA kits (Yeasen Tech. Co., Ltd, China). The reactions were stopped by the addition of SDS-PAGE loading buffer.

Microsomal membrane extractions

Microsomal membranes were extracted as described by Peng *et al.* (2002) using fresh rice stem tissues (14 g) at heading stage with minor modification. The samples were ground to a fine powder in liquid nitrogen and extracted with 70 mL of ice-cold extraction buffer (50 mM MOPS/NaOH buffer, pH 7.5, 0.25 M sucrose) containing protease inhibitors (1.0 mM PMSF, 1.0 μM pepstatin A and 1.0 μM leupeptin). The extracts were transferred to 15-mL tubes and centrifuged at 2000 *g* for 10 min at 4 °C. The resultant supernatant was filtered through two layers of gauze, and the filtrate was centrifuged at 100 000 *g* for 30 min. The remaining pellet was suspended in extraction buffer containing protease inhibitors and incubated for 30 min at 4 °C under continuous stirring in the presence of 0.05% digitonin. Finally, the homogenate was centrifuged at 5000 *g* for 15 min. The protein concentration in the supernatant was determined using the BCA kits (Yeasen Tech. Co., Ltd, China).

Immunoprecipitation and Western blot analysis

Microsomal membrane extracts were suspended in the extraction buffer containing protease inhibitors and held under continuous stirring for 30 min at 4 °C in the presence of 2% Triton X-100. The homogenates were then centrifuged at 5000 *g* for 15 min, and the extracted proteins were measured by the BCA kits as used at the same amounts in mutant and wild-type plants. 500 μL supernatants (2% Triton X-100 soluble) was mixed with 5 μL (9 μg) of anti-CESA4/7 and incubated for 1 h at 4 °C. Next, 40 μL of protein A–agarose was added into sample tubes and gently shaken for 1 h at 4 °C with end-over-end rotation. After centrifugation for 1 min at 2000 *g*, the harvested pellets were washed three times with ice-cold extraction buffer and heated in 50 μL of sampling buffer at 70 °C for 5 min, then at 100 °C for 5 min. The obtained proteins were loaded into a 10% SDS-PAGE gel.

Following electrophoresis separation, the proteins were transferred to a PVDF membrane. The membrane was blocked with TBS buffer (20 mM Tris-HCl and 500 mM NaCl, pH 7.5) plus 5% nonfat dry milk for 1.5 h, rinsed with TTBS buffer (0.05% Tween-20 in TBS) for three times and incubated with primary antibody serum (CESA4 antibody, 1 : 400 dilution; CESA7 antibody, 1 : 400 dilution; CESA9 antibody, 1 : 500 dilution) for 1 h at room temperature. Generations of CESA4-, CESA7- and CESA9-specific antibodies were described previously (Zhang *et al.*, 2009). After three times of washing with TTBS, the membrane was incubated with secondary antibody (affinity-purified phosphatase-labelled goat anti-rabbit IgG at a 1 : 10000) for 1 h at room temperature. The membrane was finally washed three times with TTBS and one time with TBS (200 mM Tris-HCl, 150 mM NaCl, pH 7.5). The reactions were detected by the ECL Plus Western Blotting Detection. The relative protein levels were calculated using Quantity One software and the RuBisCO large subunit protein (RbcL) of SDS-PAGE gel as internal reference.

Acknowledgements

This work was supported in part by grants from the National Science Foundation of China (31670296; 31571721), Fundamental Research Funds for the Central Universities of China (Program No. 2662015PY018), the National 111 Project (B08032), the National Transgenic Project (2009ZX08009-119B) and the Doctoral Scientific Research Foundation of Liaoning Province (201601107). We would like to thank Dr. Qifa Zhang for general comments on the manuscript and Dr Yihua Zhou for kindly providing partial OsCESA4, OsCESA7 and OsCESA9 antibodies. Also, we thank all colleagues in Biomass and Bioenergy Research Center for reading and discussion of the manuscript.

References

Abramson, M., Shoseyov, O. and Shani, Z. (2010) Plant cell wall reconstruction toward improved lignocellulosic production and processability. *Plant Sci.* **178**, 61–72.

Bonawitz, N.D., Kim, J.I., Tobimatsu, Y., Ciesielski, P.N., Anderson, N.A., Ximenes, E., Maeda, J. *et al.* (2014) Disruption of Mediator rescues the stunted growth of a lignin-deficient Arabidopsis mutant. *Nature*, **509**, 376–380.

Burton, R.A. and Fincher, G.B. (2014) Plant cell wall engineering: applications in biofuel production and improved human health. *Curr. Opin. Biotechnol.* **26**, 79–84.

Chen, F. and Dixon, R.A. (2007) Lignin modification improves fermentable sugar yields for biofuel production. *Nat. Biotechnol.* **25**, 759–761.

Chiniquy, D., Sharma, V., Schultink, A., Baidoo, E.E., Rautengarten, C., Cheng, K., Carroll, A. *et al.* (2012) XAX1 from glycosyltransferase family 61 mediates xylosyltransfer to rice xylan. *Proc. Natl Acad. Sci. USA*, **109**, 17117–17122.

Crook, M.J. and Ennos, A.R. (1994) Stem and root characteristics associated with lodging resistance in four winter wheat cultivars. *J. Agric. Sci.* **123**, 167–174.

Crowell, E.F., Bischoff, V., Desprez, T., Rolland, A., Stierhof, Y.D., Schumacher, K., Gonneau, M. *et al.* (2009) Pausing of Golgi bodies on microtubules regulates secretion of cellulose synthase complexes in Arabidopsis. *Plant Cell*, **21**, 1141–1154.

Ding, S.Y., Liu, Y.S., Zeng, Y., Himmel, M.E., Baker, J.O. and Bayer, E.A. (2012) How does plant cell wall nanoscale architecture correlate with enzymatic digestibility? *Science*, **338**, 1055–1060.

Dische, Z. (1962) Color reactions of carbohydrates. In *Methods in Carbohydrate Chemistry*, vol. **1** (Whistler, R.L. and Wolfrom, M.L., eds), pp. 477–512. New York: Academic Press.

Doudna, J.A. and Charpentier, E. (2014) The new frontier of genome engineering with CRISPR-Cas9. *Science*, **346**, 1258096.

Fletcher, P.J. and van Staden, J.F. (2003) Determination of ethanol in distilled liquors using sequential injection analysis with spectrophotometric detection. *Anal. Chim. Acta*, **499**, 123–128.

Fry, S.C. (1988) *The Growing Plant Cell Wall: Chemical and Metabolic Analysis.* London: Longman, pp. 95–97.

Haigler, C.H., Brown, R.M. and Benziman, M. (1980) Calcofluor white ST Alters the *in vivo* assembly of cellulose microfibrils. *Science*, **210**, 903–906.

Harris, D.M., Corbin, K., Wang, T., Gutierrez, R., Bertolo, A.L., Petti, C., Smilgies, D.M. *et al.* (2012) Cellulose microfibril crystallinity is reduced by mutating C-terminal transmembrane region residues CESA1A903V and CESA3T942I of cellulose synthase. *Proc. Natl Acad. Sci. USA*, **109**, 4098–4103.

Himmel, M.E., Ding, S.Y., Johnson, D.K., Adney, W.S., Nimlos, M.R., Brady, J.W. and Foust, T.D. (2007) Biomass recalcitrance: engineering plants and enzymes for biofuels production. *Science*, **315**, 804–807.

Huang, D., Wang, S., Zhang, B., Shang-Guan, K., Shi, Y., Zhang, D., Liu, X. et al. (2015) A Gibberellin-Mediated DELLA-NAC signaling cascade regulates cellulose synthesis in rice. *Plant Cell*, **27**, 1681–1696.

Islam, M.S., Peng, S., Visperas, R.M., Ereful, N., Bhuiya, M.S.U. and Julfiquar, A.W. (2007) Lodging-related morphological traits of hybrid rice in a tropical irrigated ecosystem. *Field. Crop. Res.* **101**, 240–248.

Kumar, R., Mago, G., Balan, V. and Wyman, C.E. (2009) Physical and chemical characterizations of corn stover and poplar solids resulting from leading pre-treatment technologies. *Bioresour. Technol.* **100**, 3948–3962.

Kurek, I., Kawagoe, Y., Jacob-Wilk, D., Doblin, M. and Delmer, D. (2002) Dimerization of cotton fiber cellulose synthase catalytic subunits occurs via oxidation of the zinc-binding domains. *Proc. Natl Acad. Sci. USA*, **99**, 11109–11114.

Li, X., Qian, Q., Fu, Z., Wang, Y., Xiong, G., Zeng, D., Wang, X. et al. (2003) Control of tillering in rice. *Nature*, **422**, 618–621.

Li, F., Ren, S., Zhang, W., Xu, Z., Xie, G., Chen, Y., Tu, Y. et al. (2013) Arabinose substitution degree in xylan positively affects lignocellulose enzymatic digestibility after various NaOH/H$_2$SO$_4$ pretreatments in *Miscanthus*. *Bioresour. Technol.* **130**, 629–637.

Li, M., Si, S., Hao, B., Zha, Y., Wan, C., Hong, S., Kang, Y. et al. (2014) Mild alkali-pretreatment effectively extracts guaiacyl-rich lignin for high lignocellulose digestibility coupled with largely diminishing yeast fermentation inhibitors in *Miscanthus*. *Bioresour. Technol.* **169**, 447–454.

Li, F., Zhang, M., Guo, K., Hu, Z., Zhang, R., Feng, Y., Yi, X. et al. (2015) High-level hemicellulosic arabinose predominately affects lignocellulose crystallinity for genetically enhancing both plant lodging resistance and biomass enzymatic digestibility in rice mutants. *Plant Biotechnol. J.* **13**, 514–525.

Liu, L., Shang-Guan, K., Zhang, B., Liu, X., Yan, M., Zhang, L., Shi, Y. et al. (2013) Brittle Culm1, a COBRA-like protein, functions in cellulose assembly through binding cellulose microfibrils. *PLoS Genet.* **9**, e1003704.

Olek, A.T., Rayon, C., Makowski, L., Kim, H.R., Ciesielski, P., Badger, J., Paul, L.N. et al. (2014) The structure of the catalytic domain of a plant cellulose synthase and its assembly into dimers. *Plant Cell*, **26**, 2996–3009.

Pauly, M. and Keegstra, K. (2008) Cell-wall carbohydrates and their modification as a resource for biofuels. *Plant J.* **54**, 559–568.

Pear, J.R., Kawagoe, Y., Schreckengost, W.E., Delmer, D.P. and Stalker, D.M. (1996) Higher plants contain homologs of the bacterial celA genes encoding the catalytic subunit of cellulose synthase. *Proc. Natl Acad. Sci. USA*, **93**, 12637–12642.

Peng, L., Hocart, C.H., Redmond, J.W. and Williamson, R.E. (2000) Fractionation of carbohydrates in Arabidopsis root cell walls shows that three radial swelling loci are specifically involved in cellulose production. *Planta*, **211**, 406–414.

Peng, L., Xiang, F., Roberts, E., Kawagoe, Y., Greve, L.C., Kreuz, K. and Delmer, D.P. (2001) The experimental herbicide CGA 325'615 inhibits synthesis of crystalline cellulose and causes accumulation of non-crystalline beta-1,4-glucan associated with CesA protein. *Plant Physiol.* **126**, 981–992.

Peng, L., Kawagoe, Y., Hogan, P. and Delmer, D. (2002) Sitosterol-β-glucoside as primer for cellulose synthesis in plants. *Science*, **295**, 147–150.

Segal, L., Creely, J.J., Martin, A.E. and Conrad, C.M. (1959) An empirical method for estimating the degree of crystallinity of native cellulose using the X-Ray diffractometer. *Text. Res. J.* **29**, 786–794.

Sethaphong, L., Haigler, C.H., Kubicki, J.D., Zimmer, J., Bonetta, D., DeBolt, S. and Yingling, Y.G. (2013) Tertiary model of a plant cellulose synthase. *Proc. Natl Acad. Sci. USA*, **110**, 7512–7517.

Sluiter, A., Hames, B., Ruiz, R., Scarlata, C., Sluiter, J., Templeton, D. and Crocker, D. (2008) *Determination of structural carbohydrates and lignin in biomass* (Tech. Rep. NREL/TP-510-42618, NREL, Golden, Co).

Smalle, J. and Vierstra, R.D. (2004) The ubiquitin 26S proteasome proteolytic pathway. *Annu. Rev. Plant Biol.* **55**, 555–590.

Somerville, C.R. (2006) Cellulose synthesis in higher plants. *Annu. Rev. Cell Dev. Biol.* **22**, 53–78.

Song, X.Q., Liu, L.F., Jiang, Y.J., Zhang, B.C., Gao, Y.P., Liu, X.L., Lin, Q.S. et al. (2013) Disruption of secondary wall cellulose biosynthesis alters cadmium translocation and tolerance in rice plants. *Mol. Plant*, **6**, 768–780.

Tanaka, K., Murata, K., Yamazaki, M., Onosato, K., Miyao, A. and Hirochika, H. (2003) Three distinct rice cellulose synthase catalytic subunit genes required for cellulose synthesis in the secondary wall. *Plant Physiol.* **133**, 73–83.

Taylor, N.G., Howells, R.M., Huttly, A.K., Vickers, K. and Turner, S.R. (2003) Interactions among three distinct CesA proteins essential for cellulose synthesis. *Proc. Natl Acad. Sci. USA*, **100**, 1450–1455.

Wang, L., Guo, K., Li, Y., Tu, Y., Hu, H., Wang, B., Cui, X. et al. (2010) Expression profiling and integrative analysis of the CESA/CSL superfamily in rice. *BMC Plant Biol.* **10**, 282.

Wilkerson, C.G., Mansfield, S.D., Lu, F., Withers, S., Park, J.Y., Karlen, S.D., Gonzales-Vigil, E. et al. (2014) Monolignol ferulate transferase introduces chemically labile linkages into the lignin backbone. *Science*, **344**, 90–93.

Xie, G., Yang, B., Xu, Z., Li, F., Guo, K., Zhang, M., Wang, L. et al. (2013) Global identification of multiple OsGH9 family members and their involvement in cellulose crystallinity modification in rice. *PLoS ONE*, **8**, e50171.

Xu, N., Zhang, W., Ren, S., Liu, F., Zhao, C., Liao, H., Xu, Z. et al. (2012) Hemicelluloses negatively affect lignocellulose crystallinity for high biomass digestibility under NaOH and H$_2$SO$_4$ pretreatments in *Miscanthus*. *Biotechnol. Biofuels*, **5**, 58.

Zhang, B., Deng, L., Qian, Q., Xiong, G., Zeng, D., Li, R., Guo, L. et al. (2009) A missense mutation in the transmembrane domain of CESA4 affects protein abundance in the plasma membrane and results in abnormal cell wall biosynthesis in rice. *Plant Mol. Biol.* **71**, 509–524.

Zhang, W., Yi, Z., Huang, J., Li, F., Hao, B., Li, M., Hong, S. et al. (2013) Three lignocellulose features that distinctively affect biomass enzymatic digestibility under NaOH and H$_2$SO$_4$ pretreatments in *Miscanthus*. *Bioresour. Technol.* **130**, 30–37.

Zhao, J., Wang, T., Wang, M., Liu, Y., Yuan, S., Gao, Y., Yin, L. et al. (2014) DWARF3 participates in an SCF complex and associates with DWARF14 to suppress rice shoot branching. *Plant Cell Physiol.* **55**, 1096–1109.

11

An (*E,E*)-α-farnesene synthase gene of soybean has a role in defence against nematodes and is involved in synthesizing insect-induced volatiles

Jingyu Lin[1], Dan Wang[2], Xinlu Chen[1], Tobias G. Köllner[3], Mitra Mazarei[1], Hong Guo[4], Vincent R. Pantalone[1], Prakash Arelli[5], Charles Neal Stewart Jr[1], Ningning Wang[2] and Feng Chen[1,*]

[1]*Department of Plant Sciences, University of Tennessee, Knoxville, TN, USA*
[2]*Department of Plant Biology and Ecology, College of Life Sciences, Nankai University, Tianjin, China*
[3]*Department of Biochemistry, Max Planck Institute for Chemical Ecology, Jena, Germany*
[4]*Department of Biochemistry, Cellular and Molecular Biology, University of Tennessee, Knoxville, TN, USA*
[5]*Crop Genetics Research Unit, USDA-ARS, Jackson, TN, USA*

Correspondence
email fengc@utk.edu

Keywords: sesquiterpene synthase, volatile, transgenic hairy roots.

Summary

Plant terpene synthase genes (*TPSs*) have roles in diverse biological processes. Here, we report the functional characterization of one member of the soybean *TPS* gene family, which was designated *GmAFS*. Recombinant GmAFS produced in *Escherichia coli* catalysed the formation of a sesquiterpene (*E,E*)-α-farnesene. *GmAFS* is closely related to (*E,E*)-α-farnesene synthase gene from apple, both phylogenetically and structurally. *GmAFS* was further investigated for its biological role in defence against nematodes and insects. Soybean cyst nematode (SCN) is the most important pathogen of soybean. The expression of *GmAFS* in a SCN-resistant soybean was significantly induced by SCN infection compared with the control, whereas its expression in a SCN-susceptible soybean was not changed by SCN infection. Transgenic hairy roots overexpressing *GmAFS* under the control of the CaMV 35S promoter were generated in an SCN-susceptible soybean line. The transgenic lines showed significantly higher resistance to SCN, which indicates that *GmAFS* contributes to the resistance of soybean to SCN. In soybean leaves, the expression of *GmAFS* was found to be induced by *Tetranychus urticae* (two-spotted spider mites). Exogenous application of methyl jasmonate to soybean plants also induced the expression of *GmAFS* in leaves. Using headspace collection combined with gas chromatography–mass spectrometry analysis, soybean plants that were infested with *T. urticae* were shown to emit a mixture of volatiles with (*E,E*)-α-farnesene as one of the most abundant constituents. In summary, this study showed that *GmAFS* has defence roles in both below-ground and above-ground organs of soybean against nematodes and insects, respectively.

Introduction

Soybean is a crop of global importance. Its yield can be significantly reduced due to diseases caused by microbial pathogens and infestation by herbivorous insects (Hartman *et al.*, 2011). The approaches to managing biotic agents of soybean plants, similar to other major crops, include sound cultural practices, application of synthetic pesticides and deployment of resistant cultivars (Oerke, 2006). The development of disease/insect-resistant soybean may be assisted by the mechanistic elucidation of plant natural defences, especially the isolation of defence genes. The production of secondary metabolites is one strategy of plant natural defences against pathogens and insects (Bennett and Wallsgrove, 1994; Zhao *et al.*, 2013). The most structurally diverse group of plant secondary metabolites is terpenoids, which have diverse roles in the interactions of plants with the environment, including serving as defences against pathogens and insects (Gershenzon and Dudareva, 2007). The soybean genome has been fully sequenced (Schmutz *et al.*, 2010). This valuable resource is expected to facilitate the

identification of candidate genes involved in the biosynthesis of terpenoids, especially of those that have roles in natural defences of soybean plants.

Terpene synthases (TPSs) are key enzymes for terpene biosynthesis. They catalyse the formation of terpenes from isoprenyl diphosphate substrates of various chain lengths (Degenhardt *et al.*, 2009). In flowering plants, *TPSs* form a mid-sized gene family in each species (Chen *et al.*, 2011). Over the past 10 years, we have been engaged in functional characterization of the *TPS* gene family in natural defences in several crop plants, including rice (Yuan *et al.*, 2008), sorghum (Zhuang *et al.*, 2012) and poplar trees (Danner *et al.*, 2011). We have also embarked on a project to study the *TPS* family of soybean. A recent study (Liu *et al.*, 2014) showed that the soybean *TPS* family (*GmTPSs*) consists of more than 20 members. The expression of 21 *GmTPS* genes was examined in different soybean tissues. While many genes were found to be expressed in primarily reproductive organs, twelve *GmTPS* genes also showed different expression patterns in response to mechanical wounding (Liu *et al.*, 2014). *GmTPS3* was determined to encode geraniol synthase, and transgenic tobaccos overexpressing *GmTPS3* showed increased resistance to cotton leaf worms (Liu *et al.*, 2014).

In this study, we report the functional characterization of *GmTPS21*, which we designated *GmAFS* (*G. max* α-farnesene synthase). *GmAFS* was selected for this investigation because it was identified as one of the candidate defence genes against soybean cyst nematodes (SCNs) in our previous study (Mazarei *et al.*, 2011). SCN is the most important pathogen of the soybean crop (Koenning and Wrather, 2010). Thus, it has been highly desired to identify and isolate soybean defence genes for genetic improvement of soybean for enhanced resistance against SCN. There were three objectives in this study. The first objective was to determine the biochemical function of the protein encoded by *GmAFS*. Terpene synthases can be categorized into monoterpene synthases, sesquiterpene synthases, and diterpene synthases, depending on the products they form (Chen *et al.*, 2011). We used *in vitro* biochemistry to determine the specific biochemical activity of GmAFS. The second objective was to determine whether *GmAFS* indeed has a role in SCN resistance. For this objective, transgenic hairy roots overexpressing *GmAFS* were produced and assayed for SCN resistance. The third objective was to examine whether *GmAFS* has roles in soybean defence against other pests. In many plant systems, insect herbivory can induce the biosynthesis and emission of volatile terpenoids (Shrivastava *et al.*, 2010). For the third objective, we specifically examined whether *GmAFS* has a role in soybean defence against insects in above-ground tissues.

Results

Expression of *GmAFS* is induced by SCN infection in SCN-resistant soybean

In our previous GeneChip analysis, the expression of *GmAFS* corresponding to Gma.625.1.S1_at was shown to be significantly induced by SCN infection in the SCN-resistant soybean TN02-226, whereas gene expression was unchanged in SCN-infected susceptible (TN02-275) plants (Mazarei *et al.*, 2011; see Supplemental Table 2 therein). As it is possible that false-positive results could occur in microarray experiments from cross-hybridization (Dai *et al.*, 2002), quantitative reverse-transcription PCR (qRT-PCR) experiments were performed. First, root tissues were collected from the SCN-resistant soybean line TN02-226 and the SCN-susceptible soybean line TN02-275 with (3 days post-SCN inoculation) and without SCN infection. These samples were subject to qRT-PCR analysis for *GmAFS*. No significant difference was observed in the expression of *GmAFS* in the SCN-susceptible soybean with or without SCN infection. In contrast, the expression of *GmAFS* in the SCN-resistant soybean was significantly increased (about 2.5-fold) by SCN infection in comparison with that of the control roots without SCN infection (Figure 1).

Evolutionary relatedness of GmAFS with other terpene synthases

With the confirmation of *GmAFS* expression in relation to SCN infestation (Figure 1), our next objective was to determine the evolutionary relatedness of GmAFS with other TPSs including some with known functions. From the latest version of the annotated soybean genome, 22 putative full-length TPS genes

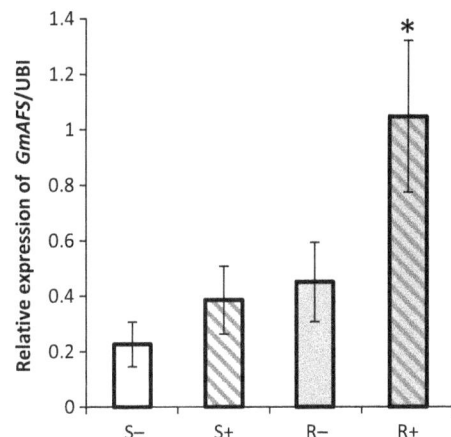

Figure 1 Expression of *GmAFS* gene in the SCN-infected (+) and noninfected control (−) roots of the SCN-resistant (R) and SCN-susceptible (S) soybean lines using quantitative RT-PCR. The expression of *GmAFS* was examined in the root tissues from TN02-226 (R) and TN02-275 (S) soybean breeding lines with (+)/without (−) the treatment of SCN HG type 1.2.5.7 (race 2). The PCR products for soybean ubiquitin-3 (*GmUBI-3*) were used to judge equality of concentration of cDNA templates in different samples. Bars represent mean values of three biological replicates with standard error. Bars with asterisks are significantly different at $P < 0.05$ as tested by Fisher's least significant difference.

including *GmAFS* (*Glyma.13G321100*) were identified. Phylogenetic trees were reconstructed using all TPSs from four sequenced dicot plants: soybean, Arabidopsis, apple and poplar. From this analysis, the soybean TPSs were determined to belong to five subfamilies: a, b, c, e/g and g (Figure 2). GmAFS is a member of the TPS-b subfamily. Most members of the TPS-b subfamily encode monoterpene synthase (Chen *et al.*, 2011), except apple (E,E)-α-farnesene synthase (MdAFS) (Green *et al.*, 2007) and poplar (E,E)-α-farnesene synthase (PtTPS2) (Danner *et al.*, 2011), which are sesquiterpene synthases. GmAFS clustered together with MdAFS and PtTPS2 (Figure 2) and showed 53% and 51% sequence identity to MdAFS and PtTPS2, respectively. Besides sharing high overall sequence similarity, GmAFS, MdAFS and PtTPS2 exhibit conserved structural features. These include the aspartate-rich DDxxD motif, NSD/DTE motif, and H-α1 loop (Figure 3). The H-α1 loop of apple MdAFS has been previously demonstrated to function in the binding of the metal ion K[+] (Green *et al.*, 2009). Evolutionary relatedness, sequence homology and conserved structural features suggested that *GmAFS* encodes (E,E)-α-farnesene synthase.

GmAFS encodes a sesquiterpene synthase producing (E,E)-α-farnesene

Under the standard assay conditions (Zhuang *et al.*, 2012), GmAFS did not show activity with either geranyl diphosphate or farnesyl diphosphate. As it has been demonstrated that the (E,E)-α-farnesene synthases from apple and poplar need K[+] in addition to magnesium for catalytic activity (Danner *et al.*, 2011; Green *et al.*, 2009), we also performed assays containing K[+]. In the presence of K[+], GmAFS converted farnesyl diphosphate into (E,E)-α-farnesene (Figure 4a). However, GmAFS was not able to convert geranyl diphosphate into monoterpenes under the same conditions (Figure 4b).

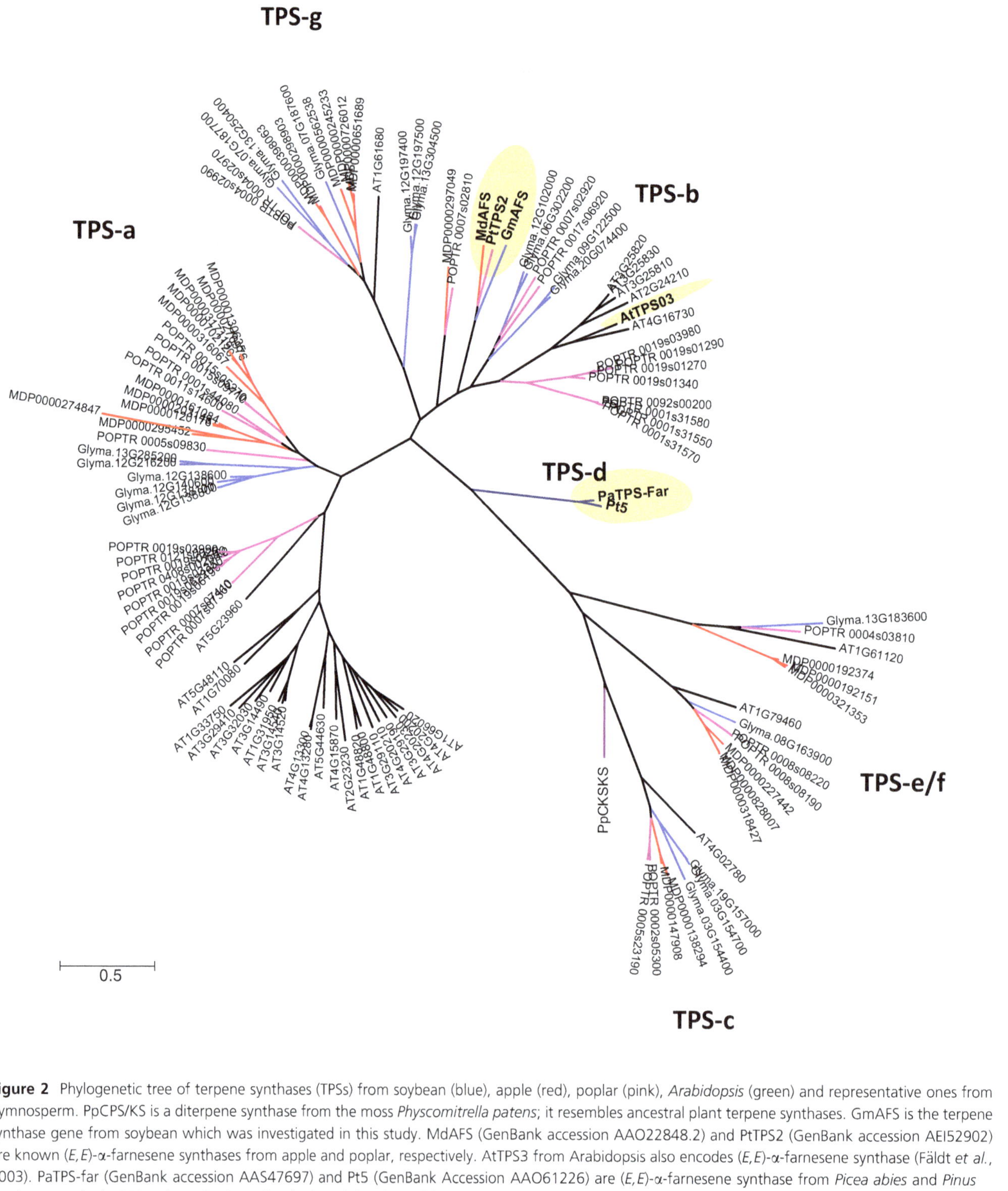

Figure 2 Phylogenetic tree of terpene synthases (TPSs) from soybean (blue), apple (red), poplar (pink), *Arabidopsis* (green) and representative ones from gymnosperm. PpCPS/KS is a diterpene synthase from the moss *Physcomitrella patens*; it resembles ancestral plant terpene synthases. GmAFS is the terpene synthase gene from soybean which was investigated in this study. MdAFS (GenBank accession AAO22848.2) and PtTPS2 (GenBank accession AEI52902) are known (*E,E*)-α-farnesene synthases from apple and poplar, respectively. AtTPS3 from Arabidopsis also encodes (*E,E*)-α-farnesene synthase (Fäldt *et al.*, 2003). PaTPS-far (GenBank accession AAS47697) and Pt5 (GenBank Accession AAO61226) are (*E,E*)-α-farnesene synthase from *Picea abies* and *Pinus taeda*, respectively. TPS-a, b, c, d, e/g and g depict subfamilies. All known (*E,E*)-α-farnesenes are highlighted in yellow.

Structural feature of the GmAFS model and comparison with that of MdAFS

With the experimental confirmation that *GmAFS* encodes (*E,E*)-α-farnesene synthase (Figure 4), next, we compared the three-dimensional structures and the active sites of GmAFS and MdAFS. Homology models of GmAFS and MdAFS were generated by use of the structure of tobacco 5-*epi*-aristolochene synthase (PDB ID: 5EAT) as template (Facchini and Chappell, 1992). The two models superpose well with a RMSD deviation of only 1.5 Å (Figure 5a). The models for GmAFS and MdAFS obtained with the X-ray structure of (+)-bornyl

```
GmAFS  : ----------MNHSYANQS-AQEVNIVTEDTRRSANYKPNIWKYDFL-QSLDSKYDEEEFVMQLNKRVT :  57
MdAFS  : MEFRVHLQADNEQKIFQNQMKPEPEASYLINQRRSANYKPNIWKNDFLDQSLISKYDGDEYRKLSEKLIE :  70
PtTPS2 : MEYKQQVQV--VQNSFQCQNNSEDID--RRQERRSANYKPNIWKYDFL-QSLSKYDEEQYRRVTEKIRE :  65

GmAFS  : EVKG-LFVQEASVIQKLELADWIQKLGLANYFQKDINEFLESILVYVKNSNINPSIEHSLHVSALCFRLL : 126
MdAFS  : EVKIYISAETMDLVAKLELIDSVRKLGLANLFEKEIKEALDSIAA-IESDNLG--TRDDLYGTALHFKIL : 137
PtTPS2 : EVKS-IFVEAVDLLAKLKLVDSVIKLGLGSYFEEEIKQSLDIIAASIKNKNLK--VEENLYVTAIRFKLL : 132

GmAFS  : RQHGYPVLPDTLSNFLDEKGKVIRKSSYVCYGKDVVELLEASHISLEGEKILDEAKNCAINSLKFGFSPS : 196
MdAFS  : RQHGYKVSQDIFGRFMDEKGTLEN--HHFAHLKGMLELFEASNLGFEGEDILDEAKASLTLALRDSGHIC : 205
PtTPS2 : RLHGYEVSQGVFNGFFD--GTSDK--SKCTDVRGLIELFEASHIAYEGEATLDDAKAFSTRILTG-INCS : 197

GmAFS  : SININRHSNLVVEKMVHALELPSHWRVQWFEVKWHVEQYKQQK-NVDPILLELTKLNFNMIQAKLQIEVK : 265
MdAFS  : YPDSN-----LSRDVHSLELPSHRRVQWFDVKWQINAYEKDICRVNATLLELAKLNFNVVQAQLQKNLR : 270
PtTPS2 : AIESD-----LAKHVVLELPSHWRVMWFDVKWHINAYENDK-QTNRHLLALAKVNFNMVQATLQKDLG : 261

GmAFS  : DLSRWWENLGIKKELSFARNRLVESFMCAAGVAFEPKYKAVRKWLTKVIIFVLIIDDVYDIHASFEELKP : 335
MdAFS  : EASRWWANLGIADNLKFARDRLVECFACAVGVAFEPEHSSFRICLTKVINLVLIIDDVYDIYGSEEELKH : 340
PtTPS2 : DVSRWWRNLGIIENLKFTRDRLVESFLCTVGLVFEPKYSSFRKWLTKVIIMILIIDDVYDVYGSLHELQQ : 331

GmAFS  : FTLAFERWDDKELEELPQYMKICVHALKDVTNEIAYEIGGENNFHSVLPYLKKAWIDFCKALYVEAKWYN : 405
MdAFS  : FTNAVDRWDSRETEQLPECMKMCFQVLYNTTCEIAREIEEENGWNQVLPQLTKVWADFCKALLVEAEWYN : 410
PtTPS2 : FTKAVSRWDTGEVQELPECMKICFQTLYDITNEMALEMQREKDGSQALPHLKKVWADFCKAMFMEAKWFN : 401

GmAFS  : KGYIPSLEEYLSNAWISSSGPVILLLSYFATMNQA--MDIDDFLHTYEDLVYNVSLIIRLCNDLGTTAAE : 473
MdAFS  : KSHIPTLEEYLRNGCISSSVSVLLVHSFFSITHEG-TKEMADFLHKNEDLLYNISLIVRLNNDLGTSAAE : 479
PtTPS2 : EGYTPSLQEYLSNAWVSSSGTVISVHSFFSVMTELETGEISNFLEKNQDLLYNISLIIRLCNDLGTSVAE : 471

GmAFS  : REKGDVASSILCYMNQKDASEEKARKHIQDMIHKAWKKINGHYCSNR-VASVEPFLTQAINAARVAHTLY : 542
MdAFS  : QERGDSPSSIVCYMREVNASEETARKNIKGMIDNAWKKVNGKCFTTNQVPFLSSFMNNATNMARVAHSLY : 549
PtTPS2 : QERGDAASSVACYMREVNVSEEVARNHINNIVKKTWKKINGHCFTKS--PTLQLLVNINTNMARVVHNLY : 539

GmAFS  : QNGDGFGIQDRDI-KKHILSLVVEPLR-------- : 568
MdAFS  : KDGDGFGDQEKGP-RTHILSLLFQPLVN------- : 576
PtTPS2 : QHGDGFGVQDRHENKKQILTLLVEPFK
```

Figure 3 Protein sequence alignment of GmAFS with (E,E)-α-farnesene synthases from apple (MdAFS) and poplar (PtTPS2). Three conserved motifs among GmAFS, MdAFS1 and PtTPS2 are boxed: the 'DDxxD' motif, the 'NSD/DTE' motif and the 'H-α1 loop'.

Figure 4 GC–MS total ion chromatograms of reaction products from a terpene synthase assay of E. coli-expressed GmAFS. (a) The terpene synthase assay of E. coli-expressed recombinant GmAFS using farnesyl diphosphate (FPP) as substrate showing the sesquiterpene synthase activity. (b) The terpene synthase assay of E. coli expressed GmAFS using geranyl diphosphate (GPP) as substrate showing no monoterpene synthase activity.

(a) Sesquiterpene synthase activity

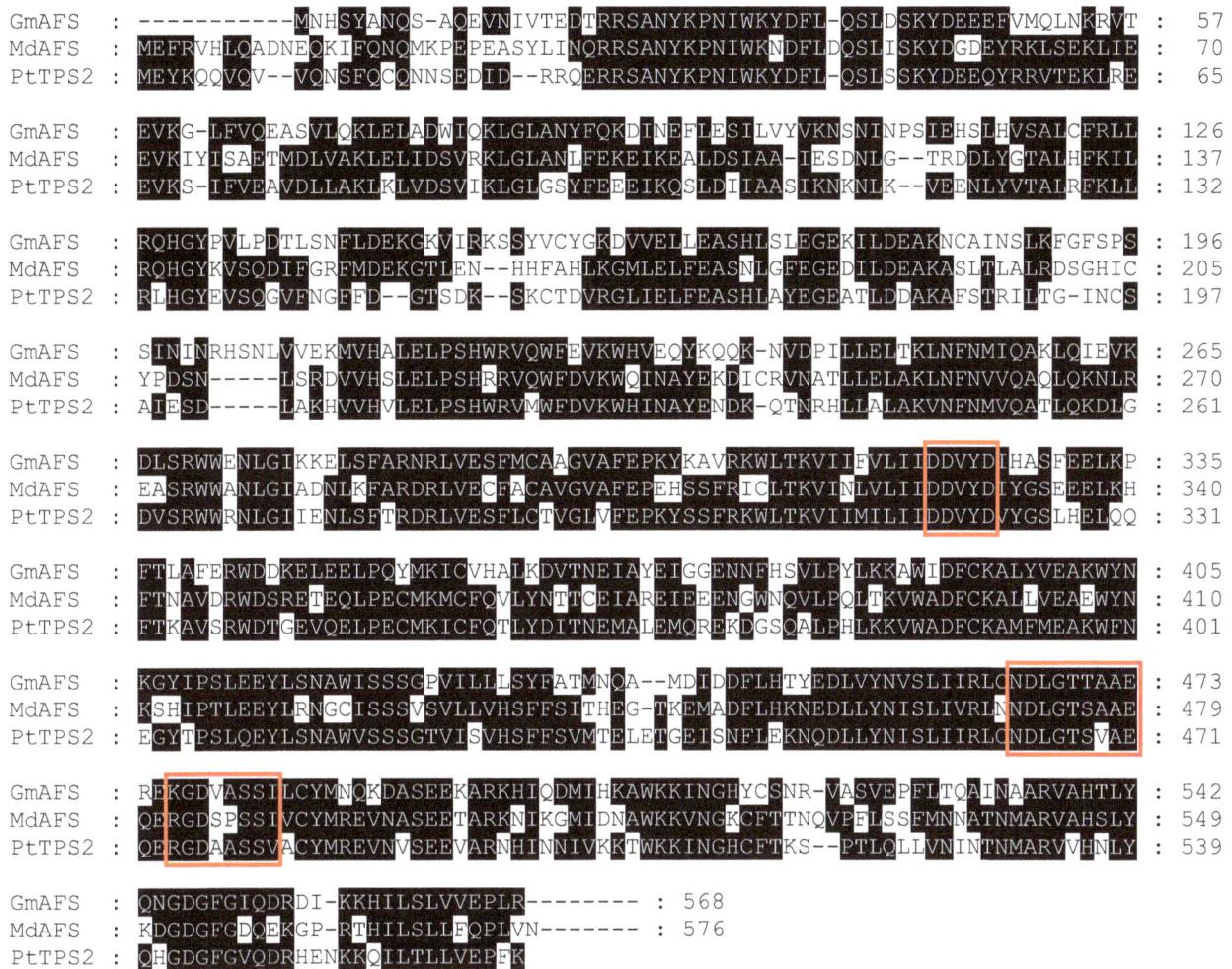

(b) Monoterpene synthase activity

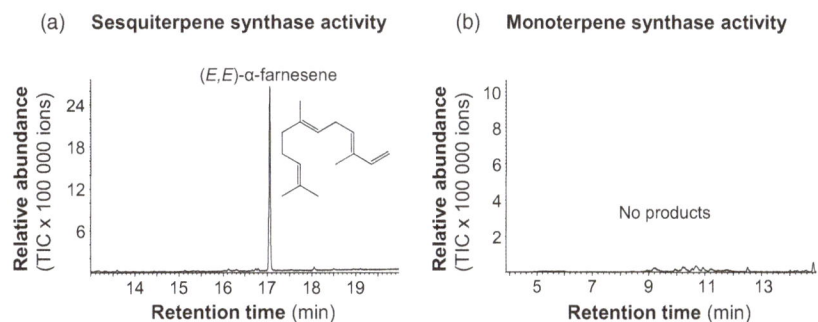

diphosphate synthase (1N20), a monoterpene synthase (Whittington et al., 2002), as the template are also similar. The active site residues for GmAFS and MdAFS were found to be conserved. These residues as well as structural motifs (e.g., H-α1 loop, DDxxD motif and H helix) in GmAFS and MdAFS were well aligned (Figure 5b). It is of interest to note that the H-α1 loops from the two models superpose well, even though the residue corresponding to P486 in MdAFS is Ala (A480) in GmAFS.

Overexpressing GmAFS in transgenic hairy roots led to enhanced resistance to soybean cyst nematode

The expression pattern of GmAFS (Figure 1) strongly suggested that GmAFS may have a role in soybean defence against SCN. To test this hypothesis, we chose to use transgenic hairy root system, which has been proved in our previous studies to be a reliable system for evaluating candidate SCN-resistant genes (Lin et al., 2013). A SCN-susceptible line of soybean (Williams 82) was used

(a)

(b)

α1 Helix

S488
S482
R468 R462   H Helix
A480
H-α1   P486   S487
loop   S481

D484
D478

E479
E473   I461
V467

S423
S428

D546   S429
D553   S424

D325   D321   C293
D330   D326   C298

DDxxD motif   F548
F555

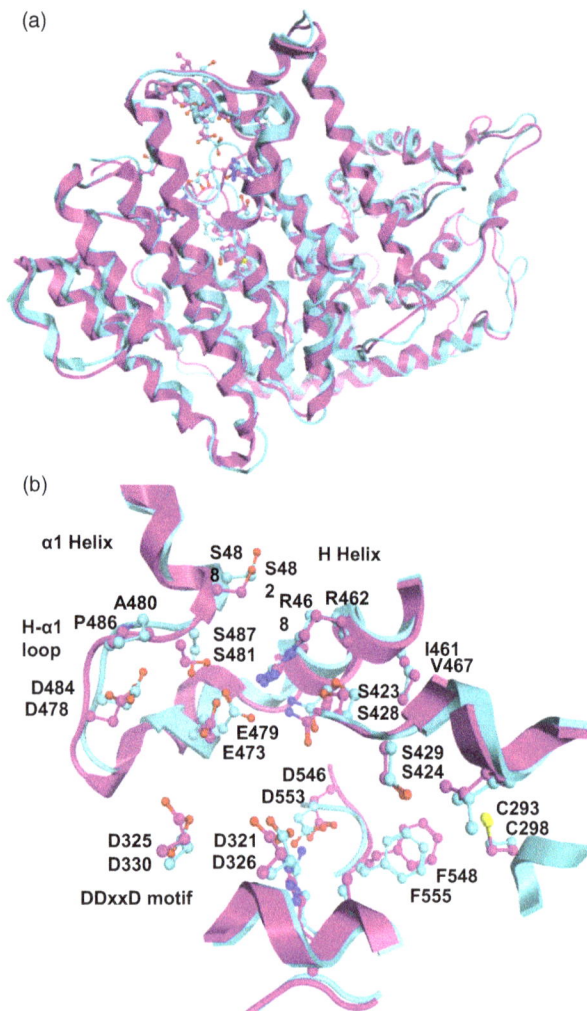

Figure 5 Structural analysis of GmAFS. (a) Supposition of the homology models for GmAFS (magenta) and MdAFS (cyan). The two models can be superposed well with a RMSD deviation of 1.5 Å. The active site residues are shown in ball and stick. (b) The active site structures and residues from the GmAFS and MdAFS models. H-α1 loop, DDxxD motif and H helix are also shown.

(a)

NOS-Ter | GmAFS | 35S-Pro | 35S-Pro | OFP | NOS-Ter

(b)

35S-Pro | OFP | NOS-Ter

(c)

Figure 6 Susceptibility of transgenic soybean hairy roots with overexpression of GmAFS to SCN race 3. (a) Schematic representation of the construct used for transgenic soybean line overexpressing GmAFS and an orange fluorescent protein (OFP) reporter gene. '35S-Pro' and 'NOS-ter' represent the CaMV 35S promoter and the NOS terminator, respectively. (b) Schematic representation of the construct used for control line containing only OFP reporter gene. (c) Ctr stands for the soybean hairy root transformed with the vector containing an orange florescence protein gene. GmAFS stands for the soybean hairy root transformed with the construct overexpressing GmAFS. Williams 82 soybean was the variety used for generating these two types of hairy roots. Bars represent mean values ($n = 20$) of the female index with standard error. Bars with asterisks are significantly different at $P < 0.05$ as tested by Fisher's least significant difference.

to produce transgenic hairy roots overexpressing GmAFS under control of CaMV35S promoter. For fast screening, transgenic hairy roots were generated to coexpress GmAFS and an orange florescence protein (OFP) reporter gene (Figure 6a) in the same cassette. As a negative control, transgenic hairy roots were also produced with a binary vector containing only the ORP reporter gene under the control of CaMV35S promoter (Figure 6b). There was no significant difference on generating the hairy roots between the GmAFS-overexpressing line and the vector control line.

The mean number of adult females and cysts for the control line was about 17.0, whereas the mean number of adult females and cysts for GmAFS-overexpressing transgenic hairy roots was about 10.6. Significantly fewer cysts were observed in transgenic soybean hairy root overexpressing GmAFS than that from control transgenic hairy roots. The female index of transgenic hairy roots overexpressing GmAFS (approximately 60) was significantly lower than that of the control (artificially set to 100) (Figure 6c), which

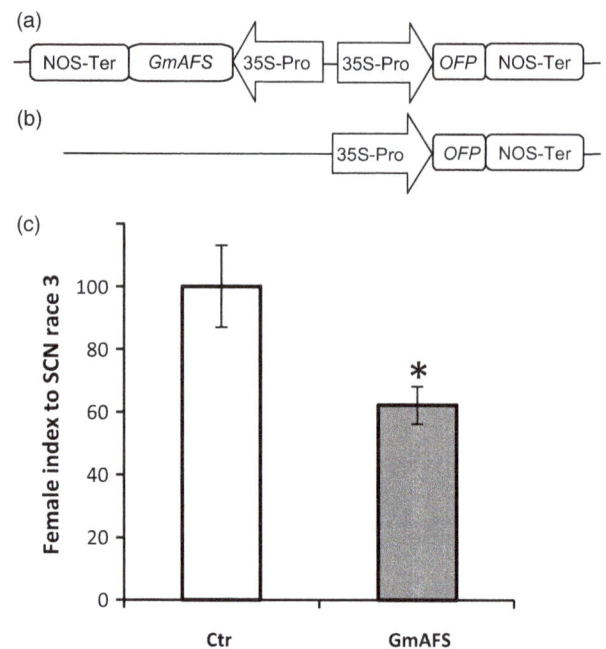

means that the transgenic soybean showed 40% decrease in susceptibility to SCN.

The expression of GmAFS in soybean leaves can be induced by herbivory and methyl jasmonate

Some terpene synthase genes are known to be expressed in multiple tissues and can be induced by multiple stresses (Fäldt et al., 2003). To understand whether GmAFS has roles in other biological processes other than defence again SCN, we examined the expression of GmAFS in leaves especially under stress conditions using qRT-PCR. The expression of GmAFS in leaves infested with Tetranychus urticae (two-spotted spider mite) was found to be 12-fold higher than its expression in control soybean leaves without T. urticae infestation (Figure 7). The jasmonate signalling pathway is essential for regulating plant defence responses to insect herbivory (War et al., 2012). To understand whether this pathway is also associated in regulating the expression of GmAFS, soybean plants were treated with methyl jasmonate and leaves were collected for gene expression analysis. The expression of GmAFS in the methyl jasmonate-treated plants was 11-fold higher than its expression in the control plants (Figure 7).

(E,E)-α-farnesene was one of the major volatile compounds emitted from soybean plants infested with *T. urticae*

Gene expression analysis of *GmAFS* in soybean leaves showed that this gene is induced by herbivory (Figure 8). To determine whether (E,E)-α-farnesene, the product of GmAFS, is released as a volatile compound from *T. urticae*-infested soybean plants, we performed dynamic headspace collection coupled with gas chromatography–mass spectrometry analysis. (E,E)-α-farnesene was among the major volatile compounds detected from the *T. urticae*-infested soybean plants. Other major compounds include Z-3-hexenyl acetate and methyl salicylate (Figure 8). Untreated control soybean plants were also analysed. While (E,E)-α-farnesene was also detected, its amount was lower than that from the *T. urticae*-infested soybean plants (Figure 8).

Discussion

GmAFS provides novel information about the evolution of terpene synthases

Terpene synthase genes form subfamilies with individual subfamilies usually associated with specific biochemical functions as monoterpene synthases, sesquiterpene synthases or diterpene synthases (Chen *et al.*, 2011). GmAFS belongs to the TPS-b subfamily, members of which generally function as monoterpene synthases. It is therefore somehow surprising to observe that GmAFS functions as a sesquiterpene synthase (Figure 4). Further phylogenetic analysis provided new insight into the evolution of the TPS-b subfamily in general and the (E,E)-α-farnesene synthases clade in particular. The (E,E)-α-farnesene synthase genes from soybean and apple are apparent orthologs, implying that their immediate ancestor gene evolved in the common ancestor of Fabidae. GmAFS and MdAFS being a pair of orthologs are also being supported by the high structural similarities of the two proteins they encode (Figure 5). There is conflicting evidence in regard to whether this gene evolved in the common ancestor of Rosid. The clustering of the poplar (E,E)-α-farnesene synthase gene with the apple and soybean (E,E)-α-farnesene synthase genes supports this hypothesis. However, the (E,E)-α-farnesene

synthase gene from Arabidopsis (AtTPS03) does not support this hypothesis. In addition to the phylogenetic evidence presented in this paper (Figure 2), the Arabidopsis (E,E)-α-farnesene synthase behaves differently at the biochemical level. While GmAFS, MdAFS and PtTPS2 all use K$^+$ and Mg^{2+} as cofactors, AtTPS03 uses only Mg^{2+} as a factor like most TPSs (Huang *et al.*, 2010). It is certainly possible that the orthologous (E,E)-α-farnesene synthase gene was lost in Arabidopsis. The analysis of the putative orthologs of this gene in other species of Rosid will provide evidence for testing this hypothesis. Prior to this study, several (E,E)-α-farnesene synthase genes have been isolated from gymnosperms (Phillips *et al.*, 2003), which belong to the TPS-d subfamily (Figure 2). Together, these results suggest that (E,E)-α-farnesene synthase genes have evolved multiple times in seed plants.

GmAFS has a role in soybean defence against SCN

The induced expression of *GmAFS* in SCN-resistant soybean suggested that GmAFS has a role in soybean defence against SCN (Figure 1). This hypothesis was supported with the overexpression of *GmAFS* in transgenic hairy roots of a soybean variety that is SCN susceptible (Figure 6). Interestingly, the expression pattern of *GmAFS* is highly similar to that of soybean salicylic acid methyltransferase gene (*GmSAMT1*) (Lin *et al.*, 2013). The expression of *GmSAMT1* in a SCN-susceptible line was not significantly changed with SCN infection. In contrast, its expression was significantly induced by SCN infection in a SCN-resistant line. Similarly, overexpression of *GmSAMT1* in a SCN-susceptible line also led to enhanced resistance to SCN (Lin *et al.*, 2013). The enhanced resistance of *GmSAMT1*-overexpressors to SCN was attributed to the changes in the salicylic acid signalling pathway (Lin *et al.*, 2013). While the defence mechanism conferred by GmAFS is unclear, it is tempting to speculate that its product (E,E)-α-farnesene has nematicidal activity. A number of terpenoids have been demonstrated to be nematicidal (Ntalli *et al.*, 2010; Oka *et al.*, 2000). It will be interesting to determine whether the defence rendered by *GmAFS* is indeed due to the toxicity of (E,E)-α-farnesene or other mechanisms. It will also be interesting to determine whether *GmAFS*, *GmSAMT1* and other SCN-resistant genes work concertedly to achieve SCN resistance.

GmAFS may be involved in indirect defence against insects

In addition to defence against SCN, *GmAFS* is suggested to have a role in soybean defence against herbivorous insects. When infested by soybean aphids, soybean plants emitted a mixture of volatile compounds including (E,E)-α-farnesene (Moraes *et al.*, 2005). Aphid-induced volatiles from soybean plants were shown to attract soybean aphid's natural enemies such as Syrphidae (Diptera), Chrysopidae (Neuroptera), and green lacewings (Mallinger *et al.*, 2011). Volatile (E,E)-α-farnesene has been shown to be active signal in attracting natural enemies. Laboratory results showed that α-farnesene was attractive to parasitic wasps including *Aphidius ervi*, *Coleomegilla maculate* and *Chrysoperla carnea* (Du *et al.*, 1998; Zhu *et al.*, 1999). James (2005) also reported the parasitic mymarid wasp, *Anagrus daanei*, was attracted to farnesene in a field study. In the current study, *T. urticae* infestation induced the expression of *GmAFS* (Figure 7) and elevated emission of (E,E)-α-farnesene (Figure 8), suggesting that (E,E)-α-farnesene may have a role in attracting the natural enemies of *T. urticae* as well. In fact, (E,E)-α-farnesene is a common compound of herbivore-induced plant volatile blends from some plants such as bean, pear, apple and poplar (Boeve

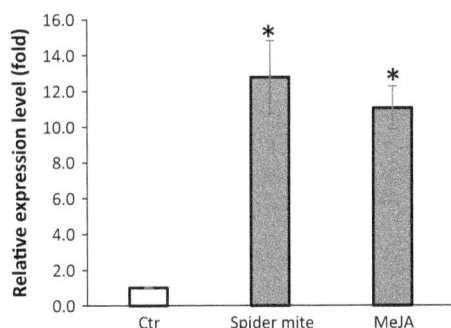

Figure 7 Expression of *GmAFS* gene in of the leaves of the Williams 82 soybean plants that were infested by *T. urticae* (spider mite), treated by methyl jasmonate (MeJA) and control plants (Ctr) using quantitative RT-PCR. The PCR products for soybean ubiquitin-3 (*GmUBI-3*) were used to judge equality of concentration of cDNA templates in different samples. Bars represent mean values of three biological replicates standard error. Bars with asterisks are significantly different at $P < 0.05$ as tested by Fisher's least significant difference.

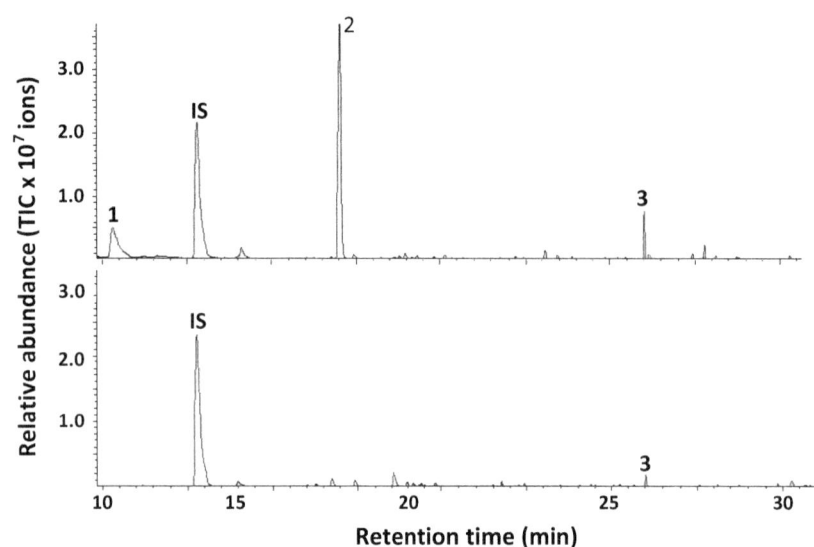

Figure 8 Volatiles emitted from Williams 82 soybean plants infested with two-spotted spider mites (*Tetranychus urticae*). The untreated Williams 82 soybean plants were analysed as a control. The upper panel shows a GC chromatogram of the volatiles from *T. urticae*-infested plants, and the lower panel shows a GC chromatogram from control plants. IS represents the internal standard. 1, *Z*-3-hexenyl acetate; 2, methyl salicylate; 3, (*E,E*)-α-farnesene. The volatile profiling experiment was repeated three times with similar results.

et al., 1996; Danner *et al.*, 2011; Du *et al.*, 1998; Scutareanu *et al.*, 2003). Therefore, (*E,E*)-α-farnesene might function in indirect defence pertaining to many insect pests for many plants, including soybean.

Other possible roles of *GmAFS* and its use for genetic improvement of soybean

In addition to its roles in defence against SCN and indirect defence against insects, GmAFS may also have other functions. For instance, the product of GmAFS may provide defence against bacterial pathogens, fungal pathogens and viruses. In a previous study, root-emitted sesquiterpene caryophyllene was shown to attract beneficial nematodes for defence against insects (Rasmann *et al.*, 2005). It will be interesting to determine whether GmAFS-produced (*E,E*)-α-farnesene has a similar function. *GmAFS* (*GmTPS21*) was found to be highly expressed in stem and mature flowers and its expression increased nearly 25-fold after 2 h of wounding (Liu *et al.*, 2014). Besides such functional studies, *GmAFS* presents a useful genetic improvement of soybean for enhanced defence against multiple biotic agents. We are in the process of producing transgenic soybean overexpressing *GmAFS*. Once produced, we will test these lines for defence function against individual pathogens and insects as well as the agronomic performance of the transgenic soybean in the field.

Experimental procedures

Plants, insects and plant treatments

Three soybean (*Glycine max*) lines were used in this study. These included 'Williams 82' and two genetically related breeding lines: TN02-226 and TN02-275, which are resistant and susceptible, respectively, to soybean cyst nematode (SCN) HG type 1.2.5.7 (race 2). TN02-226 and TN02-275 were used in our prior gene profiling to identify candidate SCN-resistant genes (Mazarei *et al.*, 2011). Williams 82 soybean was used as the plant materials for SCN HG type 0 (race 3) bioassay and gene expression examination under two-spotted spider mites (*Tetranychus urticae*) and methyl jasmonate treatment. Chlorine gas-sterilized soybean seeds were germinated on autoclaved filter paper moistened with sterile distilled water. Three soybean seedlings were grown

in a single pot under 150–200 μmol m^{-2} s^{-1} irradiance 12-h light/12-h dark cycle at 28 °C/25 °C for 21 days. For the methyl jasmonate treatment, 3-week-old Williams 82 seedlings were irrigated with 25 mL of 5 mM methyl jasmonate. After 24 h, leaves were harvested for gene expression analysis. For the insect treatment, 3-week-old Williams 82 soybean seedlings were transferred to the glasshouse for treatment with *T. urticae*. A total of 200 spider mites were used to infest one soybean plant. After 3 days of infestation, soybean plants were subjected to volatile profiling. After volatile profiling, leaves were harvested for gene expression analysis.

Database search and sequence analysis

TPS genes from soybean and apple were identified by analysing their respective genome sequences housed at Phytozome v9.1 (http://www.phytozome.net) using Blast search. The TPS genes from Arabidopsis and poplar were from a previous dataset (Chen *et al.*, 2011). Multiple protein sequence alignment was performed using ClustalX 2.1 (Larkin *et al.*, 2007). A maximum-likelihood tree was constructed using MEGA 6.0 (Tamura *et al.*, 2013) with the Jones–Taylor–Thornton (JTT) model and bootstrapping of 1000 replicates.

Isolation of full-length cDNA of *GmAFS*

A full-length cDNA of *GmAFS* was isolated via RT-PCR from soybean roots infested by SCN. Total RNA was extracted from root tissues using the RNeasy Plant Mini Kit (Qiagen, Valencia, CA), and DNA contamination was removed with DNase treatment following the manufacturers' instructions. Then purified, total RNA was reverse-transcribed into first-strand cDNA in a 15 μL reaction volume using the First-Strand cDNA Synthesis Kit (GE Healthcare, Piscataway, NJ) as previously described (Chen *et al.*, 2003). The following primers were designed for cloning and semiquantitative RT-PCR as follows: *GmAFS*-F: 5'-ATGAATCAC TCATACGCGAATCAATC-3' and *GmAFS*-R: 5'-CTATCTAAGGG GTTCAACAACCAGTG-3'. The PCR program used to amplify the target genes was performed as follows: 94 °C for 2 min followed by 35 cycles at 94 °C for 30 s, 56 °C for 30 s and 72 °C for 1 min 50 s, and a final extension at 72 °C for 10 min. PCR products were cloned into vector pEXP5/CT-TOPO and fully sequenced.

Escherichia coli expression of GmAFS and terpene synthase enzyme assay

The assays were conducted in standard assay conditions, as previously reported (Zhuang *et al.*, 2012). To study the biochemical function of GmAFS, the above-mentioned protein expression vector pEXP5/CT-TOPO harbouring *GmAFS* was transformed into the *E. coli* strain BL21 (DE3) CodonPlus (Stratagene, La Jolla, CA). Fifty millilitres of liquid cultures of the bacteria harbouring the expression constructs were grown at 37 °C to an OD600 of 0.6. Isopropyl β-D-1-thiogalactopyranoside (IPTG) with the final concentration of 500 μM was added to the culture for induction, and the cells were kept cultured for 20 h at 18 °C. Then, the cells were collected by centrifugation and disrupted by a 4×30 sec sonication treatment in chilled extraction buffer (50 mM Mopso, pH 7.0, with 5 mM $MgCl_2$, 5 mM sodium ascorbate, 0.5 mM PMSF, 5 mM dithiothreitol and 10% (v/v) glycerol). The cell fragments were removed by centrifugation at 14 000 ***g***, and the supernatant was desalted into assay buffer (10 mM Mopso, pH 7.0, 1 mM dithiothreitol, 10% (v/v) glycerol) by passage through an Econopac 10DG column (Bio-Rad, Hercules, CA).

The enzyme assays for recombinant GmAFS were performed at 30 °C for 1 h, using 50 μL of the crude enzyme and 50 μL assay buffer with 10 μM substrate (geranyl diphosphate and farnesyl diphosphate, respectively), 10 mM $MgCl_2$, 0.05 mM $MnCl_2$, 50 mM KCl, 0.2 mM $NaWO_4$ and 0.1 mM NaF in a Teflon-sealed, screw-capped 1 mL GC glass vial. A solid-phase microextraction (SPME) fibre consisting of 100 μm polydimethylsiloxane (Supelco, Bellefonte, PA) was placed in the headspace of the vial for 10 min for collecting the volatile. For analysis of the absorbed reaction products, the SPME fibre was inserted directly into the injector of the gas chromatograph. GC-MS analysis and product identification were performed as described below.

Homology models

Homology models were built for both GmAFS and apple (*E,E*)-α-farnesene (MdAFS) based on the X-ray structure of Tobacco 5-*epi*-Aristolochene Synthase (PDB ID: 5EAT) using the homology modelling program in the molecular operation environment (Molecular Operating Environment (MOE), 2015.10; Chemical Computing Group Inc., 1010 Sherbooke St. West, Suite #910, Montreal, QC, Canada, H3A 2R7, 2015).

Transcript abundance analysis of *GmAFS* using quantitative reverse-transcription PCR

The expression of *GmAFS* in the two genetically related breeding lines: TN02-226 and TN02-275, was analysed in their roots with or without the treatment with SCN HG type 1.2.5.7 (race 2) (Mazarei *et al.*, 2011). The expression of *GmAFS* in the soybean Williams 82 was analysed in leaves of the control plants, the plants treated with *T. urticae* and the plants treated with methyl jasmonate. Gene expression was measured using quantitative reverse-transcription PCR (qRT-PCR) as previously reported (Lin *et al.*, 2013). When performing quantitative RT-PCR analysis for *GmAFS*, soybean ubiquitin-3 gene (*GmUBI-3*, GenBank accession D28123) was used as a reference gene. The sequences of gene specific primers were as follows: GmAFS-rt-F 5′-GCTTGGATTTCATCTTCGGGA-3′, GmAFS-rt-R 5′-GGTCCCTAA ATCATTGCACAATCT-3′, GmUBI-3-F 5′-GTGTAATGTTGGATGTG TTCCC-3′, and GmUBI-3-R 5′-ACACAATTGAGTTCAACACAA ACCG-3′. All qRT-PCR assays were conducted in triplicate. PCR efficiencies for target and reference genes were equal among samples. Ct values and relative abundance were calculated using software supplied with the Applied Biosystems 7900 HT Fast Real-Time PCR system. The qRT-PCR data were analysed as previously described by Yuan *et al.* (2006).

Construction of binary vectors for root transformation and generation of transgenic soybean hairy roots

The pCAMBIA 1305.2 vector was used as a backbone binary vector. Our construct was built based on previously described plasmids pJL-OFP and pJL-OFP-35S:GUS (Lin *et al.*, 2013), which contained the coding sequence of an orange fluorescent protein (OFP) reporter gene, originally called as *pporRFP* (Mann *et al.*, 2012). The *GmAFS* cDNA was inserted into the BamHI and SacI sites of pJL-OFP-35S:GUS to replace the GUS gene, which resulted in the pJL-OFP-35S:GmAFS construct. The constructs including pJL-OFP and pJL-OFP-35S:GmAFS were introduced into the *Agrabacterium rhizogenes* strain K599 by the freeze-thaw method (Chen *et al.*, 1994). To test GmAFS's role in soybean cyst nematode resistance, transgenic soybean hairy root with overexpression of *GmAFS* using Williams 82 soybean was generated as previously described (Lin *et al.*, 2013). After about 4 weeks, the hairy roots grew to approximately 10 cm in length. Transgenic soybean roots were screened using dual fluorescent protein flashlight (NightSea, Lexington, MA) to detect *OFP* expression. The tap roots and hairy roots without *OFP* expression were excised off from the composite plant. The tap root and OFP-negative hairy roots and all but one transgenic hairy root were excised under the wounding site of the soybean composite plants containing OFP-positive transgenic hairy roots harbouring pJL-OFP or pJL-OFP-35S:GmAFS. The composite soybean plants with a single transgenic hairy root were subjected to SCN bioassays.

SCN treatment for breeding lines and SCN bioassay on transgenic hairy root overexpressing *GmAFS*

The cultures of SCN HG type 1.2.5.7 (race 2) and SCN HG type 0 (race 3) were maintained in Dr. Arelli's laboratory (Arelli *et al.*, 2000). SCN HG type 1.2.5.7 (race 2) was used in 6-day treatment on the soybean breeding lines, TN02-226 (SCN resistant) and TN02-275 (SCN susceptible) (Mazarei *et al.*, 2011). SCN HG type 0 (race 3) was used for in 35-day bioassay on transgenic hairy root overexpressing *GmAFS*. Active second-stage juvenile (J2) nematodes (1000 J2 SCN/plant) were used for inoculation. SCN inoculation and bioassay for transgenic hairy root overexpressing *GmAFS* were performed as previously described (Lin *et al.*, 2016). The inoculation was carried out in a growth chamber for 2 days with 16-h day length and 25 °C day/night temperature. The composite soybean plants with inoculated hairy roots were transplanted to sterile sand in 50-cm³ cone-tainers (12 cm in length, 2.5 cm inside diameter) randomly arranged within the tray (Stuewe and Sons, Tangent, OR) and maintained for 35 days in the growth chamber with 16-h day length and 22 °C day/night temperature and 100–110 μmol/cm/s light intensity. The composite plants were watered every other day and fertilized weekly with Peters Professional fertilizer (Scotts, Marysville, OH). The result was combination of four independent experiments with four to five plants analysed in each bioassay. Female index, a well-known method to compare the soybean resistance level (Niblack, 2005), was used in this study. Female index was calculated as average number of adult females and cysts for the transgenic soybean divided by average number of females and cysts for control line, multiplied by 100. The susceptibility of hairy root harbouring pJL-*OFP*-35S:*GmAFS* at the cyst stage was calculated based on female index. Each of the SCN bioassays was conducted with four replicates.

Plant volatile collection and identification

Volatiles emitted from the *T. urticae*-treated and control soybean plants were collected in an open headspace sampling system (Analytical Research Systems, Gainesville, FL). Three plants grown in a pot with root systems wrapped with aluminium foil were placed in a glass chamber (30 cm high × 10 cm diameter), with a removable O-ring snap lid with an air outlet port. The charcoal-purified air was passed into the chamber at a flow rate of 0.8 L min^{-1} from the top through a Teflon® hose. Volatiles were collected by pumping air from the chamber through a SuperQ volatile collection trap (Analytical Research Systems, Gainesville, FL). After 16-h collection, 100 µL of methylene chloride containing 1-octanol (0.003%) as an internal standard was used to elute the volatiles into a glass tube for quantification. The volatile analysis was performed in triplicate to confirm the volatile products.

Plant volatiles and volatile terpenoids from TPS enzyme assays were analysed by a Shimadzu 17A gas chromatograph coupled to a Shimadzu QP5050A quadrupole mass selective detector. Compounds' separation was performed on a Restek SHR5XLB column with 30 m × 0.25 mm internal diameter × 0.25 µm thickness (Shimadzu, Columbia, MD). Helium was used as the carrier gas at flow rate of 1.7 mL min^{-1}, and a splitless injection (injection temperature 250 °C) was used. A temperature gradient of 5 °C min^{-1} from 60 °C (6 min hold) to 300 °C was applied. Products were identified based on the National Institute of Standards and Technology (NIST) mass spectral database by comparing of retention times and mass spectra with authentic reference compounds if available. Compound quantification was performed as previously reported (Chen *et al.*, 2009). Representative single-ion peaks of each compound were integrated and compared with the equivalent response of the internal standard (single-ion method).

Statistical analysis

Statistical analysis for gene expression and female index was tested with a one-way ANOVA followed by Fisher's LSD with an alpha level of 0.05 using R software (version 3.1.0) (R Foundation for Statistical Computing, Vienna, Austria).

Acknowledgements

This work was supported by the Tennessee Soybean Promotion Board. We thank Dr. Tarek Hewezi's laboratory for sharing with us the SCN race 3. We would also like to thank Dana Pekarchick and Susan Thomas (USDA-ARS, Jackson, TN, USA) for assistance with maintaining the SCN cultures.

References

Arelli, P.R., Sleper, D.A., Yue, P. and Wilcox, J.A. (2000) Soybean reaction to Races 1 and 2 of *Heterodera glycines*. *Crop Sci.* **40**, 824–826.

Bennett, R.N. and Wallsgrove, R.M. (1994) Secondary metabolites in plant defense mechanisms. *New Phytol.* **127**, 617–633.

Boeve, J.L., Lengwiler, U., Dorn, S., Turlings, T.C.J. and Tollsten, L. (1996) Volatiles emitted by apple fruitlets infested by larvae of the European apple sawfly. *Phytochem. Lett.* **42**, 373–381.

Chen, H., Nelson, R.S. and Sherwood, J.L. (1994) Enhanced recovery of transformants of *Agrobacterium-tumefaciens* after freeze-thaw transformation and drug selection. *Biotechniques*, **16**, 664–670.

Chen, F., Tholl, D., D'Auria, J.C., Farooq, A., Pichersky, E. and Gershenzon, J. (2003) Biosynthesis and emission of terpenoid volatiles from *Arabidopsis* flowers. *Plant Cell*, **15**, 481–494.

Chen, F., Al-Ahmad, H., Joyce, B., Zhao, N., Köllner, T.G., Degenhardt, J. and Stewart, C.N. (2009) Within-plant distribution and emission of sesquiterpenes from *Copaifera officinalis*. *Plant Physiol. Biochem.* **47**, 1017–1023.

Chen, F., Tholl, D., Bohlmann, J. and Pichersky, E. (2011) The family of terpene synthases in plants: a mid-size family of genes for specialized metabolism that is highly diversified throughout the kingdom. *Plant J.* **66**, 212–229.

Dai, H.Y., Meyer, M., Stepaniants, S., Ziman, M. and Stoughton, R. (2002) Use of hybridization kinetics for differentiating specific from non-specific binding to oligonucleotide microarrays. *Nucleic Acids Res.* **30**, e86.

Danner, H., Boeckler, G.A., Irmisch, S., Yuan, J.S., Chen, F., Gershenzon, J., Unsicker, S.B. *et al.* (2011) Four terpene synthases produce major compounds of the gypsy moth feeding-induced volatile blend of *Populus trichocarpa*. *Phytochemistry*, **72**, 897–908.

Degenhardt, J., Köllner, T. and Gershenzon, J. (2009) Monoterpene and sesquiterpene synthases and the origin of terpene skeletal diversity in plants. *Phytochemistry*, **70**, 1621–1637.

Du, Y.J., Poppy, G.M., Powell, W., Pickett, J.A., Wadhams, L.J. and Woodcock, C.M. (1998) Identification of semiochemicals released during aphid feeding that attract parasitoid *Aphidius ervi*. *J. Chem. Ecol.* **24**, 1355–1368.

Facchini, P.J. and Chappell, J. (1992) Gene family for an elicitor-induced sesquiterpene cyclase in tobacco. *Proc. Natl Acad. Sci. USA*, **89**, 11088–11092.

Fäldt, J., Martin, D., Miller, B., Rawat, S. and Bohlmann, J. (2003) Traumatic resin defense in Norway spruce (*Picea abies*): methyl jasmonate-induced terpene synthase gene expression, and cDNA cloning and functional characterization of (+)-3-carene synthase. *Plant Mol. Biol.* **51**, 119–133.

Gershenzon, J. and Dudareva, N. (2007) The function of terpene natural products in the natural world. *Nat. Chem. Biol.* **3**, 408–414.

Green, S., Friel, E.N., Matich, A., Beuning, L.L., Cooney, J.M., Rowan, D.D. and MacRae, E. (2007) Unusual features of a recombinant apple alpha-farnesene synthase. *Phytochemistry*, **68**, 176–188.

Green, S., Squire, C.J., Nieuwenhuizen, N.J., Baker, E.N. and Laing, W. (2009) Defining the potassium binding region in an apple terpene synthase. *J. Biol. Chem.* **284**, 8661–8669.

Hartman, G.L., West, E.D. and Herman, T.K. (2011) Crops that feed the World 2. Soybean-worldwide production, use, and constraints caused by pathogens and pests. *Food Sec.* **3**, 5–17.

Huang, M.S., Abel, C., Sohrabi, R., Petri, J., Haupt, I., Cosimano, J., Gershenzon, J. *et al.* (2010) Variation of herbivore-induced volatile terpenes among *Arabidopsis* ecotypes depends on allelic differences and subcellular targeting of two terpene synthases, TPS02 and TPS03. *Plant Physiol.* **153**, 1293–1310.

James, D. (2005) Further field evaluation of synthetic herbivore-induced plant volatiles as attractants for beneficial insects. *J. Chem. Ecol.* **31**, 481–495.

Koenning, S.R. and Wrather, J.A. (2010) Suppression of soybean yield potential in the continental United States from plant diseases estimated from 2006 to 2009. *Plant Health Prog.*

Larkin, M.A., Blackshields, G., Brown, N.P., Chenna, R., McGettigan, P.A., McWilliam, H., Valentin, F. *et al.* (2007) Clustal W and Clustal X version 2.0. *Bioinformatics*, **23**, 2947–2948.

Lin, J., Mazarei, M., Zhao, N., Zhu, J.W.J., Zhuang, X.F., Liu, W.S., Pantalone, V.R. *et al.* (2013) Overexpression of a soybean salicylic acid methyltransferase gene confers resistance to soybean cyst nematode. *Plant Biotechnol. J.* **11**, 1135–1145.

Lin, J., Mazarei, M., Zhao, N., Hatcher, C.N., Wuddineh, W.A., Rudis, M., Tschaplinski, T.J. *et al.* (2016) Transgenic soybean overexpressing GmSAMT1 exhibits resistance to multiple-HG types of soybean cyst nematode *Heterodera glycines*. *Plant Biotechnol. J.* **14**, 2100–2109.

Liu, J., Huang, F., Wang, X., Zhang, M., Zheng, R., Wang, J. and Yu, D. (2014) Genome-wide analysis of terpene synthases in soybean: functional characterization of GmTPS3. *Gene*, **544**, 83–92.

Mallinger, R.E., Hogg, D.B. and Gratton, C. (2011) Methyl salicylate attracts natural enemies and reduces populations of soybean aphids (Hemiptera: Aphididae) in soybean agroecosystems. *J. Econ. Entomol.* **104**, 115–124.

Mann, D.G.J., LaFayette, P.R., Abercrombie, L.L., King, Z.R., Mazarei, M., Halter, M.C., Poovaiah, C.R. *et al.* (2012) Gateway-compatible vectors for high-throughput gene functional analysis in switchgrass (*Panicum virgatum* L.) and other monocot species. *Plant Biotechnol. J.* **10**, 226–236.

Mazarei, M., Liu, W., Al-Ahmad, H., Arelli, P.R., Pantalone, V.R. and Stewart, C.N. (2011) Gene expression profiling of resistant and susceptible soybean lines infected with soybean cyst nematode. *Theor. Appl. Genet.* **123**, 1193–1206.

Moraes, M.C.B., Laumann, R., Sujii, E.R., Pires, C. and Borges, M. (2005) Induced volatiles in soybean and pigeon pea plants artificially infested with the neotropical brown stink bug, *Euschistus heros*, and their effect on the egg parasitoid, *Telenomus podisi*. *Entomol. Exp. Appl.* **115**, 227–237.

Niblack, T.L. (2005) Soybean cyst nematode management reconsidered. *Plant Dis.* **89**, 1020–1026.

Ntalli, N.G., Ferrari, F., Giannakou, I. and Menkissoglu-Spiroudi, U. (2010) Phytochemistry and nematicidal activity of the essential oils from 8 Greek Lamiaceae aromatic plants and 13 terpene components. *J. Agric. Food Chem.* **58**, 7856–7863.

Oerke, E.C. (2006) Crop losses to pests. *J. Agric. Sci.* **144**, 31–43.

Oka, Y., Nacar, S., Putievsky, E., Ravid, U., Yaniv, Z. and Spiegel, Y. (2000) Nematicidal activity of essential oils and their components against the root-knot nematode. *Phytopathology*, **90**, 710–715.

Phillips, M.A., Wildung, M.R., Williams, D.C., Hyatt, D.C. and Croteau, R. (2003) cDNA isolation, functional expression, and characterization of (+)-α-pinene synthase and (−)-α-pinene synthase from loblolly pine (*Pinus taeda*): Stereocontrol in pinene biosynthesis. *Arch. Biochem. Biophys.* **411**, 267–276.

Rasmann, S., Köllner, T.G., Degenhardt, J., Hiltpold, I., Toepfer, S., Kuhlmann, U., Gershenzon, J. *et al.* (2005) Recruitment of entomopathogenic nematodes by insect-damaged maize roots. *Nature*, **434**, 732–737.

Schmutz, J., Cannon, S.B., Schlueter, J., Ma, J., Mitros, T., Nelson, W., Hyten, D.L. *et al.* (2010) Genome sequence of the palaeopolyploid soybean. *Nature*, **463**, 178–183.

Scutareanu, P., Bruin, J., Posthumus, M.A. and Drukker, B. (2003) Constitutive and herbivore-induced volatiles in pear, alder and hawthorn trees. *Chemoecology*, **13**, 63–74.

Shrivastava, G., Rogers, M., Wszelaki, A., Panthee, D.R. and Chen, F. (2010) Plant volatiles-based insect pest management in organic farming. *Crit. Rev. Plant Sci.* **29**, 123–133.

Tamura, K., Stecher, G., Peterson, D., Filipski, A. and Kumar, S. (2013) MEGA6: molecular evolutionary genetics analysis version 6.0. *Mol. Biol. Evol.* **30**, 2725–2729.

War, A.R., Paulraj, M.G., Ahmad, T., Buhroo, A.A., Hussain, B., Ignacimuthu, S. and Sharma, H.C. (2012) Mechanisms of plant defense against insect herbivores. *Plant Signal Behav.* **7**, 1306–1320.

Whittington, D.A., Wise, M.L., Urbansky, M., Coates, R.M., Croteau, R.B. and Christianson, D.W. (2002) Bornyl diphosphate synthase: structure and strategy for carbocation manipulation by a terpenoid cyclase. *Proc. Natl Acad. Sci. USA*, **99**, 15375–15380.

Yuan, J.S., Reed, A., Chen, F. and Stewart, C.N. (2006) Statistical analysis of real-time PCR data. *BMC Bioinformatics*, **7**, 85.

Yuan, J.S., Köllner, T.G., Wiggins, G., Grant, J., Degenhardt, J. and Chen, F. (2008) Molecular and genomic basis of volatile-mediated indirect defense against insects in rice. *Plant J.* **55**, 491–503.

Zhao, N., Wang, G.-D., Norris, A., Chen, X.-L. and Chen, F. (2013) Studying plant secondary metabolism in the age of genomics. *Crit. Rev. Plant Sci.* **32**, 369–382.

Zhu, J.W., Cosse, A.A., Obrycki, J.J., Boo, K.S. and Baker, T.C. (1999) Olfactory reactions of the twelve-spotted lady beetle, *Coleomegilla maculata* and the green lacewing, *Chrysoperla carnea* to semiochemicals released from their prey and host plant: electroantennogram and behavioral responses. *J. Chem. Ecol.* **25**, 1163–1177.

Zhuang, X., Köllner, T.G., Zhao, N., Li, G., Jiang, Y., Zhu, L., Ma, J. *et al.* (2012) Dynamic evolution of herbivore-induced sesquiterpene biosynthesis in sorghum and related grass crops. *Plant J.* **69**, 70–80.

12

Ectopic expression of specific GA2 oxidase mutants promotes yield and stress tolerance in rice

Shuen-Fang Lo[1,2], Tuan-Hua David Ho[2,3,4]*, Yi-Lun Liu[1,2], Mirng-Jier Jiang[1,2], Kun-Ting Hsieh[5], Ku-Ting Chen[1], Lin-Chih Yu[1], Miin-Huey Lee[6], Chi-yu Chen[6], Tzu-Pi Huang[6], Mikiko Kojima[7], Hitoshi Sakakibara[7], Liang-Jwu Chen[2,5]* and Su-May Yu[1,2,4]*

[1]Institute of Molecular Biology, Academia Sinica, Nankang, Taipei, Taiwan, ROC

[2]Agricultural Biotechnology Center, National Chung Hsing University, Taichung, Taiwan, ROC

[3]Institute of Plant and Microbial Biology, Academia Sinica, Taipei, Taiwan, ROC

[4]Department of Life Sciences, National Chung Hsing University, Taichung, Taiwan, ROC

[5]Institute of Molecular Biology, National Chung Hsing University, Taichung, Taiwan, ROC

[6]Department of Plant Pathology, National Chung Hsing University, Taichung, Taiwan, ROC

[7]RIKEN Center for Sustainable Resource Science, Yokohama, Kanagawa, Japan

*Correspondence
email sumay@imb.sinica.edu.tw (S-MY);
email ljchen@nchu.edu.tw (L-JC);
emails tho@gate.sinica.edu.tw;
ho@biology2.wustl.edu (T-HDH)

Keywords: rice, gibberellin, GA 2 oxidase 6, plant architecture, yield, stress tolerance, photosynthesis rate.

Summary

A major challenge of modern agricultural biotechnology is the optimization of plant architecture for enhanced productivity, stress tolerance and water use efficiency (WUE). To optimize plant height and tillering that directly link to grain yield in cereals and are known to be tightly regulated by gibberellins (GAs), we attenuated the endogenous levels of GAs in rice via its degradation. GA 2-oxidase (GA2ox) is a key enzyme that inactivates endogenous GAs and their precursors. We identified three conserved domains in a unique class of C_{20} GA2ox, GA2ox6, which is known to regulate the architecture and function of rice plants. We mutated nine specific amino acids in these conserved domains and observed a gradient of effects on plant height. Ectopic expression of some of these GA2ox6 mutants moderately lowered GA levels and reprogrammed transcriptional networks, leading to reduced plant height, more productive tillers, expanded root system, higher WUE and photosynthesis rate, and elevated abiotic and biotic stress tolerance in transgenic rice. Combinations of these beneficial traits conferred not only drought and disease tolerance but also increased grain yield by 10–30% in field trials. Our studies hold the promise of manipulating GA levels to substantially improve plant architecture, stress tolerance and grain yield in rice and possibly in other major crops.

Introduction

Rice is a major staple crop feeding more of the human population than any other crop, and increase in rice yield is crucial for meeting the world's demand for food production in the next several decades. However, rice production has plateaued in major rice growing countries (IRRI, 2010), and global climate changes, such as rising temperature and water scarcity, further aggravate the stability of rice production. The development of new strategies for breeding rice that maintains high productivity in an adverse environment remains a major challenge.

The grain yield potential in rice is determined by both genetic and environmental factors. Plant architectures such as height, tiller number and root system are important target traits for rice breeding. The rice tiller is a specialized grain-bearing branch that normally arises from the axil of each leaf and grows independently of the mother stem (culm) with its own adventitious roots. GAs control germination, plant height, tillering, root growth, flowering and seed production (Hussien et al., 2014; Lo et al., 2008; Yamaguchi, 2008). Thus, maintenance of optimal levels of bioactive GAs is important for plant growth and development. Slight reductions in GA levels result in semi-dwarfed plants that are more lodging-resistant in association with an improvement of harvest index (HI) (Khush, 1999). Manipulation of two such genes, *Reduced height 1* (*Rht1*), encoding a wheat GA signalling

factor, and *semi-dwarf* (*sd1*), encoding a rice GA biosynthesis enzyme, combined with N-fertilizer application, led to a quantum leap of yield in semi-dwarf cultivars of the respective plants. This provided the basis for the so-called Green Revolution between the 1960s and 1990s (Botwright et al., 2005; Peng et al., 1999).

Genetic, biochemical and structural studies have significantly advanced our knowledge on biochemical pathways of GA biosynthesis and catabolism, genes and enzymes involved in these pathways, and the molecular mechanism of GA signalling in plants (Hedden and Thomas, 2012; Yamaguchi, 2008). Bioactive GA (GA_1, GA_3, GA_4 and GA_7) concentrations are maintained by the balanced activities of GA 3-oxidases (GA3oxs) and GA 20-oxidases (GA20oxs), essential enzymes regulating GA biosynthesis, and GA 2-oxidases (GA2oxs), necessary for GA inactivation (Sun, 2008; Yamaguchi, 2008). The levels of bioactive GAs can be reduced by the commonly found class C_{19} GA2oxs, which hydroxylate the C-2 of active C_{19}-GAs (GA_1 and GA_4) or C_{19}-GA precursors (GA_9 and GA_{20}), and the class C_{20} GA2oxs, which specifically hydroxylate C_{20}-GA precursors (GA_{12} and GA_{53}) (Lo et al., 2008; Sakamoto et al., 2003; Yamaguchi, 2008). Class C_{20} GA2oxs contain three unique and conserved amino acid motifs that are absent in class C_{19} GA2oxs (Lee and Zeevaart, 2005; Lo et al., 2008), and the function of these domains is not well understood. Ectopic expression of a C_{20} GA2ox and GA2ox6 mutants deleting the conserved domain III generates semi-dwarf plants with increased tillers and root system in transgenic rice (Lo et al., 2008).

Ectopic expression of GA biosynthesis and catabolism enzymes has been used to control the endogenous bioactive GA level in transgenic plants. Reduction in GA level could be accomplished by attenuating either the GA biosynthesis enzymes or those involved in its degradation. It has recently been reported that ectopic expression of a cytochrome P450 monooxygenase gene from Populus trichocarpa leads to the suppression of GA biosynthesis genes causing the reduction in shoot growth and improvement in tolerance to salt stress in transgenic rice (Wang et al., 2016). However, knocking down of GA biosynthesis enzymes does not always lead to substantial reduction in GA level due to the presence of multiple gene families for these biosynthesis enzymes (Coles et al., 1999; Rieu et al., 2008). On the other hand, simple over-expression of a GA degradation enzyme can achieve the goal of effectively reducing GA level (Lo et al., 2008). However, constitutive overexpression of GA2ox enzymes often leads to severe dwarfism and sterility in transgenic plants (Sakai et al., 2003; Sakamoto et al., 2004). By overex-pressing GA2ox1 under the control of a GA-inducible GA biosynthesis gene (GA3ox2) promoter, semi-dwarf transgenic rice with normal flowering and grain development can be obtained (Sakamoto et al., 2003).

Tillering is a crucial agronomic trait directly linked to the number of panicles, which in turn is one of the most important determinants of grain yield (Yang and Hwa, 2008). Plant height is also an important agronomic trait that is tightly associated with the HI and yield potential (Yang and Hwa, 2008). Manipulation of GA levels holds a promise for further improvement of plant architecture and increase in grain yield, as limitation of plant height usually coincides with the increase in the number of dwarfed tillers that bear panicles and grains (Lo et al., 2008). However, excessive tillering diverts nutrients, and tillers emerge late in the growing season may have high frequency of incompletely filled grains. Additionally, plant height is negatively correlated with tiller number, which in turn exhibits a trade-off with filled grains. Consequently, to maximize the grain yield, optimum tiller number and plant height are important targets of rice breeding. However, there is no available information concerning the manipulation of plant height and tiller number to maximize the grain yield in cereals by modulating the level of endogenous GA.

The Green Revolution has had a profound impact in world agriculture in the last few decades, and it is credited with saving over a billion people from starvation. However, the success of Green Revolution was achieved through laborious breeding and selection for high-yield semi-dwarf varieties, and most currently cultivated crops have not yet reached their optimal architectures, stress tolerances and grain yields. In this study, we were able to adjust the architecture by ectopic expression of modified GA2ox6. We discovered that point mutation of certain amino acids of GA2ox6 conferred extraordinary beneficial traits such as semi-dwarfism, increase in grain yield and enhancement of abiotic and biotic stress tolerance in transgenic rice.

Results

Key amino acids for functions of C_{20} GA2oxs

A total of 18 putative C_{20} GA2ox genes identified from 11 different plant species encode proteins containing three con-served motifs (Figures 1a and S1, Table S1). Sixteen amino acids in these motifs were identical, including motif I, $^{S}/_{P}YRWG$; motif

II, $xS^{W}/_{V}SEA^{F}/_{Y}H^{I}/_{V}P/_{I}^{L}/_{M}$; and motif III, DVxxxGxKxGLxxF (x represents less conserved amino acid) (Figure 1a). A total of 7 identical and 2 less conserved amino acid residues were selected for point mutations. Residues were substituted with alanine (A) except in the case of alanine 141, which was replaced with glutamate (E) (Figure 1b). Individual GA2ox6 mutants were expressed under the control of the Ubi promoter, and RT-PCR analyses showed that they were expressed at similar levels in independent transgenic rice lines (Figure S2).

The relative plant height was then scored to estimate the impact of GA2ox6 mutations on the plant architecture. The impact by the wild-type GA2ox6 (GA2ox6-WT) was set as 100%, and GA2ox6 mutants were calculated relative to this value. Mutations of GA2ox6 at amino acid residues Y123A and H143A allowed nearly full recovery of plant heights, thus almost totally abolishing the impact on plant height; mutations at E140A, A141E and G343A partially restored plant heights with 42, 55 and 66% impacts, respectively; mutations at D338A and V339A had small effects on dwarfed plant heights, with 80–82% impacts, and mutations at two other amino acid residues, W138A and T341A, resulted in even shorter plant heights, compared to GA2ox6-WT (Figure 1b). These results indicate that among the seven amino acids identified in the class C_{20} GA2ox, Y123 and H143 are most important, E140, A141 and G343 are second important, and D338 and V339 are less important for the function of GA2ox6 in controlling plant height. The discovery of the importance of Y123, E140, A141 and H143 in conserved motif I and II of GA2ox6 is consistent with a hypothesis that they could be involved in substrate binding (Lee and Zeevaart, 2005).

GA2ox6 regulates plant architecture

Five transgenic lines expressing GA2ox6 mutants Y123A, E140A, A141E, H143A and G343A exhibited different degrees of reduction in plant heights from seedling to adult stages (Figures 2a and S3a, Table 1). In contrast, the tiller number increased inversely to plant height in lines E140A, A141E, G343A and GA2ox6-WT (Figure 2b, Table 1). Germination rates in all transgenic lines, except GA2ox6-WT, were similar to nontrans-formed control (NT) (Figure S3b). Quantitative analysis showed that levels of GA precursors such as GA_{53}, GA_{44}, GA_{19} and GA_{20} were significantly decreased in shoots of A141E and G343A seedlings (Figure 3). However, it is unclear why the level of GA_1 was below the limit of detection in either NT or transgenic rice plants. To demonstrate that the different degrees of reduction in plant heights in lines A141E, G343A and GA2ox6-WT were caused by different levels of GA deficiency; we further treated the 17-day-old seedlings with exogenous GA_3 at 5 μM. The shoots of NT, A141E, G343A and GA2ox6-WT significantly elongated after 3 days of treatment; and the transgenic plants reached to similar plant height as NT after 5 days of GA treatment (Figure S4). These observations indicate that the transgenic plants A141E, G343A and GA2ox6-WT are likely to have different levels of endogenous GA, which can be compensated by the treatment of a high level of exogenous GA.

E140A and A141E mutants of GA2ox6 enhance grain yield

Three field trials in Spring and Fall of 2011, and Fall of 2013 indicated that the grain yield of line A141E was higher than other lines and NT. We also found line G343A displayed increased tolerance to various abiotic stresses compared to NT and the other transgenic lines. Consequently, transgenic lines A141E,

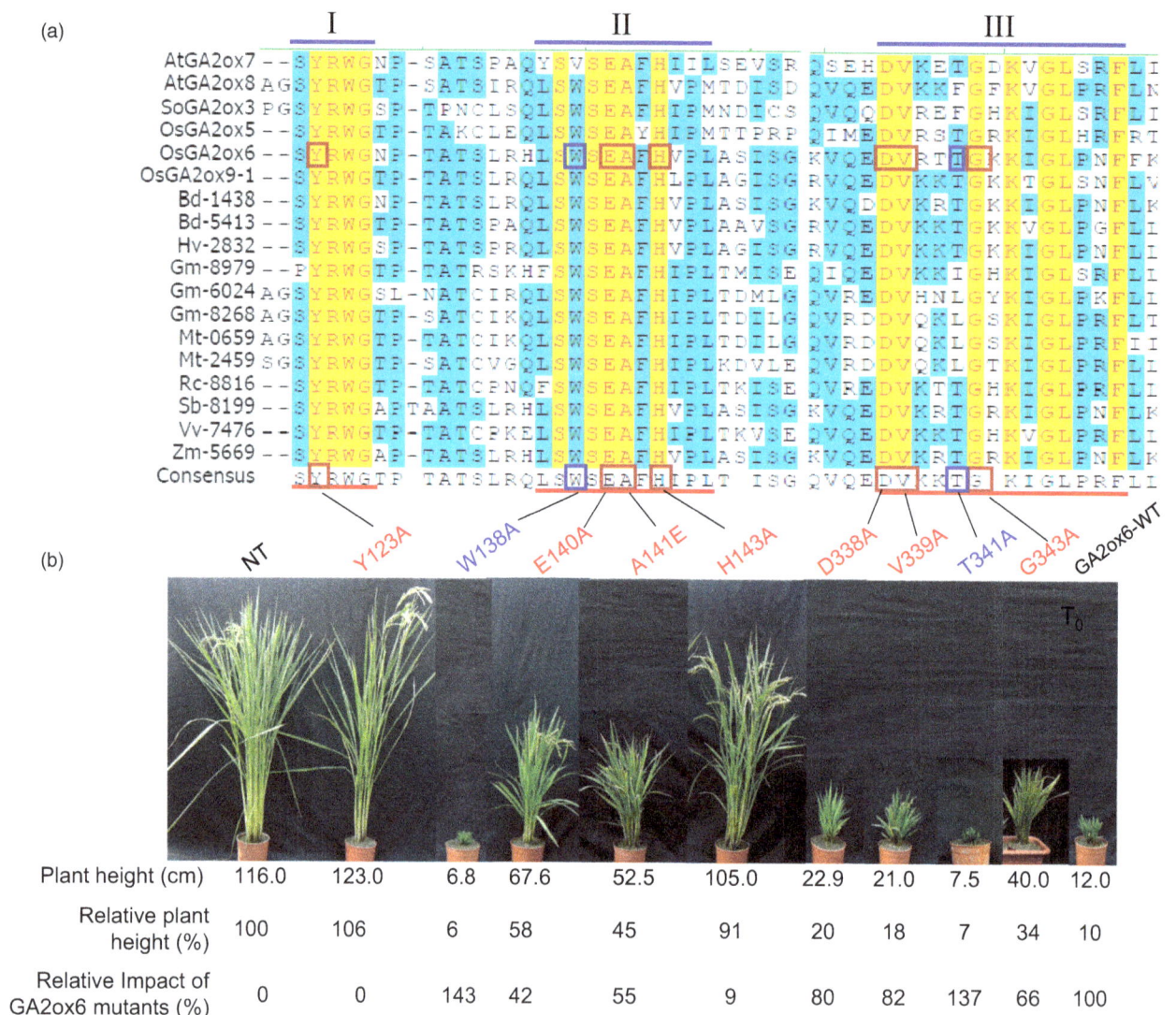

Figure 1 Three conserved motifs essential for function of C_{20} GA2oxs in controlling plant height. (a) Amino acid sequence alignment of C_{20} GA2oxs from different plant species. Roman numerals above the sequences indicate the three unique and conserved motifs present in C_{20} GA2oxs. Identical and conserved amino acid residues are highlighted in yellow and blue, respectively. Red underlines denote the conserved 30 amino acids in motifs I, II and III of C_{20} GA2oxs. Point mutations were introduced into the rice GA2ox6 (OsGA2ox6). Mutations that reduced or enhanced GA2ox6 impacts in transgenic rice are marked by red or blue squares, respectively. (b) Three-month-old T0 plants of nontransformed control (NT) and transgenic lines overexpressing various GA2ox6 mutants. The impact of GA2ox6-WT on plant height in transgenic lines was set as 100%, and the impact of other GA2ox6 mutants was calculated relative to this value.

G343A and GA2ox6-WT were compared with NT for essential traits associated with grain yield and stress tolerance. The ratio of shoot to root dry weights decreased in both young and mature transgenic plants (Figures 4a,b and S3c). WUE also increased in transgenic young plants (Figure 4c). Importantly, the average grain yield of line A141E increased significantly by 17–32%, in contrast to those of lines G343A, which were reduced to 78–90%, as compared with NT in three separate field trials conducted during different seasons (Figures 5a and S5, Table 1). Transgenic line E140A, which was analysed later, was also found to exhibit increased average grain yield by 19% in one field trial (Figure S5). The HIs of both lines A141E and G343A were significantly increased over that of NT (Figure 5b, Table 1).

Examinations of other traits revealed that total chlorophyll, photosynthesis rate and the number of productive tillers bearing seeds were higher in transgenic lines; however, the number of filled seeds was significantly increased only in line A141E and reduced in line GA2ox6-WT (Figure 6, Table 1). The grain and panicle weights and panicle length of line A141E were similar to NT but significantly reduced in lines G343A and GA2ox6-WT (Figure S6, Table 1), indicating that more carbon assimilates were produced and partitioned into grains on increased tillers in line A141E. Grain morphology showed no difference in all lines (Figure S6). Line E140A displayed similar traits to line A141E mentioned above.

A141E and G343A mutants of GA2ox6 enhance abiotic stress tolerance

A141E and G343A transgenic rice lines were tested for tolerance to various abiotic stresses. Both lines were significantly more

Figure 2 Ectopic expression of GA2ox6 and its mutants reduces plant height but increases tiller numbers in transgenic rice. Ninety-five-day-old T2 transgenic plants were used in this study. (a) Morphology of NT and transgenic plants. Scale bar = 10 cm. Plant heights were quantified. (b) Tiller number. n = 18 for each line.

tolerant to dehydration, and line G343A also had greater tolerance to salt and heat than NT (Figure 7a).

Accumulation of the amino acid proline is induced in many plant species in response to environmental stresses and that has been proposed to play an important role in the adjustment to osmotic and oxidative stresses caused by salt and drought (Sperdouli and Moustakas, 2012; Szabados and Savoure, 2010). As shown in Figure 7b, proline levels and CAT and APX activities were all greater in lines A141E and G343A than in NT under normal growth conditions. These factors were also increased in NT but were much greater in transgenic lines under dehydration conditions, and were generally higher in line G343A than in line A141E regardless of dehydration. In contrast, the two transgenic lines had lower total peroxide levels than NT under both normal and dehydration conditions. These findings demonstrated enhanced physiological adaptation to abiotic stresses in rice over-expressing GA2ox6 mutants.

Leaf shape plays a crucial role in photosynthesis and plant development by affecting light interception, leaf temperature and water loss (O'Toole and Cruz, 1980). Leaves are shorter and wider in transgenic lines in contrast to long and narrow leaves in NT (Figure 8a). Leaves rolled and wilted significantly in NT but were less wilted in transgenic lines after dehydration (Figures 8a and

S7a). After rehydration in water, leaves opened slowly in NT but rapidly in transgenic lines (Video S1). The two top young leaves of NT and lines A141E and G343A wilted during osmotic stress treatment with PEG; after recovery in water, these leaves in lines A141E and G343A expanded and continued to grow, while in NT never expanded and ultimately died (Video S2). Lines A141E and G343A grown in soil were also dehydrated. Leaves of both lines recovered completely from dehydration after re-watering, but those of NT were severely damaged and survival rate was low (Figure S7b).

Bulliform cells (BCs) are large, bubble-shaped epidermal cells that are present in groups on the upper surface of leaves of many grasses, and are thought to play a role in providing mechanical strength in the unfolding of developing leaves and in the rolling and unrolling of mature leaves in response to alternating dehydration and well-watered conditions, respectively (Moore et al., 1998). Histological examination of leaf cross sections revealed that BCs were more prominent and extended deeper into the leaf interior (Figures 8b and S8), and the number and volume of BCs were also increased in transgenic lines as compared to NT (Figure 8c). Expanded BCs may retain more water facilitating leaf unrolling after dehydration. Indeed, transgenic lines contain more water in shoots than NT after dehydration for 3.5 h (Figure S9). In addition to shoot, the entire plants also contain more water in transgenic lines (Figure 8d).

E140A, A141E and G343A mutants of GA2ox6 enhance disease resistance

Diseases are a major constraint for achieving optimal yields. Underexpression of a GA biosynthesis enzyme, GA20ox3, reduces plant height but enhances resistance to pathogens *Magnaporthe oryzae* (causing rice blast) and *Xanthomonas oryze pv. oryzae* (causing bacterial blight) in rice (Qin et al., 2013). We found that E140A, A141E and G343A mutants displayed smaller lesion sizes after infection with *X. oryzae pv. oryzae*, maintained greater seedling weight after infection with *Pythium arrhenomanes* (causing seedling blight), and restricted the movement of *Fusarium fujikuroi* (causing the foolish-seedling disease) in shoots as compared with NT (Figure 9).

Reprogramming GA-regulated transcriptional networks

GA-deficient mutants A141E and G343A exhibited pleiotropic alterations in morphology and physiology. To better understand how the GA regulation network controlling plant growth and development, the genomewide transcriptomics profiling was performed for shoots and roots in A141E and G343A in comparison with NT. Primary responsive genes were identified and validated by microarray and RT-PCR analyses (Figures 10 and S10–S12 and Tables S2 and S3). Based on stringent statistics and filtering (P values < 0.05, signal ratio changes > threefold), transcription was found to be significantly reprogrammed in two transgenic lines, with total 765 and 688 genes being up- and down-regulated in roots, and 588 and 382 genes being up- and down-regulated in shoots, respectively, in A141E and G343A (Figure S10a). Large sets of genes were specifically or commonly up- and down-regulated in shoots and roots in A141E and G343A, with more genes being affected in G343A than in A141E and more in roots than in shoots (Figure S10b). Genes involved in abiotic and biotic stress responses and signalling were highly enriched in shoots and roots of the two lines (Figures 10a and S10c, Tables S2 and S3).

Table 1 Comparison of traits in nontransformed control (NT) and transgenic lines overexpress A141E, and G343A and GA2ox6-WT

Traits	NT	A141E	G343A	GA2ox6-WT
Seedling shoot length (cm) (16 DAI)	13.1 ± 0.2* (100)[†]	8.3 ± 0.2 (63)	6.2 ± 0.1 (47)	4.9 ± 0.2 (37)
Seedling tiller number (16 DAI)	1.0 ± 0.0 (100)	1.24 ± 0.4 (124)	1.38 ± 0.1 (138)	1.48 ± 0.2 (148)
Seedling root number (16 DAI)	13.4 ± 0.7 (100)	15.7 ± 0.4 (117)	17.4 ± 0.4 (130)	17.4 ± 0.8 (130)
Plant height (cm) (95 DAI)	63.3 ± 0.6 (100)	40.7 ± 0.3 (64)	25.2 ± 0.8 (40)	12.3 ± 0.9 (19)
Tiller number per plant (95 DAI)	13.3 ± 0.4 (100)	16.7 ± 0.4 (125)	19.4 ± 0.9 (146)	20.6 ± 0.8 (155)
Productive tiller number per plant	10.6 ± 0.3 (100)	13.5 ± 0.3 (127)	16.2 ± 0.3 (152)	12.9 ± 0.4 (122)
Panicle length (cm)	19.8 ± 0.4 (100)	18.3 ± 0.2 (93)	15.3 ± 0.3 (78)	10.8 ± 0.2 (55)
Weight per panicle (g)	2.55 ± 0.12 (100)	2.51 ± 0.08 (98)	1.46 ± 0.07 (57)	0.61 ± 0.03 (24)
Weight of 1000 grains (g)	26.1 ± 0.2 (100)	25.8 ± 0.4 (99)	24.1 ± 0.3 (92)	23.2 ± 0.03 (89)
Fertility (%)	90.8 ± 0.01	91.6 ± 0.03	93.2 ± 0.01	93.6 ± 0.02
Total shoot weight (g)	47.1 ± 1.4 (100)	38.8 ± 0.6 (82)	25.1 ± 0.4 (53)	NA
Grain number per plant	900 ± 0.4 (100)	1130 ± 0.8 (125)	870 ± 0.7 (97)	470 ± 0.4 (53)
Grain yield (Ton/ha, 2013-Fall)	4.22 ± 0.2 (100)	4.92 ± 0.1 (117)	3.28 ± 0.2 (78)	NA
Harvest Index (2013-Fall) (Grain weight/shoot weight/plant)	0.55 ± 0.01 (100)	0.77 ± 0.02 (141)	0.89 ± 0.03 (161)	NA

NA, not available; DAI, days after imbibition.

*SE; $n \geqq 18$ for NT, A141E, G343A, GA2ox6-WT.

[†]Values in parentheses indicate % of NT.

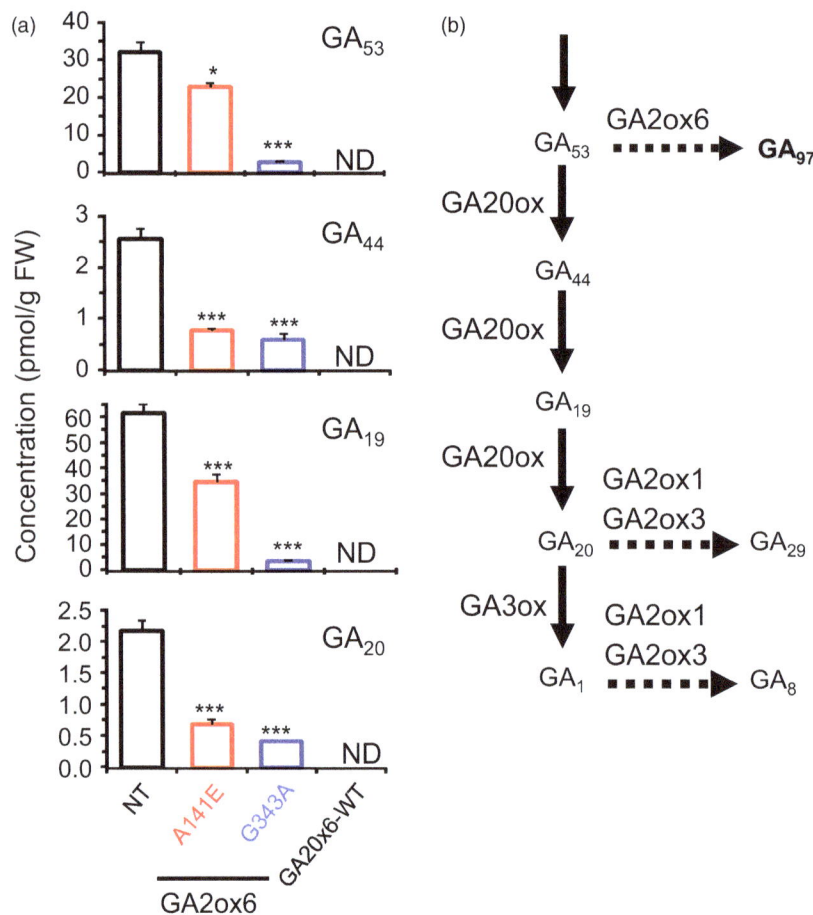

Figure 3 Ectopic expression of GA2ox6 and its mutants reduces the accumulation of GA precursors in transgenic lines. (a) Concentrations of GA precursors in NT and transgenic rice over-expressing wild-type GA2ox6 and its mutants. ND: nondetectable. (b) GA2ox6 is known to inactivate GA_{53} in the GA biosynthesis pathway. $n = 7, 8, 8, 8$ for NT and lines A141E, G343A, and GA2ox6-WT, respectively.

We found that up-regulation of six groups of genes in shoots and roots of A141E and G343A may be responsible for enhanced tillering, root growth, biotic and abiotic stress tolerance and grain yield (Figure 10): (1) Group 1 genes encoding auxin response factors (ARFs), auxin efflux transporter PINs and type B Arabidopsis response regulator (ARR), which are known to promote tillering and root development in cereals (Hussien et al., 2014; Orman-Ligeza et al., 2013), were up-regulated, and the repressor

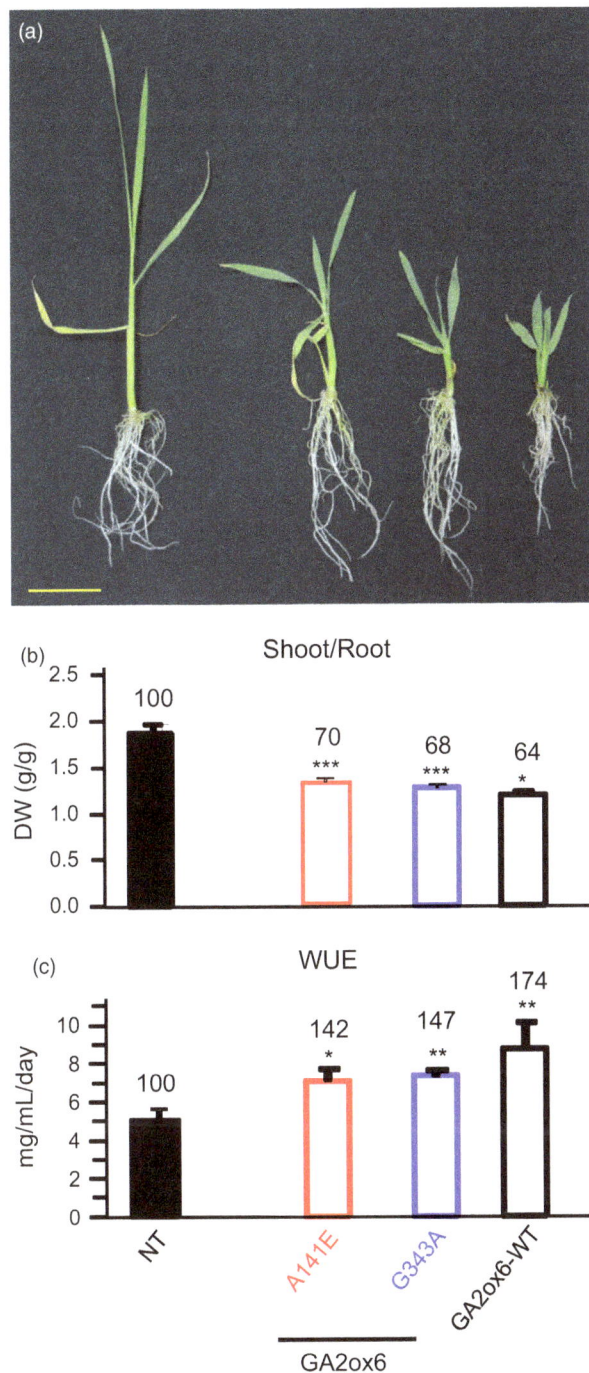

Figure 4 Root growth and WUE are significantly enhanced in GA-deficient transgenic rice. (a) Morphology of 25-day-old plants, scale bar = 5 cm. (b) Shoot/root ratio. (c) WUE. n = 21 for each line.

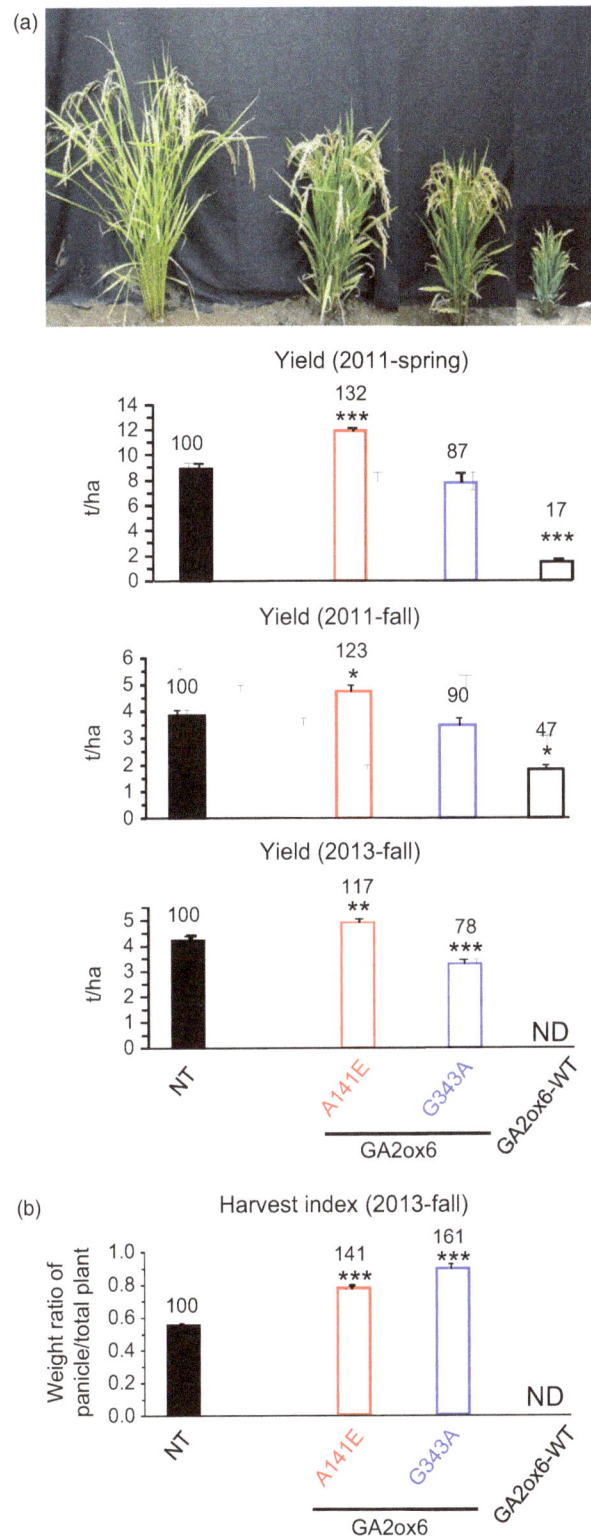

Figure 5 Grain yield is significantly enhanced in GA-deficient transgenic rice. (a) Morphology of rice plants near harvest and grain yield in ton per hectare in spring and fall, 2011, and in fall, 2013. (b) Harvest index in fall, 2013; 2011-Spring: n = 10 for each line; 2011-Fall: n = 18, 24, 20, 27 for NT and lines A141E, G343A and GA2ox6-WT, respectively; 2013-Fall: n = 32, 55, 55, 30 for NT and lines A141E, G343A and GA2ox6-WT, respectively. ND: not-determined.

protein Aux/IAA was down-regulated. (2) Group 2 genes encoding RuBisCo large subunit and proteins involved in light reaction of photosynthesis, such as cytochrome b6/f, NADPH Q-oxidoreductase and PSII 10 kDa polypeptide (PSII 10), were up-regulated in A141E and G343A shoots, which correlates with the enhanced photosynthesis rates seen in these plants. It should be noted that the expression level of photosynthesis related genes is actually reduced in A141E and G343A in roots (Figure 10a). However, the significance of this reduction is not yet clear. (3)

Group 3 genes encoding proteins involved in abiotic stress tolerance, such as AP2/EREBPs, WRKYs and NACs. (4) Group 4 genes encoding proteins responsible for ROS scavenging, such as glutaredoxins (GRXs) and superoxide dismutase (SOD), which could also be regulated by the DELLA-dependent pathway. (5) Group 5 genes encoding numerous signalling receptor kinases, such as DUF26 kinases, cell wall (CW)-associated kinases and leucine-rich repeat (LRR) kinases. (6) Group 6 genes encoding proteins involved in plant defence, such as xylanase inhibitor proteins (XIPs), polygalacturonase inhibitor proteins (PGIPs), resistance (R) proteins and pathogenesis-related (PR) proteins (Juge, 2006; Pogorelko et al., 2013), were highly up-regulated.

Discussion

To optimize plant architecture for enhancing productivity as well as stress tolerance or water saving capacity in crops is one of major challenges in modern agricultural biotechnology. In the present study, constitutive ectopic expression of GA2ox6 mutants controlling endogenous GA levels has successfully manipulated plant height, tiller number, shoot-to-root ratio, WUE, photosynthesis rate, stress tolerance, pathogen resistance, harvest index and grain yield in transgenic rice.

Five amino acids are important for functions of C_{20} GA2oxs

Although the function of the class C_{20} GA2oxs in deactivation of GA precursors is known in both dicots and monocots (Lee and Zeevaart, 2005; Lo et al., 2008), the function of the three unique and conserved domains in this enzyme family is not well understood. Only motif II of the three domains has been proposed to function as a substrate-binding site based on amino acid sequence similarity to the GA biosynthetic enzyme GA20ox (Lee and Zeevaart, 2005; Lo et al., 2008). Within these three domains, 16 of 30 amino acids are identical in GA2oxs from 11 different species, indicating their functional conservation throughout evolution (Figure 1a). However, individuals of these identical amino acids seem to play distinct as well as common functions. For example, amino acids Y123 and H143 were most important, and E140, A141 and G343 were second important, while D338 and V339 were less important, for the function of GA2ox6 in controlling plant height and tillering (Figure 1b). Interestingly, mutations in the less conserved amino acids W138 and T341 enhanced the function of GA2ox6 by causing even more dwarfed plants. Mutations of GA2ox6 at E140, A141 and G343 conferred ideal plant architecture and physiology in transgenic rice in terms of productivity and stress tolerance.

Moderate semi-dwarf rice mutants possess high-yield potential

Three seasons of field trials demonstrated that transgenic line A141E significantly increased grain yield by 17–32%, and one field trial showed that line E140A also increased grain yield by 19% (Figures 5 and S5, Table 1). Yield potential is known to be controlled by complicated multiple factors. The so-called New Plant Type (NPT) of rice, including lower tillering (9–10 tillers/plant), no unproductive tillers, 200–250 grains/panicle, dark green, thick and erect leaves, and vigorous and deep root system, has been conceptualized for further improving grain yield potential for rice post the green revolution era (Jeon et al., 2011). Although the architecture in lines E140A and A141E only

meet certain criteria of NPT, a few newly acquired traits may contribute to the increase in its grain yield as compared with NT: first, the ratio of shoot to root dried biomass was reduced and a more vigorous root system was maintained (Figures 4a and S3c, S9), which got better access to nutrients and water in soil and thus had a potential of improving the HI (Figure 5b, Table 1). Second, dark green and erect leaves (Figure S8) with increased total chlorophyll content were developed and photosynthesis rate was enhanced. Third, more productive tillers were generated for grain production and filling (Figure 6, Table 1). No obvious negative phenotype was associated with line A141E and E140A, except slightly delayed germination rate (1-day delay) (Figure S3b) and heading date (1-week delay). Overexpression of an ethylene-responsive AP2/ERF factor, EATB, suppresses the expression of GA20ox2, also leading to reduced plant height, increased tillering and higher grain yield in transgenic rice (Qi et al., 2011). Our studies demonstrate that lines E140A and A141E, with heights 75 and 64% of NT at adult stage, respectively (Figure 2a), possess combined traits favourable for plant growth and represent ideotypes of rice with high-yield potentials.

Moderate GA-deficient mutant is more stress tolerant

GA-deficient rice is also significantly more tolerant to dehydration and slightly more tolerant to high salinity, heat and cold than NT (Figure 7a). GA deficiency has been implicated in enhancing salt tolerance in Arabidopsis, presumably by overexpressing a stress-responsive transcription factor encoded by the dwarf and delayed-flowering 1 (ddf1), which activates stress-related genes such as RD29A/COR78 (Magome et al., 2004), and in promoting cold and pathogen tolerance in a gibberellin-insensitive dwarf 1 (gid1) rice mutant but with unknown mechanism (Tanaka et al., 2006). Several factors might contribute to the capacity of abiotic stress tolerance in lines A141E and G343A. First, abiotic stresses induce the formation of toxic ROS that causes protein and membrane damages, and efficient scavenging of ROS by catalase (CAT) and ascorbate peroxidase (APX) is essential for osmotic tolerance in plants (Apel and Hirt, 2004; Hasegawa et al., 2000). Increase in proline levels and APX and CAT activities and decrease in ROS or peroxide accumulation may account for enhanced abiotic stress tolerance in these lines (Figure 7b). The notion is supported by a study showing that under abiotic or biotic stress, the GA-repressible DELLA proteins are induced to activate the expression of ROS-detoxification enzymes, thus reducing ROS levels and conferring stress tolerance in Arabidopsis (Achard et al., 2008). Second, higher chlorophyll levels enhance drought tolerance (Figure 6), as chlorophyll content is regarded one of several measures for capacity of dehydration tolerance in rice (Luo, 2010). Normally, chlorophyll contents declined in plants under drought conditions. Chlorophyll content, Fv/Fm which represents the maximum quantum yield of PSII, was reduced to less extents in drought-tolerant than in drought-sensitive barley varieties (Guo et al., 2009). Third, the higher water content in roots and leaves (Figures 8d and S9) enhance drought tolerance. Plants with higher water status could maintain physiological functions under drought conditions (Luo, 2010). Fourth, less water was consumed for the same production as NT, leading to significant increase in WUE in transgenic lines (Figure 4c). WUE is defined as the economic production per unit water consumption, but it may or may not related to drought resistance (Luo, 2010). GA-deficient rice turns out to have both enhanced WUE and drought resistance.

Total chlorophyll

Photosynthesis rate

Productive tiller

Filled seeds

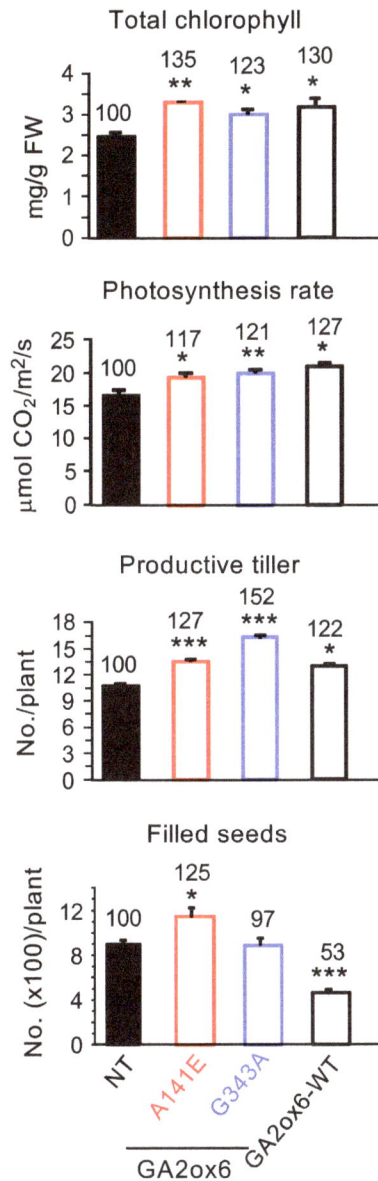

Figure 6 Moderate GA deficiencies improve multiple agronomic traits. Measured parameters: total chlorophyll content in 25-day-old plants, maximal photosynthesis rate, number of productive tillers and numbers of fertile (filled) seeds per plant before harvest. $n = 21$ for each line.

It is worthwhile noting that leaf blades of transgenic lines remained relatively open, in contrast to leaves rolled adaxially quickly in NT after dehydration. The leaf water potential, and in particular, loss of turgor pressure in BCs is tightly associated with the degree of leaf rolling (O'Toole and Cruz, 1980). The expanded BC volume may account for the unrolled leaf phenotypes after dehydration, which allows photosynthesis to continue in transgenic lines under stress conditions. The number and size of BCs also affect leaf shape, as loss of function of a *Narrow Leaf 7* (*NAL7*) gene, which controls auxin biosynthesis, results in reduced size and number of BCs and leaf width in rice (Fujino *et al.*, 2008). The GA-deficient transgenic rice has shorter and wider leaf blades than NT, and such phenotype correlates with the increase in volume and number of BCs (Figure 8b,c), indicating crosstalk

between GA and auxin signals may control the development of BCs and leaf width.

Although the productivity of line G343A was decreased by 10–22% in three field trials, it is more tolerant to dehydration, salinity and heat (Figure 7a). Under severe abiotic stress, enhanced stress tolerance may offset slight reduction in grain yield potential in line G343A as compared with line A141E. Transgenic line GA2ox6-WT also exhibited similar traits to lines A141E and G343A; however, the grain yield was significantly lower; we attribute this to overproduction of tillers that compete for carbon assimilates and mineral nutrients. Further, insufficient activity of GA may also impair reproductive growth and attenuate seed production in the GA2ox-WT line. Greater tiller number (Figure 2) and lower GA levels (Figure 3) may also explain the reduced grain yields in line G343A as compared with line A141.

Moderate GA deficiency reprograms transcriptional networks

GA is an essential hormone regulating growth and development throughout the entire life cycle of plants. Altered expression of genes regulated by GA may impact many aspects of plant growth and developmental processes. In the present study, we mainly focused on transcriptomics relevant to grain yield, abiotic and biotic stress tolerance (Figure 10), as these traits have been foci of many research.

Balance and crosstalk of hormones control root development, for example, auxin promotes lateral root growth (Orman-Ligeza *et al.*, 2013), while GA deficiency promotes adventitious root growth (Lo *et al.*, 2008). We showed that mutations at A141 and G343 resulted in down-regulation of Aux/IAA but up-regulation of ARFs and ARR, which may increase auxin biosynthesis and re-direct its transportation, leading to increase in bud activity and cytokinin biosynthesis (Domagalska and Leyser, 2011; Hussien *et al.*, 2014; Muller and Leyser, 2011; Orman-Ligeza *et al.*, 2013) and correlating with enhanced tiller and root growth in these two mutants.

Lower endogenous GA levels could reduce the destruction of DELLA protein (Sun, 2008), which in turn directly or indirectly induces a set of abiotic stress-related transcription factors (Gallego-Bartolome *et al.*, 2011; Qi *et al.*, 2014; Sun, 2010), such as those encoded by the group 3 genes, AP2/EREBPs, WRKYs and NACs, which are indeed up-regulated in shoots and roots of A141E and G343A transgenic lines. We also found that some genes encoding proteins responsible for ROS scavenging, such as SOD, APX, peroxiredoxin and glutaredoxin (GRX) were up-regulated in A141E and G343A (Tables S2 and S3), which could be related to ROS scavenging as well as tolerance to biotic and abiotic stresses (Figure 7).

GA deficiency also redirected important transcriptional networks related to biotic stress responses. One set of up-regulated genes includes signalling molecules and receptors, such as group 5 genes encoding DUF26 kinases, CW-associated kinases and LRR kinases. These genes together with other up-regulated group 6 genes encoding XIPs, PGIPs, R proteins and PR proteins could be responsible for the enhanced plant defence. It is intriguing that some PGIPs and most of XIPs are strongly induced in A141E and G343A transgenic lines, as these cell wall degrading enzyme inhibitors have been shown to be induced by pathogen infection and important in plant defence as well as in cell elongation (Juge, 2006; Pogorelko *et al.*, 2013). Conceivably, some PGIPs and XIPs inhibit fungal pectic enzymes and xylanases, respectively, which

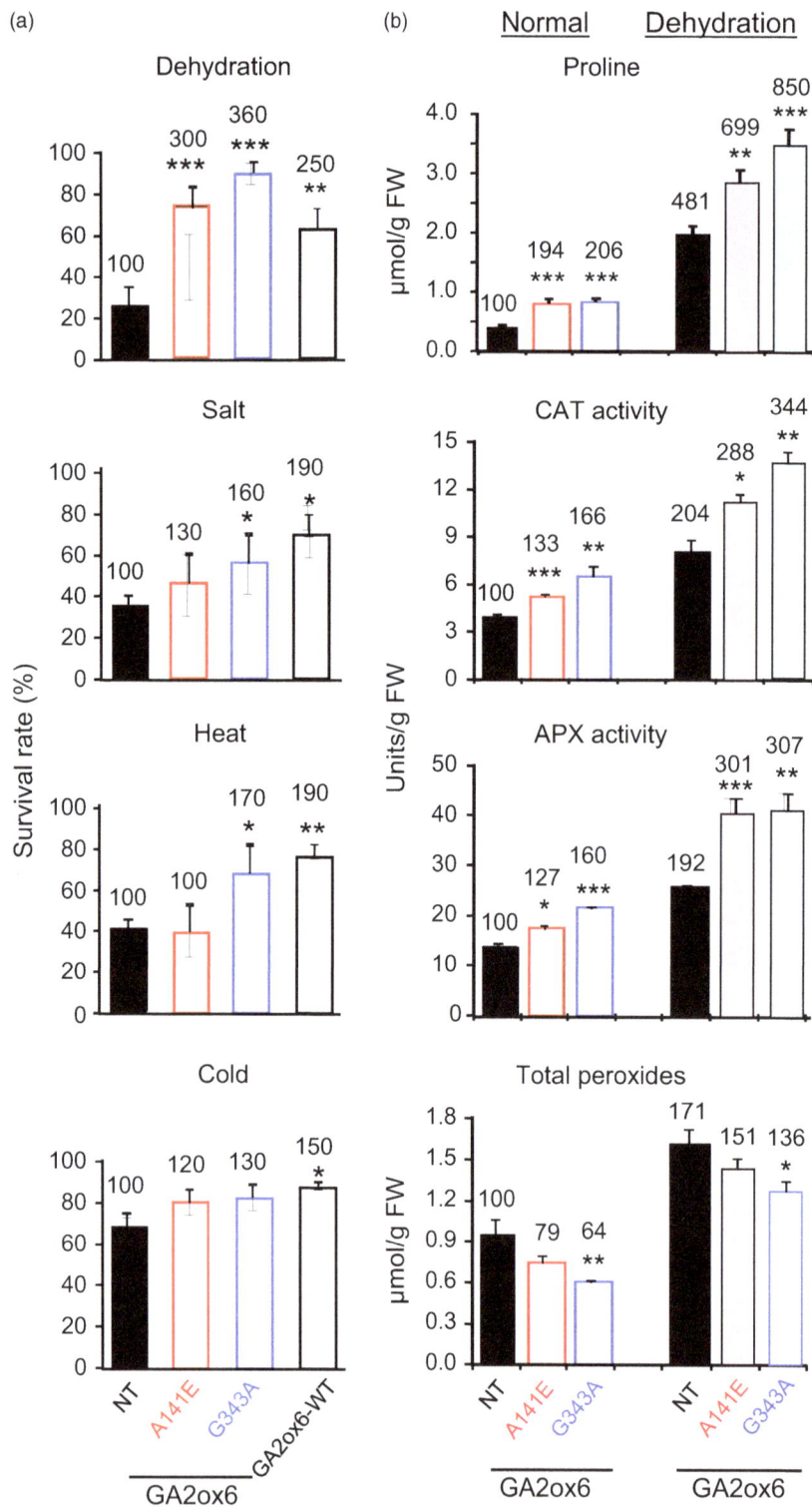

Figure 7 Abiotic stress tolerance, proline contents and ROS scavenging enzymes are enhanced in GA-deficient transgenic rice plants. Fourteen-day-old plants were used in following experiments. (a) Survival rates after recovery from various stress treatments, n = 120, 49, 64 and 73 for drought treatment; n = 124, 74, 58 and 85 for salt treatment; n = 117, 44, 58 and 85 for cold treatment; and n = 134, 74, 69 and 56 for heat treatment, for NT and lines A141E, G343A and GA2ox6-WT, respectively. (b) Proline content, catalase (CAT) and ascorbate peroxidase (APX) activities and total peroxide contents in plants treated with or without dehydration. n = 6 for each line.

are needed to degrade plant cell walls for a successful invasion, and other PGIPs and XIPs inhibit endogenous pectic enzymes and xylanase activity leading to lower cell wall extensibility and dwarfism. Additionally, we observed up-regulation of genes associated with PAMP-triggered immunity (PTI) and effector-triggered immunity (ETI) systems (Jones and Dangl, 2006), both of which correlate with the enhanced disease resistance in A141E and G343A.

Lines A141E and G343A behave very similarly, but also have major differences in some traits. For example, line G343A is shorter and has more tiller number and bulliform cell volume and greater abiotic stress tolerance, but with reduced grain yield as compared to line A141E. These differences in traits could be resulted from more significantly reduced levels of some GA precursors in line G343A (Figure 3), which is consistent with the observation that the expression of more genes was affected in

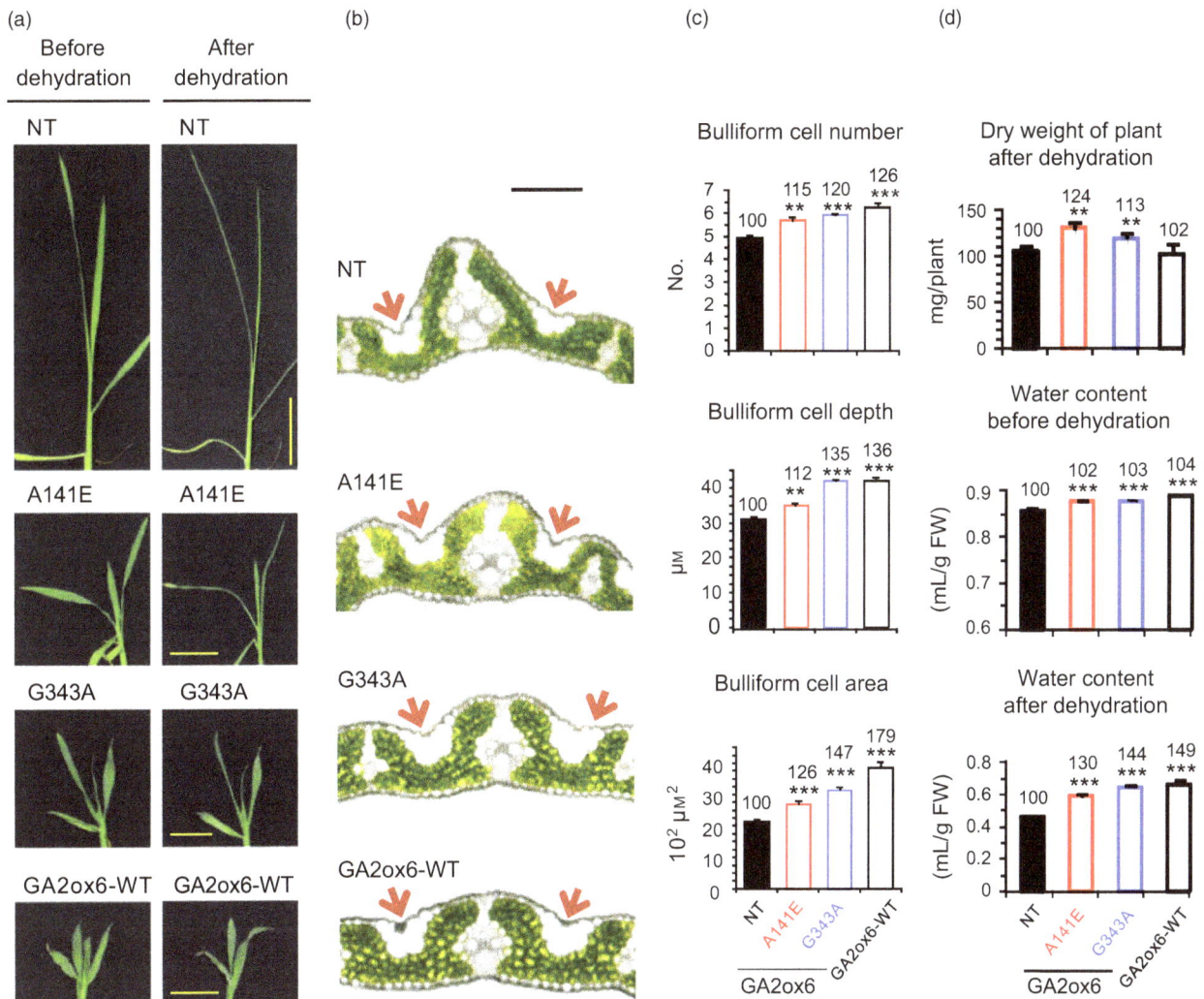

Figure 8 The elevated abiotic stress tolerance is associated with enhanced bulliform cell volume and water contents in GA-deficient plants. Fourteen-day-old plants were used in following experiments. (a) Leaf morphology before and after 3.5-h dehydration. Scale bar = 3 cm. Live images are also shown in Video S1. (b) Cross section of first fully expanded leaf under normal growth conditions. Scale bar = 100 μm. Arrow indicates bulliform cells. More leaf sections are also shown in Fig. S7. (c) Quantification of bulliform cell number, depth and area, n = 18, 14, 22, 14 for NT, A141E, G343A and GA2ox6-WT, respectively. (d) Plant dry weight after dehydration and water contents before and after 3.5-h dehydration. n = 21 for each line.

both shoots and roots in line G343A than in line A141E (Figure S10, Tables S2 and S3) as compared to the NT. Among these genes, 7.6 and 13.9% in shoots and roots of A141E, and 10.4 and 14.1% in shoots and roots of G343A, respectively, are known to be involved in abiotic and biotic stress responses and signalling. The fold induction of ARFs and ARR, AP2/EREBPs, WRKYs and NACs involved in abiotic stress tolerance was also significantly higher in G343A than in A141A. Overexpression of stress-related proteins may also attribute to the yield penalty in G343A as compared with A141E.

Taken together, several factors might contribute to the tolerance for water-deficit stress in GA-deficient transgenic lines. Elevated proline levels and ROS scavenging activities protect plants from damage under dehydration stress, and higher chlorophyll contents maintain photosynthesis under dehydration. Higher water contents may also allow plants to maintain metabolic functions under drought conditions. Finally, an increase in BC cell volume that store water may facilitate leaf unfolding

and rapid recovery from dehydration. Accumulation of ROS has also been shown to correlate with the aggressiveness of pathogens, and resistant cultivars had weaker symptoms, less ROS accumulation and higher activity of the ROS scavenging system in plants (El-Komy, 2014; Govrin and Levine, 2000). It appears that multiple defence mechanisms against pathogens are turned on in the GA-deficient transgenic rice.

In summary, although the Green Revolution has drastically enhanced productivity of wheat and rice, our studies demonstrated that ectopic expression of GA2ox mutants can further modulate endogenous GA levels, leading to partitioning of even greater proportions of plant biomass into grains. In addition, these manipulations resulted in enhanced tolerance to both biotic and abiotic stresses. This advanced version of Green Revolution could be applied to other crops for further optimization of plant architecture and function so that more stress- and disease-tolerant varieties requiring less water but with higher yields could be produced.

(a)

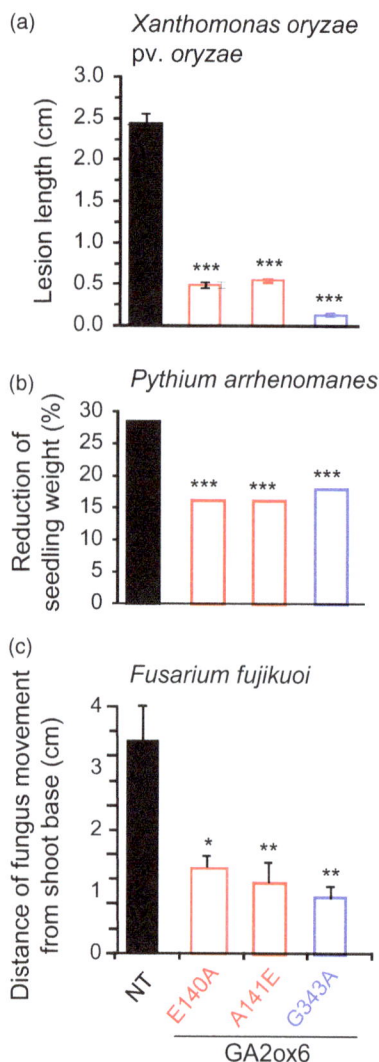

(b)

(c)

Figure 9 GA-deficient transgenic rice is more resistant to pathogens. (a) Lesion expansion on leaves after infection by *Xanthomonas oryze pv. oryzae*. (b) Seedling weight after infection by *Pythium arrhenomanes*. (c) Upward migration of *F. fujikuoi* from shoot base. $n = 10$ for each line in each treatment.

Materials and methods

Plant materials

The rice cultivar *Oryza sativa* L. cv Tainung 67 was used in this study. Transgenic and NT seeds were surface-sterilized with 2.5% NaOCl and placed on MS agar medium (Murashige and Skoog Basal Medium, Sigma, St. Louis, MO, USA) and incubated at 28 °C with 16-h light and 8-h dark for 14–20 days. Plants were transplanted to a GM-field with bird-free screen.

Database searching and phylogenetic analysis of C_{20} *GA2oxs*

Database searches for GA2oxs from different plant species and identification of C_{20} *GA2oxs* using the 30 amino acids present in the three unique conserved motifs in rice GA2ox6 were carried out as described (Lo *et al.*, 2008). Deduced amino acid sequences of all C_{19} and C_{20} GA2oxs were aligned as described (Lo *et al.*, 2008).

Site-directed mutation of rice GA2ox6 and rice transformation

To generate point mutations in three conserved domains of *GA2ox6*, amino acid residues Y123, W138, E140, H143, D338, V339, T341, G343 were substituted with alanine (A), and A141 was substituted with glutamate (E). Point mutations were conducted as described (Kunkel, 1985). *Ubi:GA2ox6* in plasmid pAHC18 (Lo *et al.*, 2008) was used as the template for point mutations of *Ubi:GA2ox6* by QuikChange® Site-directed Mutagenesis Kit (Stratagene, http://www.stratagene.com/) according to the manufacturer's instructions. Primers for mutagenesis are listed in Table S4. All plasmids were linearized with *Hind*III and inserted into the same site in pCAMBIA1301 (Hajdukiewicz *et al.*, 1994). Resulting binary vectors were transferred into *Agrobacterium tumefaciens* strain EHA105 and subsequently used for rice transformation as described (Ho *et al.*, 2000). Correct point mutations in recombinant GA2ox6 were confirmed by nucleotide sequencing of genomic DNAs isolated from transgenic lines.

Gibberellin quantification

To measure levels of endogenous GAs, ~200 mg of shoot tissues was collected from 21-day-old seedlings. Tissues were placed in a 2.0-mL Eppendorf tube (Labcon, Petaluma, CA, USA), lyophilized and ground to powder by TissueLyser (QIAGEN, Hilden, Germany) in liquid nitrogen. The concentrations of endogenous GA compounds were analysed as described previously (Ayano *et al.*, 2014; Kojima *et al.*, 2009). Two biological repeats were carried out in this measurement.

Water consumption and water use efficiency (WUE)

Eighteen-day-old seedlings were weighed, transferred to plastic tubes containing 50 mL water, and tube openings were sealed with parafilm. Water consumption was measured, and water was re-filled, every 2 days for up to 8 days. The increase in fresh weight of plants during the 8-day period, and final fresh and dried weights of plants were determined. The WUE of each plant was calculated by dividing the average increase in fresh weight by the average water consumption per day.

Yield evaluation

Homozygosity of T3 transgenic lines was confirmed by hygromycin selection. Three-week-old seedlings were transplanted to the open GM-field at National Chung-Hsing University and grown under natural conditions. At least two repeated blocks for each transgenic and NT lines, 24 plants in each block with a layout of 3 rows X 8 lines, and the space for each plant is 25 cm × 25 cm, were designed for field tests. The plant height, matured tiller number, biomass, panicle number and total yield data were collected from 18 plants, excluding the two marginal lines, in each block.

Chlorophyll content and photosynthetic rate

Fresh leaves were collected from the first expanded leaf of 80-day-old plants in the field. Leaves were ground with liquid nitrogen in a mortar and pestle. Pigments were extracted with 95% ethanol, and light absorptions at 648.6 and 664.2 nm were determined using a UV/visible spectrophotometer (Biowave II; Biochrom Ltd., Holliston, MA, USA). Concentrations of total chlorophylls were calculated as described (Lichtenthaler, 1987).

Leaf photosynthetic rate was measured from 9:30 to 16:00 using the LI-6400 Portable Photosynthesis System with a leaf chamber fluorometer attached (Model 6400-40; LICOR Inc.,

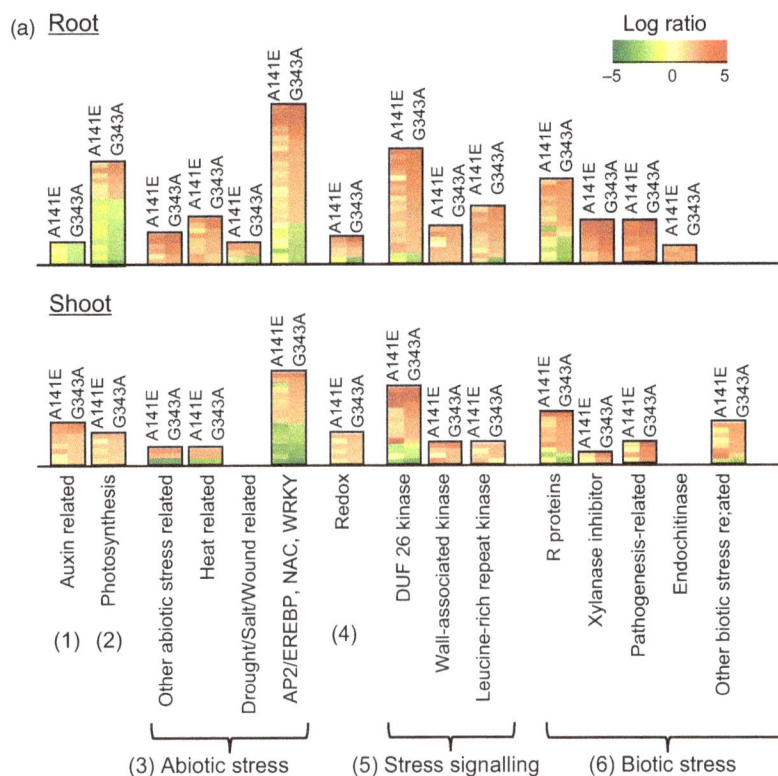

Figure 10 Genes that are significantly up-regulated contribute to increase in grain yield and abiotic and biotic stress tolerance in GA-deficient transgenic rice. (a) Expression profiles of six groups of genes (number in parenthesis labelled in blue) related to grain yield and abiotic and biotic stress tolerance were compared in roots and shoots between A141E and G343A mutants and NT. Clustered genes up- and down-regulated are marked in red and green, respectively. For detailed lists of genes in each cluster and extent of changes, see Tables S2 and S3. (b) Coordinated events and pathways leading to increase in grain yield and abiotic and biotic stress tolerance. Numbers denote up-regulated genes corresponding to gene clusters in A. Red and green fonts indicate representative up- and down-regulated genes, respectively. For abbreviation of gene names, see Materials and Methods. Small upward and downward arrowheads indicate up- and down-regulation of genes. Open arrowheads indicate suggested sequences of events.

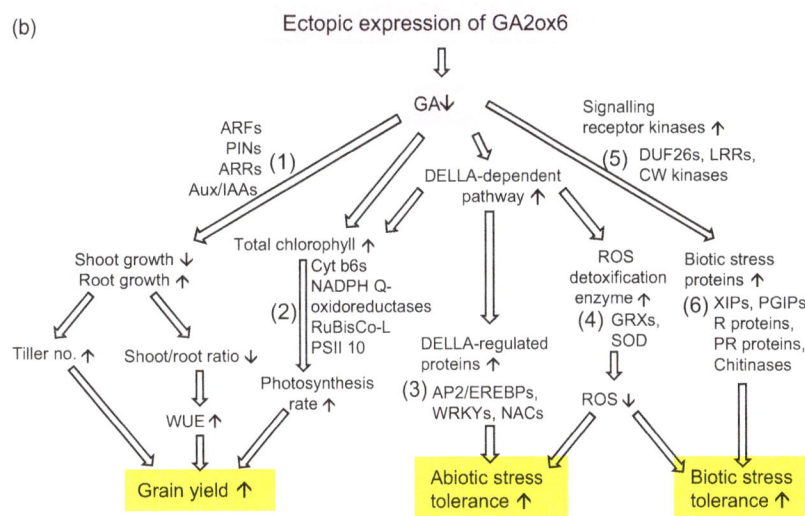

Lincoln, NE, USA); parameters used included: CO_2 flux, 500 μM/S; block temperature, 28 °C; photosynthetically active radiation, 800 μmol photons m^2/s. Triplicate measurements of photosynthetic assimilation rate (μM CO_2 m^2/S) were recorded after equilibration to a steady state (~20 min) and with three first fully expanded leaves for each plant.

Stress treatments

Fourteen-day-old plants were transferred to distilled water for one day and then incubated in a cold (4 °C) or warm (42 °C) incubator or in 200 mM NaCl solution for 2 days, or dehydrated (air-dried) on the bench at room temperature (25–27 °C) for 6 h. All stressed plants were allowed to recover in water for 6 days in a 28 °C incubator, and survival rates were determined.

Videos of plant responses to and recovery from stress treatments

Twenty five-day-old plants were treated with dehydration or 30% PEG6000 on the bench at room temperature (25–27 °C) for 3.5 h. Plants were then allowed to recover in water or roots were covered with 0.8% agar gel replenished with water at intervals for 20–24 h. Photographs were taken at an interval of 2 min. Time-lapse videos were organized by the Windows Movie Maker v2.6.

Quantification of proline and total peroxide content

Shoots of 14-day-old seedlings with or without 3-h dehydration were weighed. Proline was extracted with a ninhydrin reagent and quantified by the absorbance at 520 nm. Proline concentrations were calculated by a calibration curve and expressed as µmol proline g-1 fresh weight (Bates et al., 1973). Total peroxides of shoots of 15-day-old seedlings were extracted with 5% (w/v) trichloroacetic acid (TCA) according to the method described by Sagisaka (1976).

Activity assay of antioxidant enzymes

Proteins in shoots of 14-day-old seedlings with or without 3-h dehydration were extracted using the sodium phosphate buffer (50 mM, pH6.8). Activities of catalase and ascorbate peroxidase were determined by spectrophotometric methods as described by Kato and Shimizu (1985) and Nakano and Asada (1981), respectively.

Leaf structure examination

The first fully expanded leaves of 25-day-old plants were collected. Cross sections of leaf blades were made with Microtome Leica VT1200 (Leica Microsystems GmbH, Wetzlor, Germany) and examined with a light microscope (Upright microscope ECLIPSE Ni-U, Nikon Co., Tokyo, Japan).

Water content

Seventeen-day-old plants were used in this experiment. For determination of water contents, shoots and roots were blotted dry with paper towels and weighed before and after dehydration (air dry) for 3.5 h. For determination of dried weight, plant materials were dried in 80 °C oven for 2 days, and dried weights of shoots and roots were measured. The water content before dehydration was calculated by subtracting dried weight from fresh weight before dehydration, and the water content after dehydration was calculated by subtracting dried weight from fresh weight after dehydration.

Pathogen infection

Transgenic and NT seeds were surface-sterilized with 2.5% NaOCl and dipped in running water for 3 days. Germinated seeds were infected with the following pathogens. X. oryzae pv. oryzae: germinated seeds were transferred to pots filled with peat moss, incubated at 28 °C with 16-h light and 8-h dark for 14 days, and transferred to pots containing field soils and peat moss (1:1) for another 21 days. Plants were then infected with bacteria at a concentration of 1×10^8 cfu/mL using sterilized scissors (Kauffman et al., 1973). Length of lesion developed was measured. P. arrhenomanes: germinated seeds were transferred to 0.5X Kimura solution, pH 5.0, containing the fungus with concentration of absorbance 1.5 at OD_{600} and incubated at 28 °C with 16-h light and 8-h dark for 10 days. The reduction in total seedling fresh weight was measured. F. fujikuroi: germinated seeds were submerged in the spore suspension with concentration of 1×10^6 spores/mL for 1 h. Infected seeds were transferred to 0.5X Kimura solution, pH 5.0, and incubated at 28 °C with 16-h light and 8-h dark for 14 days. The upward migration distance of the fungus from shoot base was measured.

Microarray analysis

Total RNA was purified from rice shoots and roots of 17-day-old seedlings using Trizol® reagent (Invitrogen, Waltam, Massachusets, USA), and further purified with RNeasy Mini Kit (QIAGEN). RNA quality assessment and array experiment were conducted in the Affymetrix Gene Expression Service Lab of Academia Sinica,

using the GeneChip® Rice Genome Array (Affymetrix, Taipei, Taiwan). The Affymetrix CEL files were imported into GeneSpring 12.6 software (Agilent Technol., Santa Clara, CA, USA) for data normalization, generating the MAS5.0 algorithm. The flags of detection calls as 'Present (P)' or 'Margin (M)' were subjected to further analysis (Pepper et al., 2007). Genes with signal ratio greater than threefold changes were used for GeneOntology significance analysis with the AgriGO database (http://bioinfo.cau.edu.cn/agriGO), and functional classification was performed with the MapMan software (3.6.0RC1) (Thimm et al., 2004).

RT-PCR analyses

Total RNA was purified from rice leaves or roots, and RT-PCR analyses were conducted as described (Lo et al., 2008)

Primers

Nucleotides for all primers used for PCR and RT-PCR analyses are provided in Table S4.

Statistical analysis

All numerical data are presented as mean ± SEM (error bars indicate standard error of the mean). Statistical analyses were carried out by comparing the raw data of all individuals of each transgenic line with those of NT with the Student's t-test using the SigmaPlot software (version 11.0; Systat Software, Inc., San Jose, CA, USA).

For the relative levels shown above bars in figures, the value of each parameter in NT was set as 100, and the value of transgenic lines was calculated relative percentage to this value. Difference was compared between transgenic lines and NT. Significance levels were determined with the t-test: * $0.05 > P \geqq 0.01$, ** $0.01 > P \geqq 0.001$, *** $0.001 > P$.

Acknowledgement

We thank Dr. John O'Brien for critical review of the manuscript; and Dr. Paul Wei-Che Hsu for microarray data analysis. This work was supported by grants (94F005-1 and 94S1502 to Su-May Yu and 94F005-2 to Tuan-Hua David Ho) from Academia Sinica and the National Science Council (NSC101-2321-B001-035 to Su-May Yu and NSC101-2313-B-005-004 to Liang-Jwu Chen) of the Republic of China, and in part by the Ministry of Education, Taiwan, R.O.C., under the ATU plan to Su-May Yu, Tuan-Hua David Ho and Liang-Jwu Chen. None of the authors have any conflict of interest to declare.

Authors' contributions

SMY, THDH and SFL involved in the designing of the project; S-MY, T-HDH, LJC, M-HL, C-YC, T-PH and HS involved in the supervision of the project; S-FL, K-TH, K-TC, M-JJ and MK performed the experiments; M-JJ and L-CY provided materials and other support; S-FL, S-MY and T-HDH involved in the preparation of manuscript.

References

Achard, P., Renou, J.P., Berthome, R., Harberd, N.P. and Genschik, P. (2008) Plant DELLAs restrain growth and promote survival of adversity by reducing the levels of reactive oxygen species. Curr. Biol. **18**, 656–660.

Apel, K. and Hirt, H. (2004) Reactive oxygen species: metabolism, oxidative stress, and signal transduction. Annu. Rev. Plant Biol. **55**, 373–399.

Ayano, M., Kani, T., Kojima, M., Sakakibara, H., Kitaoka, T., Kuroha, T., Angeles-Shim, R.B. et al. (2014) Gibberellin biosynthesis and signal transduction is essential for internode elongation in deepwater rice. Plant, Cell Environ. 37, 2313–2324.

Bates, L.S., Waldren, R.P. and Teare, I.D. (1973) Rapid determination of free proline for water-stress studies. Plant Soil 39, 205–207.

Botwright, T.L., Rebetzke, G.J., Condon, A.G. and Richards, R.A. (2005) Influence of the gibberellin-sensitive Rht8 dwarfing gene on leaf epidermal cell dimensions and early vigour in wheat (Triticum aestivum L.). Ann. Bot. 95, 631–639.

Coles, J.P., Phillips, A.L., Croker, S.J., Garcia-Lepe, R., Lewis, M.J. and Hedden, P. (1999) Modification of gibberellin production and plant development in Arabidopsis by sense and antisense expression of gibberellin 20-oxidase genes. Plant J. 17, 547–556.

Domagalska, M.A. and Leyser, O. (2011) Signal integration in the control of shoot branching. Nat. Rev. Mol. Cell Biol. 12, 211–221.

El-Komy, M.H. (2014) Comparative analysis of defense responses in chocolate spot-resistant and -susceptible faba bean (Vicia faba) cultivars following infection by the necrotrophic fungus Botrytis fabae. Plant Pathol. J. 30, 355–366.

Fujino, K., Matsuda, Y., Ozawa, K., Nishimura, T., Koshiba, T., Fraaije, M. and Sekiguchi, H. (2008) NARROW LEAF 7 controls leaf shape mediated by auxin in rice. Mol. Genet. Genomics 279, 499–507.

Gallego-Bartolome, J., Alabadi, D. and Blazquez, M.A. (2011) DELLA-induced early transcriptional changes during etiolated development in Arabidopsis thaliana. PLoS ONE 6, e23918.

Govrin, E.M. and Levine, A. (2000) The hypersensitive response facilitates plant infection by the necrotrophic pathogen Botrytis cinerea. Curr. Biol. 10, 751–757.

Guo, P., Baum, M., Grando, S., Ceccarelli, S., Bai, G., Li, R., von Korff, M. et al. (2009) Differentially expressed genes between drought-tolerant and drought-sensitive barley genotypes in response to drought stress during the reproductive stage. J. Exp. Bot. 60, 3531–3544.

Hajdukiewicz, P., Svab, Z. and Maliga, P. (1994) The small, versatile pPZP family of Agrobacterium binary vectors for plant transformation. Plant Mol. Biol. 25, 989–994.

Hasegawa, P.M., Bressan, R.A., Zhu, J.K. and Bohnert, H.J. (2000) Plant cellular and molecular responses to high salinity. Annu. Rev. Plant Physiol. Plant Mol. Biol. 51, 463–499.

Hedden, P. and Thomas, S.G. (2012) Gibberellin biosynthesis and its regulation. Biochemical J. 444, 11–25.

Ho, S.L., Tong, W.F. and Yu, S.M. (2000) Multiple mode regulation of a cysteine proteinase gene expression in rice. Plant Physiol. 122, 57–66.

Hussien, A., Tavakol, E., Horner, D.S., Muñoz-Amatriaín, M., Muehlbauer, G.J. and Rossini, L. (2014) Genetics of tillering in rice and barley. Plant Genome, (accessed 13 January 2014). doi: 10.3835/plantgenome2013.10.0032.

IRRI (2010) Rice Policy - Why is it happening? http://beta.irri.org/solutions/index.php?option=com_content&task=view&id=15.

Jeon, J.-S., Jung, K.-H., Kim, H.-B., Suh, J.-P. and Khush, G. (2011) Genetic and molecular insights into the enhancement of rice yield potential. J. Plant Biol. 54, 1–9.

Jones, J.D.G. and Dangl, J.L. (2006) The plant immune system. Nature, 444, 323–329.

Juge, N. (2006) Plant protein inhibitors of cell wall degrading enzymes. Trends Plant Sci. 11, 359–367.

Kato, M. and Shimizu, S. (1985) Chlorophyll metabolism in higher plants VI. Involvement of peroxidase in chlorophyll degradation. Plant Cell Physiol. 26, 1291–1301.

Kauffman, H., Reddy, A., Hsieh, S. and Merca, S. (1973) Improved technique for evaluating resistance of rice varieties to Xanthomonas oryzae. Plant Dis. Rep. 57, 537–541.

Khush, G.S. (1999) Green revolution: preparing for the 21st century. Genome, 42, 646–655.

Kojima, M., Kamada-Nobusada, T., Komatsu, H., Takei, K., Kuroha, T., Mizutani, M., Ashikari, M. et al. (2009) Highly sensitive and high-throughput analysis of plant hormones using MS-probe modification and liquid chromatography-tandem mass spectrometry: an application for hormone profiling in Oryza sativa. Plant Cell Physiol. 50, 1201–1214.

Kunkel, T.A. (1985) Rapid and efficient site-specific mutagenesis without phenotypic selection. Proc. Natl Acad. Sci. USA 82, 488–492.

Lee, D.J. and Zeevaart, J.A. (2005) Molecular cloning of GA 2-oxidase3 from spinach and its ectopic expression in Nicotiana sylvestris. Plant Physiol. 138, 243–254.

Lichtenthaler, H.K. (1987) [34] Chlorophylls and carotenoids: Pigments of photosynthetic biomembranes. In Methods in Enzymol (Lester Packer, R.D. ed.), pp. 350–382. Academic Press, Elsevier, Amsterdam, Netherlands: Academic Press.

Lo, S.F., Yang, S.Y., Chen, K.T., Hsing, Y.I., Zeevaart, J.A., Chen, L.J. and Yu, S.M. (2008) A novel class of gibberellin 2-oxidases control semidwarfism, tillering, and root development in rice. Plant Cell 20, 2603–2618.

Luo, L.J. (2010) Breeding for water-saving and drought-resistance rice (WDR) in China. J. Exp. Bot. 61, 3509–3517.

Magome, H., Yamaguchi, S., Hanada, A., Kamiya, Y. and Oda, K. (2004) dwarf and delayed-flowering 1, a novel Arabidopsis mutant deficient in gibberellin biosynthesis because of overexpression of a putative AP2 transcription factor. Plant J. 37, 720–729.

Moore, R., Clark, W.D., Vodopich, D.S., Stern, K.R. and Lewis, R. (1998) Botany. Dubuque IA USA: McGraw-Hill College.

Muller, D. and Leyser, O. (2011) Auxin, cytokinin and the control of shoot branching. Ann. Bot. 107, 1203–1212.

Nakano, Y. and Asada, K. (1981) Hydrogen peroxide is scavenged by ascorbate-specific peroxidase in Spinach Chloroplasts. Plant Cell Physiol. 22, 867–880.

Orman-Ligeza, B., Parizot, B., Gantet, P.P., Beeckman, T., Bennett, M.J. and Draye, X. (2013) Post-embryonic root organogenesis in cereals: branching out from model plants. Trends Plant Sci. 18, 459–467.

O'Toole, C. and Cruz, R.T. (1980) Response of leaf water potential, stomatal resistance, and leaf rolling to water stress. Plant Physiol. 65, 428–432.

Peng, J., Richards, D.E., Hartley, N.M., Murphy, G.P., Devos, K.M., Flintham, J.E., Beales, J. et al. (1999) 'Green revolution' genes encode mutant gibberellin response modulators. Nature 400, 256–261.

Pepper, S.D., Saunders, E.K., Edwards, L.E., Wilson, C.L. and Miller, C.J. (2007) The utility of MAS5 expression summary and detection call algorithms. BMC Bioinformatics, 8, 273.

Pogorelko, G., Lionetti, V., Bellincampi, D. and Zabotina, O. (2013) Cell wall integrity: targeted post-synthetic modifications to reveal its role in plant growth and defense against pathogens. Plant Signal. Behav. 8, e25435.

Qi, W., Sun, F., Wang, Q., Chen, M., Huang, Y., Feng, Y.Q., Luo, X. et al. (2011) Rice ethylene-response AP2/ERF factor OsEATB restricts internode elongation by down-regulating a gibberellin biosynthetic gene. Plant Physiol. 157, 216–228.

Qi, T., Huang, H., Wu, D., Yan, J., Qi, Y., Song, S. and Xie, D. (2014) Arabidopsis DELLA and JAZ Proteins Bind the WD-Repeat/bHLH/MYB complex to modulate gibberellin and jasmonate signaling synergy. Plant Cell 26, 1118–1133.

Qin, X., Liu, J.H., Zhao, W.S., Chen, X.J., Guo, Z.J. and Peng, Y.L. (2013) Gibberellin 20-oxidase gene OsGA20ox3 regulates plant stature and disease development in rice. Mol. Plant-Microbe Interact. 26, 227–239.

Rieu, I., Ruiz-Rivero, O., Fernandez-Garcia, N., Griffiths, J., Powers, S.J., Gong, F., Linhartova, T. et al. (2008) The gibberellin biosynthetic genes AtGA20ox1 and AtGA20ox2 act, partially redundantly, to promote growth and development throughout the Arabidopsis life cycle. Plant J. 53, 488–504.

Sagisaka, S. (1976) The occurrence of peroxide in a perennial plant, Populus gelrica. Plant Physiol. 57, 308–309.

Sakai, M., Sakamoto, T., Saito, T., Matsuoka, M., Tanaka, H. and Kobayashi, M. (2003) Expression of novel rice gibberellin 2-oxidase gene is under homeostatic regulation by biologically active gibberellins. J. Plant. Res. 116, 161–164.

Sakamoto, T., Morinaka, Y., Ishiyama, K., Kobayashi, M., Itoh, H., Kayano, T., Iwahori, S. et al. (2003) Genetic manipulation of gibberellin metabolism in transgenic rice. Nat. Biotechnol. 21, 909–913.

Sakamoto, T., Miura, K., Itoh, H., Tatsumi, T., Ueguchi-Tanaka, M., Ishiyama, K., Kobayashi, M. et al. (2004) An overview of gibberellin metabolism enzyme genes and their related mutants in rice. Plant Physiol. 134, 1642–1653.

Sperdouli, I. and Moustakas, M. (2012) Interaction of proline, sugars, and anthocyanins during photosynthetic acclimation of *Arabidopsis thaliana* to drought stress. *Plant Physiol.* **169**, 577–585.

Sun, T.P. (2008) Gibberellin metabolism, perception and signaling pathways in Arabidopsis. *Arabidopsis Book*, **6**, e0103.

Sun, T.P. (2010) Gibberellin-GID1-DELLA: a pivotal regulatory module for plant growth and development. *Plant Physiol.* **154**, 567–570.

Szabados, L. and Savoure, A. (2010) Proline: a multifunctional amino acid. *Trends Plant Sci.* **15**, 89–97.

Tanaka, N., Matsuoka, M., Kitano, H., Asano, T., Kaku, H. and Komatsu, S. (2006) gid1, a gibberellin-insensitive dwarf mutant, shows altered regulation of probenazole-inducible protein (PBZ1) in response to cold stress and pathogen attack. *Plant, Cell Environ.* **29**, 619–631.

Thimm, O., Blasing, O., Gibon, Y., Nagel, A., Meyer, S., Kruger, P., Selbig, J. *et al.* (2004) MAPMAN: a user-driven tool to display genomics data sets onto diagrams of metabolic pathways and other biological processes. *Plant J.* **37**, 914–939.

Wang, C., Yang, Y., Wang, H., Ran, X., Li, B., Zhang, J. and Zhang, H. (2016) Ectopic expression of a cytochrome P450 monooxygenase gene PtCYP714A3 from *Populus trichocarpa* reduces shoot growth and improves tolerance to salt stress in transgenic rice. *Plant Biotech. J.* **14**, 1838–1851.

Yamaguchi, S. (2008) Gibberellin metabolism and its regulation. *Annu. Rev. Plant Biol.* **59**, 225–251.

Yang, X.C. and Hwa, C.M. (2008) Genetic modification of plant architecture and variety improvement in rice. *Heredity* **101**, 396–404.

13

Improvement in phosphate acquisition and utilization by a secretory purple acid phosphatase (OsPAP21b) in rice

Poonam Mehra, Bipin Kumar Pandey and Jitender Giri*

National Institute of Plant Genome Research, New Delhi, India

*Correspondence
email jitender@nipgr.ac.in

Keywords: acid phosphatase, organic phosphates, P use efficiency, overexpression, RNAi, root, secretory protein.

Summary

Phosphate (Pi) deficiency in soil system is a limiting factor for rice growth and yield. Majority of the soil phosphorus (P) is organic in nature, not readily available for root uptake. Low Pi-inducible purple acid phosphatases (PAPs) are hypothesized to enhance the availability of Pi in soil and cellular system. However, information on molecular and physiological roles of rice PAPs is very limited. Here, we demonstrate the role of a novel rice PAP, OsPAP21b in improving plant utilization of organic-P. OsPAP21b was found to be under the transcriptional control of OsPHR2 and strictly regulated by plant Pi status at both transcript and protein levels. Biochemically, OsPAP21b showed hydrolysis of several organophosphates at acidic pH and possessed sufficient thermostability befitting for high-temperature rice ecosystems with acidic soils. Interestingly, OsPAP21b was revealed to be a secretory PAP and encodes a distinguishable major APase (acid phosphatase) isoform under low Pi in roots. Further, OsPAP21b-overexpressing transgenics showed increased biomass, APase activity and P content in both hydroponics supplemented with organic-P sources and soil containing organic manure as sole P source. Additionally, overexpression lines depicted increased root length, biomass and lateral roots under low Pi while RNAi lines showed reduced root length and biomass as compared to WT. In the light of these evidences, present study strongly proposes OsPAP21b as a useful candidate for improving Pi acquisition and utilization in rice.

Introduction

Given its key role in metabolism and signalling, phosphorus (P) is essential for plant growth and development. However, plant-available P (Pi) is often a limiting factor for crop production in many world soils. About 20 mha of upland area under rice cultivation is Pi deficient (Neue et al., 1990). In major rice-producing areas such as India, ~60% soils have low to medium Pi availability (Murumkar et al., 2015). Application of phosphatic fertilizers can ameliorate soil Pi deficiency. Unfortunately, the source of Pi fertilizers, rock phosphate is finite, rapidly depleting and concentrated only in few regions worldwide (Cordell et al., 2009). Further, applied Pi is quickly fixed into insoluble inorganic or organic forms due to its high reactivity and microbial action (Pandey et al., 2013; Richardson and Simpson, 2011). As rice is one of the major consumers of Pi fertilizers, enhancement of P use efficiency is highly desired for sustainable rice production.

Owing to its high reactivity, there exist more than 170 mineral or inorganic forms of P in soil (Holford, 1997). However, only ionic forms (orthophosphates; $H_2PO_4^-$ and HPO_4^{2-}) of P in soil solution constitute the 'plant-available' P. Further, a large fraction of soil P (50%–80%) occurs as organic-P and constitutes the bound P pool, not readily available for root uptake (Wang et al., 2009). These organic forms majorly include phosphate monoesters (derivatives of inositol hexakisphosphate) and to some extent phosphate diesters (nucleic acids, phospholipids, cyclic phosphates) and phosphonates (George and Richardson, 2008; Turner, 2008). Consequently, despite having significant amount of P in the form of organic complexes in soil, plants experience its

deficiency. Plants have devised several strategies to solubilize these P forms which involve exudation of organic acids and protons, modification of root system architecture, association with mycorrhiza and secretion of APases and phytases (Shen et al., 2011).

PAPs are involved in Pi acquisition and utilization in plants (Kuang et al., 2009). These enzymes represent the largest group of APases (E.C. 3.1.3.2) and are characterized by their pink or purple colour in water solution due to a charge transfer transition between Tyr residue to Fe (III) in binuclear metal centre (Oddie et al., 2000; Wang et al., 2015). PAPs contain five conserved blocks of seven amino acids (**D**XG/G**D**XX**Y**/**G**NH(D/E)/VXX**H**/**G**H**X**H) which coordinate the metal binding at binuclear centre (Li et al., 2002). PAPs are present in diverse organisms including bacteria, fungi, plants and mammals (Kuang et al., 2009). Plant PAPs exist as multigene family in Arabidopsis (Li et al., 2002), rice (Zhang et al., 2011), soybean (Li et al., 2012a) and maize (González-Muñoz et al., 2015). These PAPs catalyse the hydrolysis of several P-containing organic compounds at acidic pH (pH 4–7). Intracellular PAPs remobilize Pi from cellular reserves, whereas secreted ones hydrolyse P compounds in apoplast and rhizosphere. Several PAPs have been reported to be induced by Pi deficiency in Arabidopsis and crop plants (González-Muñoz et al., 2015; Li et al., 2002; Mehra et al., 2016; Zhang et al., 2011). Overexpression of few of them could improve the plant growth on organic-P-supplemented media (reviewed in Tian and Liao, 2015). In Arabidopsis, 29 putative PAPs have been reported, of which 11 are low Pi inducible (Wang et al., 2014). Some of these PAPs, AtPAP17, AtPAP26, AtPAP12, AtPAP25, AtPAP15 and

AtPAP10, have been well characterized for their ability to hydrolyse organic-P (Del Vecchio *et al.*, 2014; Kuang *et al.*, 2009; del Pozo *et al.*, 1999; Tran *et al.*, 2010b; Wang *et al.*, 2011). In rice, 26 PAPs have been identified, of which ten are induced significantly under Pi deficiency (Mehra *et al.*, 2016; Zhang *et al.*, 2011). Recently, rice PAPs, *OsPAP10a* and *OsPAP10c*, the low Pi-inducible rice homologues of *AtPAP10*, were shown to increase ATP hydrolysis when overexpressed in rice (Lu *et al.*, 2016; Tian *et al.*, 2012). However, except for these two PAPs, no detailed study of any of the rice PAPs has been carried out so far using rice as a system. Further, studies encompassing biochemical properties, transcriptional regulation and loss of function of rice PAPs are largely missing.

Here, we investigated role of a novel PAP, OsPAP21b in improving Pi acquisition and utilization in rice through elaborate biochemical, molecular and functional characterization. Our results revealed that OsPAP21b is a secretory protein and plays key roles in organic-P utilization in rice.

Results

OsPAP21b is a low Pi-induced gene and regulated by OsPHR2

In our previous transcriptome study of two rice genotypes under Pi deficiency, *OsPAP21b* was highly induced especially, in low Pi-tolerant genotype (Mehra *et al.*, 2016). In the present study, we found relatively higher up-regulation of *OsPAP21b* in roots as compared to shoot tissues under Pi deficiency (Figure 1a). This indicates that *OsPAP21b* is a root-preferential phosphate starvation response (PSR) gene. We further found that up-regulation is specific to Pi starvation as prolonged exposure to different nutrient deficiencies led to the induction of *OsPAP21b* in roots under Pi deficiency only (Figure 1c). Although slight up-regulation of *OsPAP21b* was also observed after 7 days of N and K deficiency, after 15 days, it was down-regulated. This suggests that *OsPAP21b* is primarily responsive to Pi deficiency. We next analysed the transcriptional regulation of *OsPAP21b* in 15-day-old Pi-starved seedlings resupplied with +Phi (320 μm Phosphite) or +P (320 μm NaH$_2$PO$_4$). Phi is a non-metabolizable form of Pi which cannot substitute for Pi in plants, and is known to interfere or suppress the expression of many low Pi-inducible genes (Varadarajan *et al.*, 2002). However, in our study, *OsPAP21b* was found to be largely non-responsive to Phi treatment as *OsPAP21b* was consistently up-regulated even after 48 h of Phi supply (Figure 1b). On the other hand, Pi resupply to Pi-starved seedlings suppressed the expression of *OsPAP21b* within 1 h in roots and 2 h in shoot (Figure 1b). Notably, this suppression was accompanied with simultaneous increase in total P content in these tissues. This implies that induction of *OsPAP21b* strongly depends on Pi status of plants and not by local availability of Pi in media.

OsPHR2 physically interacts with the promoter of OsPAP21b

Most of the low Pi-induced molecular responses in rice are regulated by a MYB transcription factor, OsPHR2. To test whether *OsPAP21b* is also transcriptionally regulated by OsPHR2-dependent pathway, we scanned 2 kb promoter region of *OsPAP21b* and found one potential OsPHR2 binding site (P1BS element) between −421 and −414 upstream of ATG. The slow migrating protein–DNA complexes in EMSA gel indicated binding of OsPHR2 with *OsPAP21b* promoter (Figure 1d). Further, competitive EMSA with 400-fold excess of unlabelled *OsPAP21b* promoter

(competitor) confirmed specificity of this physical interaction (Figure 1d).

Phylogeny of OsPAP21b with other plant PAPs of Ib subgroup

PAPs of Ib subgroup were identified in eight different plants including rice and Arabidopsis using NCBI blast search. On the basis of multiple sequence alignments, neighbour-joining tree was constructed which subdivided all Ib subgroup PAPs into four clades (Figure 2a). Interestingly, OsPAP21b showed closest homology with PAPs of other monocots (*Zea mays* and *Sorghum bicolor*) and formed one clade (Clade I). On the other hand, Arabidopsis homologue of OsPAP21b, AtPAP21 grouped with Ib PAPs of other dicots and formed a distinct clade, clade IV. Other 1b rice PAPs, OsPAP18 and OsPAP20 separated into two different clades (III and II, respectively) which revealed further divergence among rice Ib subgroup PAPs.

OsPAP21b is a functional acid phosphatase

OsPAP21b had all seven conserved amino acids required for its catalytic activity. To test its activity *in vitro*, we purified recombinant 6xHis-OsPAP21b protein by immobilized metal-ion chromatography (IMAC) and analysed on SDS-PAGE (Figure S1). A band of expected size, i.e. 51.24 kDa, was detected which was further confirmed by immunoblotting with anti-OsPAP21b antibody (Figure S1d). Enzyme activity assays showed that OsPAP21b can release Pi from different P-containing organic and inorganic substrates revealing that OsPAP21b is a functional phosphatase with broad substrate specificity (Figure 2b). The highest activity of OsPAP21b was detected with generic substrate pNPP followed by inorganic substrate PPi. Analysis of kinetic parameters revealed that OsPAP21b possesses sufficient specific activity with pNPP (2.0246 ± 0.0932 units/mg protein) and ADP (1.8363 ± 0.2684 units/mg protein) with K_m 0.09 ± 0.01 mm and 0.077 ± 0.004 mm, respectively. Interestingly, OsPAP21b showed fairly high activity with phosphorylated amino acids (p-Ser, p-Thr and p-Tyr) and ADP. Our analysis further confirmed APase nature of OsPAP21b as highest activity was observed in acidic pH (pH 5.0) (Figure 2c). Additionally, OsPAP21b was found to be moderately thermostable with highest activity at 65 °C (Figure 2d). We further investigated the effect of different anions and divalent cations on the activity of OsPAP21b (Figure 2e). Notably, activity of OsPAP21b was completely inhibited by high concentration of Pi and showed 50% activity inhibition (IC$_{50}$) at 5.09 ± 1.34 mm concentration of Pi. Further, Co^{2+}, Mn^{2+} and Ni^{2+} were found to be preferred cofactors for OsPAP21b activity.

Overexpression of *OsPAP21b* improved plant growth on organic-P substrates

To elucidate the functional roles of *OsPAP21b*, full-length cDNA of *OsPAP21b* along with 5′ and 3′ UTRs was constitutively overexpressed under *ZmUbi1* promoter in rice (OE lines; Figures S2, S3). All OE lines showed significant overexpression of *OsPAP21b* as compared to WT (wild type) at both transcript and protein levels (Figure 3). To assess the effects of *OsPAP21b* overexpression, three independent T3 homozygous lines and WT were grown under +P (320 μm NaH$_2$PO$_4$), −P (1 μm NaH$_2$PO$_4$) and +ATP (15-day-old −P grown seedlings recovered with 320 μm ATP for the next 15 days) for 30 days. Notably, OE lines and WT showed increased accumulation of both transcripts and protein under −P conditions in roots as compared to their

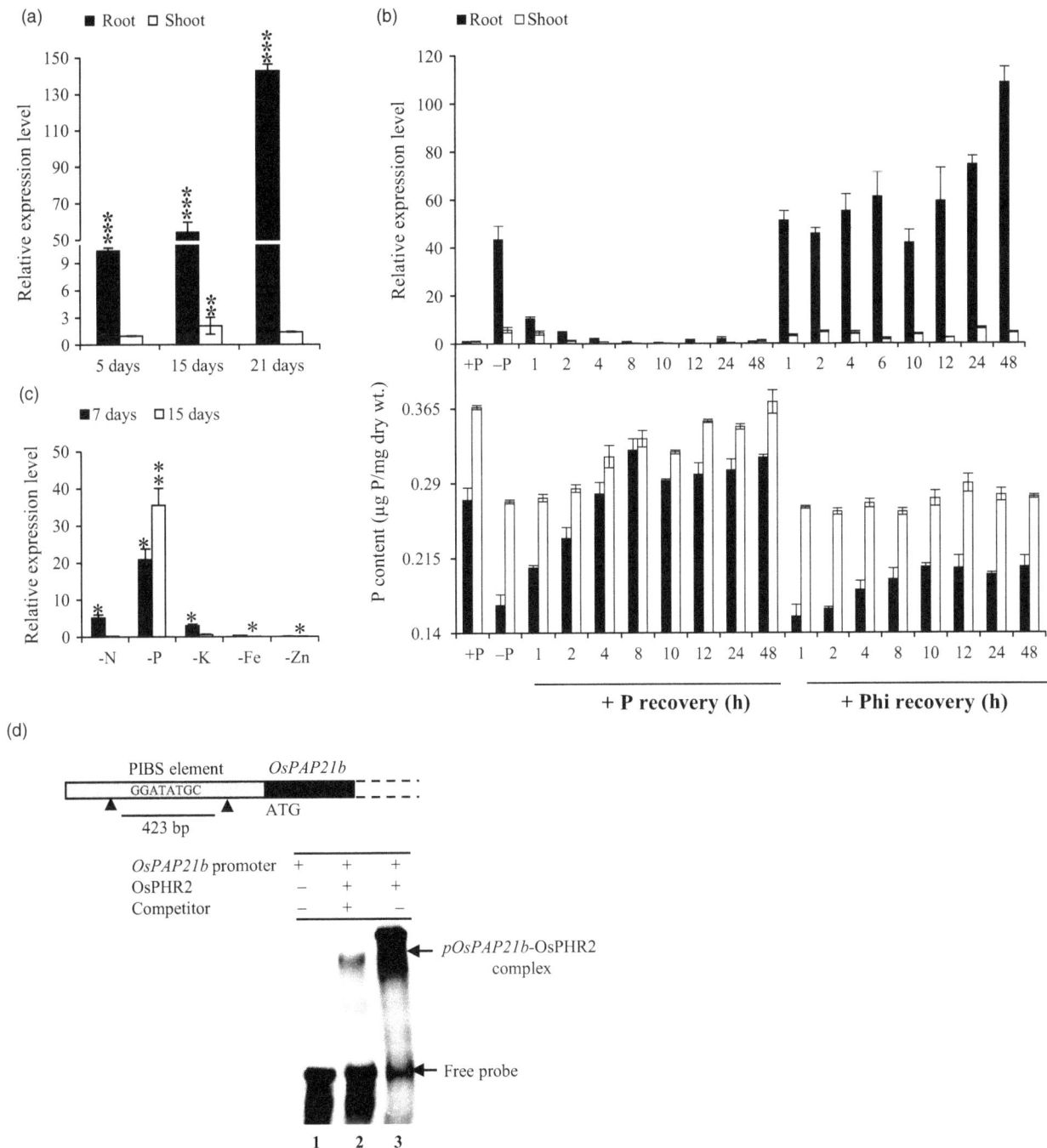

Figure 1 Transcriptional regulation of *OsPAP21b* under Pi deficiency. (a) Expression of *OsPAP21b* in rice under Pi deficiency. Relative expression under Pi deficiency was evaluated with respect to Pi sufficient conditions at 5, 15 and 21 days in WT. (b) Expression of *OsPAP21b* (upper panel) and total P content (lower panel) in roots and shoots of 15-day-old seedlings under +P, −P and after recovery of Pi-starved seedlings with either Pi or Phi. (c) Expression profiling of *OsPAP21b* after 7 and 15 days of nitrogen (−N), phosphorus (−P), potassium (−K), iron (−Fe) and zinc (−Zn) deficiency in roots. Gene expression levels under nutrient-deficient conditions with respect to corresponding sufficient conditions were determined by qRT-PCR. *P value <0.05; **P value <0.01; ***P value <0.001 were determined by Student's t-test. (d) Binding of *OsPAP21b* promoter with OsPHR2 by EMSA. 423-bp promoter region of *OsPAP21b* containing one P1BS element (−421 to −414 bp) was radiolabeled with [α^{32}P]CTP and used for binding assays with recombinant OsPHR2 protein. Slow migrating protein–DNA complexes and free probe are indicated by arrow at the top and bottom of the PAGE gel, respectively.

corresponding +P condition (Figure 3a, d). Moreover, relatively higher protein levels were observed in roots as compared to shoots in OE lines under −P condition (Figure 3c). This was consistent with higher up-regulation of *OsPAP21b* in roots as compared to shoots in OE lines as compared to WT (Figure 3a,

b). Further, distinct band of OsPAP21b could not be detected in WT by western blot, indicating low expression of OsPAP21b (Figure 3c, d). However, with higher protein load, a faint band could be visualized in WT under −P condition in roots (Figure 3d).

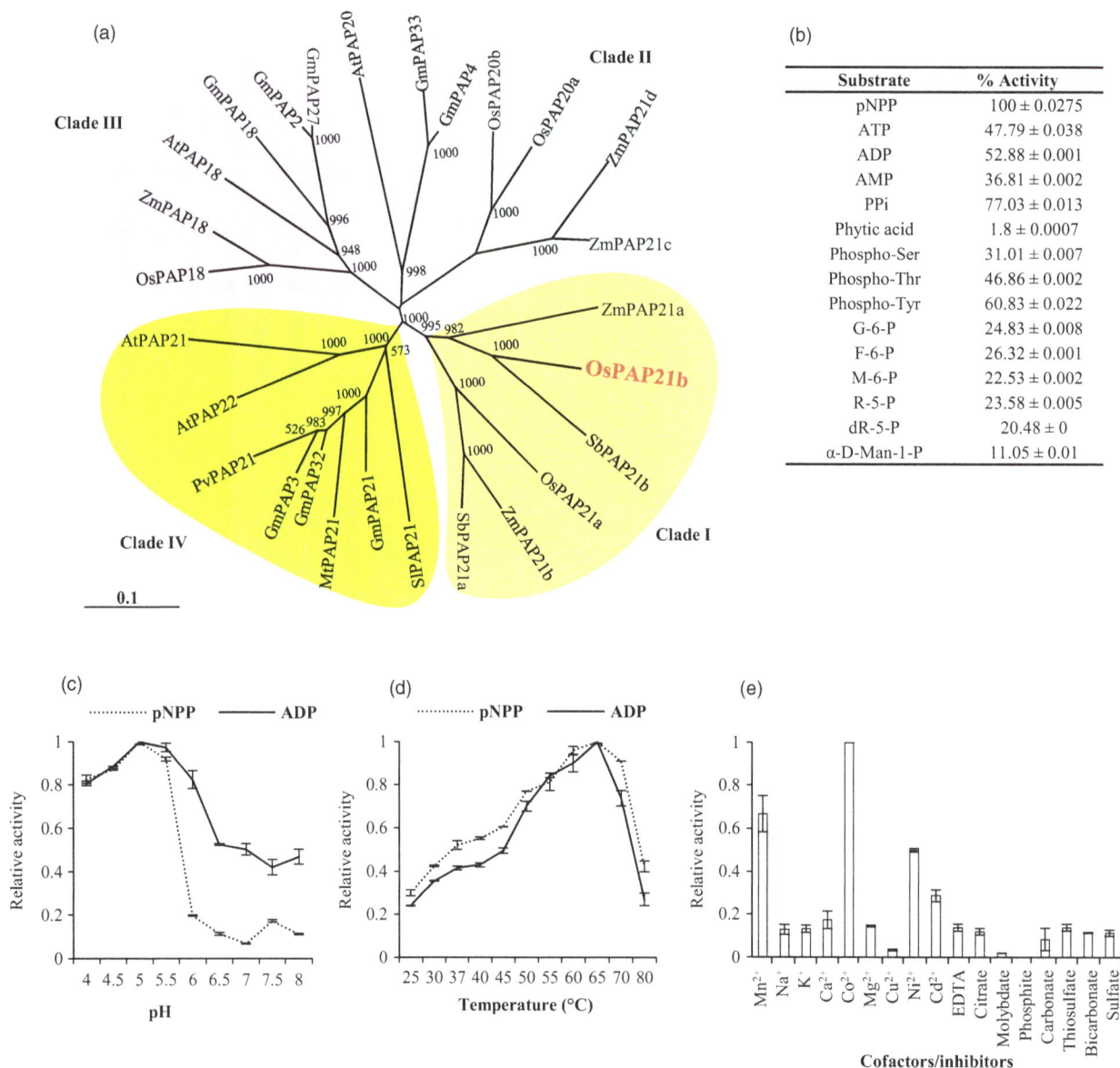

Substrate	% Activity
pNPP	100 ± 0.0275
ATP	47.79 ± 0.038
ADP	52.88 ± 0.001
AMP	36.81 ± 0.002
PPi	77.03 ± 0.013
Phytic acid	1.8 ± 0.0007
Phospho-Ser	31.01 ± 0.007
Phospho-Thr	46.86 ± 0.002
Phospho-Tyr	60.83 ± 0.022
G-6-P	24.83 ± 0.008
F-6-P	26.32 ± 0.001
M-6-P	22.53 ± 0.002
R-5-P	23.58 ± 0.005
dR-5-P	20.48 ± 0
α-D-Man-1-P	11.05 ± 0.01

Figure 2 Phylogeny and biochemical properties of OsPAP21b. (a) NJ tree representing phylogenetic relationship of OsPAP21b with other PAPs of Ib subgroup from *Oryza sativa* (OsPAPs), *Arabidopsis thaliana* (AtPAPs), *Zea mays* (ZmPAPs), *Glycine max* (GmPAPs), *Phaseolus vulgaris* (PvPAPs), *Solanum lycopersicum* (SlPAPs), *Medicago truncatula* (MtPAPs) and *Sorghum bicolor* (SbPAPs). Tree was generated in ClustalX 2.0.11 with 1000 bootstrap and viewed with TreeView 1.6. Scale bar represents rate of amino acid substitutions. (b) Relative APase activity of OsPAP21b on different P-containing substrates. APase activity with pNPP was considered as 100%. (c) Effect of pH and (d) temperature on APase activity of OsPAP21b using pNPP and ADP as substrates. (e) Influence of different anions and divalent cations on APase activity of OsPAP21b using pNPP as substrate. For calculating relative APase activities, maximum activity was considered equal to 1. Values are means from three independent experiments with standard deviations.

Interestingly, there was a drastic decrease in the level of transcripts in all ATP supplied OE lines (recovery) as compared to +P (Figure 3a, b). This indicates probable degradation or down-regulation of *OsPAP21b* transcripts under ATP recovery and points towards post-transcriptional regulation of *OsPAP21b* in OE lines in Pi status-dependent manner.

Morphologically, total lateral root length in OE lines was increased by almost 1.6–2 times as compared to WT under −P (Figure S4). Additionally, lateral length/cm of root was significantly increased by 1.2–1.5 times in OE line in relation to WT (Figure S4). These results indicate that *OsPAP21b* influences root system architecture under Pi deficiency by increasing lateral length. Further analysis revealed ~6%–15% increase in root

length in OE lines than WT under −P (Figure S5b). However, noticeable advantage (8%–11%) in shoot length was recorded only in ATP recovered Pi-starved OE lines (Figure S5c). Similarly, OE lines produced higher root and shoot biomass under +ATP as compared to WT (Figure 4c, f). Marked increase in root biomass was also observed in OE lines as compared to WT under −P condition (Figure 4b, e). However, no significant differences in plant biomass were found between WT and OE lines under +P (Figure 4a, d). These results suggest that constitutive overexpression of *OsPAP21b* does not affect the normal plant growth and development under sufficient Pi supply; however, it plays important role in improving growth on organic-P substrate through better Pi uptake and utilization. We further investigated growth

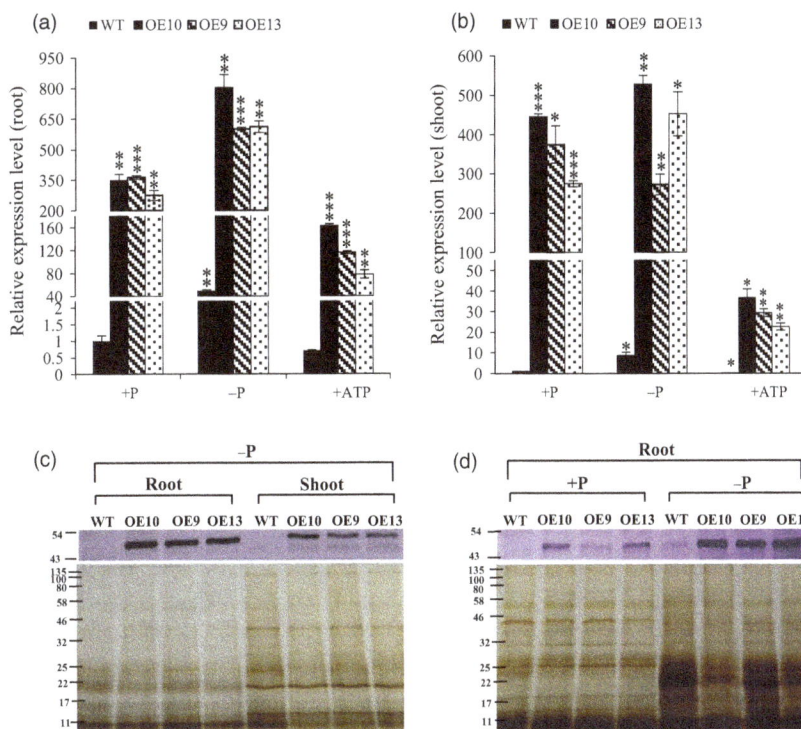

Figure 3 Analysis of *OsPAP21b* transcript and protein levels in OE lines (OE10, 9, 13). (a) Relative expression levels of *OsPAP21b* in roots and (b) shoots of OE lines as compared to WT. All relative expression levels were calculated with respect to WT +P. Plants were raised for 30 days under +P, −P and +ATP (15-day-old Pi-starved seedlings were recovered with ATP for another 15 days) conditions. *P value <0.05; **P value <0.01; ***P value <0.001. (c) Western blot showing increased protein levels of OsPAP21b in 30-day-old OE lines relative to WT in roots and shoots under −P condition. (d) Western blot of OsPAP21b in roots of WT and OE lines under +P and −P conditions. 24 µg of total protein was loaded in each lane and resolved on 12% SDS-PAGE. OsPAP21b accumulation was probed with anti-OsPAP21b antibody. Silver-stained protein gels show the equal amount of protein loading in each lane.

performance of OE lines on organophosphates other than ATP (Figure S6). Similar to ATP recovery, OE lines showed significant increase in root and shoot length and biomass on ADP and p-Ser as compared to WT (Figure S6). However, on phytate-supplemented media, significant differences in shoot biomass were observed only in OE9 as compared to WT. Apart from hydroponics, Pi-starved *OsPAP21b* OE lines also showed better recovery than WT in soil system supplemented with only organic manure as P source (Figure 5a). About 47%–68% increase in plant biomass (Figure 5b) and 54%–87% in P content (Figure 5c) was observed in OE lines as compared to WT. Interestingly, all OE lines also showed early flowering as compared to WT (Figure S7). These results suggest high potential of OsPAP21b in utilizing natural organic-P sources.

Overexpression of *OsPAP21b* enhanced APase activity and P content

Total APase activity of 30-day-old seedlings was measured under +P, −P and +ATP conditions (Figure 6a, b). Irrespective of treatments, APase activity was fairly high in root as compared to shoot. Under +P, total APase activity was significantly increased by ~2.5- to 3.8-fold in roots and ~5.5- to 8.7-fold in shoots of OE lines as compared to WT. APase activity was further increased in all OE lines and WT in −P condition relative to +P condition. However, in response to ATP recovery, APase activity was significantly decreased in OE lines as compared to both +P and −P conditions. This again reveals Pi-dependent regulation of OsPAP21b. To assess the effect of increased APase activity on Pi

uptake, total P content per plant was quantitated (Figure 6c, d). Significant increase was observed in roots (35%–69%) and shoots (42%–63%) of ATP recovered OE lines as compared to WT. About twofold to 2.5-fold increase in total P content was also observed in roots of Pi-starved OE lines in relation to WT. However, no significant increase in total P content was found in OE lines as compared to WT under +P.

Expression of other PAPs and Pi transporters is altered in *OsPAP21b* transgenics

As OE lines showed higher P accumulation under −P, we studied expression of other low Pi-inducible PAPs and Pi transporters in roots of WT and OE lines grown under +P, −P and +ATP. Our analysis revealed significant up-regulation of *OsPAP3b*, *OsPAP10a*, *OsPAP10c*, *OsPAP23* and *OsPAP27a* under Pi deficiency in OE lines as compared to WT (Figure S8). Similarly, significant up-regulation of Pi transporters *OsPT2*, *OsPT4* and *OsPT9* was also observed in OE lines as compared to WT under Pi deficiency (Figure S9). These results indicate potential signalling roles of *OsPAP21b* and also explain high P accumulation in transgenics.

OsPAP21b encodes a secretory PAP

As predicted by SignalP 3.0, OsPAP21b contains a signal peptide at its N terminal end and therefore may be a secretory PAP (Zhang *et al.*, 2011). To confirm this, we did multiple experiments. First, secretory APase activity of OsPAP21b was tested by staining of roots using BCIP as substrate. Intense blue colour precipitate was

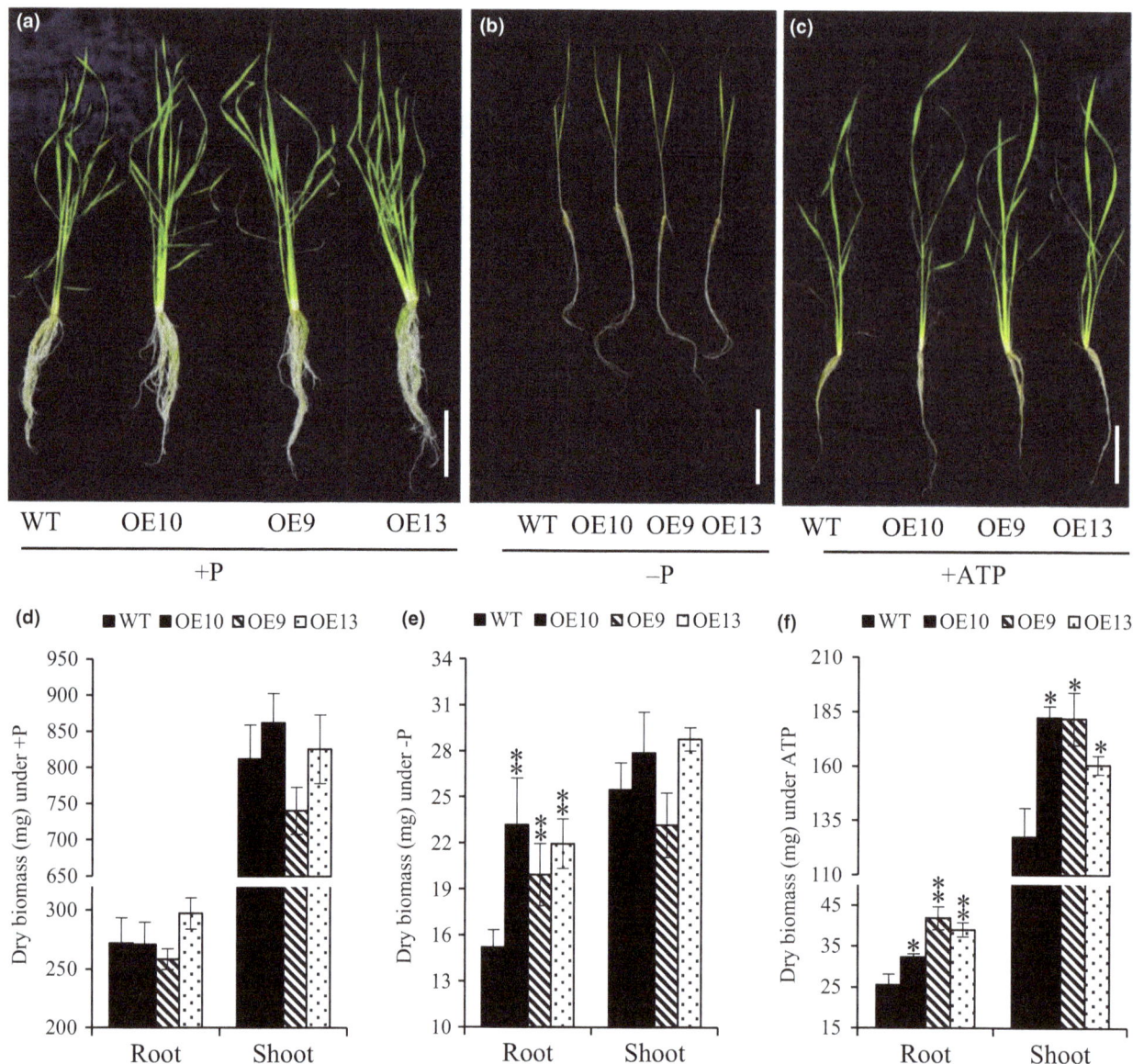

Figure 4 Growth of WT and *OsPAP21b* OE lines (OE10, 9, 13) under different P treatments. (a–c) Phenotype of WT and *OsPAP21b* OE lines under +P, −P and +ATP recovery conditions. scale bar = 10 cm. (d–f) Root and shoot dry biomass. Plants were raised hydroponically for 30 days under +P, −P and +ATP recovery conditions (Pi-starved seedlings for 15 days were recovered with ATP for 15 subsequent days). Each values represents mean of 15 seedlings in three replicates with standard error. *P value <0.05; **P value <0.01.

observed on root surfaces of OE lines as compared to WT under +P (Figure 7a). This indicates increased hydrolysis of BCIP into stable coloured indolyl derivative by secreted OsPAP21b in OE lines. However, under −P conditions, apparent differences in activity staining could not be observed between OE lines and WT. This may be due to rapid saturation of colour intensity by diverse APases secreted in response to Pi deficiency. Second, secretory APase activity was measured in plant growth media under +P and −P conditions. Again all OE lines showed 18%–44% increase in APase activity as compared to WT under +P (Figure 7b). Under −P conditions, increase in secretory APase activity was observed as high as 73% in OE10 as compared to WT (Figure 7b). Observed difference in total secretory APase activity could also be attributed to differences in total root surface area. To overrule this, secretory APase activity was also determined with equal amount of concentrated protein from growth media. Notably, all OE lines

showed significantly higher (42%–49%) secretory APase activity as compared to WT under +P (Figure 7c). However, under −P, only 6%–10% increase in APase activity could be observed in OE lines as compared to WT (Figure 7c). We further confirmed these results by western blotting of concentrated secreted protein with anti-OsPAP21b antibody (Figure 7e). Lastly, we confirmed OsPAP21b secretion by plasmolysis of onion epidermal cells overexpressing YFP-OsPAP21b fusion protein. YFP signals were clearly noticed in apoplast of plasmolysed cells (Figure 7d). Overexpression of YFP-OsPAP21b in onion cells showed dotted fluorescence pattern all over the cell which indicated its localization in endomembrane system (Figure S10). All this clearly revealed the secretory nature of OsPAP21b. To further determine the localization of OsPAP21b, co-localization assays were performed with various organelle markers tagged with mCherry (Figure S10). Although perfect overlap between fluorescence

Figure 5 Growth and P content of soil-grown WT and *OsPAP21b* OE lines. (a) Phenotype of 2-month-old WT and *OsPAP21b* OE lines in soils supplemented with manure as organic-P source. Scale bar =30 cm. (b) Root and shoot dry biomass and (c) total P content per plant of *OsPAP21b* OE lines and WT. Each bar represents average of 3 replicates (*n* = 10) with standard error. ****P* value <0.001.

signals obtained from OsPAP21b and markers could not be obtained, the pattern of YFP-OsPAP21b seems to be similar to Golgi marker.

OsPAP21b encodes a major APase isoform

To investigate the effect of *OsPAP21b* overexpression on APase profiles, we performed in-gel APase assays with total root and shoot proteins of WT and OE lines (Figure 8). Three isoforms could be clearly identified and were named as E1, E2 and E3. Similar isoforms were also identified earlier in rice by Tian *et al.* (2012). However, in our study a fourth isoform, named as E4 was spotted in WT root only under −P condition. This suggests that E4 is a low Pi-inducible isoform that is predominantly induced in root tissues. Interestingly, intense overexpression of E4 isoform was observed in all OE lines as compared to WT, irrespective of treatment and tissue indicating E4 is indeed encoded by *OsPAP21b* (Figure 8a, b). Further, coomassie-stained protein bands (from another non-reducing PAGE gel) corresponding to E4 isoform were identified as OsPAP21b by mass spectrometry with a significantly high Mascot score (128) (Figure S11). Finally, reduction in this form in RNAi lines as compared to WT further validates its identity (Figure 8c, d). Taken together, these results confirm that OsPAP21b encodes a major low Pi-inducible isoform in root.

Effects of *OsPAP21b* silencing on plant growth

Realizing the positive influence of *OsPAP21b* overexpression on low Pi tolerance, we raised RNAi lines (Ri) of *OsPAP21b* to appraise its contribution under Pi deficiency. Similar to OE lines, Ri lines were also raised under +P, −P and +ATP conditions. Expression analysis of Ri lines revealed significant down-regulation of *OsPAP21b* in all three conditions (Figure S12). Morphological analysis of Ri lines revealed significant decrease in root biomass in relation to WT under ATP recovery (Figure 9e). Notably, Ri9 with highest down-regulation of *OsPAP21b* showed reduction in root biomass under all conditions as compared to WT (Figure 9c–e). However, no significant differences in shoot biomass were evident between Ri lines and WT under any treatments except for Ri9 which showed ~20% reduction in shoot biomass under +ATP (Figure 9e). Further, significant differences in root length were also observed only in Ri9 line as compared to WT under all three conditions, indicating significant role of *OsPAP21b* in affecting root architecture (Figure S13).

Silencing of OsPAP21b decreased APase activity and P content

Significant decrease in APase activity was observed in roots and shoots of RNAi lines grown in −P and ATP-supplemented media as compared to WT (Figure 9f, g). Notably, about 20%–35%

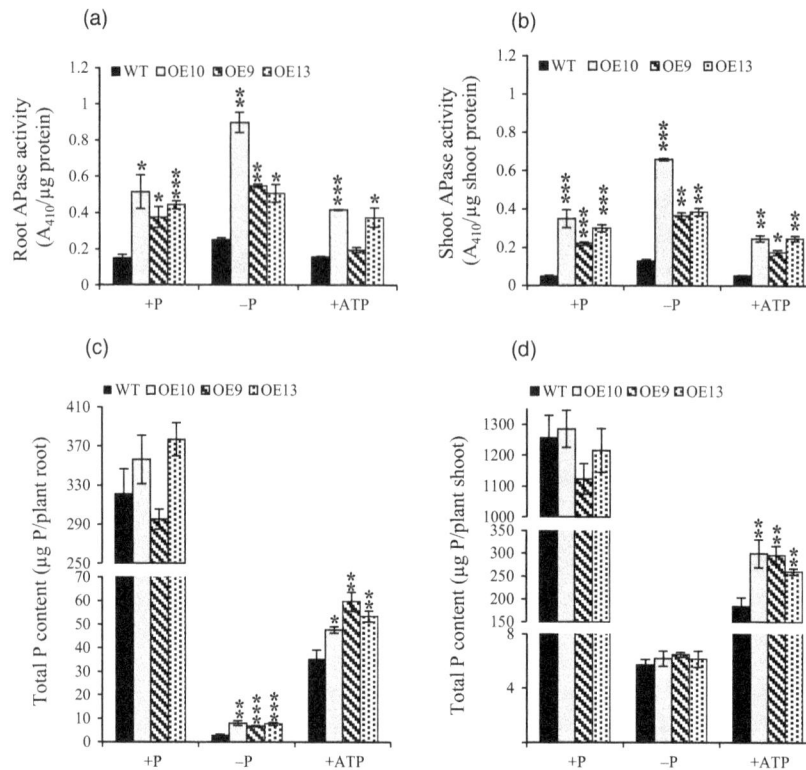

Figure 6 Total APase activity and P content of WT and *OsPAP21b* OE lines. (a) Total APase activity in roots and (b) shoots of 30-day-old seedlings raised under +P, −P and +ATP conditions. Activity was determined with 1 µg of total root protein using pNPP as substrate. (c) Total P content in roots and (d) shoots of 30-day-old seedlings under +P, −P and +ATP conditions. Each bar represents means of three replicates (*n* = 5) with standard error. *P value <0.05; **P value <0.01; ***P value <0.001.

decrease in APase activity was observed under −P in roots of RNAi lines. This suggests significant contribution of OsPAP21b in total plant APase activity under −P. Further, quantitation of total P content of Ri lines revealed significant decrease in P content in roots under +ATP (Figure 9h). However, except for Ri9, no significant differences in root P content were observed in Ri lines under −P and +P conditions. Notably, any significant differences in shoot P content could not be observed between WT and Ri lines under any treatment except for Ri9 which showed significant reduction under +ATP (Figure 9i). Collectively, these results suggest the importance of OsPAP21b in hydrolysis of organic-P compounds and improving Pi utilization in rice.

Discussion

Organophosphates constitute ~80% of total soil P; however, remain unavailable for root uptake before mineralization. PAPs, especially secretory ones, are emerging as major plant enzymes for releasing Pi from these sources. Twenty-six rice PAPs are classified into three main groups (I, II and III) and seven subgroups (Ia, Ib, Ic, IIa, IIb, IIIa, IIIb) (Zhang *et al.*, 2011). Despite being a large gene family, only one PAP of subgroup Ic (OsPAP23) and two PAPs of subgroup Ia (OsPAP10a and OsPAP10c) have been shown to enhance the extracellular Pi utilization from organic-P sources (Li *et al.*, 2012b; Lu *et al.*, 2016; Tian *et al.*, 2012). Interestingly, both rice and Arabidopsis PAPs of subgroup Ic are reported to possess phytase activity (Li *et al.*, 2012b; Wang *et al.*, 2009; Zhang *et al.*, 2011), whereas PAPs from subgroup Ia are major secretory PAPs in both rice and Arabidopsis (Lu *et al.*, 2016;

Tian *et al.*, 2012; Tran *et al.*, 2010b; Wang *et al.*, 2011). Here, we have provided detailed characterization of one of the major low Pi-responsive rice PAPs, *OsPAP21b* that belongs to a different subgroup, Ib which phylogenetically lies between Ia and Ic.

Several transcriptome studies have commonly reported a high induction of *OsPAP21b* under Pi deficiency across diverse rice genotypes suggesting its important role(s) in low Pi response (Li *et al.*, 2010; Mehra *et al.*, 2016; Pariasca-Tanaka *et al.*, 2009; Secco *et al.*, 2013; Takehisa *et al.*, 2013; Zhang *et al.*, 2011). Notably, majority of these studies were conducted in roots. We also found preferential induction of *OsPAP21b* in roots which might be due to the active roles of roots in soil P solubilization and acquisition. Additionally, magnitude of *OsPAP21b* up-regulation was progressively enhanced with prolonged exposure to Pi deficiency. These evidences indicate spatio-temporal regulation of *OsPAP21b* in response to low Pi. Further, several PSR genes are positively controlled by MYB transcription factor, OsPHR2 during Pi starvation (Zhou *et al.*, 2008). Earlier study by Zhang *et al.* (2011) indicated the role of OsPHR2 in regulation of PAPs by showing increased accumulation of *OsPAP21b* and other PAPs transcripts in *OsPHR2* overexpression lines. Our study confirmed that OsPHR2 indeed binds with the promoter of *OsPAP21b* and regulates its expression.

Several lines of evidences indicate a strict Pi status-dependent regulation of OsPAP21b at both transcription and translational levels, and therefore, its key role in Pi acquisition and utilization. These are: (i) suppression of *OsPAP21b* accumulation in Pi-starved WT within 1 h of Pi resupply in roots. (ii) Resupply of ATP to Pi-starved *OsPAP21b* constitutive OE lines repressed both

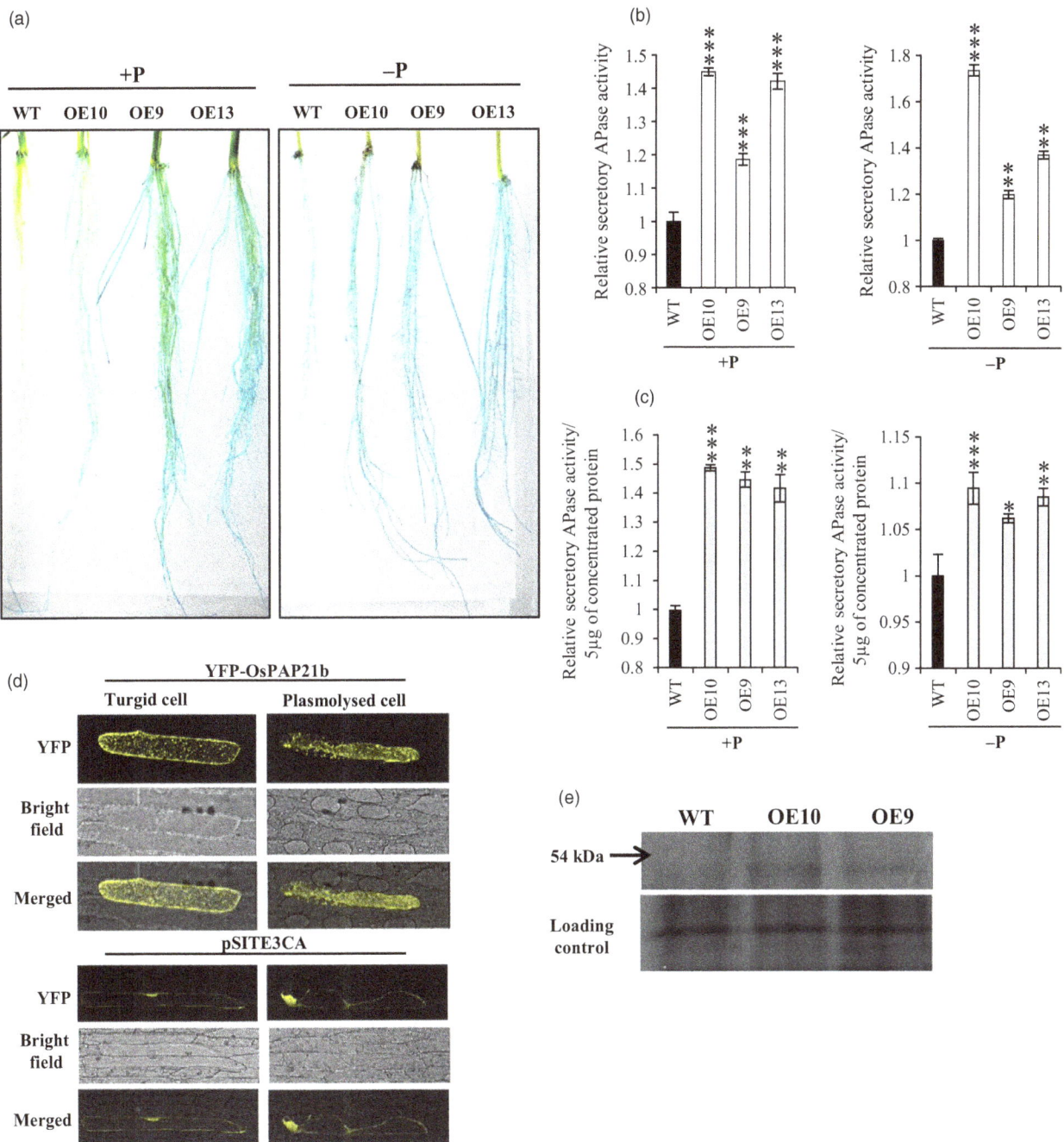

Figure 7 *OsPAP21b* encodes a secretory PAP. (a) Activity staining of roots of 15-day-old WT and *OsPAP21b* OE lines under +P and −P. Roots were overlaid with 0.015% BCIP in 0.5% agar and blue colour shows APase activity. (b) Quantitation of secretory APase activity in growth medium of 30-day-old seedlings under +P and −P conditions. (c) Quantitation of secretory APase activity in 30-day-old seedlings under +P and −P conditions using 5 μg of concentrated total protein secreted into medium. Relative activity was determined with respect to APase activity in WT which was considered equal to one. *P value <0.05; **P value <0.01; ***P value <0.001. (d) YFP fluorescence in turgid and plasmolysed onion epidermal cells overexpressing YFP-OsPAP21b or only YFP. (e) Detection of OsPAP21b under −P conditions in concentrated media protein of 30-day-old WT and OE lines by western blotting. Silver-stained protein gel shows equal protein load in each lane.

accumulation of *OsPAP21b* transcripts and total APase activity as compared to −P grown seedlings. Notably, suppression of *OsPAP21b* under Pi recovery was accompanied with simultaneous increased P accumulation in seedlings. (iii) Pi deficiency further enhanced the accumulation of *OsPAP21b* transcripts and protein in OE lines. This can be attributed to the overexpression of full-length cDNA which contains full 5′ and 3′ UTRs (Figure S2). 5′ UTR

is known to possess regulatory elements that can regulate gene expression. Further, the presence of a long 3′ UTR can potentially affect the stability of *OsPAP21b* transcripts in OE lines in response to Pi status. Such post-transcriptional regulations of PAPs have also been reported for AtPAP10 (Wang *et al.*, 2011). Similarly, AtPAP10 overexpression transgenics also showed increased PAP protein accumulation under −P condition despite having

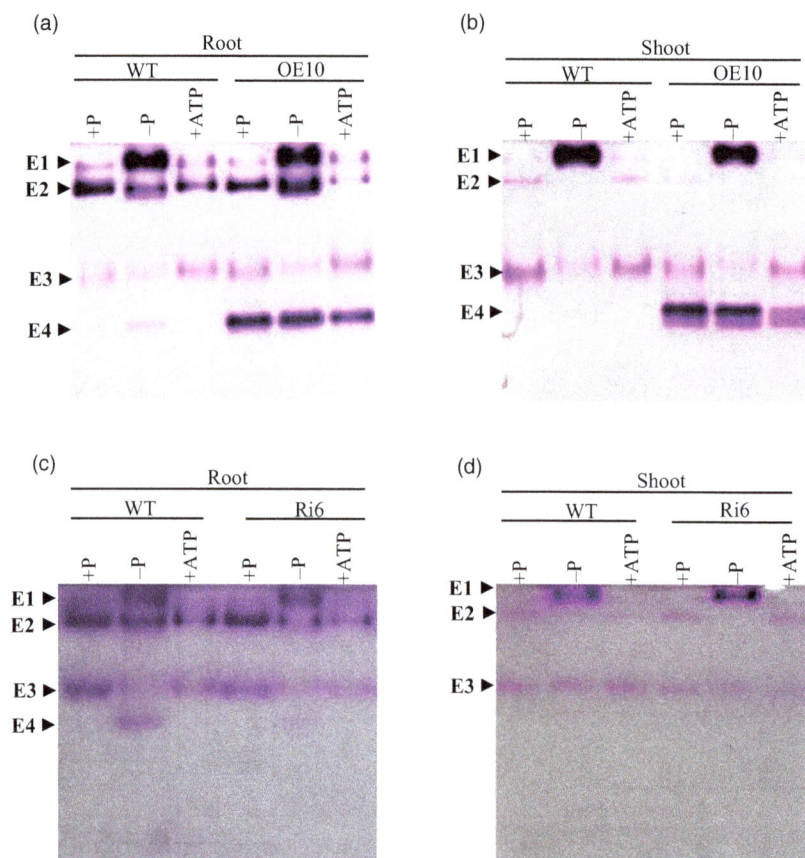

Figure 8 APase profiles of *OsPAP21b* transgenics and WT. (a) APase profiles of total root and (b) shoot proteins in 30-day-old WT and OE10 seedlings grown under +P, −P and +ATP conditions. (c) APase profiles of total root and (d) shoot proteins in 30-day-old WT and Ri6 under +P, −P and +ATP conditions. 10 μg of total root and shoot proteins were separated on 10% non-reducing SDS-PAGE. Gels were stained with fast black potassium salt and β-naphthyl acid phosphate. Different APase isoforms E1, E2, E3 and E4 are indicated by arrow heads.

constitutive promoter (Zhang *et al.*, 2014). (iv) In-gel APase assays also confirmed increase and subsequent decrease in activity of E4 isoform (corresponding to OsPAP21b) under −P and ATP recovery, respectively. (v) Lastly, enzyme activity assays with recombinant OsPAP21b revealed complete inhibition of APase activity at a concentration of Pi (10 mM) which corresponds to cellular Pi levels. All these points indicate some post-transcriptional or post-translational regulation, which influences levels of either constitutively expressing *OsPAP21b* transcripts or protein in response to elevated Pi status. Such direct link between Pi status and OsPAP21b levels clearly proved its importance in Pi acquisition and utilization.

Plant microRNAs and Pi resupply-inducible serine proteases have been proposed for such regulation (reviewed in Tran *et al.*, 2010a). However, any direct evidences for these mechanisms are largely missing and needs further investigations. Interestingly, native western blotting of *OsPAP21b* OE lines with anti-OsPAP21b antibody showed high molecular weight oligomer formations by OsPAP21b (Figure S14). Further, immunoblotting of non-reducing SDS-PAGE gel of total protein from OE10 also showed complexes of OsPAP21b. Interestingly, most of these complexes perfectly aligned with E1, E2 and E4 APase isoforms (Figure S15). Y2H assays (Figure S16); however, excluded any probable oligomerization of OsPAP21b protein with itself as suggested for several high molecular weight (HMWs ~55 kDa) PAPs (Li *et al.*, 2002) or other co-expressed PAPs. It seems that

these putative OsPAP21b-associated proteins may govern its protein level regulations. Moreover, transcriptional regulation of PAPs via PHR2 has also been proposed to be influenced by SPX proteins (Zhang *et al.*, 2011). However, more in-depth investigations of such regulations need to be carried out in future. Furthermore, while *OsPAP21b* transcript is repressed by Pi; however, even 48 h of Phi treatment did not have any significant effect on *OsPAP21b* expression. This indicates systemic regulation of *OsPAP21b*. Recently, Jost *et al.* (2015) also showed that transcripts of *AtPAP17* and *AtPAP1* are non-responsive to Phi treatments which suggest that these PAPs could discriminate well between Pi and Phi.

Although overexpression of *OsPAP21b* led to increased protein accumulation and hence increased APase activity under +P conditions in OE lines as compared to WT, no significant or negative effect on plant growth performance under +P conditions was observed in OE lines. This could be due to no significant change in plant P content between WT and OE transgenics under +P conditions. Previous studies with *AtPAP10* also showed increased root-associated APase activity of OE lines; however, no significant differences were detected between WT and transgenics phenotypically under +P conditions (Wang *et al.*, 2011). Under −P conditions, OE lines showed significant increase in APase activity and P content of roots as compared to WT which suggests that OsPAP21b can also hydrolyse the intracellular organophosphates. Analysis of Ri lines and in-gel APase isoforms

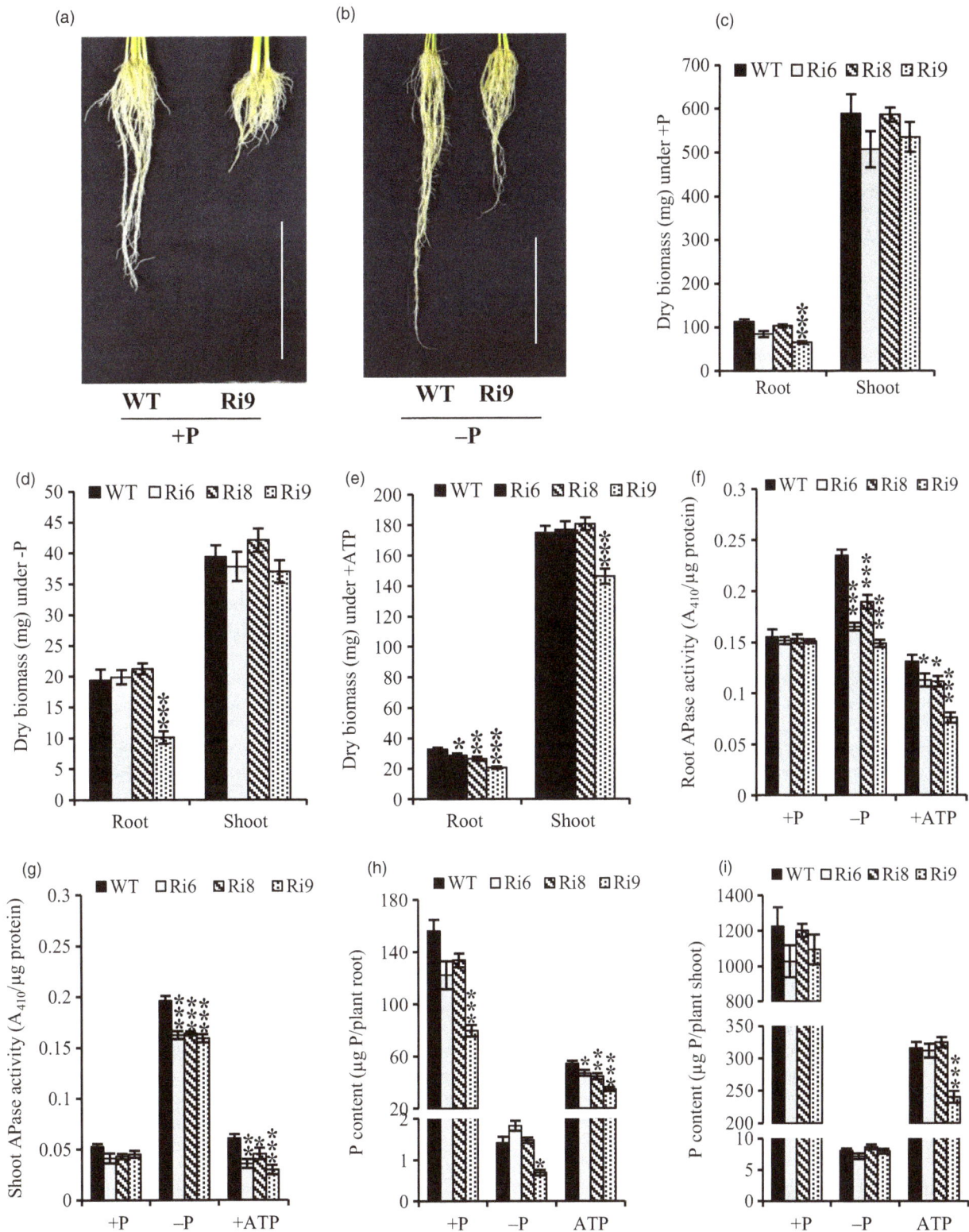

Figure 9 Analysis of RNAi lines of *OsPAP21b* for root phenotype, biomass, APase activity and P content. (a) Root growth of 30-day-old WT and Ri9 under +P and (b) −P (*n* = 5). Scale bar = 10 cm. (c–e) Root and shoot dry biomass under +P, −P and ATP recovery conditions. (f–g) Total APase activity and (h–i) total P content in roots and shoots of seedlings raised under +P, −P and +ATP conditions. Each value represents mean of three replicates with standard error. *P* value <0.05; **P* value <0.01; ***P* value <0.001.

also confirmed major contribution of OsPAP21b in total APase activity under −P. Down-regulation of *OsPAP21b* in Ri9 led to a decrease in APase activity leading to decrease in P content and

root biomass as compared to WT under −P. Noticeably, the most important role of OsPAP21b was observed in extracellular Pi utilization from different organic sources. Increased intracellular and secretory APase activity in OE lines sufficiently enhanced plant P content upon recovery with ATP which ultimately led to faster recovery of Pi-starved OE lines as compared to WT. Unlike OE lines, Ri lines of *OsPAP21b* showed decreased APase activity and loss of root biomass under recovery conditions as compared to WT. Down-regulation of *OsPAP21b* also led to loss of shoot biomass in Ri under ATP recovery indicating its major role in extracellular ATP utilization.

Notably, nucleic acids form a large portion of organic-P in soil and need to be hydrolysed by phosphatases before root uptake (Turner, 2008). Additionally, it has been reported that plant cells secrete ATP in extracellular matrix to maintain cell viability (Chivasa *et al.*, 2005). These secreted ATPs could also act as source of organic-P for OsPAP21b. Apart from ATP, Pi-starved OE lines also showed improved recovery on a variety of other organic-P substrates. APase assay of recombinant OsPAP21b protein also confirmed its broad substrate specificity for variety of organophosphates. Further, OE lines of *OsPAP21b* also showed better growth as compared to WT in soil containing manure as organic-P source. Taken all together, OsPAP21b with both intra- and extracellular APase activities proved to be an ideal candidate for improving low Pi tolerance in rice. Interestingly, none of the PAPs of Ib or any other subgroups other than Ia were earlier shown to be of secretory in nature. Identification and characterization of OsPAP21b also upholds the future possibility of identifying more potent secretory PAPs in rice. Nevertheless, moderate thermostability and acidic range pH optima of OsPAP21b makes it a suitable candidate for high-temperature rice ecosystems especially low Pi acidic soils.

Our results further revealed a potential role of rice PAPs in modulating root system architecture under low Pi. Studies with *AtPAP10* and *NtPAP12* have also indicated role of PAPs in modulating RSA by hydrolysing cell wall bound enzymes involved in cell wall biosynthesis (Kaida *et al.*, 2010; Wang *et al.*, 2011). Arabidopsis mutants of several PAPs also show poor root growth as compared to WT (Wang *et al.*, 2014). As several metabolic intermediates or signalling molecules are also substrate of OsPAP21b, an indirect role of OsPAP21b in controlling RSA is quite possible.

Rice, the staple food crop for more than half of the world's population, has only ~25% P use efficiency (Vinod and Heuer, 2012). Therefore, developing rice genotypes for improved P use efficiency has been recognized as a critical step towards its sustainable production. The present work is an extension of such efforts and identifies a novel PAP, OsPAP21b that can substantially enhance the solubilization and scavenging of abundant natural organic-P sources. Further, Pi fertilizer loss to waterbodies poses great threat to aquatic life due to eutrophication. Such studies would also help to reduce the use of Pi fertilizers and protect environment.

Experimental procedures

Plant material and growth conditions

For hydroponic experiments, rice (*Oryza sativa* cv. PB1) seeds were surface-sterilized and germinated in the dark as described previously (Mehra *et al.*, 2016). Evenly germinated seeds were placed on nylon mesh floating over Yoshida growth medium, pH 5.0–5.5. Experiments were carried out in green house at 30/28 °C (day/night) temperature, 70% relative humidity and 16/8-h photoperiod. For Pi sufficient and deficient treatments, nutrient media containing 320 μM (+P) and 1 μM NaH$_2$PO$_4$ (−P) were

supplied, respectively. Different nutrient deficiency conditions (−N, −P, −K, −Fe and −Zn) were created as described (Mehra and Giri, 2016).

For +Phi/Pi treatments, seedlings were raised hydroponically under +P and −P conditions for 15 days, after which −P raised plants were supplied with 320 μM of NaH$_2$PO$_4$ or 320 μM of Phi (Na$_2$HPO$_3$·5H$_2$O) for different time intervals, and tissues were collected for P content and gene expression analyses. Relative expression levels of *OsPAP21b* under −P and after Phi/Pi recovery were calculated with respect to expression under +P condition.

OsPAP21b overexpression and RNAi lines were grown in −P or +P media for 15 days. Afterwards, half of the Pi-starved seedlings were recovered with 320 μM ATP (Sigma) for another 15 days. Subsequently, all 1-month-old +P, −P and ATP recovered −P seedlings (+ATP) were phenotyped. For studying performance of *OsPAP21b* OE lines on different organic substrates, 15-day-old −P grown seedlings were recovered with 320 μM of different organic substrates (ATP, ADP, p-Ser, phytate) for another 15 days. For analysis of growth in soil system, 15-day-old hydroponically raised −P seedlings were recovered in soil supplemented with only organic manure (2 sand: 1 soilrite: 1 vermiculite: 1 organic manure) as P source (43 ± 0.7 mg/kg total P). Plants were raised in five replicates in greenhouse till 2 months of age before harvesting. Unless stated otherwise, experiments were performed in three biological replicates (*n* = 10–15).

Biochemical characterization of OsPAP21b

Coding region of *OsPAP21b* was cloned into *NdeI* and *EcoRI* sites of pET28a vector (Novagen) and transformed in *E. coli*, BL21(DE3) pLysS cells. Transformed cells were treated with 0.3 mM IPTG at 15 °C for 18 h to induce the expression of 6XHis-tagged OsPAP21b. Recombinant OsPAP21b protein was isolated and purified by Ni^{2+}-affinity chromatography as reported earlier (Mehra and Giri, 2016). Activity assays were performed with 1.5 μg of recombinant OsPAP21b protein in 100 μL reaction mixture containing 50 mM sodium acetate buffer pH 5.0, 5 mM MgCl$_2$ and 10 mM pNPP (*p*-nitrophenol phosphate) as standard substrate. Reactions were incubated at 37 °C for 30 min, and released Pi was measured by yellow vanadomolybdate method (Kitson and Mellon, 1944) by adding 100 μL of vanadate–molybdate reagent. For measurement of enzyme activity with different P-containing substrates, all substrates were used at final concentrations of 10 mM in the reaction describe above. For measurement of activity at different pH, 50 mM of sodium acetate (pH 4–6) and Tris-maleate buffer (6.5–8.0) were used with pNPP and ADP as substrates. Optimum temperature was determined by incubating reactions at different temperatures in Veriti™ thermal cycler (Applied Biosystems). For determination of cofactors and inhibitors of OsPAP21b, 10 mM chloride salts of different cations and sodium salts of different anions were used in reaction mixture with 10 mM pNPP as substrate. For calculation of IC$_{50}$, 0 to 25 mM of NaH$_2$PO$_4$ was incubated with 10 mM of pNPP as substrate. Kinetics constant, K_m and V_{max} were estimated from Lineweaver–Burk plot of enzyme activity at different concentrations of pNPP and ADP (0.1–100 mM).

Vector construction and rice transformation

To generate *OsPAP21b*-overexpressing (OE) rice transgenics, full-length cDNA (Figure S2) was cloned in Gateway-compatible binary vector pANIC6B (Mann *et al.*, 2012) under the transcriptional control of maize ubiquitin promoter (*pZmUbi1*). For raising RNAi transgenics of *OsPAP21b* (Ri), 307-bp region of *OsPAP21b* cDNA was amplified and cloned in pANIC8b vector (Mann *et al.*, 2012) by Gateway Technology (Invitrogen). *Agrobacterium*-mediated transformation of rice genotype, PB1 was carried out as described (Toki *et al.*, 2006). Resultant transformants (T0) were confirmed for the presence of transgene by PCR with gene-specific primers of hygromycin phosphotransferase (*hpt*; Table S1, Figure S3) and histochemical GUS assay (Jefferson *et al.*, 1987). All experiments were subsequently performed with T3 generation homozygous transgenic lines. Overexpression and down-regulation of *OsPAP21b* in transgenic lines was confirmed by qRT-PCR using primers listed in Table S1.

For plant phenotyping, qRT-PCR, EMSA, protein extraction, quantitation of total plant and secretory APase activity, in-gel APase profiling and mass spectrometry, activity staining of root surface-associated APases, yeast two-hybrid assays, generation of anti-OsPAP21b antibody, immunoblot analysis and total P content analysis, please see Supplementary methods.

Acknowledgements

This work was supported by research grant, BT/PR3299/AGR/2/813/2011 of DBT, Government of India, and NIPGR core grant. P.M. and B.K.P. acknowledge research fellowships by CSIR and DBT, respectively.

References

Chivasa, S., Ndimba, B.K., Simon, W.J., Lindsey, K. and Slabas, A.R. (2005) Extracellular ATP functions as an endogenous external metabolite regulating plant cell viability. *Plant Cell*, **17**, 3019–3034.

Cordell, D., Drangert, J.O. and White, S. (2009) The story of phosphorus: global food security and food for thought. *Glob. Environ. Chang.* **19**, 292–305.

Del Vecchio, H.A., Ying, S., Park, J., Knowles, V.L., Kanno, S., Tanoi, K., She, Y.M. *et al.* (2014) The cell wall-targeted purple acid phosphatase AtPAP25 is critical for acclimation of *Arabidopsis thaliana* to nutritional phosphorus deprivation. *Plant J.* **80**, 569–581.

George, T.S. and Richardson, A.E. (2008) Potential and limitations to improving crops for enhanced phosphorus utilization. In *Ecophysiology of Plant-Phosphorus Interactions* (White, P.J. and Hammond, J.P., eds), pp. 247–270. Dordrecht: Springer.

González-Muñoz, E., Avendaño-Vázquez, A.-O., Chávez Montes, R.A., de Folter, S., Andrés-Hernández, L., Abreu-Goodger, C. and Sawers, R.J.H. (2015) The maize (*Zea mays* ssp. *mays* var. B73) genome encodes 33 members of the purple acid phosphatase family. *Front. Plant Sci.* **6**, 341.

Holford, I.C.R. (1997) Soil phosphorus: its measurement, and its uptake by plants. *Aust. J. Soil Res.* **35**, 227–239.

Jefferson, R.A., Kavanagh, T.A. and Bevan, M.W. (1987) GUS fusions: betaglucuronidase as a sensitive and versatile gene fusion marker in higher plants. *EMBO J.* **6**, 3901–3907.

Jost, R., Pharmawati, M., Lapis-Gaza, H.R., Rossig, C., Berkowitz, O., Lambers, H. and Finnegan, P.M. (2015) Differentiating phosphate-dependent and phosphate-independent systemic phosphate-starvation response networks in *Arabidopsis thaliana* through the application of phosphite. *J. Exp. Bot.* **66**, 2501–2514.

Kaida, R., Serada, S., Norioka, N., Norioka, S., Neumetzler, L., Pauly, M., Sampedro, J. *et al.* (2010) Potential role for purple acid phosphatase in the dephosphorylation of wall proteins in tobacco cells. *Plant Physiol.* **153**, 603–610.

Kitson, R.E. and Mellon, M.G. (1944) Colorimetric determination of phosphorus as molybdivanadophosphoric acid. *Ind. Eng. Chem. Anal. Ed.* **16**, 379–383.

Kuang, R., Chan, K.H., Yeung, E. and Lim, B.L. (2009) Molecular and biochemical characterization of AtPAP15, a purple acid phosphatase with phytase activity, in Arabidopsis. *Plant Physiol.* **151**, 199–209.

Li, D., Zhu, H., Liu, K., Liu, X., Leggewie, G., Udvardi, M. and Wang, D. (2002) Purple acid phosphatases of *Arabidopsis thaliana*: comparative analysis and differential regulation by phosphate deprivation. *J. Biol. Chem.* **277**, 27772–27781.

Li, L., Liu, C. and Lian, X. (2010) Gene expression profiles in rice roots under low phosphorus stress. *Plant Mol. Biol.* **72**, 423–432.

Li, C., Gui, S., Yang, T., Walk, T., Wang, X. and Liao, H. (2012a) Identification of soybean purple acid phosphatase genes and their expression responses to phosphorus availability and symbiosis. *Ann. Bot.* **109**, 275–285.

Li, R.J., Lu, W.J., Guo, C.J., Li, X.J., Gu, J.T. and Xiao, K. (2012b) Molecular characterization and functional analysis of OsPHY1, a purple acid phosphatase (PAP)-type phytase gene in rice (*Oryza sativa* L.). *J. Integr. Agric.* **11**, 1217–1226.

Lu, L., Qiu, W., Gao, W., Tyerman, S.D., Shou, H. and Wang, C. (2016) *OsPAP10c*, a novel secreted acid phosphatase in rice, plays an important role in the utilization of external organic phosphorus. *Plant Cell Environ.* **39**, 2247–2259.

Mann, D.G., LaFayette, P.R., Abercrombie, L.L., King, Z.R., Mazarei, M., Halter, M.C. *et al.* (2012) Gateway-compatible vectors for high-throughput gene functional analysis in switchgrass (*Panicum virgatum* L.) and other monocot species. *Plant Biotechnol. J.* **10**, 226–236.

Mehra, P. and Giri, J. (2016) Rice and chickpea *GDPDs* are preferentially influenced by low phosphate and CaGDPD1 encodes an active glycerophosphodiester phosphodiesterase enzyme. *Plant Cell Rep.* **35**, 1699–1717.

Mehra, P., Pandey, B.K. and Giri, J. (2016) Comparative morphophysiological analyses and molecular profiling reveal Pi-efficient strategies of a traditional rice genotype. *Front. Plant Sci.* **6**, 1184.

Murumkar, S.B., Pawar, G.R. and Naiknaware, M.D. (2015) Effect of different sources and solubility of phosphorus on growth, yield and quality of *Kharif* rice (*Oryza sativa* L.). *Int. J. Trop. Agric.* **33**, 245–249.

Neue, H.U., Lantin, R.S., Cayton, M.T.C. and Autor, N.U. (1990) Screening of rices for adverse soil tolerance. In *International Symposium on Genetic Aspects of Plant Mineral Nutrition* (Bassam, N.E., Dambroth, M. and Loughman, B.C., eds), pp. 523–532. The Netherlands: Kluwer Academic Publishers.

Oddie, G.W., Schenk, G., Angel, N.Z., Walsh, N., Guddat, L.W., de Jersey, J. *et al.* (2000) Structure, function, and regulation of tartrate-resistant acid phosphatase. *Bone*, **27**, 575–584.

Pandey, B.K., Mehra, P. and Giri, J. (2013) Phosphorus starvation response in plants and opportunities for crop improvement. In *Climate Change and Plant Abiotic Stress Tolerance* (Tuteja, N. and Gill, S.S., eds), pp. 991–1012. Weinheim: Wiley-VCH Verlag GmbH and Co. KGaA.

Pariasca-Tanaka, J., Satoh, K., Rose, T., Mauleon, R. and Wissuwa, M. (2009) Stress response versus stress tolerance: a transcriptome analysis of two rice lines contrasting in tolerance to phosphorus deficiency. *Rice*, **2**, 167–185.

del Pozo, J.C., Allona, I., Rubio, V., Leyva, A., de la Pena, A., Aragoncillo, C. and Paz-Ares, J. (1999) A type 5 acid phosphatase gene from *Arabidopsis thaliana* is induced by phosphate starvation and by some other types of phosphate mobilizing/oxidative stress conditions. *Plant J.* **19**, 579–589.

Richardson, A.E. and Simpson, R.J. (2011) Soil microorganisms mediating phosphorus availability update on microbial phosphorus. *Plant Physiol.* **156**, 989–996.

Secco, D., Jabnoune, M., Walker, H., Shou, H., Wu, P., Poirier, Y. and Whelan, J. (2013) Spatio-temporal transcript profiling of rice roots and shoots in response to phosphate starvation and recovery. *Plant Cell*, **25**, 4285–4304.

Shen, J., Yuan, L., Zhang, J., Li, H., Bai, Z., Chen, X., Zhang, W. *et al.* (2011) Phosphorus dynamics: from soil to plant. *Plant Physiol.* **156**, 997–1005.

Takehisa, H., Sato, Y., Antonio, B.A. and Nagamura, Y. (2013) Global transcriptome profile of rice root in response to essential macronutrient deficiency. *Plant Signal. Behav.* **8**, e24409.

Tian, J. and Liao, H. (2015) The role of intracellular and secreted purple acid phosphatases in plant phosphorus scavenging and recycling. In *Annual Plant Reviews, Volume 48, Phosphorus Metabolism in Plants* (Plaxton, W.C. and Lambers, H., eds), pp. 265–287. John Wiley & Sons: Hoboken.

Tian, J., Wang, C., Zhang, Q., He, X., Whelan, J. and Shou, H. (2012) Overexpression of OsPAP10a, a root-associated acid phosphatase, increased extracellular organic phosphorus utilization in rice. *J. Integr. Plant Biol.* **54**, 631–639.

Toki, S., Hara, N., Ono, K., Onodera, H., Tagiri, A., Oka, S. and Tanaka, H. (2006) Early infection of scutellum tissue with Agrobacterium allows high-speed transformation of rice. *Plant J.* **47**, 969–976.

Tran, H.T., Hurley, B.A. and Plaxton, W.C. (2010a) Feeding hungry plants: the role of purple acid phosphatases in phosphate nutrition. *Plant Sci.* **179**, 14–27.

Tran, H.T., Qian, W., Hurley, B.A., She, Y.M., Wang, D. and Plaxton, W.C. (2010b) Biochemical and molecular characterization of AtPAP12 and AtPAP26: the predominant purple acid phosphatase isozymes secreted by phosphate-starved *Arabidopsis thaliana*. *Plant Cell Environ.* **33**, 1789–1803.

Turner, B.L. (2008) Resource partitioning for soil phosphorus: a hypothesis. *J. Ecol.* **96**, 698–702.

Varadarajan, D.K., Karthikeyan, A.S., Matilda, P.D. and Raghothama, K.G. (2002) Phosphite, an analog of phosphate, suppresses the coordinated expression of genes under phosphate starvation. *Plant Physiol.* **129**, 1232–1240.

Vinod, K.K. and Heuer, S. (2012) Approaches towards nitrogen- and phosphorus-efficient rice. *AoB Plants*, **2012**, pls028.

Wang, X.R., Wang, Y.X., Tian, J., Lim, B.L., Yan, X.L. and Liao, H. (2009) Overexpressing *AtPAP15* enhances phosphorus efficiency in soybean. *Plant Physiol.* **151**, 233–240.

Wang, L., Li, Z., Qian, W., Guo, W., Gao, X., Huang, L. *et al.* (2011) The Arabidopsis purple acid phosphatase AtPAP10 is predominantly associated with the root surface and plays an important role in plant tolerance to phosphate limitation. *Plant Physiol.* **157**, 1283–1299.

Wang, L., Lu, S., Zhang, Y., Li, Z., Du, X. and Liu, D. (2014) Comparative genetic analysis of *Arabidopsis* purple acid phosphatases AtPAP10, AtPAP12, and AtPAP26 provides new insights into their roles in plant adaptation to phosphate deprivation. *J. Integr. Plant Biol.* **56**, 299–314.

Wang, J., Si, Z., Li, F., Xiong, X., Lei, L., Xie, F. *et al.* (2015) A purple acid phosphatase plays a role in nodule formation and nitrogen fixation in *Astragalus sinicus*. *Plant Mol. Biol.* **88**, 515–529.

Zhang, Q., Wang, C., Tian, J., Li, K. and Shou, H. (2011) Identification of rice purple acid phosphatases related to phosphate starvation signalling. *Plant Biol.* **13**, 7–15.

Zhang, Y., Wang, X., Lu, S. and Liu, D. (2014) A major root-associated acid phosphatase in Arabidopsis, AtPAP10, is regulated by both local and systemic signals under phosphate starvation. *J. Exp. Bot.* **65**, 6577–6588.

Zhou, J., Jiao, F., Wu, Z., Li, Y., Wang, X., He, X., Zhong, W. *et al.* (2008) OsPHR2 is involved in phosphate-starvation signaling and excessive phosphate accumulation in shoots of plants. *Plant Physiol.* **146**, 1673–1686.

14

The defence-associated transcriptome of hexaploid wheat displays homoeolog expression and induction bias

Jonathan J. Powell[1,2,]*, Timothy L. Fitzgerald[1], Jiri Stiller[1], Paul J. Berkman[1], Donald M. Gardiner[1], John M. Manners[3], Robert J. Henry[2] and Kemal Kazan[1,2]

[1]Commonwealth Scientific and Industrial Research Organisation Agriculture, St Lucia, Queensland, Australia
[2]Queensland Alliance for Agriculture and Food Innovation, University of Queensland, St Lucia, Queensland, Australia
[3]Commonwealth Scientific and Industrial Research Organisation Agriculture, Black Mountain, Australian Capital Territory, Australia

*Correspondence
email Jonathan.Powell@csiro.au

Keywords: biotic stress, homoeolog expression bias, polyploidy, RNA-seq, wheat.

Summary

Bread wheat (*Triticum aestivum* L.) is an allopolyploid species containing three ancestral genomes. Therefore, three homoeologous copies exist for the majority of genes in the wheat genome. Whether different homoeologs are differentially expressed (homoeolog expression bias) in response to biotic and abiotic stresses is poorly understood. In this study, we applied a RNA-seq approach to analyse homoeolog-specific global gene expression patterns in wheat during infection by the fungal pathogen *Fusarium pseudograminearum*, which causes crown rot disease in cereals. To ensure specific detection of homoeologs, we first optimized read alignment methods and validated the results experimentally on genes with known patterns of subgenome-specific expression. Our global analysis identified widespread patterns of differential expression among homoeologs, indicating homoeolog expression bias underpins a large proportion of the wheat transcriptome. In particular, genes differentially expressed in response to *Fusarium* infection were found to be disproportionately contributed from B and D subgenomes. In addition, we found differences in the degree of responsiveness to pathogen infection among homoeologous genes with B and D homoeologs exhibiting stronger responses to pathogen infection than A genome copies. We call this latter phenomenon as 'homoeolog induction bias'. Understanding how homoeolog expression and induction biases operate may assist the improvement of biotic stress tolerance in wheat and other polyploid crop species.

Introduction

The vast majority of extant plants species either currently exist in a state of polyploidy (neopolyploidy) or have been affected by polyploidization events during their evolutionary history (paleopolyploidy; Blanc and Wolfe, 2004; Wood *et al.*, 2009). Polyploidy occurs via whole-genome duplication events in the case of autopolyploids or by one or more interspecific hybridization events between different species in the case of allopolyploids (Adams and Wendel, 2005). Polyploid species often display distinct characteristics such as larger seeds (Beaulieu *et al.*, 2007) and leaves (Sugiyama, 2005) and more vigorous growth (Ni *et al.*, 2009) than their progenitor species. In addition, species with higher states of ploidy often possess better abiotic and biotic stress tolerance than their progenitors (Comai, 2005). Mechanisms of increased biotic and abiotic stress tolerance in polyploid species may involve polyploidy-contributed heterosis and expression dosage due to increased gene copy number (Chen, 2007).

Polyploidization with its accompanying genomic flux has an enormous impact on global transcriptional regulation relative to patterns of gene expression observed in progenitor species (Chen and Ni, 2006). Accordingly, plants undergo dramatic alterations to global gene expression immediately after a polyploidization event followed by a gradual reversion on an evolutionary timescale to a diploid state (Feldman and Levy, 2009). Without such correction, increased dosage may also be detrimental to plant fitness in newly formed polyploids due to the risk of

unbalancing the fine-tuned regulation of many biological functions in progenitor species (Bekaert *et al.*, 2011; Birchler *et al.*, 2005). Postpolyploidization events may result in genome asymmetry as homoeologs may be silenced (Sehrish *et al.*, 2014) or lost (Schnable *et al.*, 2011) or homoeolog expression bias occurs where homoeologs show expression that is different from an assumed equal parental expression ratio (Feldman and Levy, 2009). This process has been shown to occur in a nonrandom fashion, resulting in uneven contribution of particular biological processes and molecular functions from specific subgenomes, a phenomenon termed functional compartmentalization or subfunctionalization (Bekaert *et al.*, 2011).

Homoeolog expression biases have been shown to impact on plant growth, development and stress responses in several polyploid species. In allopolyploid cotton, widespread, nonadditive expression patterns and expression partitioning have been identified for homoeologs responding to various abiotic stresses (Dong and Adams, 2011; Liu and Adams, 2007). In newly formed allopolyploid *Arabidopsis*, nonadditive gene expression between homoeologs inherited from *Arabidopsis thaliana* and *A. aerenosa* correlated with transcript instability (Kim and Chen, 2011). Genes with nonadditive expression and associated transcript instability were found to have a strong association with biotic and abiotic stress response gene ontologies (Kim and Chen, 2011). Recent work highlighted a high degree of expression bias and expression partitioning in hexaploid wheat during drought and heat stress (Liu *et al.*, 2015).

Comparative transcriptome analyses in polyploids have been utilized to study polyploidy-associated phenomena in several plant species including the tetraploid cotton (*Gossypium hirsutum*; Flagel *et al.*, 2012; Yoo *et al.*, 2013) and *Arabidopsis arenosa* (Ng *et al.*, 2012; Pignatta *et al.*, 2010), hexaploid (bread) wheat (*Triticum aestivum*; Akhunova *et al.*, 2010; Leach *et al.*, 2014) and dodecaploid common cordgrass (*Spartina anglica*; Chelaifa *et al.*, 2010). Akhunova *et al.* (2010) observed a greater contribution to gene expression from the A and B subgenomes compared to the D subgenome in wheat using a homoeolog distinguishing microarray; however, the method utilized in this study was unable to distinguish between A and B homoeologs. Leach *et al.* (2014) utilized RNA-seq to observe homoeolog expression bias under basal growth conditions in root and shoot tissue for genes occurring on group 1 and group 5 chromosomes of hexaploid wheat. Overall, this study indicated homoeolog expression bias affects a large proportion of the wheat genome but individual subgenomes do not contribute disproportionately to overall homoeolog expression bias, a phenomenon termed 'balanced homoeolog expression bias'. Approximately 45% of homoeolog triplets displayed expression of all three homoeologs. However, a single homoeolog copy appears to predominantly contribute to the overall transcript abundance (Leach *et al.*, 2014). Pfeifer *et al.* (2014) found a high degree of subgenome specialization within the wheat grain transcriptome with particular subgenomes contributing disproportionately to various biological functions including gene expression and translation (A subgenome), cellular macromolecule metabolism (B subgenome) and transport, secretion and communication/signalling (D subgenome).

Previous work has suggested polyploid species tend to exhibit increased tolerance against pathogen attack (Peng *et al.*, 2003) and this could at least partly explain early success of neopolyploid species (Oswald and Nuismer, 2007). However, the association between polyploidy and disease resistance can be complex and is not easy to directly test, especially in crop plants where polyploid species have been subjected to artificial selection for resistance while diploid progenitors have not. In other cases, polyploidy may result in increased susceptibility to pathogens if one of the progenitors involved in the polyploid species contains a suppressor or disease susceptibility locus that interfere with the expression of resistance (Kerber, 1991). To date, relatively little work has been performed to assess the effect of biotic stress on genome asymmetry and homoeolog expression bias in polyploid plant species. Previous work in wheat has implicated a primary role for the B subgenome contributing towards biotic stress responses based on distribution of QTL for disease resistance on B subgenome-specific chromosomes (Feldman *et al.*, 2012). The genes involved in the biosynthesis pathway for 2-benzoxazolinone (BOA), an important phytoalexin, are deployed predominantly from the B subgenome (Nomura *et al.*, 2005). Nomura *et al.* (2005) also demonstrated that hexaploid wheat progenitor species (i.e. diploids and tetraploids) are each able to synthesize BOA, suggesting silencing of A and D homoeologs postpolyploidization in hexaploid wheat, resulting in subfunctionalization of BOA synthesis to the B subgenome. Additionally, little work has been performed how genes induced or repressed during a biotic stress response might be biased in responsiveness between homoeologous gene copies. Identifying such 'homoeolog induction biases' is also critical to better understanding how each subgenome contributes to biotic stress responses and may shed light on which processes contribute to the success of polyploids against stresses.

In this work, we investigated homoeolog-specific gene expression patterns of bread wheat infected with *F. pseudograminearum*, a necrotrophic fungal pathogen (Akinsanmi *et al.*, 2006), to determine whether different subgenomes respond to pathogen attack differently. *F. pseudograminearum* is the predominant cause of crown rot (Chakraborty *et al.*, 2006), a disease with economic significance (Murray and Brennan, 2009) and a highly quantitative basis of resistance (Li *et al.*, 2010). Resistance to crown rot in wheat has been previously shown to vary with ploidy level with tetraploid wheat (*T. durum*) displaying greater susceptibility to this pathogen than hexaploid wheat (Liu *et al.*, 2012). Previous work has explored wheat responses during *F. pseudograminearum* infection using microarrays (Desmond *et al.*, 2008); however, this approach was not able to infer expression patterns in a homoeolog-specific manner. Our analyses suggest that individual wheat subgenomes contribute disproportionately to the overall response to *F. pseudograminearum* with B and D subgenomes displaying a greater contribution than the A subgenome. Potential implications of this phenomenon on wheat breeding are also discussed.

Results

Expanding the known set of homoeolog triplets from the wheat chromosome survey genome sequence using a reciprocal best BLAST analysis

Correctly identifying the full complement of homoeologs in polyploids is essential for homoeolog expression analysis. However, this remains a technical challenge within hexaploid wheat since gene sequence collections are relatively incomplete or contain redundant sequence copies. In order to comprehensively assess potential homoeolog expression bias within the wheat transcriptome, we first aimed to identify as many homoeologous sequences as possible for each wheat gene within the International Wheat Genome Sequencing Consortium (IWGSC) chromosomal survey sequence (CSS) CDS reference (Mayer *et al.*, 2014). To do this, we used an approach similar to the one employed by Pfeifer *et al.* (2014) but modified BLAST parameters in an attempt to identify a larger set of homoeologous genes. As explained in Materials and Methods, homoeologous triplets were identified as reciprocal best BLAST (RBB) hits (Moreno-Hagelsieb and Latimer, 2008) between subgenome-specific CDS. In order for a homoeolog triplet to be identified, consistent agreement of RBB hits between each subgenome (i.e. A to B, B to D and D to A) is required.

Here, homoeologous triplets were inferred from the global CDS library for 38 889 genes derived from the chromosome arm assemblies (Mayer *et al.*, 2014) to form 12 963 triplets corresponding to approximately 39% of CDS in the reference and 28% of total predicted protein coding genes. However, not all homoeolog triplets could be identified mostly due to the absence of a complete reference with gene models. In addition, potential gene deletions and gene duplications producing highly similar sequences could also confound an RBB strategy used for homoeolog identification. Here, we were able to identify 98.7% of all homoeologs previously identified by Pfeifer *et al.* (2014). Furthermore, our analysis identified significantly more homoeolog triplets (12 963) than the analysis in Pfeifer *et al.* (2014) (6576) mainly because we did not apply a minimum BLAST score cut-off. However, we performed additional analyses to ensure homoeolog triplets were correctly inferred, including confirming that homoeolog chromosomal locations (i.e. 1AL/1BL/1DL) were conserved unless such locations were affected by known translocation events. Overall, we concluded that the number of homoeologs triplets we could identify would be sufficient for global analysis of homoeolog expression patterns in wheat.

Homoeolog-specific alignment of RNA-seq reads validated using differing alignment protocols

To assess homoeolog expression bias in bread wheat during response to biotic stress, an established laboratory infection assay (Yang et al., 2010) was performed to infect wheat seedlings with F. pseudograminearum. Four biological replicates of F. pseudograminearum (Fp)-inoculated and noninoculated (mock) wheat plants were sampled as described in Materials and Methods. We then used RNA-seq to characterize the transcriptional response in a homoeolog-specific manner. For each of the approaches described below, reads were aligned to the global coding sequence reference and read counts were then extracted for genes within the inferred homoeolog triplets described in the previous section. For the purpose of testing whether the alignment stringency we used was adequate to differentiate between homoeologs, independent alignments were performed using different methods that employ distinct aligner algorithms. No significant differences in read count estimates were found between heuristic seeding (Bowtie2; Langmead and Salzberg, 2012) versus exhaustive k-mer (Biokanga; Stephen et al., 2012) alignment methods (Appendix S1). No significant differences in read count estimates were observed between random assignment or sloughing of reads which aligned equally well to multiple locations on the reference. To test whether the degree to which homoeologs overlap influences read alignment accuracy, blocks of overlapping coding sequence within homoeolog triplets were retrieved and used as a reference for alignment. From these observations, we concluded that alignment biases should not significantly affect expression estimates for homoeolog triplets where all three sequences are present in the reference used for alignment.

Homoeolog-specific alignment stringency validated using for benzoxazolinone biosynthesis pathway

For validating the stringency of the method for estimating transcript abundance in a subgenome-specific fashion, expression of genes for a well-characterized phytoalexin biosynthesis pathway with known subgenome-specific expression was used. As stated above, benzoxazolinone compounds, 2-benzoxazolinone (BOA) and 6-methoxy-benzoxazolinone (MBOA), have been characterized in wheat (Nomura et al., 2002). All homoeologous copies of the TaBx1-TaBx5 genes involved in BOA biosynthesis in wheat have been identified previously and Bx gene expression shown to be predominantly contributed from the B subgenome (Nomura et al., 2005). To determine whether we could confirm this finding within our data set, we first used BLAST to identify which sequences within CSS reference correspond to previously characterized wheat Bx genes (Appendix S1). Perfect matches were found for all A and D subgenome Bx copies and all B subgenome copies except TaBx3. The TaBx3B sequence (AB042628.1) was added to the reference and read alignment and differential expression analysis were performed again.

This analysis showed that all five Bx genes (TaBx1B, TaBx2B, TaBx3B, TaBx4B and TaBx5B) are highly expressed within mock-and pathogen-infected samples (Appendix S2) with TaBx4B and TaBx5B significantly repressed by infection (~two-fold). This observation indicates the RNA-seq analysis we used was able to distinguish between homoeologs when estimating expression, even for genes such as Bxs sharing a high degree of sequence similarity (Nomura et al., 2005). From this, we concluded that alignment biases should not significantly affect homoeolog expression estimates provided that all three sequences are present in the reference used for alignment.

A large degree of homoeolog expression bias occurs during infection

To assess the degree to which homoeolog expression bias occurs within the transcriptome, read counts for A, B and D homoeologs under mock and infected conditions were retrieved from binary alignment map files and used as inputs for DESeq. Genes for which expression of A, B and D homoeologs were not significantly different from each other (A = B = D) were placed into 'Category 1'. Homoeolog triplets where only one of the homoeolog was differentially expressed (i.e. A > B = D, B > A = D, D > A = B A < B = D, B < A = D or D < A = B) were placed into 'Category 2' (Figure 1a). Homoeolog triplets for which expression was significantly different for each homoeolog (i.e. A > B > D, A > D > B, B > A > D, D > A > B, B > D > A and D > B > A) were placed into 'Category 3' (Figure 1a). Comparing numbers of homoeolog triplets within Categories 2 and 3 to those in Category 1 provides a measure of the degree of homoeolog expression bias.

For mock-treated samples, 3855 homoeolog triplets (31%) were assigned to Category 1 while for Fp-treated samples, 3592 homoeolog triplets (29%) were assigned to Category 1 (Figure 1b). Remaining triplets were placed into Category 2 or Category 3; as such, homoeolog expression bias was detected within 69% and 71% of the homoeolog triplets in mock- and Fp-treated samples, respectively. The majority of non-Category 1 homoeolog triplets (5197 mock and 5292 infected) exhibited expression bias towards a single homoeolog (Category 2; Figure 1b). Fewer homoeolog triplets exhibited an unequal expression bias towards two homoeologs (Category 3) with 3294 and 3462 homoeolog triplets identified in mock and infected samples, respectively.

Patterns of homoeolog expression bias are mostly conserved under infected and uninfected conditions

Understanding how homoeolog biases contribute to responses during infection requires observation of the degree to which homoeolog expression bias patterns are fixed between basal growth and infected conditions and the way in which patterns change during application of a biotic stress. To do this, patterns of expression bias within individual homoeolog triplets were compared under mock and infected conditions (Figure 1). This analysis revealed 2992 homoeologs in total displayed Category 1 expression patterns under both conditions. The majority of triplets showing homoeolog-specific expression (3801) displayed Category 2 expression patterns under both mock and infected conditions. Significantly fewer homoeologs retained Category 3 expression under both infected and mock conditions (2606; χ^2 distribution test $P < 0.01$) compared with the number of homoeologs retaining Category 1 and Category 2 expression. For 1672 triplets, the complexity of expression increased under pathogen infection (i.e. 816 triplets from Category 1 to 2, 809 triplets from Category 2 to 3 and 47 triplets Category 1 to 3)

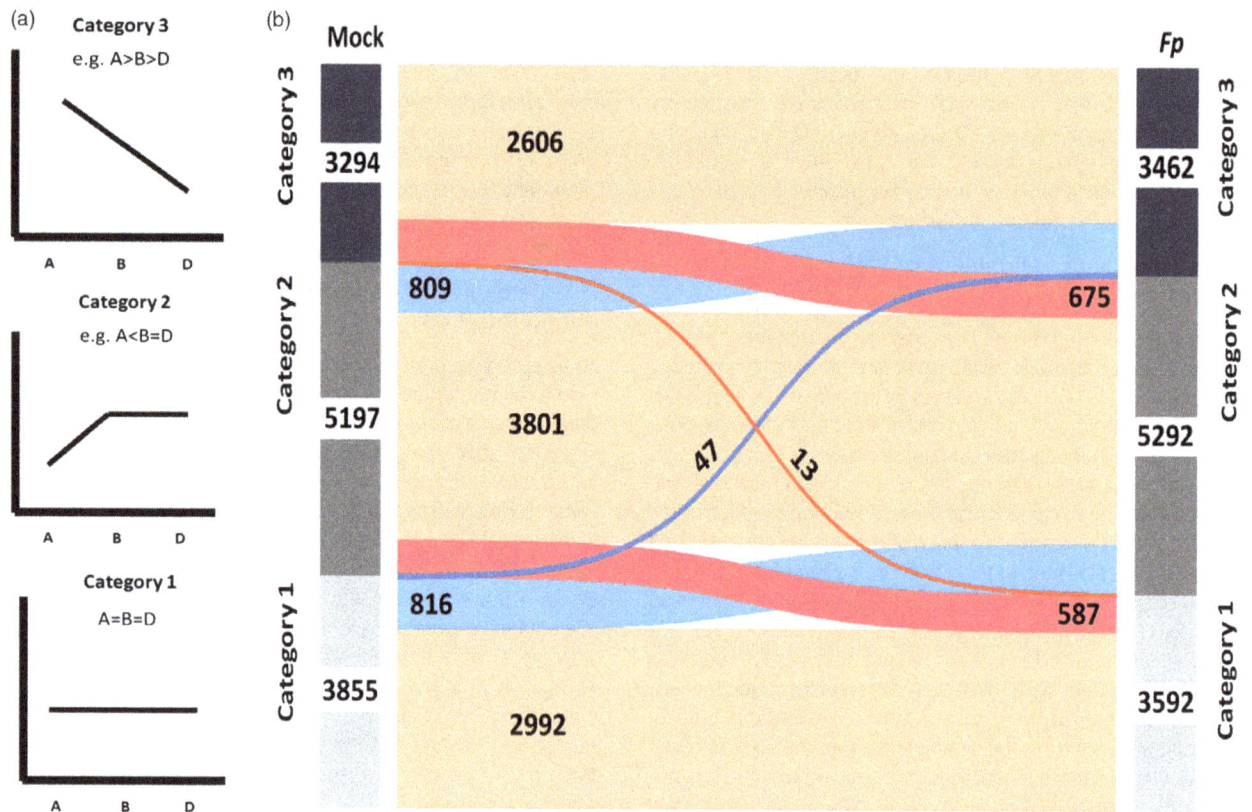

Figure 1 Homoeolog Expression Bias during Biotic Stress. (a) It illustrates three categories of expression pattern identified in this study. Category 1 denotes triplets within which all three homoeologs were expressed to an equivalent level. Category 2 denotes triplets for which one homoeolog was expressed to a significantly different degree compared to both other homoeologs (e.g. A > B = D). Category 3 denotes triplets in which all three homoeologs were significantly differently expressed from each other (e.g. A > B > D). (b) Sankey diagram showing patterns of transition for homoeolog triplets between mock (left side)- and Fp-infected (right side) conditions. Beige flows represent homoeolog triplets which retained the same expression pattern under both conditions. Blue flows represent homoeolog triplets which displayed increased expression bias under infected condition relative to mock. Red flows represent homoeolog triplets which displayed reduced expression bias under infected condition relative to mock.

while for 1275 triplets, the complexity of expression was decreased under pathogen treatment (587 triplets from Category 2 to 1, 675 triplets from Category 3 to 2 and 13 triplets from Category 3 to 1). Interestingly, where the type of homoeolog expression bias is conserved between mock and pathogen infection (i.e. Category 2 or Category 3 retention), the pattern of expression was also generally conserved (i.e. if A > B > D under mock conditions, then A > B > D under infected conditions as well). A similar trend was observed for homoeolog triplets changing between categories of subgenome expression bias (Category 2 to Category 3 transitions and vice versa) in that the same homoeolog would retain expression bias (i.e. for A > B = D under mock transitioning to A > B > D or A > D > B under infection; Table S1).

Homoeolog induction bias is a primary driver of subgenomic specificity in the wheat transcriptome

To infer which genes were induced during *Fusarium* infection, differential expression analysis was performed using DESeq. In total, 2755 genes showed significant differential expression in response to *Fusarium* infection at 3 dpi, representing altered expression of approximately 2.8% of the annotated transcriptome (~99K genes). Total differentially expressed genes were comprised of 1867 up-regulated genes with fold changes ranging

Table 1 Table displaying counts of differentially expressed genes globally and for each subgenome specifically in *Triticum aestivum* L. Observations in rows marked by asterisk showed bias from expected proportions as determined by χ^2 test ($P < 0.01$)

	Global	A genome	B genome	D genome
Up-regulated genes*	1867	559	639	669
Down-regulated genes	888	275	306	307
Total*	2755	834	945	976

from 232 to 1.17. In addition, 888 were down-regulated under infection (Table 1).

Within the total set of 1867 up-regulated genes, 1526 occurred within identified homoeologous triplets. After consolidating homoeologous differentially expressed genes (considering differential expression of multiple homoeolog copies as response of a single gene), the canonical transcriptomic response consisted of 944 genes showing differential expression (62% of all differentially expressed homoeologous triplets). Patterns of differential expression among identified homoeologs revealed 139 triplets where all homoeologs were differentially expressed (417 gene copies in total). We denote instances where one or two homoeologs were differentially expressed between mock and infected conditions as cases of 'homoeolog induction bias'. Homoeolog induction bias was found to impact a large proportion of the biotic stress-induced transcriptome. Of 591 genes differentially expressed between mock- and pathogen-inoculated plants in a homoeolog-specific manner, 54 triplets displayed differential expression of A and B homoeologs only (108 gene copies), 82 from B and D homoeologs only (164 gene copies) and 73 from both A and D homoeologs only (146 gene copies; Figure 2a). A high proportion of biotic stress-responsive genes were contributed from a single homoeolog copy with 177, 206 and 208 genes and were differentially expressed either from the A, B or D subgenome alone, respectively (Figure 2a). Analysis indicated homoeologs differentially expressed from a single subgenome were disproportionately high (χ^2 distribution test $P < 0.01$) compared to genes induced from multiple subgenomes. Consequently, single homoeolog expression events also contributed most of the observed functional diversity within the induced transcriptome.

Another potential bias occurring within differentially expressed homoeolog triplets is a bias in the magnitude of induction between homoeologs during infection. To observe whether co-induced homoeologs differed in magnitude of induction, pairwise comparisons (A vs. B, B vs. D and A vs. D subgenomes) of DE gene fold change values for inferred homoeologs were performed using Spearman's ranking analysis (Figure 2b). Expression fold change values were found to be highly correlated across all three comparisons with an R^2 value of 0.94 for A subgenome versus B correlation, 0.91 for B versus D correlation and 0.93 for A versus D correlation. When only considering those genes with greater than twofold DE, the degree of correlation was reduced but still significant ($R^2 = 0.72$ for A vs. B; 0.66 for B vs. D and 0.61 for A vs. D) with high confidence ($P < 0.001$) across comparisons (Figure 2b). These observations indicate subgenome specificity for response to biotic stress is primarily driven by homoeolog induction bias (i.e. which homoeologs are induced) rather than homoeolog expression bias (i.e. driving the magnitude of induction) when multiple homoeologs in a triplet are differentially expressed.

Homoeolog expression bias and induction bias impact on biotic stress-related genes and pathways and establish subgenome specificity

For the purpose of determining whether observed expression and induction biases would involve genes commonly implicated in biotic stress responses, gene descriptions for differentially expressed homoeologs were retrieved using BLAST2GO (Conesa et al., 2005). This analysis allowed assignment of functional descriptions for genes that are well known to be involved in pathogen responses such as pathogenesis-related proteins, leucine-rich repeat proteins (LRR) and leucine-rich receptor-like kinases (LRKs), ABC transporters and pleiotropic drug resistance proteins, germin-like proteins and glutathione-S-transferases (Tables S2 and S3). Contribution of biotic stress-induced homoeologs was disproportionately contributed (χ^2 test $P < 0.05$) with pathogenesis-related proteins and leucine-rich receptors and kinases contributed more from the D genome compared to A and B and much less frequently from all three subgenomes equally (Table S2).

The plant defence-associated hormone jasmonic acid is produced from alpha-linolenate via a series of reactions. In total, 13 homoeolog triplets encoding enzymes in the jasmonate biosynthesis pathway were identified in our analysis. Interestingly, the expression of the gene encoding the lipoxygenase (LOX) enzyme that catalyses the first step in the jasmonate biosynthesis pathway showed a biased expression towards the A subgenome both under mock- and pathogen-inoculated conditions (see inset in Figure 3). The D subgenome also contributes to the LOX expression, but no LOX expression could be detected from the B subgenome under either mock- or pathogen-inoculated conditions (Figure 3 inset). In addition, we identified three paralogous genes (tentatively named as OPR1, OPR2 and OPR3), most likely encoding different isoforms of the enzyme 12-oxophytodienoate reductase (OPR) within each subgenome. Of these three

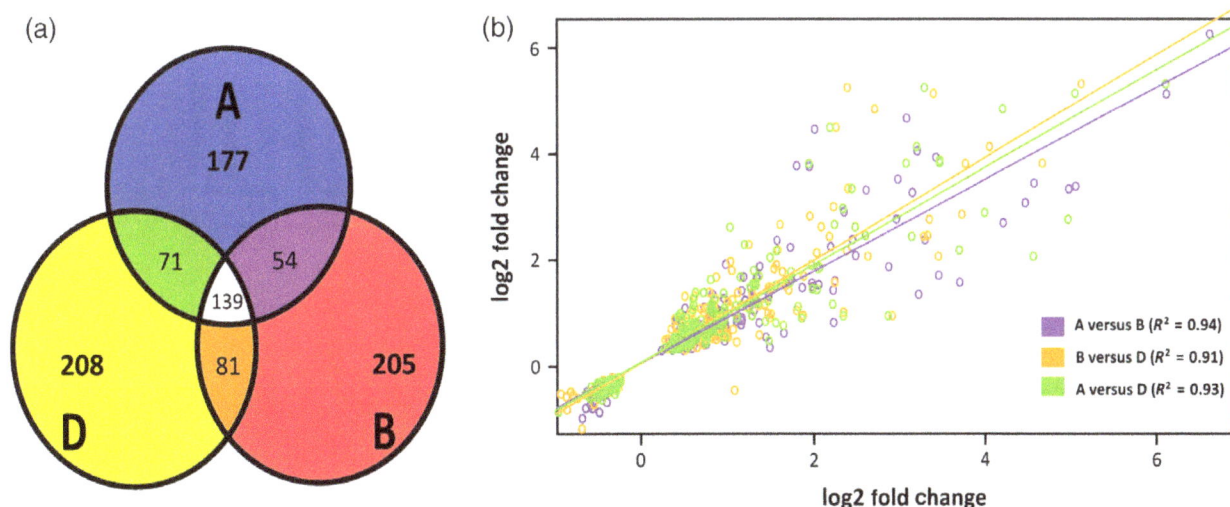

Figure 2 Homoeolog Induction Bias during Biotic Stress. (a) Venn diagram showing counts of differentially expressed genes (*Fusarium* induced) within identified homoeologous triplets. Counts represent triplets where one (no intersection), two (intersection of two circles) or all three (intersection of three circles) genes were differentially expressed. The first two descriptions represent cases of homoeolog induction bias, with the disproportionate indicating homoeolog induction bias strongly underpins the biotic stress-induced transcriptome. (b) Pairwise correlation expressions for homoeologs differentially expressed from A and B subgenome copies (orange), B and D subgenome copies (purple) and A and D subgenome copies (green).

OPR genes, *OPR1* expression showed a bias towards the A subgenome while *OPR2* and *OPR3* showed a bias towards B and D subgenomes. In addition, while the biased expression pattern of OPR1 from the A subgenome was not different between pathogen- versus mock-treated samples, B and D homoeolog copies of OPR2 and OPR3 showed a strong induction bias following pathogen inoculation. Similarly, four paralogous genes encoding different isoforms of the enzyme enoyl-CoA hydratase could be identified within each subgenome. The homoeologs of these paralogous genes showed biased expression towards

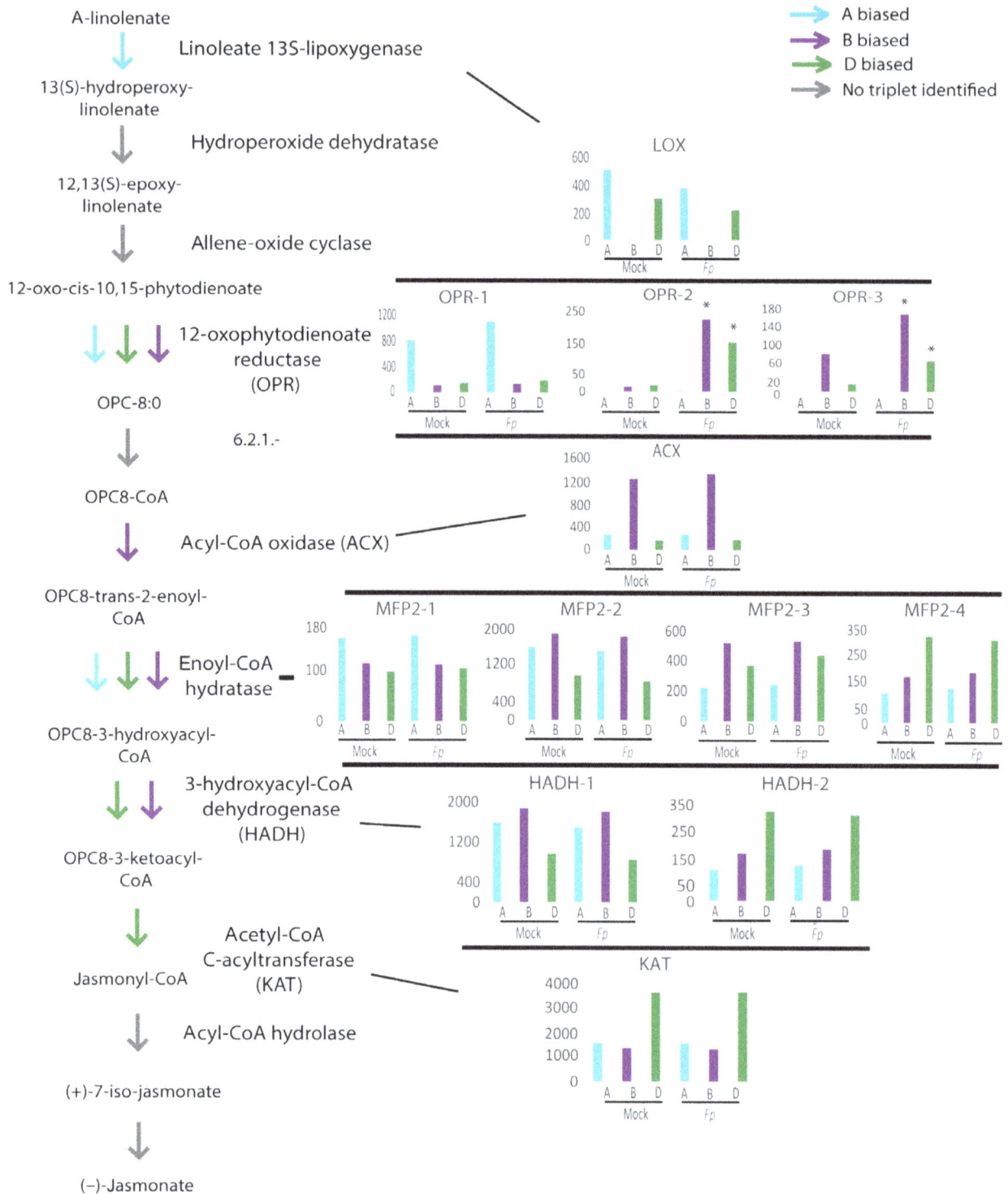

Figure 3 Homoeolog expression bias and induction bias within the jasmonate biosynthesis pathway. Cyan, purple and green arrows represent steps encoded by triplets displaying an expression bias towards the A, B and D subgenome homoeologs, respectively. Grey arrows represent enzymatic steps for which no triplets could be identified. Histograms display read counts for homoeologs under mock- and *Fp*-inoculated conditions and are grouped by corresponding enzyme identity. Homoeologs significantly induced during infection are marked with asterisks.

different subgenomes. Finally, expression patterns of homoeolog genes encoding the enzyme 3-ketoacyl-CoA thiolase (KAT) showed a bias towards the D subgenome (Figure 3).

In addition to these defence-related genes, we identified two defence-related biosynthetic pathways showing a strong bias towards contribution from B and D subgenomes. Folate (also known as vitamin B) has been shown to be important in SA-mediated systemic immunity (Wittek *et al.*, 2015) and folate starvation has been demonstrated as an important resistance strategy in soya bean against soya bean cyst nematodes (Liu *et al.*, 2012). In plants, folate biosynthesis initiates with production of tetrahydrofolate from either guanosine triphosphate or chorismate (Ravanel *et al.*, 2001). Tetrahydrofolate then undergoes a series of transformations into five distinct folate derivatives. We were able to identify homoeolog triplets for genes that encode enzymes catalysing the twenty-four steps in the tetrahydrofolate and folate transformations pathway, while the homoeolog triplets encoding enzymes for the remaining seven steps could not be identified. The genes encoding five of these enzymatic steps showed expression bias towards the B subgenome and eight showed expression bias towards the D subgenome (Appendix S3). Also, for four of these steps, multiple paralogous genes encoding these enzymes within each subgenome were identified. Interestingly, these paralogous genes for a given subgenome homoeolog were found, and these shared the same expression bias pattern. For instance, several paralogous genes (e.g. *MTHFD1.1*, *MTHFD1.2* and *MTHFD1.3*) encoding the enzyme methylenetetrahydrofolate dehydrogenase (EC 1.5.1.5) showed biased expression towards the D subgenome (Appendix S3)

Discussion

In this study, we utilized an RNA-seq-based approach to observe global expression patterns during infection by a pathogen in an unbiased manner and to provide the sensitivity to distinguish between homoeologous copies during read alignment. Past studies examining homoeolog expression patterns in wheat have been restricted by technical limitations inherent to probe-based methods (Akhunova *et al.*, 2010; Leach *et al.*, 2014) with more recent approaches adopting RNA-seq methods to overcome some of these challenges (Pont *et al.*, 2011; Pfeifer *et al.*, 2014; Nussbaumer *et al.*, 2015). RNA-seq presents a technical improvement over probe-based transcriptomic analyses since it overcomes the limitation of a finite probe set, an inherent limitation in hybridization-based detection (Wang *et al.*, 2009). In addition, applying RNA-seq-based methods to observe the transcriptome in polyploid plants increases the likelihood of detecting homoeolog-specific polymorphisms, particularly when longer reads are generated (Buggs *et al.*, 2012; Higgins *et al.*, 2012). Finally, the availability of the IWGSC assembled chromosome sequences in wheat provides a reference for RNA-seq alignment to allow estimation of gene expression in a chromosome- and subgenome-specific manner (Mayer *et al.*, 2014). However, accurate alignment of reads to reference sequence where highly similar sequences exist can still present a technical challenge for any RNA-seq application. The challenge is compounded in polyploid species by the existence of widespread multiple gene copies with high sequence similarity.

While previous work analysing the transcriptomes of polyploid species has focussed on understanding patterns of expression bias (i.e. the ratio of expression levels for homoeologs) or homoeolog expression dominance (i.e. the level of overall expression

compared with the level in progenitor species; Grover *et al.*, 2012), there is a paucity of information regarding the degree to which induction profiles across homoeologous copies of biotic stress-responsive genes differ and whether these induction biases favour particular subgenomes globally. We therefore aimed to examine subgenome-specific gene expression patterns in bread wheat during fungal infection.

A high degree of homoeolog expression bias underpins the wheat transcriptome

For differential expression analysis, misalignment biases should affect mock and treated samples equally. However, the accuracy of expression estimates could be affected substantially by misalignment of reads between homoeologs, potentially leading to incorrect inference of homoeolog expression bias. We tested alignment methods with various stringency and ambiguous read handling parameters with results suggesting Burrows–Wheeler transform-based aligners such as Bowtie2 are able to reliably distinguish between homoeolog copies, consistent with previous findings (Pfeifer *et al.*, 2014).

Global observation of expression patterns revealed that homoeolog expression bias underpins a substantial proportion of the wheat transcriptome under both basal growth conditions and increasingly so during infection. Recent work has demonstrated the D subgenome contributes disproportionately to the transcriptional response during infection by *Fusarium graminearum* suggesting the D genome may play a predominant role in responding to this pathogen (Nussbaumer *et al.*, 2015). Contribution of homoeolog expression was found to be significantly biased towards B and D subgenomes under both mock and infection conditions consistent with previous findings. This bias may help explain the disproportionate contribution of B and D subgenomes to biotic stress-responsive genes observed since actively expressed genes are more likely to be induced. In contrast to our findings, previous transcriptome analyses in wheat suggested overall expression bias is balanced across subgenomes (Leach *et al.*, 2014; Pfeifer *et al.*, 2014). However, in previous studies only a limited number of chromosomes (Group 1 and 5; Leach *et al.*, 2014) or a small set of homoeologous loci were examined (Pfeifer *et al.*, 2014). Inadequate statistical power of analyses to identify subtle biases may have also been contributed to this discrepancy. Further work is needed to confirm whether the bias we detected towards B and D subgenomes is consistently maintained across environmental conditions and genetic backgrounds. Availability of more powerful statistical analysis methods and the total wheat coding sequence repertoire would allow a greater proportion of homoeolog triplets to be successfully inferred. The retention of homoeolog expression patterns between noninfected and infected conditions suggests for the majority of genes, patterns of expression bias are relatively stable. Therefore, observing homoeolog expression bias under basal conditions may provide an indication of which homoeologs are predominantly expressed under stress-induced conditions.

Overall, the higher proportion of triplets showing increased expression bias under infection (χ^2 distribution test $P < 0.01$) suggests biotic stress increases the overall degree of expression bias in the transcriptome; however, a large proportion of triplets displayed the same pattern of expression bias under mock and inoculated conditions. This suggests patterns of subgenome expression bias within homoeolog triplets are generally fixed, but the magnitude of difference tends to increase when biotic stress is applied.

Subgenome specificity in pathogen response is underpinned by homoeolog induction bias

We describe a new concept for transcriptome analysis in polyploid species which we term 'homoeolog induction bias'. Homoeolog induction bias differs from homoeolog expression bias as the former considers which homoeolog copies are more responsive during stress conditions rather than differences in magnitude of expression between homoeologs under the same condition. Homoeolog induction bias also differs from expression partitioning (Liu and Adams, 2007) since it does not attempt to explain patterns of induction in the light of shared biological functions or molecular processes, although expression partitioning may often be a strong explanatory factor for biases in homoeolog induction patterns. In this study, the inherent genomic complexity is highlighted in the varied degree to which homoeolog expression bias and homoeolog induction bias were observed within the wheat transcriptome. However, amidst the complexity, evidence for a greater contribution of B and D subgenomes for biotic stress responses emerged.

Here, we demonstrate that a large proportion of the transcriptome diversity in the molecular response to a necrotrophic fungal pathogen is contributed through induction of a single homoeolog. Together with the degree of expression bias favouring a single homoeolog observed, the trend for homoeolog induction bias evinces progress towards functional diploidization in bread wheat (Pfeifer et al., 2014). For homoeolog triplets in which two or more homoeologs are actively expressed or induced, it will be interesting to consider why expression of multiple homoeologs is retained in some cases while lost in others. Patterns of retention may be a guided process where increased expression dosage that provides beneficial effects may have been selected for artificially (e.g. through breeding) or naturally. Genes in which expression is only contributed from a single homoeolog are highly attractive targets for knockout, knockdown or mutagenesis approaches to aid functional characterization since this avoids the need to stack multiple altered homoeologs, a time-consuming and laborious process (Fitzgerald et al., 2010, 2015).

Impact of homoeolog expression bias and induction bias on biotic stress-related genes

Homoeolog induction for genes associated with biotic stress was generally biased towards B and D subgenomes. The potentially greater importance of the B subgenome in response to pathogens is consistent with the higher proportion of QTL for pathogen resistance occurring on the B subgenome chromosomes than on the other two subgenomes (Feldman et al., 2012). In addition, results from our study also suggest a greater contribution of the D subgenome to stress responses than the A subgenome. This suggestion is consistent with the view that the incorporation of the D subgenome in wheat has been a primary driver for the dispersal of bread wheat across temperate agro-ecological zones (Berkman et al., 2013).

The polyploidization history of wheat provides some explanation for the predominant role of B and D subgenomes in biotic stress responses (Marcussen et al., 2014). The predominance of the B subgenome over the A subgenome may have resulted through changes to genetic regulation during the period where the progenitor genomes existed in a tetraploid state (Lai et al., 2015) driving genome asymmetry in a function-specific manner. In tetraploid wheat, global transcriptomic analysis revealed the A subgenome to be dominant over the B subgenome in terms of

genomic stability (Pont et al., 2011). The event in which tetraploid wheat hybridized with the D genome progenitor occurred relatively recently on an evolutionary timescale (Marcussen et al., 2014). Thus, codominance of B and D subgenomes in biotic stress response has been maintained over the comparatively short period of hexaploidy. Recent work has demonstrated genome asymmetry patterns vary between natural and synthetic tetraploid wheat genotypes (Wang et al., 2016) perhaps suggesting emergence of genome asymmetry following polyploidization occurs in a stochastic rather than directed manner. Further use of synthetic polyploid lines may reveal how genome asymmetry and subfunctionalization between subgenomes occurs.

Conclusions

Understanding homoeolog expression and induction bias in polyploid crops has critical implications for their genetic improvement, since identification of actively expressed/induced homoeologs will allow targeted inactivation of active homoeologs. Further work studying expression and induction biases across infection time points, tissue types and developmental stages to determine whether expression biases are temporally or spatially determined will provide a more comprehensive understanding of polyploidy-associated gene expression patterns. Better understanding of how polyploid species utilize their genomic repertoire to endure conditions of stress may enable new strategies to improve agronomic traits in polyploid crops.

Experimental procedures

Crown rot infection assay

A soilless infection assay was performed using the commercial wheat (*Triticum aestivum* L.) cultivar 'Chara' to observe global transcriptional change during infection by *F. pseudograminearum* isolate CS3427 (CSIRO *Fusarium* collection). *F. pseudograminearum* spores were produced in flask culture using V8 broth (Gardiner et al., 2012) by inoculating and incubating on an orbital shaker at room temperature (~22 °C) for 1 week. Spores were harvested by filtering culture through Miracloth (Calbiochem, San Diego, CA) and centrifuging the filtrate in 50-mL Falcon tubes using a Sigma 4K15 benchtop centrifuge (6000× *g*) to pellet spores. Spores were resuspended in distilled water to a final concentration of ~1 × 10⁶ spores/mL and stored at −20 °C until required. Seedlings (3 days postgermination) were immersed in *F. pseudograminearum* spores (1 × 10⁶ spores/mL) and incubated for 3 min. Four biological replicates consisting of approximately 12 plants per replicate were included for each treatment to correct for inherent biological variation between plants during infection. To observe transcriptomic change at a relatively early point during infection (prior to visible symptom development), tissue was harvested 3 days postinoculation (dpi) and coleoptile sheath enclosed shoot tissue for each plant was excised and immediately immersed in liquid nitrogen. This was performed to observe response to infection within the crown region with 3 dpi selected as a time point to observe transcriptomic change in response to *F. pseudograminearum* in line with previous work (Desmond et al., 2006; Appendix S1).

Validation of successful infection was performed in two ways: firstly, infection replicates were included in the trial and were observed at 14 dpi for development of symptoms (Figure 1). Secondly, cDNA synthesis was performed on aliquots of RNA and relative expression of marker genes for defence responses was assessed using RT-PCR (Appendix S1). Having observed a strong molecular response at this time point, RNA samples were sent to the Ramaciotti Centre (Sydney, Australia) for library preparation and sequencing.

Homoeoallele triplet identification

We applied reciprocal best BLAST (RBB; Moreno-Hagelsieb and Latimer, 2008) between subgenome-specific coding sequence (CDS) subsets derived from the published wheat chromosome arm assemblies (Mayer et al., 2014). RBB hits were identified for each comparison of CDS sets from each of the A vs B, A vs D and B vs D subgenomes. A homoeolog triplet was identified as a set of three genes displaying agreement of genes present in the RBB hits between each pairwise comparison of the subgenome (i.e. A to B, B to D and D to A). Using a custom python script, we compared the homoeoallele triplets identified by our approach with those identified by Pfeifer et al. (2014) to determine the level of agreement between the two methods and the number of novel triplets identified by our approach.

RNA extraction and quality control

RNA extraction was performed using a Qiagen RNeasy extraction kit as per the manufacturer's instructions with the option for on-column DNase I (Qiagen) digestion. RNA concentration was initially determined using a Nanodrop 2100 spectrophotometer. Integrity of RNA samples was determined using an Agilent Bioanalyser (performed by Australian Genome Sequencing Facility) with all samples having a RIN score >8.5. cDNA synthesis was performed using Invitrogen Superscript III cDNA synthesis kit using oligo-dT primers to promote transcription of whole mRNA molecules according to manufacturer instructions.

RNA-seq library preparation and sequencing

Library preparation and sequencing were performed at the Ramaciotti Centre as described below. RNA was quantified and integrity was assessed a second time prior to sequencing. Sequencing libraries were prepared using standard Illumina library preparation methods. An Illumina HiSeq 2000 platform was used to generate 100-base pair (bp) paired-end (PE) reads from isolated mRNA extracted from F. pseudograminearum and mock-infected plants pooled into four biological reps yielding approximately 175 million reads (35gb) in total. Given the estimated size of the bread wheat transcriptome, this represented an average total sequence coverage of ~230x or approximately ~30x per biological sample. Sequence files were deposited to the National Centre for Biotechnology Information (NCBI) Sequence Read Archive under BioProject ID PRJNA297822.

RNA-seq analysis

To exclude sequencing errors where possible, sequence quality was analysed using SolexaQA (Cox et al., 2010) and paired-end reads were trimmed to ensure PHRED score >30 prior to alignment (minimum read length 70 bp). Reads were aligned to the T. aestivum Chromosomal Survey Sequence (Mayer et al., 2014) cDNA collection using Bowtie2 (2.2.3; Langmead and Salzberg, 2012) obtained http://plants.ensembl.org/index.html on 14 May 2014. Paired-end reads were utilized to help solve ambiguous read alignments. On average, 66% of total reads were successfully aligned to reference across samples. Appendix S1 delineates the process and command lines used within this analysis. Analysis of differential expression was performed using DESeq (Anders, 2010). Homoeolog expression

bias was inferred to triplets for which A, B or D homoeologs were differentially expressed under the same condition (adjusted P value <0.05 with Bonferroni corrected FDR) and homoeolog induction bias was inferred to triplets in which one or two homoeologs were differentially expressed between mock-inoculated and Fusarium-inoculated conditions (adjusted P value <0.05 with Bonferroni corrected FDR.

A graphical overview of the overall analysis pipeline is provided in Appendix S4.

Acknowledgements

Funding for this work was provided by the Grains Research and Development Corporation (GRDC) under CSP00155. JJP also acknowledges the GRDC for provision of a graduate research scholarship (GRS10532). We gratefully acknowledge BioPlatforms Australia for providing funding for sequencing.

References

Adams, K.L. and Wendel, J.F. (2005) Polyploidy and genome evolution in plants. Curr. Opin. Plant Biol. **8**, 135–141.

Akhunova, A.R., Matniyazov, R.T., Liang, H. and Akhunov, E.D. (2010) Homoeolog-specific transcriptional bias in allopolyploid wheat. BMC Genom. **11**, 505.

Akinsanmi, O., Backhouse, D., Simpfendorfer, S. and Chakraborty, S. (2006) Genetic diversity of Australian Fusarium graminearum and F. pseudograminearum. Plant. Pathol. **55**, 494–504.

Anders, S. (2010) Analysing RNA-Seq data with the DESeq package. Mol. Biol. **43**, 1–17.

Beaulieu, J.M., Moles, A.T., Leitch, I.J., Bennett, M.D., Dickie, J.B. and Knight, C.A. (2007) Correlated evolution of genome size and seed mass. New Phytol. **173**, 422–437.

Bekaert, M., Edger, P.P., Pires, J.C. and Conant, G.C. (2011) Two-phase resolution of polyploidy in the Arabidopsis metabolic network gives rise to relative and absolute dosage constraints. Plant Cell, **23**, 1719–1728.

Berkman, P.J., Visendi, P., Lee, H.C., Stiller, J., Manoli, S., Lorenc, M.T., Lai, K. et al. (2013) Dispersion and domestication shaped the genome of bread wheat. Plant Biotech. J. **11**, 564–571.

Birchler, J.A., Riddle, N.C., Auger, D.L. and Veitia, R.A. (2005) Dosage balance in gene regulation: biological implications. Trends Genet. **21**, 219–226.

Blanc, G. and Wolfe, K.H. (2004) Widespread paleopolyploidy in model plant species inferred from age distributions of duplicate genes. Plant Cell, **16**, 1667–1678.

Buggs, R.J., Renny-Byfield, S., Chester, M., Jordon-Thaden, I.E., Viccini, L.F., Chamala, S., Leitch, A.R. et al. (2012) Next-generation sequencing and genome evolution in allopolyploids. Am. J. Bot. **99**, 372–382.

Chakraborty, S., Liu, C., Mitter, V., Scott, J., Akinsanmi, O., Ali, S., Dill-Macky, R. et al. (2006) Pathogen population structure and epidemiology are keys to wheat crown rot and Fusarium head blight management. Austral. Plant Pathol. **35**, 643–655.

Chelaifa, H., Monnier, A. and Ainouche, M. (2010) Transcriptomic changes following recent natural hybridization and allopolyploidy in the salt marsh species Spartina × townsendii and Spartina anglica (Poaceae). New Phytol. **186**, 161–174.

Chen, Z.J. (2007) Genetic and epigenetic mechanisms for gene expression and phenotypic variation in plant polyploids. Ann. Rev. Plant Biol. **58**, 377–401.

Chen, Z.J. and Ni, Z. (2006) Mechanisms of genomic rearrangements and gene expression changes in plant polyploids. BioEssays, **28**, 240–252.

Comai, L. (2005) The advantages and disadvantages of being polyploid. *Nat. Rev. Genet.* **6**, 836–846.

Conesa, A., Götz, S., García-Gómez, J.M., Terol, J., Talón, M. and Robles, M. (2005) Blast2GO: A universal tool for annotation, visualization and analysis in functional genomics research. *Bioinformatics*, **21**, 3674–3676.

Cox, M.P., Peterson, D.A. and Biggs, P.J. (2010) SolexaQA: At-a-glance quality assessment of Illumina second-generation sequencing data. *BMC Bioinformatics*, **11**, 485.

Desmond, O.J., Edgar, C.I., Manners, J.M., Maclean, D.J., Schenk, P.M. and Kazan, K. (2006) Methyl jasmonate induced gene expression in wheat delays symptom development by the crown rot pathogen *Fusarium pseudograminearum*. *Physiol. Mol. Plant Pathol.* **67**, 171–179.

Desmond, O.J., Manners, J.M., Schenk, P.M., Maclean, D.J. and Kazan, K. (2008) Gene expression analysis of the wheat response to infection by *Fusarium pseudograminearum*. *Physiol. Mol. Plant Pathol.* **73**, 40–47.

Dong, S. and Adams, K.L. (2011) Differential contributions to the transcriptome of duplicated genes in response to abiotic stresses in natural and synthetic polyploids. *New Phytol.* **190**, 1045–1057.

Feldman, M. and Levy, A.A. (2009) Genome evolution in allopolyploid wheat— a revolutionary reprogramming followed by gradual changes. *J. Genet. Genomics* **36**, 511–518.

Feldman, M., Levy, A.A., Fahima, T. and Korol, A. (2012) Genomic asymmetry in allopolyploid plants: wheat as a model. *J. Exp. Bot.* **63**, 5045–5059.

Fitzgerald, T.L., Kazan, K., Li, Z., Morell, M.K. and Manners, J.M. (2010) A high-throughput method for the detection of homoeologous gene deletions in hexaploid wheat. *BMC Plant Biol.* **10**, 264.

Fitzgerald, T.L., Powell, J.J., Stiller, J., Weese, T.L., Abe, T., Zhao, G., Jia, J. *et al.* (2015) An assessment of heavy ion irradiation mutagenesis for reverse genetics in wheat (*Triticum aestivum* L.). *PLoS ONE*, **10**, e0117369.

Flagel, L.E., Wendel, J.F. and Udall, J.A. (2012) Duplicate gene evolution, homoeologous recombination, and transcriptome characterization in allopolyploid cotton. *BMC Genom.* **13**, 302.

Gardiner, D.M., McDonald, M.C., Covarelli, L., Solomon, P.S., Rusu, A.G., Marshall, M., Kazan, K. *et al.* (2012) Comparative pathogenomics reveals horizontally acquired novel virulence genes in fungi infecting cereal hosts. *PLoS Pathog.* **8**, e1002952.

Grover, C., Gallagher, J., Szadkowski, E., Yoo, M., Flagel, L. and Wendel, J. (2012) Homoeolog expression bias and expression level dominance in allopolyploids. *New Phytol.* **196**, 966–971.

Higgins, J., Magusin, A., Trick, M., Fraser, F. and Bancroft, I. (2012) Use of mRNA-seq to discriminate contributions to the transcriptome from the constituent genomes of the polyploid crop species *Brassica napus*. *BMC Genom.* **13**, 247.

Kerber, E. (1991) Stem-rust resistance in 'Canthatch' hexaploid wheat induced by a nonsuppressor mutation on chromosome 7DL. *Genome*, **34**, 935–939.

Kim, E.-D. and Chen, Z.J. (2011) Unstable transcripts in Arabidopsis allotetraploids are associated with nonadditive gene expression in response to abiotic and biotic stresses. *PLoS ONE*, **6**, e24251.

Lai, K., Lorenc, M.T., Lee, H.C., Berkman, P.J., Bayer, P.E., Visendi, P., Ruperao, P. *et al.* (2015) Identification and characterization of more than 4 million intervarietal SNPs across the group 7 chromosomes of bread wheat. *Plant Biotech. J.* **13**, 97–104.

Langmead, B. and Salzberg, S.L. (2012) Fast gapped-read alignment with Bowtie 2. *Nat. Methods*, **9**, 357–359.

Leach, L.J., Belfield, E.J., Jiang, C., Brown, C., Mithani, A. and Harberd, N.P. (2014) Patterns of homoeologous gene expression shown by RNA sequencing in hexaploid bread wheat. *BMC Genom.* **15**, 276.

Li, H.B., Xie, G.Q., Ma, J., Liu, G.R., Wen, S.M., Ban, T., Chakraborty, S. *et al.* (2010) Genetic relationships between resistances to Fusarium head blight and crown rot in bread wheat (*Triticum aestivum* L.). *Theor. Appl. Genet.* **121**, 941–950.

Liu, Z. and Adams, K.L. (2007) Expression partitioning between genes duplicated by polyploidy under abiotic stress and during organ development. *Curr. Biol.* **17**, 1669–1674.

Liu, Y., Ma, J., Yan, W., Yan, G., Zhou, M., Wei, Y., Zheng, Y. *et al.* (2012) Different tolerance in bread wheat, durum wheat and barley to Fusarium crown rot disease caused by Fusarium pseudograminearum. *J. Phytopathol.* **160**, 412–417.

Liu, Z., Xin, M., Qin, J., Peng, H., Ni, Z., Yao, Y. and Sun, Q. (2015) Temporal transcriptome profiling reveals expression partitioning of homeologous genes contributing to heat and drought acclimation in wheat (*Triticum aestivum* L.). *BMC Plant Biol.* **15**, 152.

Marcussen, T., Sandve, S.R., Heier, L., Spannagl, M., Pfeifer, M., Jakobsen, K.S., Wulff, B.B. *et al.* (2014) Ancient hybridizations among the ancestral genomes of bread wheat. *Science*, **345**, 1250092.

Mayer, K.F., Rogers, J., Doležel, J., Pozniak, C., Eversole, K., Feuillet, C., Gill, B. *et al.* (2014) A chromosome-based draft sequence of the hexaploid bread wheat (*Triticum aestivum*) genome. *Science*, **345**, 1251788.

Moreno-Hagelsieb, G. and Latimer, K. (2008) Choosing BLAST options for better detection of orthologs as reciprocal best hits. *Bioinformatics*, **24**, 319–324.

Murray, G.M. and Brennan, J.P. (2009) Estimating disease losses to the Australian wheat industry. *Austral. Plant Pathol.* **38**, 558–570.

Ng, D.W., Zhang, C., Miller, M., Shen, Z., Briggs, S. and Chen, Z. (2012) Proteomic divergence in Arabidopsis autopolyploids and allopolyploids and their progenitors. *Heredity*, **108**, 419–430.

Ni, Z., Kim, E.-D., Ha, M., Lackey, E., Liu, J., Zhang, Y., Sun, Q. *et al.* (2009) Altered circadian rhythms regulate growth vigour in hybrids and allopolyploids. *Nature*, **457**, 327–331.

Nomura, T., Ishihara, A., Imaishi, H., Endo, T.R., Ohkawa, H. and Iwamura, H. (2002) Molecular characterization and chromosomal localization of cytochrome P450 genes involved in the biosynthesis of cyclic hydroxamic acids in hexaploid wheat. *Mol. Genet. Genomics*, **267**, 210–217.

Nomura, T., Ishihara, A., Yanagita, R.C., Endo, T.R. and Iwamura, H. (2005) Three genomes differentially contribute to the biosynthesis of benzoxazinones in hexaploid wheat. *Proc. Natl Acad. Sci. USA*, **102**, 16490–16495.

Nussbaumer, T., Warth, B., Sharma, S., Ametz, C., Bueschl, C., Parich, A., Pfeifer, M. *et al.* (2015) Joint transcriptomic and metabolomic analyses reveal changes in the primary metabolism and imbalances in the subgenome orchestration in the bread wheat molecular response to *Fusarium graminearum*. *G3 Genes Genomes Genet.*, **5**, 2579–2592.

Oswald, B.P. and Nuismer, S.L. (2007) Neopolyploidy and pathogen resistance. *Proc. R. Soc. Lond. B Biol. Sci.* **274**, 2393–2397.

Peng, J., Ronin, Y., Fahima, T., Röder, M.S., Li, Y., Nevo, E. and Korol, A. (2003) Domestication quantitative trait loci in Triticum dicoccoides, the progenitor of wheat. *Proc. Natl Acad. Sci. USA*, **100**, 2489–2494.

Pfeifer, M., Kugler, K.G., Sandve, S.R., Zhan, B., Rudi, H., Hvidsten, T.R., Mayer, K.F. *et al.* (2014) Genome interplay in the grain transcriptome of hexaploid bread wheat. *Science*, **345**, 1250091.

Pignatta, D., Dilkes, B.P., Yoo, S.Y., Henry, I.M., Madlung, A., Doerge, R.W., Jeffrey Chen, Z. *et al.* (2010) Differential sensitivity of the *Arabidopsis thaliana* transcriptome and enhancers to the effects of genome doubling. *New Phytol.* **186**, 194–206.

Pont, C., Murat, F., Confolent, C., Balzergue, S. and Salse, J. (2011) RNA-seq in grain unveils fate of neo-and paleopolyploidization events in bread wheat (*Triticum aestivum* L.). *Genome Biol.* **12**, R119.

Ravanel, S., Cherest, H., Jabrin, S., Grunwald, D., Surdin-Kerjan, Y., Douce, R., and Rebeille, F. (2001) Tetrahydrofolate biosynthesis in plants: molecular and functional characterization of dihydrofolate synthetase and three isoforms of folylpolyglutamate synthetase in *Arabidopsis thaliana*. *Proc. Natl Acad. Sci. USA*, **98**, 15360–15365.

Schnable, J.C., Springer, N.M. and Freeling, M. (2011) Differentiation of the maize subgenomes by genome dominance and both ancient and ongoing gene loss. *Proc. Natl Acad. Sci. USA*, **108**, 4069–4074.

Sehrish, T., Symonds, V.V., Soltis, D.E., Soltis, P.S. and Tate, J.A. (2014) Gene silencing via DNA methylation in naturally occurring *Tragopogon miscellus* (Asteraceae) allopolyploids. *BMC Genom.* **15**, 701.

Stephen, S., Cullerne, D., Spriggs, A., Helliwell, C., Lovell, D. and Taylor, J. (2012) *BioKanga: a suite of high performance bioinformatics applications.* Available at https://sourceforge.net/projects/biokanga/.

Sugiyama, S.-I. (2005) Polyploidy and cellular mechanisms changing leaf size: comparison of diploid and autotetraploid populations in two species of Lolium. *Ann. Bot.* **96**, 931–938.

Wang, Z., Gerstein, M. and Snyder, M. (2009) RNA-Seq: a revolutionary tool for transcriptomics. *Nat. Rev. Genet.* **10**, 57–63.

Wang, X., Zhang, H., Li, Y., Zhang, Z., Li, L. and Liu, B. (2016) Transcriptome asymmetry in synthetic and natural allotetraploid wheats, revealed by RNA-sequencing. *New Phytologist*, **209**, 1264–1277.

Wittek, F., Kanawati, B., Wenig, M., Hoffmann, T., Franz-Oberdorf, K., Schwab, W., Schmitt-Kopplin, P. *et al.* (2015) Folic acid induces salicylic acid-dependent immunity in Arabidopsis and enhances susceptibility to *Alternaria brassicicola*. *Mol. Plant Pathol.* **16**, 616–622.

Wood, T.E., Takebayashi, N., Barker, M.S., Mayrose, I., Greenspoon, P.B. and Rieseberg, L.H. (2009) The frequency of polyploid speciation in vascular plants. *Proc. Natl Acad. Sci. USA*, **106**, 13875–13879.

Yang, X., Ma, J., Li, H., Ma, H., Yao, J. and Liu, C. (2010) Different genes can be responsible for crown rot resistance at different developmental stages of wheat and barley. *Eur. J. Plant Pathol.* **128**, 495–502.

Yoo, M., Szadkowski, E. and Wendel, J. (2013) Homoeolog expression bias and expression level dominance in allopolyploid cotton. *Heredity*, **110**, 171–180.

15

The *Lr34* adult plant rust resistance gene provides seedling resistance in durum wheat without senescence

Amy Rinaldo[1], Brian Gilbert[1], Rainer Boni[2], Simon G. Krattinger[2], Davinder Singh[3], Robert F. Park[3], Evans Lagudah[1] and Michael Ayliffe[1,*]

[1]*CSIRO Agriculture, Canberra, ACT, Australia*
[2]*Department of Plant and Microbial Biology, University of Zurich, Zurich, Switzerland*
[3]*Plant Breeding Institute, University of Sydney, Narellan, NSW, Australia*

*Correspondence
e-mail michael.ayliffe@csiro.au

Summary

The hexaploid wheat (*Triticum aestivum*) adult plant resistance gene, *Lr34/Yr18/Sr57/Pm38/Ltn1*, provides broad-spectrum resistance to wheat leaf rust (*Lr34*), stripe rust (*Yr18*), stem rust (*Sr57*) and powdery mildew (*Pm38*) pathogens, and has remained effective in wheat crops for many decades. The partial resistance provided by this gene is only apparent in adult plants and not effective in field-grown seedlings. *Lr34* also causes leaf tip necrosis (*Ltn1*) in mature adult plant leaves when grown under field conditions. This D genome-encoded bread wheat gene was transferred to tetraploid durum wheat (*T. turgidum*) cultivar Stewart by transformation. Transgenic durum lines were produced with elevated gene expression levels when compared with the endogenous hexaploid gene. Unlike nontransgenic hexaploid and durum control lines, these transgenic plants showed robust seedling resistance to pathogens causing wheat leaf rust, stripe rust and powdery mildew disease. The effectiveness of seedling resistance against each pathogen correlated with the level of transgene expression. No evidence of accelerated leaf necrosis or up-regulation of senescence gene markers was apparent in these seedlings, suggesting senescence is not required for *Lr34* resistance, although leaf tip necrosis occurred in mature plant flag leaves. Several abiotic stress-response genes were up-regulated in these seedlings in the absence of rust infection as previously observed in adult plant flag leaves of hexaploid wheat. Increasing day length significantly increased *Lr34* seedling resistance. These data demonstrate that expression of a highly durable, broad-spectrum adult plant resistance gene can be modified to provide seedling resistance in durum wheat.

Keywords: *Puccinia*, *Triticum*, rust, ABC transporter, *Blumeria*.

Introduction

Wheat rust diseases caused by *Puccinia graminis* f. sp. *tritici* (stem rust), *P. striiformis* f. sp. *tritici* (stripe/yellow rust) and *P. triticina* (leaf rust) remain a major threat to world production of hexaploid bread wheat (*Triticum aestivum* L., $2n = 6x$ = AABBDD) and tetraploid durum (pasta) wheat (*T. turgidum* L. subsp. *durum*, $2n = 4x = 28$, AABB) (Chen *et al.*, 2014; Huerta-Espino *et al.*, 2011; Kolmer, 2005; Singh *et al.*, 2011a). Resistance to these wheat rust pathogens is achieved most economically using resistance genes derived from wheat landraces and wild relative species. However, most wheat rust resistance genes are ultimately overcome by pathogen evolution to virulence.

Wheat rust resistance genes have been broadly categorized into two groups, all-stage or seedling resistance genes and adult plant resistance (APR) genes. All-stage resistance genes, function at all stages of plant development and although often race specific, can provide high levels of resistance. Nine cloned all-stage wheat rust resistance genes each encode a nucleotide-binding site leucine-rich repeat protein (NLR), a large class of disease resistance proteins characterized in numerous plant species (Cloutier *et al.*, 2007; Feuillet *et al.*, 2003; Huang *et al.*, 2003; Liu *et al.*, 2014; Mago *et al.*, 2015; Periyannan *et al.*, 2013; Saintenac *et al.*, 2013; Steuernagel *et al.*, 2016). These proteins recognize pathogen molecules (effectors) introduced

into plant cells or alternatively effector-mediated modifications of host proteins, leading to defence activation (reviewed by Dodds and Rathjen, 2010). Pathogen effector genes can rapidly evolve to avoid plant recognition, making NLR resistance genes ineffective.

Unlike all-stage resistance, APR occurs in mature wheat plants only and tends to provide partial resistance, although specific APR gene combinations show additive resistance effects (Singh *et al.*, 2011b). Some APR genes provide resistance to all isolates of a pathogen species (broad spectrum) and in some cases resistance to multiple pathogen species. For example, the *Lr34/Yr18/Sr57/Pm38/Ltn1* gene (hereafter called *Lr34*) provides resistance to *P. triticina* (*Lr*), *P. striiformis* f. sp. *tritici* (*Yr*), *P. graminis* f. sp. *tritici* (*Sr*) and *Blumeria graminis* (*Pm*). This gene also confers a leaf tip necrosis phenotype (Ltn) on flag leaves when plants are field-grown (Dyck, 1991; Hulbert *et al.*, 2007; Lagudah *et al.*, 2006; Risk *et al.*, 2012; Shah *et al.*, 2011). Similarly, the *Lr67/Yr46/Sr56/Pm39/Ltn3* gene (hereafter referred to as *Lr67*) shows broad-spectrum, partial resistance to rust and mildew pathogens (Herrera-Foessel *et al.*, 2014). In contrast, the *Yr36* APR gene provides *P. striiformis* f. sp. *tritici* resistance only (Uauy *et al.*, 2005).

The molecular basis of APR is poorly understood when compared with NLR-mediated resistance. The three APR genes described above have been cloned and encode an ABC-type transporter (*Lr34*), a hexose transporter (*Lr67*) and a protein kinase fused to a START domain (*Yr36*) (Fu *et al.*, 2009; Krattinger *et al.*, 2009; Moore *et al.*, 2015). The Lr67 protein may act as a dominant negative regulator of hexose transport (Moore *et al.*, 2015), while the Yr36 protein increases chloroplast H_2O_2 accumulation by phosphorylation of a thylakoid-associated ascorbate peroxidase (Guo *et al.*, 2015). The mode of action of the Lr34 transporter and molecules it transports are unknown. The deletion of a single phenylalanine codon in the D genome-encoded *lr34*-susceptible allele converts it to a functional *Lr34* resistance gene (Chauhan *et al.*, 2015). This mutation is believed to have occurred after the domestication of wheat (Krattinger *et al.*, 2013).

Lr34 has been used extensively in hexaploid wheat cultivation for many decades and remained durable to all pathogen races over this long period of time (Dyck *et al.*, 1966; Kolmer *et al.*, 2008). It provides partial resistance that is insufficient to prevent yield losses from rust diseases unless supplemented with additional resistance genes. *Lr34* partial resistance occurs in mature plants (60 days postgermination) and is often associated with Ltn (Dyck, 1991; Singh, 1992a). Ltn, which appears to be accelerated leaf senescence, is particularly apparent in flag leaves (Krattinger *et al.*, 2009; Risk *et al.*, 2012) and has been used as a phenotypic marker for *Lr34* identification (Shah *et al.*, 2011). Ltn, however, is influenced by both the environment and genetic background (Shah *et al.*, 2010, 2011; Singh, 1992a). *Lr34* resistance occurs in mature flag leaves after the onset of Ltn (Krattinger *et al.*, 2009) with leaf tips showing the greatest resistance (Hulbert *et al.*, 2007). Flag leaf tips show up-regulation of abiotic stress-responsive genes in the absence of pathogen infection and higher levels of pathogenesis-related (PR) protein expression upon *P. triticina* infection (Hulbert *et al.*, 2007). This apparent abiotic stress response in uninfected, mature *Lr34* tissue is suggested to prime the plant defence response for elevated expression upon rust pathogen challenge (Hulbert *et al.*, 2007).

Expression of *Lr34* in transgenic barley seedlings results in a very deleterious phenotype due to the induction of rapid, developmental leaf senescence (Risk *et al.*, 2013). Unlike *Lr34* wheat plants, these barley plants show constitutive induction of defence pathways in the absence of pathogen infection (Chauhan *et al.*, 2015) and are resistant to pathogens at the seedling and adult plant stage (Risk *et al.*, 2013). Barley does not contain orthologous *Lr34* sequences (Krattinger *et al.*, 2011), and co-expression of the *lr34*-susceptible allele appears to help attenuate negative effects in this species (Chauhan *et al.*, 2015).

In contrast to barley, rice encodes an *Lr34* orthologue (Krattinger *et al.*, 2011). Expression of the wheat *Lr34* gene in rice was associated with early leaf tip necrosis and deleterious pleiotropic effects in most cases, although not as extreme as observed in barley (Krattinger *et al.*, 2016). In some lines, leaf tip necrosis occurred in seedlings at the two-leaf stage and plants subsequently showed a severe negative impact on axillary shoot formation, plant vigour and spikelet production (Krattinger *et al.*, 2016). However, a single line with low seedling expression was recovered that was only marginally compromised. Remarkably, this gene also provided resistance to *Magnaporthe oryzae*, the hemibiotrophic causal agent of rice blast disease. Deletion of the critical phenylalanine codon present in the rice *Lr34* orthologue did not result in disease resistance (Krattinger *et al.*, 2016).

Given the remarkable durability of *Lr34*, we have introduced this hexaploid wheat (ABD) D genome-encoded gene into durum wheat (AB) by transgenesis as a potentially useful source of disease resistance. Amongst these durum transgenics, several lines showed obvious seedling resistance to leaf rust, stripe rust and powdery mildew diseases, a phenotype not associated with the endogenous *Lr34* gene. A strong correlation between seedling resistance and transgene expressions level was observed. Unlike barley and rice, no deleterious accelerated senescence or developmental phenotypes were observed in these seedlings. *Lr34* is therefore potentially of significant benefit for durum wheat germplasm improvement and, with elevated expression, can provide seedling resistance that is not conferred by the endogenous hexaploid wheat gene.

Results

Lr34-mediated leaf rust resistance occurs in hexaploid wheat seedlings when grown at a constant 10 °C throughout *P. triticina* challenge (Risk *et al.*, 2012). However, these seedlings are not resistant when grown at higher temperatures (Risk *et al.*, 2012; Rubiales and Niks, 1995; Singh and Gupta, 1992). *Lr34* cold-induced seedling resistance was exploited to screen transgenic durum cultivar Stewart plants containing a *Lr34* transgene (Figure S1) for *P. triticina* resistance. T1 seedlings from 12 lines were infected with *P. triticina* and after inoculation grown at a constant 10 °C with a 16-h light/8-h dark photoperiod. Potentially resistant progeny were observed in 10 T1 families. On susceptible plants, pustules were present after 20 days postinoculation (dpi), which increased in size by 36 dpi. Resistant seedlings were not immune to *P. triticina*, but had an obvious reduction in pustule size (Figure S2). Homozygous *Lr34* lines were identified using PCR and confirmed by DNA blot analysis of DNAs from 25 T2 seedlings using a restriction enzyme/probe combination that also determined transgene copy number (Figure S3). Four independent homozygous T1 lines were produced that contained a single transgene and potentially showed weak (36-4), moderate (17-1, 41-2) and high (39-2) levels of seedling leaf rust resistance at 10 °C.

Homozygous T2 progeny were re-screened for seedling leaf rust resistance at 10 °C. For each line, *P. triticina* growth was quantified 30 dpi (Risk *et al.*, 2012) by pooling equivalent leaves from 10 to 15 seedlings and the relative amount of fungal chitin present per gram of fresh tissue determined using a chitin assay (Ayliffe *et al.*, 2013). Chitin levels were expressed as fluorescence units of bound WGA-FITC, a fluorophore-conjugated lectin that specifically binds chitin. A clear reduction in fungal biomass was apparent in homozygous *Lr34* families compared with Stewart control plants (Figure 1a, black columns). Hexaploid wheat cultivar Thatcher carrying the endogenous D genome *Lr34* gene (Th+Lr34) had significantly less rust disease compared with Thatcher control plants, as expected (Figure 1a).

These genotypes were then tested for *P. triticina* resistance when plants were grown at 22 °C with a 16-h light/8-h dark photoperiod. Under these conditions, Stewart control plants were moderately susceptible (3C on the Stakman scale, where 0 is immune and 4 is highly susceptible (Stakman *et al.*, 1962)) to the *P. triticina* isolate used, with obvious macroscopic rust growth and sporulation occurring (Figure 2a). Remarkably, an obvious increase in seedling resistance occurred in T2 seedlings of these durum lines compared with nontransgenic control plants 14 dpi in three replicated experiments (Figures 1a, 2a, 5a, 6a). As expected, Th+34 seedlings had high levels of infection when grown at 22 °C, albeit less than the Tc control (Figures 1a, 2a). The relative levels of rust growth on transgenic seedlings at 22 °C

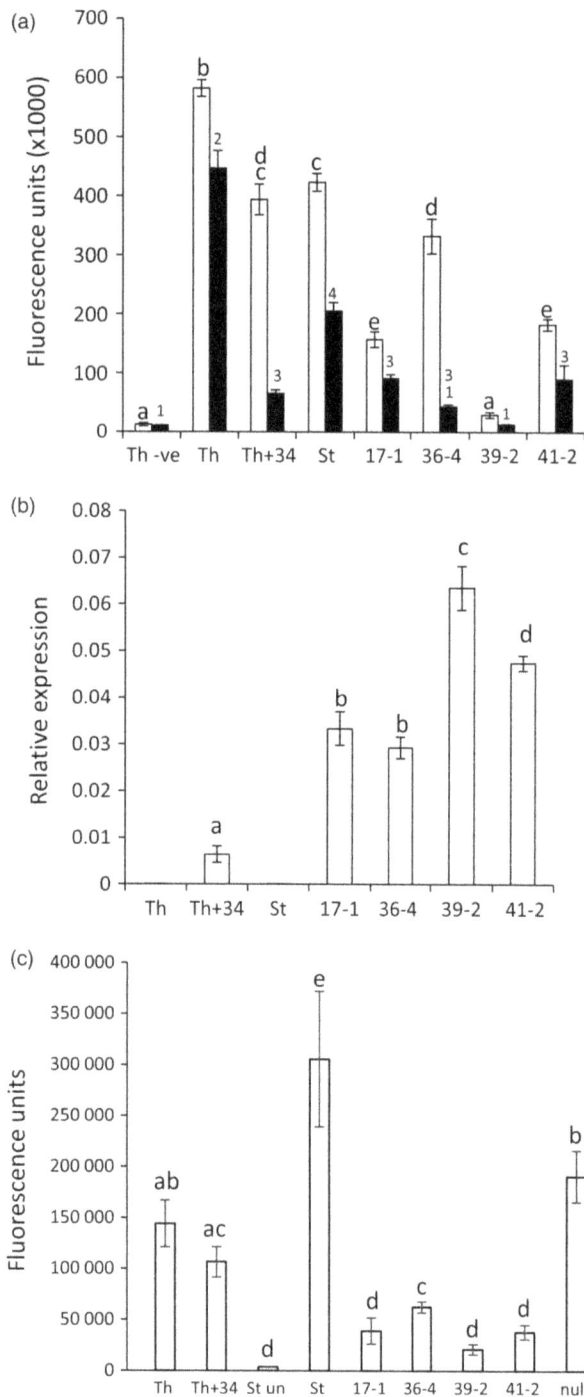

(a)

(b)

(c)

Figure 1 *Lr34* durum resistance corresponds to transgene expression levels. (a) Chitin assay quantification of *P. triticina* growth on wheat seedlings (three- to four-leaf stage) of hexaploid cultivar Thatcher (Th), a near-isogenic *Lr34* Thatcher line (Th+34), Stewart (St), *Lr34* lines 17-1, 36-4, 39-2, 41-2 and uninfected Thatcher (Th-ve). Seedlings were grown at 10 °C (black columns) or 22 °C (white columns) and harvested at 30 and 14 dpi, respectively. Common letters (white columns) or numbers (black columns) indicate data not significantly different (ANOVA, $P < 0.05$), throughout. Each value is the average of four chitin measurements from 10 to 15 pooled seedlings. (b) *Lr34* expression was quantified by Q-PCR and normalized relative to *GAPDH* in uninfected, three- to four-leaf seedlings of genotypes described in (a). Each data point is derived from three biological replicates, each with three technical replicates. (c) Chitin assay quantification of *P. striiformis* f. sp. *tritici* growth (14 dpi) on 10-15 seedlings per genotype described in (a) at the three- to four-leaf stage. Using pairwise *t*-tests ($P < 0.05$), line 39-2 had significantly less pathogen growth than lines 17-1 and 41-2, which each had less growth than line 36-4. A nontransgenic Stewart line regenerated from tissue culture (null) is included.

expression levels and disease resistance in these durum lines (Figure 1b). All four lines had significantly greater (4.5- to 10-fold) seedling *Lr34* expression compared with the endogenous Th+34 gene (Figure 1b), presumably explaining the durum seedling resistance observed for this APR gene under glasshouse conditions. Increased *Lr34* expression in these lines is likely due to transgene integration into more transcriptionally favourable regions of the genome.

As *Lr34* also provides APR to *P. striiformis* f. sp. *tritici* (wheat stripe rust disease) (Singh, 1992b) and *Blumeria graminis* (wheat powdery mildew disease) (Spielmeyer *et al.*, 2005), *Lr34* durum lines were screened for seedling resistance to these pathogens (22 °C, 16-h light). An obvious reduction in seedling stripe rust growth was observed on transgenic lines both macroscopically and by relative fungal biomass quantification in three replicated experiments (Figures 1c, 2b, S4). The levels of disease resistance again largely correlated with transgene expression levels of each line (Figure 1b). Similarly, significantly less *B. graminis* growth occurred on two *Lr34* durum lines tested with this pathogen using 14 biological replicates per genotype (Figure 3a).

To further investigate the *Lr34* resistance of hexaploid wheat seedlings grown at 10 °C, transcript levels were quantified. Seedlings were grown at 22 °C (16-h light) and then half transferred to 10 °C growth conditions with the same light regime. After 3 days of acclimation, half of the seedlings in each cabinet were infected with *P. striiformis* f. sp. *tritici* and tissues harvested from infected and uninfected seedlings 3 dpi. A fourfold induction of *Lr34* expression occurred in Tc+34 seedlings grown at 10 °C after *P. triticina* infection (Figure 3b) compared with uninfected seedlings. No equivalent increase in *Lr34* expression occurred in Tc+34-infected seedlings grown at 22 °C. These data are consistent with similar studies that showed *Lr34* induction in *P. triticina*-infected wheat seedlings grown at 10 °C, but not in infected seedlings grown under higher temperatures (Risk *et al.*, 2012). The ability of *Lr34* to provide seedling resistance in hexaploid wheat at 10 °C therefore appears due to elevated gene expression upon pathogen infection, although additional effects from these growth conditions (e.g. reduced pathogen growth rate and potential cold acclimation response) may contribute to resistance.

Microscopic analyses were undertaken on transgenic durum lines 17-1, 36-4, 39-2 and 41-2 and control Stewart seedlings

largely mirrored what was seen at 10 °C with line 39-2 again showing the greatest resistance (Figures 1a, 2a, 5a, 6a), although line 36-4 was an obvious exception in Figure 1a.

To determine whether a correlation existed between *Lr34* transgene expression levels and the resistance observed in each line, Q-PCR was undertaken on uninfected leaf tissue of T2 seedlings grown at 22 °C (16-h light). The PCR primers used amplified transcripts from only Lr34 and not from related homoeologues present on the wheat A genome (no *Lr34* homologue exists on the B genome), or the chromosome 7D *lr34* susceptibility allele of hexaploid wheat (Krattinger *et al.*, 2009). A strong correlation occurred between transgene

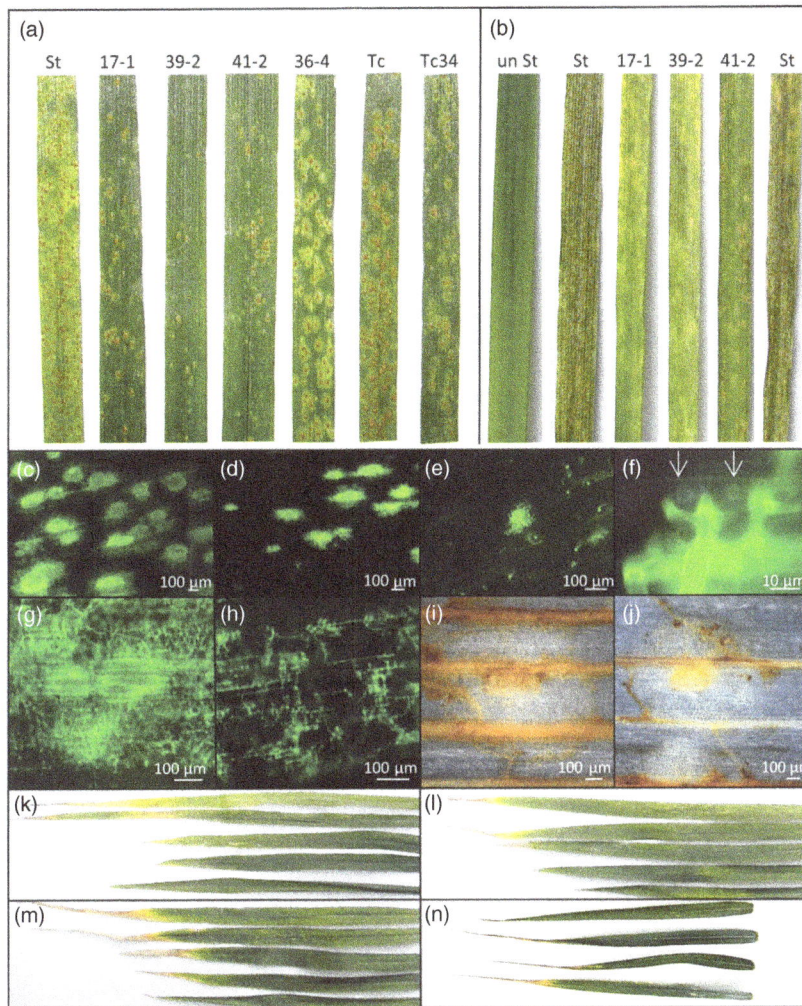

Figure 2 Phenotypic analysis of *Lr34* durum seedlings. (a) *P. triticina* growth (22 °C, 16-h light) 14 dpi on leaves of seedling of Stewart, *Lr34* lines 17-1, 39-2, 41-2 and 36-4, Thatcher hexaploid wheat and *Lr34* Thatcher. (b) *P. striiformis* f. sp. *tritici* growth (22 °C, 16-h light) 14 dpi on *Lr34* seedlings of 17-1, 39-2 and 41-2, uninfected Stewart and infected Stewart. Microscopy of *P. triticina* growth (12 dpi) on Stewart (c) and *Lr34* line 39-2 (d) stained with WGA-FITC. (e) *P. triticina* growth (12dpi) on *Lr34* line 39-2 showing one moderate and numerous small infection sites. (f) *P. triticina* haustoria (arrows) in cells of line 39-2. (g, h) *P. striiformis* f. sp. *tritici* growth on Stewart (g) and *Lr34* line 39-2 (h) (WGA-FITC stained). 3', 3'-Diaminobenzidine staining of *P. triticina* infection sites on Stewart (i) and line 39-2. (j). A brown precipitate shows H_2O_2 accumulation in vascular tissue and cells surrounding uredinia (white circle of cells with yellow centre) in each line. Leaves, in descending order of age (i.e. flag leaf at bottom), from glasshouse-grown Stewart (k), 39-2 (l) and 41-2 (m) plants. (n) Flag leaves from field-grown (top to bottom) Stewart and lines 17-1, 39-2 and 41-2. Tissues in panels c–j were grown at 22 °C, 16-h light.

after infection with either *P. triticina* or *P. striiformis* f. sp. *tritici*. On all lines, a mixture of infection sites occurred, ranging from small sites to large sporulating uredinia. Control Stewart seedlings showed extensive growth of both pathogens with infection sites usually producing sporulating uredinia (Figure 2c, g), while transgenic durum lines showed less hyphal growth, fewer uredinia and generally smaller infection sites albeit with haustoria (Figure 2d–f, h). Autofluorescent cells were uncommon in all lines infected with either pathogen, suggesting that cell death was not a predominant feature of the resistance response. These data are consistent with previous analysis of adult hexaploid *Lr34* wheat plants after *P. triticina* infection where reduced rust growth occurred without obvious cell death (Risk *et al.*, 2012).

As rust growth does still occur on *Lr34* transgenic seedlings, the effect of additional plant growth following *P. striiformis* infection was examined 26 dpi. By this time, infected seedling

leaves of all plants were becoming chlorotic and senescent, suggesting that little more growth of this biotrophic pathogen would occur due to leaf ageing (Figure S5). Relative fungal biomass quantification again showed significantly less stripe rust pathogen growth (Figure 3c) on *Lr34* durum seedlings that correlated with transgene expression levels (Figure 1b). These data suggest that rust pathogen growth on these seedlings does not ever reach that observed on nontransgenic control lines.

Expression of *Lr34* in barley causes strong seedling leaf necrosis that appears to be an accelerated senescence response (Risk *et al.*, 2013). To assess potential deleterious effects of *Lr34* expression in durum wheat, T3 plants were grown to maturity in the glasshouse. No difference in plant tiller number, tiller height and seed weight yield occurred between control and transgenic durum lines (Figure S6). During plant development, no dramatic differences in leaf senescence rates were apparent (Figure 2k–m).

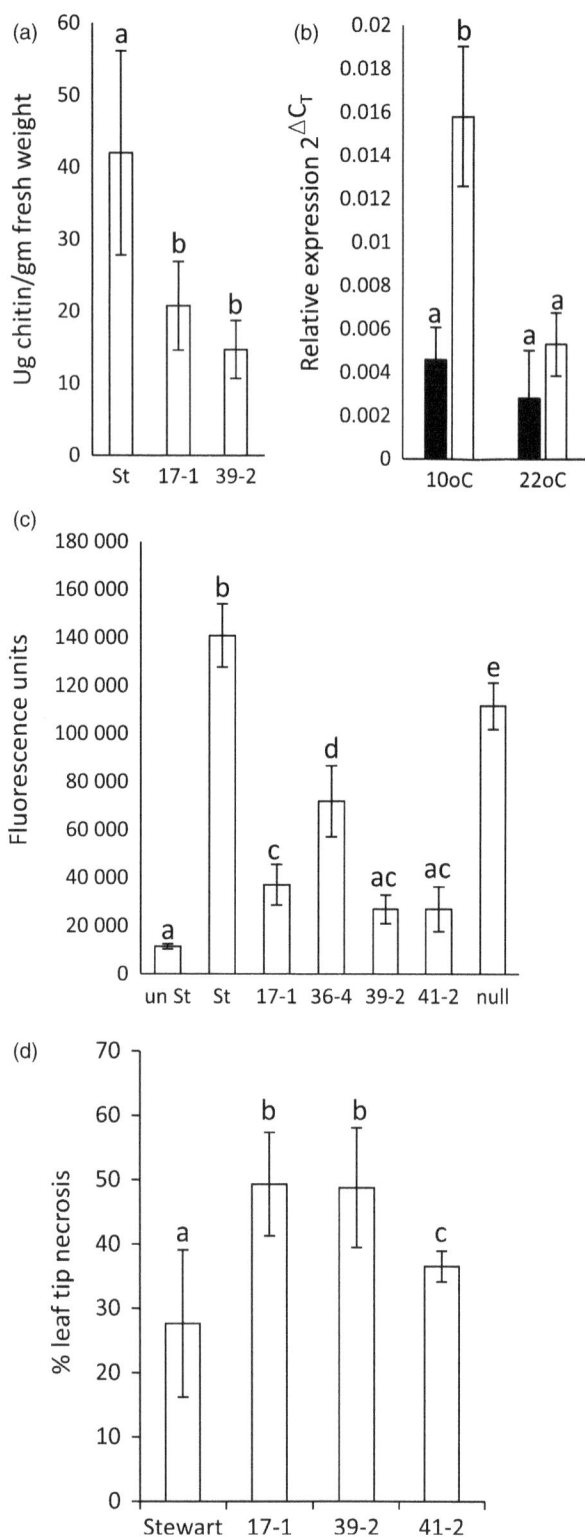

Figure 3 Further characterization of *Lr34* durum lines. (a) Chitin assay quantification of *Blumeria graminis* growth (7 dpi) on leaves of Stewart, 17-1 and 39-2. Each data point is from 14 plants and three technical replicates. Fluorescence was converted to ug of chitin/gm of tissue using a standard curve. Data were compared using the Student's *t*-test, $P < 0.05$. (b) Relative *Lr34* expression in uninfected (black columns) and *P. striiformis* f. sp. *tritici*-infected (white columns) hexaploid wheat seedlings grown at 10 or 22 °C. Each data point was from pooled leaf tissue (6-12 seedlings) and triplicate Q-PCRs. (c) *P. striiformis* chitin assay quantification (26 dpi) of Stewart, 17-1, 36-4, 39-2, 41-2 and a nontransgenic Stewart line regenerated from tissue culture (null). Seedlings infected at the three- to four-leaf stage were grown at 22 °C (16-h light) and tissue, when harvested, was undergoing age-related senescence (Figure S5). Uninfected Stewart seedlings (*un St*) were included. Each data point was from pooled leaf tissue of 12 seedlings and four technical replicates. (d) Ltn on flag leaves of field-grown *Lr34* durum lines and Stewart. The length of Ltn was divided by total flag leaf length to calculate % Ltn; 10–26 flag leaves were measured per line (see image 2n).

was undertaken. Two senescence genes were targeted, *S40* (Krupinska *et al.*, 2002) and *CP-MIII* (*serine carboxypeptidase*) (Parrott *et al.*, 2010), with *S40* previously shown to be highly induced in Th+Lr34 flag leaves (Krattinger *et al.*, 2009). No significant difference in *S40* expression occurred between Stewart and *Lr34* transgenic lines (Figure 4a). No consistent change in *Cp-MIII* expression occurred with the most resistant line, 39-2, and line 36-4 showing no difference to the Stewart control, while lines 17-1 and 41-2 showed a modest increase in expression (two- to threefold) (Figure 4b). These data are not consistent with elevated senescence gene expression in *Lr34* durum seedlings and suggest that this seedling resistance is not associated with senescence induction.

An interesting feature of *Lr34* expression in mature bread wheat plants is the up-regulation of abiotic stress-responsive genes, predominantly in the tips of flag leaves, in the absence of pathogen infection (Hulbert *et al.*, 2007). To determine whether similar gene expression occurred in *Lr34* durum seedlings, expression of abiotic stress-responsive genes *rab15* (Kosova *et al.*, 2014; Tsuda *et al.*, 2000) and *HSP90* (Shi *et al.*, 2012) was quantified in leaf RNAs of uninfected seedlings grown at 22 °C (16-h light). A 20- to 150-fold induction of *rab15* occurred in *Lr34* durum seedlings compared with nontransgenic Stewart (Figure 4c), while a more modest four- to six-fold increase in *HSP90* expression occurred (Figure 4d). The expression of *Lr34* in durum seedlings therefore appears to induce an apparent abiotic stress-related response in the absence of pathogen infection.

Higher *PR* gene expression levels were reported in *P. triticina*-infected flag leaves of *Lr34* hexaploid plants compared with control lines (Hulbert *et al.*, 2007). Transgenic and control durum seedlings were infected with *P. triticina* at the three- to four-leaf stage (22 °C) and tissues harvested 12 dpi. Tissue samples were quantified for both *PR* gene expression and relative fungal biomass. Again, significantly less rust growth occurred in transgenic lines compared with Stewart seedlings (Figure 5a), which correlated with *Lr34* expression levels (Figure 1b). However, no increased expression of *PR1*, *PR2* or *PR3* occurred in *Lr34* durum lines relative to control plants upon rust infection (Figure 5b and Figure S7). These data, however, are complicated by rust infection *per se* strongly inducing *PR* gene expression (Figure 5b and Figure S7: compare uninfected and infected Stewart seedlings). *PR* expression was therefore normalized relative to fungal

However, under field conditions, transgenic plants did show some increased leaf tip necrosis of flag leaves compared with non-transgenic controls (Figures 2n, 3d). Leaf tip necrosis of mature *Lr34* hexaploid wheat plants is a common phenotype.

To determine whether a weak (i.e. nonvisible) senescence response may be occurring in these transgenic durum seedlings, the first leaves (eldest) were harvested from seedlings, at the fourth leaf stage, and quantification of senescence marker genes

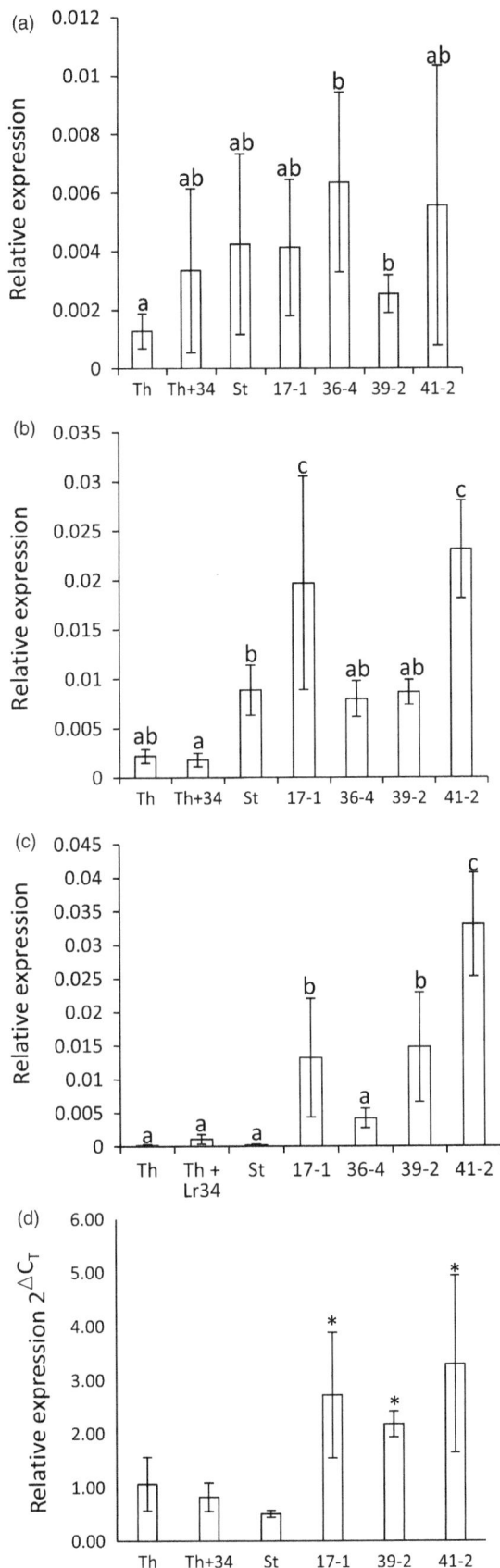

Figure 4 *Lr34* induced gene expression changes in uninfected wheat seedlings. Q-PCR analyses on RNA from uninfected wheat seedlings (four- to five-leaf stage) of hexaploid wheat cultivar Thatcher (Th), a near-isogenic *Lr34* Thatcher line (Th+Lr34), Stewart and *Lr34* lines 17-1, 36-4, 39-2 and 41-2. Panels a–d show relative expression of (a) *S40*, (b) *Cp-MIII*, (c) *Rab15* and (d) *HSP90*, each normalized relative to *GAPDH*. Data were derived from three biological replicates per genotype and three technical replicates per sample. In panel (d), transgenic lines 17-1, 39-2 and 41-2 had significantly higher levels of gene expression (*t*-test, $P < 0.05$) than the Stewart control, indicated by an asterisk above each column.

obvious increase in *PR* expression in response to rust infection compared with the control (Figure 5c).

To examine the influence of photoperiod on *Lr34* resistance, transgenic seedlings were grown at 22 °C (16-h light) to the three- to four-leaf stage and then inoculated with *P. triticina*. Half of the infected plants were then grown under continuous light, while the remaining seedlings were maintained under a 16-h light/8-h dark photoperiod. All *Lr34* lines had significantly less pathogen growth 10 dpi under continuous light when compared with 16-h light (Figure 6a). Under both conditions, rust resistance levels again correlated with the transgene expression levels (Figure 1b). In contrast, control seedlings showed equivalent or increased rust growth under constant light compared with a 16-h light/8-h dark light regime (Figure 6a). Under these highly controlled growth cabinet conditions, increasing photoperiod therefore resulted in increased levels of *Lr34* resistance. It is noteworthy that light has previously been speculated to promote *Lr34* resistance in field-grown plants (Singh and Gupta, 1992).

Given the light association of *L34* resistance and previous observation that *Yr36* APR results in chloroplast H_2O_2 accumulation (Guo *et al.*, 2015), H_2O_2 accumulation was also examined. Durum seedlings were grown at 22 °C (16-h light) and H_2O_2 content determined in infected and uninfected plants using Amplex Red (Invitrogen). While a clear increase in H_2O_2 content occurred upon *P. triticina* infection, no differences in accumulation of this reactive oxygen occurred in resistant and susceptible durum genotypes (Figure 6b). These observations were consistent with 3,3' diamino benzidine (DAB) staining of H_2O_2 in rust-infected leaf tissue with no obvious difference in H_2O_2 accumulation apparent (Figure 2i, J).

Discussion

Several rust resistance genes have been transferred from durum wheat into hexaploid wheat due to the relative simplicity of crossing these species and selecting fertile hexaploid lines or alternatively producing synthetic hexaploid wheat by crossing tetraploid wheat (AABB) with *Aegilops tauschii* (DD). Germplasm development in tetraploid wheat using interspecies crosses, while achievable (Huguet-Robert *et al.*, 2001; Klindworth *et al.*, 2012; Morris *et al.*, 2011), is more difficult due to poor vigour and low fertility in tetraploid backgrounds (Ceoloni *et al.*, 1996; Klindworth *et al.*, 2012). Of particular difficulty is the introduction of D genome genes of hexaploid wheat or *Ae. tauschii* due to the absence of homologous chromosomes in AABB tetraploids and only a few examples have been reported (Ceoloni *et al.*, 1996; Han *et al.*, 2016; Joppa *et al.*, 1998; Liu *et al.*, 1996; Luo *et al.*, 1996). Given the broad, multipathogen effectiveness and durability of *Lr34*, transfer of this gene to durum wheat by

biomass. Relative to rust growth, strong induction of all three *PR* genes occurred in the most resistant transgenic, line 39-2, while the remaining more intermediate resistant lines showed no

(a)

(b)

(c)

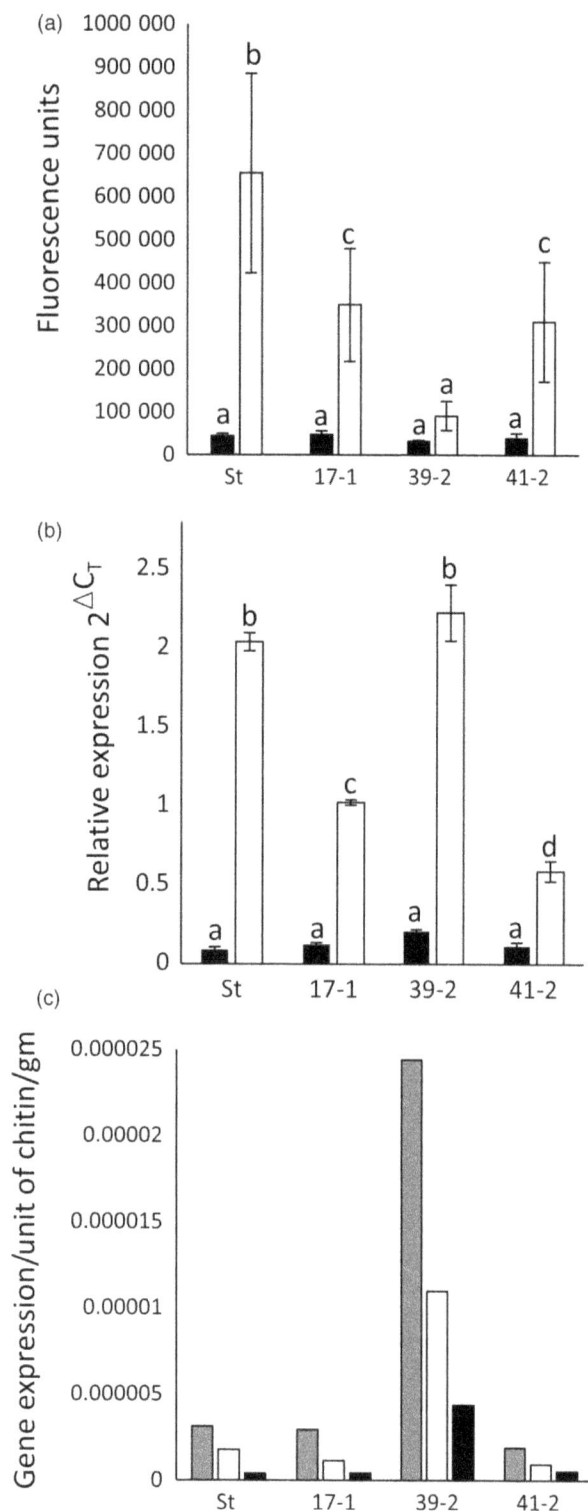

Figure 5 Pathogenesis-related (*PR*) gene expression in *Lr34* durum seedlings. (a) Chitin assay quantification of uninfected (black columns) and *P. triticina*-infected (12 dpi) (white columns) Stewart seedlings (St) and lines 17-1, 39-2 and 41-2 (10–15 seedlings per genotype). The same plants were used for RNA extraction in (b). (b) *PR-1* gene expression in wheat seedlings normalized relative to the *GAPDH*. RNA was extracted from uninfected (black columns) or *P. triticina*-infected (12 dpi) seedlings (three- to four-leaf stage) described in (a). Three biological replicates were used per genotype and three technical replicates per sample. The same RNAs were quantified for *PR-2* and *PR-3* expression (Figure S7). (c) Relative *PR* gene expression values (in B and Figure S7) were divided by chitin biomass values of rust-infected material shown in A. *PR1*, *PR2* and *PR3* values are shown as grey, white and black columns, respectively, for Stewart (St) and *Lr34* transgenic lines 17-1, 39-2 and 41-2.

leaves of hexaploid Th+34 plants when compared with seedlings (Risk *et al.*, 2012, 2013), consistent with this hypothesis. Importantly, elevated gene expression was not associated with negative pleiotropic effects in durum seedlings and no increased senescence was apparent either macroscopically or by quantification of senescence marker gene transcripts. This seedling resistance further enhances the agronomic potential of *Lr34* in durum wheat cultivation. Seedling resistance is particularly important for protecting wheat from stripe rust disease because it occurs early in the growing season.

Previous introduction of *Lr34* as a transgene into hexaploid wheat cultivar Bobwhite (BW26 AUS) also led to elevated expression levels, with some seedlings showing 10-fold higher expression; however, seedling resistance was not observed (Risk *et al.*, 2012). These lines, however, did show typical *Lr34* APR and minor leaf tip necrosis (Risk *et al.*, 2012). Why elevated *Lr34* expression does not provide seedling resistance in hexaploid wheat is unknown. In contrast, *Lr34* transgenics made in a second wheat line, BW26SU, did show seedling resistance to *P. triticina*. This line, however, was not fully rust susceptible and resistance in these lines did not correlate with transgene expression levels. One highly resistant BWSUI line showed only a 2.4-fold increase in *Lr34* expression relative to Th+34 seedlings. The authors concluded that background resistance present in BWSUI significantly enhanced *Lr34* effects in this line (Risk *et al.*, 2012). The *Lr34* gene does show additive effects with minor resistance genes (Singh *et al.*, 2011b) and several all-stage resistance genes (German and Kolmer, 1992).

We cannot exclude the possibility that the *Lr34* resistance in Stewart durum seedlings may be associated with minor gene affects. However, we feel this is improbable. Rust assays showed that Stewart is moderately susceptible to the *P. triticina* (3C rating) and *P. striiformis* f. sp. *tritici* (3, 3+C rating) isolates used in this study (Figure 2a, b), suggesting that minor resistance genes do exist in this background. However, *Lr34* durum lines also had increased powdery mildew disease resistance, meaning minor background genes effective against all three pathogen species would be needed in the Stewart background that showed additive effects with *Lr34*. In addition, resistance levels in these durum lines directly correlated with *Lr34* expression levels consistent with transgene expression being the predominant factor in this seedling resistance.

Elevated *Lr34* expression in durum wheat is well tolerated with only mild leaf tip necrosis in mature plants, although higher expression levels could possibly be deleterious. Hexaploid wheat also tolerates elevated *Lr34* expression, although cold-grown

transgenesis is potentially of use in controlling rust and mildew diseases in this crop species.

It was unexpected that this APR gene would also show high levels of seedling resistance in durum wheat which we attribute to transgene expression levels being five to ten times greater than the endogenous gene in hexaploid wheat seedlings, when uninfected seedlings grown at 22 °C were compared. A five- to 10-fold increase in *Lr34* expression also occurs in resistant flag

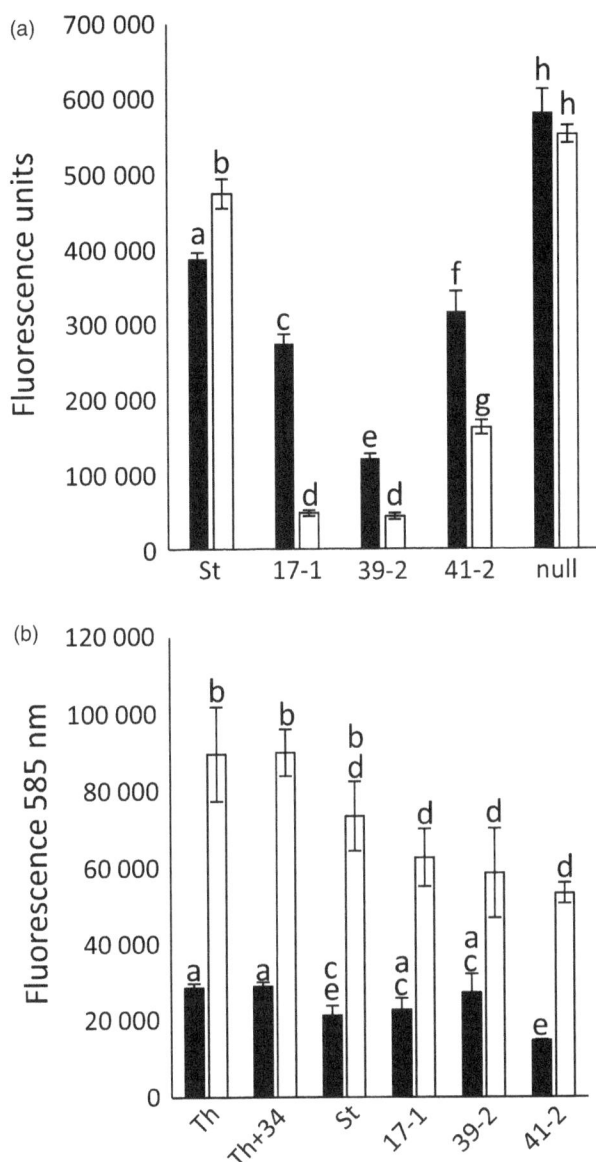

Figure 6 *Lr34* resistance is influenced by photoperiod, but not associated with H_2O_2 accumulation. (a) Quantification of *P. triticina* growth (14 dpi) on Stewart (St), *Lr34* lines 17-1, 39-2, 41-2 and a nontransgenic tissue culture regenerant (null) under two photoperiod regimes. Seedlings (four-leaf stage), grown at 22 °C with a 16-h light/8-h dark photoperiod), were infected with *P. triticina* and either returned to these growth conditions (black bars) or grown under constant light (white bars), using the same temperature regime. Each data point is from pooled tissue of approximately six seedlings. (b) Relative H_2O_2 levels, determined by Amplex Red assay, in hexaploid wheat cultivar Thatcher (Th), near-isogenic *Lr34* Thatcher (Th+34), Stewart (St) and *Lr34* durum lines 17-1, 39-2 and 41-2 from infected (white column) or uninfected (black column) plants. Each data point is derived from individual H_2O_2 measurements of three biological replicates per genotype.

necrosis and negative developmental effects were also seen in most *Lr34* rice lines (Krattinger *et al.*, 2016).

While *Lr34* resistance usually coincides with accelerated senescence in most plant tissues, including transgenic durum flag leaves, the absence of visible senescence or senescence gene up-regulation in *Lr34* durum seedlings suggests that senescence is not required for resistance. Consistent with this hypothesis, in rice the amount of leaf tip necrosis in *Lr34* lines did not necessarily correlate with resistance levels (Krattinger *et al.*, 2016), implying these two phenotypes are not directly correlated. Ltn is not always apparent in field-grown, adult *Lr34* plants being dependent on both the environment and genetic background. It is unknown, however, whether plants under these circumstances show senescence up-regulation without visible necrosis or, alternatively, show no altered senescence response. Hexaploid *Lr34* seedlings grown at 10 °C during rust infection also show resistance, presumably due to elevated *Lr34* expression, without visible accelerated senescence.

This *Lr34* durum seedling resistance, however, is associated with up-regulation of abiotic stress genes such as *rab15* and *HSP90*, suggesting that this response is required for resistance. Only some of the responses induced by *Lr34* may provide disease resistance and additional effects such as accelerated senescence may be pleiotropic. Common signalling pathways between senescence and abiotic stress responses are well established (Gepstein and Glick, 2013). Some evidence of modest up-regulated of *PR* expression was observed in rust-infected durum seedlings although it was confounded by rust infection *per se* causing *PR* induction. Other analyses have shown limited *PR* induction in *Lr34* hexaploid wheat (Risk *et al.*, 2012), which differs from the observations of Hulbert *et al.* (2007).

These data raise several possible models for *Lr34* function in durum wheat. In the first scenario, *Lr34* resistance and senescence are mechanistically related, but additional factors, regulated by plant development and the environment, are required for the latter response to occur (Figure 7a). In the case of barley, which does not have an *Lr34* ortholog and may therefore lack appropriate regulatory control of *Lr34*-mediated processes, these additional factors are inappropriately produced or recognized during initiation of age-dependent seedling leaf senescence resulting in accelerated leaf necrosis. In hexaploid wheat, a minimum transcriptional threshold needed for *Lr34* resistance is not reached until later in plant development at which time both resistance and senescence occur concomitantly. Consistent with this model, elevated expression of *Lr34* in durum seedlings and in cold-grown, infected, hexaploid seedlings results in resistance without leaf senescence (Figure 7a). In the second scenario (Figure 7b), the resistance mechanism conferred by *Lr34* is independent of an *Lr34*-mediated senescence response, which only occurs in tissues of mature plants after reaching a specific developmental age.

Lr34 durum lines that show robust seedling resistance when grown under field-like conditions will be a valuable resource to further investigate the mechanistic basis underlying resistance. These transgenic lines demonstrate a clear photoperiod effect on *Lr34* resistance. These observations were of particular interest given the proposed mechanism of the Yr36 START-kinase protein that phosphorylates a chloroplast thylakoid-associated ascorbate peroxidise (Guo *et al.*, 2015). Elevated H_2O_2 levels occur in *Yr36* transgenic plants and accelerated leaf senescence of older leaves (Guo *et al.*, 2015). However, no increased H_2O_2

seedlings showed leaf tip necrosis upon rust infection (which was not observed in durum wheat) and one transgenic line showed chlorotic spotting and reduced seed set (Risk *et al.*, 2012). In contrast, similar *Lr34* expression levels in barley result in strong seedling senescence with phenotype severity related to the transgene expression level (Risk *et al.*, 2013). Seedling leaf tip

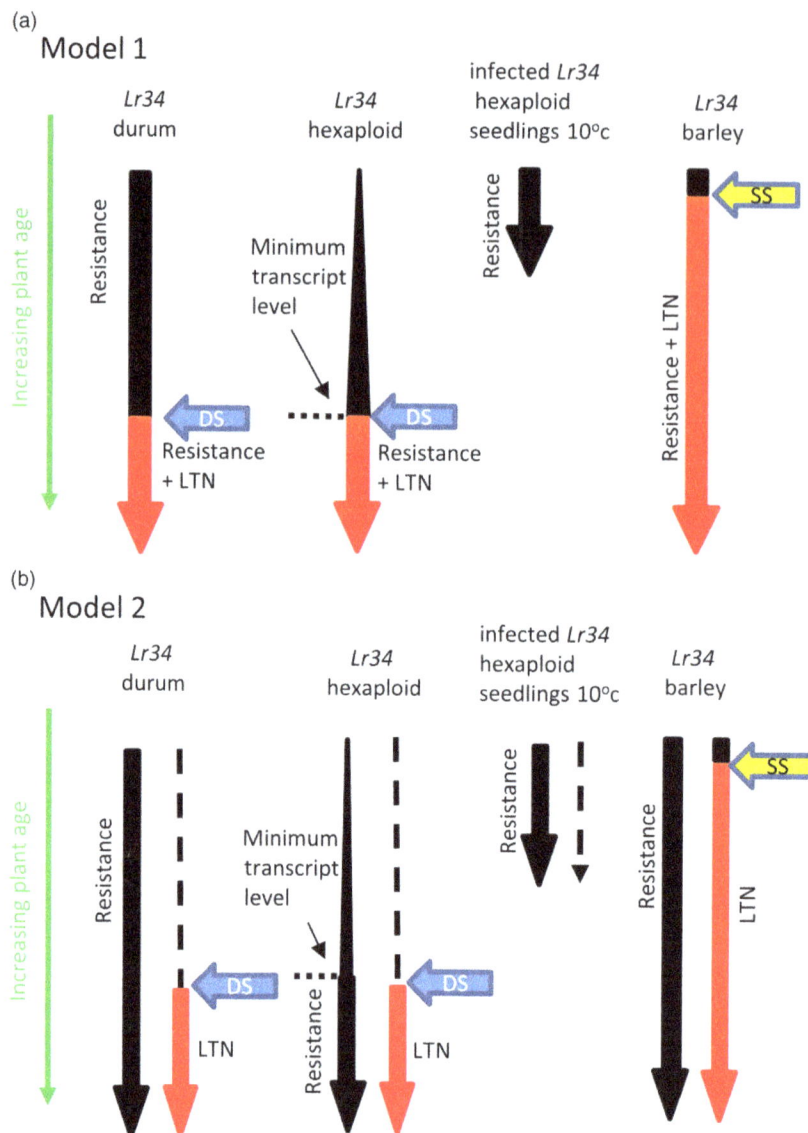

Figure 7 *Lr34* expression and phenotype models. (a) *Lr34* resistance requires a minimum transcriptional threshold (expression levels are depicted by arrow width), which is not reached in hexaploid wheat until later in plant maturity. In contrast, *Lr34* durum seedlings, *Lr34* barley seedlings and cold-treated, infected hexaploid *Lr34* seedlings are resistant due to higher *Lr34* expression levels. *Lr34* senescence, or leaf tip necrosis (Ltn) (shown in red), is mechanistically the same pathway as resistance but also dependent upon developmental signals (blue arrow labelled DS) that occur later in plant maturation. Hence, cold-grown, infected *Lr34* hexaploid seedlings and *Lr34* durum seedlings do not show leaf senescence, while adult plants of the same genotypes do. In *Lr34* barley seedlings, signalling during normal developmental senescence of seedling leaves (yellow arrow SS) is sufficient to induce accelerated necrosis due to the heterologous nature of the wheat *Lr34* gene. (b) This model is based upon the same assumptions except that *Lr34* senescence is considered an independent pathway to resistance.

accumulation occurred in either uninfected or rust-infected *Lr34* durum seedlings, suggesting a potentially different mode of action.

In summary, *Lr34* resistance in durum seedlings is not associated with necrosis or accelerated senescence. In contrast, induction of abiotic stress-response genes occurred in the absence of pathogen infection in these seedlings as previously observed in adult hexaploid wheat plants (Hulbert *et al.*, 2007). Photoperiod had a significant effect on *Lr34* phenotypes by as yet undefined mechanisms. Manipulation of APR gene expression can enhance disease resistance without associated negative pleiotropic effects in some instances, which is of potential agronomic benefit.

Experimental procedures

Plant and pathogen growth conditions

Wheat plants were grown in growth cabinets at 22 °C, 16-h light/8-h dark unless otherwise specified. Wheat seedlings were infected with *Puccinia striiformis* f. sp. *tritici* isolate accession number 821559 (pathotype 104 E137 A-) and *Puccinia triticina* isolate accession number 020281 (pathotype 104-1,2,3,(6),(7),11 + Lr37) obtained from the Plant Breeding Institute, NSW, Australia. Plants were inoculated with *P. striiformis* and *P. triticina* urediniospores and incubated in a humid chamber overnight at 10 or 22 °C, respectively. Infected plants were then transferred to growth cabinets (22 °C, 16-h light/8-h dark) for growth of rust pathogens. Extended growth times were used for experiments at 10 °C due to significantly slower rust growth (Risk *et al.*, 2012). Rust inoculum was propagated on wheat cultivar Morocco and urediniospores collected from infected plants by shaking plants over aluminium foil. For infection with *B. graminis*, durum leaves were harvested from 26-day-old seedlings and placed on MS salt media and infected with *B. graminis* isolate ISR208; 7 dpi *B. graminis* growth was measured by chitin assay (Ayliffe *et al.*, 2013) using 14 biological replicates per genotype.

Generation of homozygous *Lr34* transgenic durum wheat plants

Transgenic Stewart durum wheat plants were generated by Agrobacterium-mediated transformation of cultivar Stewart (Ishida *et al.*, 2013; Richardson *et al.*, 2014). The *Lr34* transgene (Figure S1A) was cloned into binary plasmid pWBVec8 (Murray *et al.*, 2004; Wang *et al.*, 1998), which encodes a hygromycin phosphotransferase gene. Transgenic plants were selected using 30–50 µg/mL of hygromycin. To identify Stewart plants containing at least one complete *Lr34* transgene, DNA blot analysis was undertaken on T0 plants as previously described (Ayliffe *et al.*, 2000). DNAs were restricted with *NotI* and hybridized with a probe encoding 2 kb of the 3′ terminus of the *Lr34* ORF (Table S1). A predicted 16-kb fragment with homology was identified in lines containing a complete transgene (Figure S1B). T1 plants were screened for potential *Lr34* transgene homozygosity by PCR analysis of 20 individual T2 seeds using *Lr34*-specific primers (Table S1, ABCTF4N and Lr34plusR). Transgene copy number in each family was then determined, and homozygosity confirmed, by DNA blot analysis of 25 *DraI*-restricted T2 plant DNAs hybridized with a probe complementary to a 481-bp fragment of the *Lr34* 3′ untranslated region (3′UTR) (Table S1, Figure S3). Homozygous families were identified for four independent, single-copy *Lr34* transgenic events, that is 17-1, 36-4, 39-2 and 41-2.

Fungal biomass assays

Seedlings (10–15) at the three- to four-leaf stage from either homozygous *Lr34* transgenic lines or nontransgenic control lines were infected with rust urediniospores. Rust-infected leaves were harvested 10–14 days postinoculation, and relative chitin biomass per mg fresh weight of harvested tissue was determined by pooling seedling leaves of the same genotype and measuring the binding of wheat germ agglutinin–fluorescein isothiocyanate (WGA-FITC), as previously described (Ayliffe *et al.*, 2013). Relative rust biomass/gm fresh weight of leaf tissue was expressed as fluorescence units of bound WGA-FITC. Four chitin measurements were undertaken on each seedling leaf tissue pool, and average value was presented with standard deviation.

Quantitative RT-PCR

RNA was extracted from frozen, ground wheat leaf tissue using a Spectrum Plant Total RNA Kit (Sigma-Aldrich) and On-column DNase I Digestion Kit (Sigma-Aldrich) for genomic DNA removal. cDNA was synthesized using a reverse transcriptase kit (Phusion RT-PCR Kit; Finnzymes) and Q-PCR undertaken using a CFX96 real time system and C100 touch thermocycler (Bio-Rad). Target gene sequences were normalized relative to the wheat *glyceraldehyde-3-phosphate dehydrogenase* gene (*GAPDH*) using the comparative C_T method (Schmittgen and Livak, 2008) or alternatively target sequence concentration was determined using a standard curve derived from a target fragment linear dilution series, followed by normalization relative to *GAPDH* (Rinaldo *et al.*, 2015). Three replicate reactions were undertaken per RNA sample. Primer sequences used for amplification are shown in Table S1. Melting curves for primer pairs used in Q-PCR analyses are shown in Figure S8.

Microscopy and H_2O_2 quantification

Fungal-infected tissue was stained with wheat germ agglutinin–fluorescein isothiocyanate (Sigma-Aldrich, St. Louis) and visualized under blue light (Ayliffe *et al.*, 2011). Histochemical detection of H_2O_2 in infected leaf tissue was undertaken by 3,3′-diaminobenzidine (DAB) staining (Sigma-Aldrich, St. Louis) (Thordal-Christensen *et al.*, 1997). H_2O_2 levels were determined in ground leaf tissues using an Amplex Red H_2O_2 assay kit as described by the manufacturer (ThermoFisher).

Data analysis

All data were analysed by ANOVA using an online calculator (http://statistica.mooo.com/) unless otherwise stated in figure legends. Significantly different data points had a $P < 0.05$ unless indicated otherwise in the text. Standard deviations are indicated on graphs. Graph columns with common annotation were not statistically different (ANOVA, $P < 0.05$), throughout.

Acknowledgements

The authors wish to acknowledge the Grains Research and Development Corporation for financial support and Soma Chakraborty, Melanie Soliveres, Shamsul Hoque, Dhara Bhatt, Smitha Louis and Terese Richardson for technical support.

Author contributions

AR undertook experiments described in Figures 1a, c, 2, 3c, d, 4, 5, 6a. BG undertook experiments in Figures 1c, 2c. RB and SK produced data in Figure 3a. DS and RP screened rust collections to identify and provide a *P. triticina* isolate virulent on durum cultivar Stewart. EL undertook field trial results in Figure 3d and provided experimental design planning. MA contributed images in Figure 2, undertook experiments in Figures 2l–o and 6b in addition to providing experimental design.

References

Ayliffe, M.A., Collins, N.C., Ellis, J.G. and Pryor, A. (2000) The maize *rp1* rust resistance gene identifies homologues in barley that have been subjected to diversifying selection. *Theor. Appl. Genet.* **100**, 1144–1154.

Ayliffe, M.A., Devilla, R., Mago, R., White, R., Talbot, M., Pryor, A. and Leung, H. (2011) Non-host resistance of rice to rust pathogens. *Mol. Plant Microbe Interact.* **24**, 1143–1155.

Ayliffe, M., Periyannan, S., Feechan, A., Dry, I., Schumann, U., Wang, M.-B., Pryor, A. *et al.* (2013) A simple method for comparing fungal biomass in infected plant tissues. *Mol. Plant Microbe Interact.* **26**, 658–667.

Ceoloni, C., Biagetti, M., Ciaffi, M., Forte, P. and Pasquini, M. (1996) Wheat chromosome engineering at the 4x level: the potential of different alien gene transfers into durum wheat. *Euphytica*, **89**, 87–97.

Chauhan, H., Boni, R., Bucher, R., Kuhn, B., Buchmann, G., Sucher, J., Selter, L.L. *et al.* (2015) The wheat resistance gene *Lr34* results in constitutive induction of multiple defense pathways in transgenic barley. *Plant J.* **84**, 202–215.

Chen, W., Wellings, C., Chen, X., Kang, Z. and Liu, T. (2014) Wheat stripe (yellow) rust caused by *Puccinia striiformis* f. sp. *tritici*. *Mol. Plant Pathol.* **15**, 433–446.

Cloutier, S., McCallum, B.D., Loutre, C., Banks, T.W., Wicker, T., Feuillet, C., Keller, B. *et al.* (2007) Leaf rust resistance gene *Lr1*, isolated from bread wheat (*Triticum aestivum* L.) is a member of the large *psr567* gene family. *Plant Mol. Biol.* **65**, 93–106.

Dodds, P.N. and Rathjen, J.P. (2010) Plant immunity: towards an integrated view of plant-pathogen interactions. *Nat. Rev. Genet.* **11**, 539–548.

Dyck, P.L. (1991) Genetics of adult-plant leaf rust resistance in Chinese Spring and sturdy wheats. *Crop Sci.* **31**, 309–311.

Dyck, P.L., Samborsk, D.J. and Anderson, R.G. (1966) Inheritance of adult-plant leaf rust resistance derived from common wheat varieties Exchange and Frontana. *Can. J. Genet. Cytol.* **8**, 665–671.

Feuillet, C., Travella, S., Stein, N., Albar, L., Nublat, A. and Keller, B. (2003) Map-based isolation of the leaf rust disease resistance gene *Lr10* from the hexaploid wheat (*Triticum aestivum* L.) genome. *Proc. Natl Acad. Sci. USA*, **100**, 15253–15258.

Fu, D., Uauy, C., Distelfeld, A., Blechl, A., Epstein, L., Chen, X., Sela, H. *et al.* (2009) A kinase-START gene confers temperature-dependent resistance to wheat stripe rust. *Science*, **323**, 1357–1358.

Gepstein, S. and Glick, B.R. (2013) Strategies to ameliorate abiotic stress-induced plant senescence. *Plant Mol. Biol.* **82**, 623–633.

German, S.E. and Kolmer, J.A. (1992) Effect of *Lr34* in the enhancement of resistance to leaf rust of wheat. *Theoret. Appl. Genet.* **84**, 97–105.

Guo, J.-Y., Li, K., Wang, X., Lin, H., Cantu, D., Uauy, C., Dobon-Alonso, A. *et al.* (2015) Wheat stripe rust resistance protein WKS1 reduces the ability of the thylakoid-associated ascorbate peroxidise to detoxify reactive oxygen species. *Plant Cell*, **27**, 1755–1770.

Han, C., Zhang, P., Ryan, P.R., Rathjen, T.M., Yan, Z. and Delhaize, E. (2016) Introgression of genes from bread wheat enhances the aluminium tolerance of durum wheat. *Theor. Appl. Genet.* **129**, 729–739.

Herrera-Foessel, S.A., Singh, R.P., Lillemo, M., Huerta-Espino, J., Bhavani, S., Singh, S., Lan, C. *et al.* (2014) *Lr67/Yr46* confers adult plant resistance to stem rust and powdery mildew in wheat. *Theor. Appl. Genet.* **127**, 781–789.

Huang, L., Brooks, S.A., Li, W., Fellers, J.P., Trick, H.N. and Gill, B.S. (2003) Map-based cloning of leaf rust resistance gene *Lr21* from the large and polyploid genome of bread wheat. *Genetics*, **164**, 655–664.

Huerta-Espino, J., Singh, R.P., German, S., McCallum, B.D., Park, R.F., Chen, W.Q., Bhardwaj, S.C. *et al.* (2011) Global status of wheat leaf rust caused by *Puccinia triticina*. *Euphytica*, **179**, 143–160.

Huguet-Robert, V., Dedryver, F., Röder, M.S., Korzun, V., Abélard, P., Tanguy, A.M., Jaudeau, B. *et al.* (2001) Isolation of a chromosomally engineered durum wheat line carrying the *Aegilops ventricosa Pch1* gene for resistance to eyespot. *Genome*, **44**, 345–349.

Hulbert, S.H., Bai, J., Fellers, J.P., Pacheco, M.G. and Bowden, R.L. (2007) Gene expression patterns in near isogenic lines for wheat rust resistance gene *Lr34/Yr18*. *Phytopathology*, **97**, 1083–1093.

Ishida, Y., Tsunashima, M., Hiei, Y. and Komari, T. (2013) Wheat (*Triticum aestivum* L.) transformation using immature embryos. *Methods Mol. Biol. Agrobacterium Protocols* **1223**, 189–198.

Joppa, L.R., Klindworth, D.L. and Hareland, G.A. (1998) Transfer of high molecular weight glutenins from spring wheat to durum wheat. In *Proceedings of the 9th International Wheat Genetics Symposium*, vol. **1**, Saskatoon, Saskatchewan, Canada (Slinkard, A.E., ed), pp. 257–260. University of Saskatchewan Extension Press: Saskatoon, SK, Canada.

Klindworth, D.L., Niu, Z., Chao, S., Friesen, T.L., Jin, Y., Faris, J.D., Cai, X. *et al.* (2012) Introgression and characterisation of a goatgrass gene for a high level of resistance to Ug99 stem rust in tetraploid wheat. *G3 Genes\Genomes\Genet.* **2**, 665–673.

Kolmer, J.A. (2005) Tracking wheat rust on a continental scale. *Curr. Opin. Plant Biol.* **8**, 441–449.

Kolmer, J.A., Singh, R.P., Garvin, D.F., Viccars, L., William, H.M., Huerta-Espino, J., Ogbonnaya, F.C. *et al.* (2008) Analysis of the *Lr34/Yr18* rust resistance region in wheat germplasm. *Crop Sci.* **48**, 1841–1852.

Kosova, K., Vitamvas, P. and Prasil, I.T. (2014) Wheat and barley dehydrins under cold, drought and salinity – what can LEA-II proteins tell us about plant stress response? *Front. Plant Sci.* **5**, 343.

Krattinger, S.G., Lagudah, E.S., Spielmeyer, W., Singh, R.P., Huerta-Espino, J., McFadden, H., Bossolini, E. *et al.* (2009) A putative ABC transporter confers durable resistance to multiple fungal pathogens in wheat. *Science*, **323**, 1360–1363.

Krattinger, S.G., Lagudah, E.S., Wicker, T., Risk, J.M., Ashton, A.R., Selter, L.L., Matsumoto, T. *et al.* (2011) *Lr34* multi-pathogen resistance ABC transporter: molecular analysis of homoeologous and orthologous genes in hexaploid wheat and other grass species. *Plant J.* **65**, 392–403.

Krattinger, S.G., Jordan, D.R., Mace, E.S., Raghavan, C., Luo, M.-C., Keller, B. and Lagudah, E.S. (2013) Recent emergence of the wheat *Lr34* multi-pathogen resistance; insights from haplotype analysis in wheat, rice, sorghum and *Aegilops tauschii*. *Theor. Appl. Genet.* **126**, 663–672.

Krattinger, S.G., Sucher, J., Selter, L.L., Chauhan, H., Zhou, B., Tang, M., Upadhyay, N.M. *et al.* (2016) The wheat durable, multipathogen resistance gene *Lr34* confers partial blast resistance in rice. *Plant Biotechnol. J.* **14**, 1261–1268.

Krupinska, K., Haussuhl, K., Schafer, A., van der Kooij, T.A.W., Leckband, G., Lorz, H. and Falk, J. (2002) A novel nucleus-targeted protein is expressed in barley leaves during senescence and pathogen infection. *Plant Phys.* **130**, 1172–1180.

Lagudah, E.S., McFadden, H., Singh, R.P., Huerta-Espino, J., Bariana, H.S. and Spielmeyer, W. (2006) Molecular genetic characterization of the *Lr34/Yr18* slow rusting resistance gene region in wheat. *Theor. Appl. Genet.* **114**, 21–30.

Liu, C.-Y., Shepherd, K.W. and Rathjen, A.J. (1996) Improvement of durum wheat pastamaking and breadmaking qualities. *Cereal Chem.* **73**, 155–166.

Liu, W., Frick, M., Huel, R., Nykiforuk, C.L., Wang, X., Gaudet, D.A., Eudes, F. *et al.* (2014) The stripe rust resistance gene *Yr10* encodes an evolutionary-conserved and unique CC-NBS-LRR sequence in wheat. *Mol. Plant* **7**, 1740–1755.

Luo, M.-C., Dubcovsky, J., Goyal, S. and Dvořák, J. (1996) Engineering of interstitial foreign chromosome segments containing the K+/Na+ selectivity gene *Kna1* by sequential homoeologous recombination in durum wheat. *Theor. Appl. Genet.* **93**, 1180–1184.

Mago, R., Zhang, P., Vautrin, S., Šimková, H., Bansal, U., Luo, M.-C., Rouse, M. *et al.* (2015) The wheat *Sr50* gene reveals rich diversity at a cereal disease resistance locus. *Nat. Plant* **1**, 15186. doi:10.1038/nplants.2015.186.

Moore, J.W., Herrera-Foessel, S., Lan, C., Schnippenkoetter, W., Ayliffe, M., Huerta-Espino, J., Lillemo, M. *et al.* (2015) Recent evolution of a hexose transporter variant confers resistance to multiple pathogens in wheat. *Nat. Genet.* **47**, 1494–1498.

Morris, C.F., Simeone, M.C., King, G.E. and Lafiandra, D. (2011) Transfer of soft kernel texture from *Triticum aestivum* to durum wheat, *Triticum turgidum* ssp. *durum*. *Crop Sci.* **51**, 114–122.

Murray, F., Brettell, R., Matthews, P., Bishop, D. and Jacobsen, J. (2004) Comparison of *Agrobacterium*-mediated transformation of four barley cultivars using the *GFP* and *GUS* reporter genes. *Plant Cell Rep.* **22**, 397–402.

Parrott, D.L., Martin, J.M. and Fischer, A.M. (2010) Analysis of barley (*Hordeum vulgare*) leaf senescence and protease gene expression: a family of C1A cysteine protease is specifically induced under conditions characterized by high carbohydrate, but low to moderate nitrogen levels. *New Phytol.* **187**, 313–331.

Periyannan, S., Moore, J., Ayliffe, M., Bansal, U., Wang, X., Huang, L., Deal, K. *et al.* (2013) The gene *Sr33*, an ortholog of barley *Mla* genes, encodes resistance to wheat stem rust race Ug99. *Science*, **341**, 786–788.

Richardson, T., Thistleton, J., Higgins, T.J., Howitt, C. and Ayliffe, M. (2014) Efficient Agrobacterium transformation of elite wheat germplasm without selection. *Plant Cell Tissue Organ Cult.* **119**, 647–659.

Rinaldo, A.R., Cavallini, E., Jia, Y., Moss, S.M.A., McDavid, D.A.J., Hooper, L.C., Robinson, S.P. *et al.* (2015) A grapevine anthocyanin acyltransferase, transcriptionally regulated by *VvMYBA*, can produce most acylated anthocyanins present in grape skins. *Plant Physiol.* **169**, 1897–1916.

Risk, J.M., Selter, L.L., Krattinger, S.G., Viccars, L.A., Richardson, T.M., Buesing, G., Herren, G. *et al.* (2012) Functional variability of the *Lr34* durable resistance gene in transgenic wheat. *Plant Biotechnol. J.* **10**, 477–487.

Risk, J.M., Selter, L.L., Chauhan, H., Krattinger, S.G., Kumklehn, J.K., Hensel, G., Viccars, L.A. *et al.* (2013) The wheat *Lr34* gene provides resistance against multiple fungal pathogens in barley. *Plant Biotechnol. J.* **11**, 847–854.

Rubiales, D. and Niks, R.E. (1995) Characterization of *Lr34*, a major gene conferring nonhypersensitive resistance to wheat leaf rust. *Plant Dis.* **79**, 1208–1212.

Saintenac, C., Zhang, W., Salcedo, A., Rouse, M.N., Trick, H.N., Akhunov, E. and Dubcovsky, J. (2013) Identification of wheat gene *Sr35* that confers resistance to Ug99 stem rust race group. *Science*, **341**, 783–786.

Schmittgen, T.D. and Livak, K.J. (2008) Analyzing real-time PCR data by the comparative C_T method. *Nat. Protoc.* **3**, 1101–1108.

Shah, S.J.A., Imtiaz, M. and Hussain, S. (2010) Phenotypic and molecular characterization of wheat for slow rusting resistance against *Puccinia striiformis* Westend. f. sp. *tritici*. *J. Phytopathol.* **158**, 393–402.

Shah, S.J.A., Hussain, S., Ahmad, M., Farhatullah, A.I. and Ibrahim, M. (2011) Using leaf tip necrosis as a phenotypic marker to predict the presence of durable rust resistance gene pair *Lr34/Yr18* in wheat. *J. Gen. Plant Pathol.* **77**, 174–177.

Shi, Z.-S., Li, Z.-Y., Chen, Y., Chen, M., Li, L.-C. and Ma, Y.-Z. (2012) Heat shock protein 90 in plants: molecular mechanisms and roles in stress responses. *Int. J. Sci.* **13**, 15706–15723.

Singh, R.P. (1992a) Association between gene *Lr34* for leaf rust resistance and leaf tip necrosis in wheat. *Crop Sci.* **32**, 874–878.

Singh, R.P. (1992b) Genetic association of leaf rust resistance gene *Lr34* with adult-plant resistance to stripe rust in bread wheat. *Phytopathology*, **82**, 835–838.

Singh, R.P. and Gupta, A.K. (1992) Expression of wheat leaf rust resistance gene *Lr34* in seedlings and adult plants. *Plant Dis.* **76**, 489–491.

Singh, R.P., Hodson, D.P., Huerta-Espino, J., Jin, Y., Bhavani, S., Njau, P., Herrera-Foessel, S. *et al.* (2011a) The emergence of Ug99 races of the stem rust fungus is a threat to world wheat production. *Ann. Rev. Phytopathol.* **49**, 465–481.

Singh, R.P., Huerta-Espino, J., Bhavani, S., Herrera-Foessel, S.A., Singh, D., Singh, P.K., Velu, G. *et al.* (2011b) Race non-specific resistance to rust diseases in CIMMYT spring wheats. *Euphytica*, **179**, 175–186.

Spielmeyer, W., McIntosh, R.A., Kolmer, J. and Lagudah, E.S. (2005) Powdery mildew resistance and *Lr34/Yr18* genes for durable resistance to leaf and stripe rust cosegregate at a locus on the short arm of chromosome 7D of wheat. *Theor. Appl. Genet.* **111**, 731–735.

Stakman, E.C., Stewart, D.M. and Loegering, W.Q. (1962) *Identification of physiologic races of Puccinia graminis var. tritici*. United States Department of Agriculture, Agricultural Research Service, Washington.

Steuernagel, B., Periyannan, S.K., Hernández-Pinzón, I., Witek, K., Rouse, M.N., Yu, G., Hatta, A. *et al.* (2016) MutRenSeq; three-step cloning of resistance genes from hexaploid wheat using mutagenesis and sequence capture. *Nat. Biotechnol.* **34**, 652–655. doi:10.1038/nbt.3543.

Thordal-Christensen, H., Zhang, Z., Wei, Y. and Collinge, D.B. (1997) Subcellular localisation of H_2O_2 in plants. H_2O_2 accumulation in papillae and hypersensitive response during the barley powdery mildew interaction. *Plant J.* **11**, 1187–1194.

Tsuda, K., Tsvetanov, S., Takumi, S., Mori, N., Atanassov, A. and Nakamura, C. (2000) New members of a cold-responsive group-3 *Lea/Rab* related *Cor* gene family from common wheat (*Triticum aestivum* L.). *Genes Genet. Syst.* **75**, 179–188.

Uauy, C., Brevis, J.C., Chen, X., Khan, I., Jackson, L., Chicaiza, O., Distelfeld, A. *et al.* (2005) High-temperature adult-plant (HTAP) stripe rust resistance gene *Yr36* from *Triticum turgidum* ssp. *dicoccoides* is closely linked to the grain protein content locus *Gpc-B1*. *Theor. Appl. Genet.* **112**, 97–105.

Wang, M.B., Li, Z., Matthews, P.R., Upadhyaya, N.M. and Waterhouse, P.M. (1998) Improved vectors for *Agrobacterium tumefaciens*-mediated transformation of monocot plants. *Acta Hortic.* **461**, 401–407.

16

Does *Bt* rice pose risks to non-target arthropods? Results of a meta-analysis in China

Cong Dang[1,†], Zengbin Lu[1,2,†], Long Wang[1], Xuefei Chang[1], Fang Wang[1], Hongwei Yao[1], Yufa Peng[3], David Stanley[4] and Gongyin Ye[1,*]

[1]*State Key Laboratory of Rice Biology & Key Laboratory of Agricultural Entomology of Ministry of Agriculture, Institute of Insect Sciences, Zhejiang University, Hangzhou, China*

[2]*Institute of Plant Protection, Shandong Academy of Agricultural Sciences, Jinan, China*

[3]*State Key Laboratory for Biology of Plant Diseases and Insect Pests, Institute of Plant Protection, Chinese Academy of Agricultural Sciences, Beijing, China*

[4]*Biological Control of Insects Research Laboratory, USDA/Agricultural Research Service, Columbia, MO, USA*

*Correspondence

email chu@zju.edu.cn

[†]These two authors contributed equally to this work.

Keywords: *Bt* rice, meta-analysis, non-target, functional guilds.

Summary

Transgenic *Bt* rice expressing the insecticidal proteins derived from *Bacillus thuringiensis* Berliner (*Bt*) has been developed since 1989. Their ecological risks towards non-target organisms have been investigated; however, these studies were conducted individually, yielding uncertainty regarding potential agroecological risks associated with large-scale deployment of *Bt* rice lines. Here, we developed a meta-analysis of the existing literature to synthesize current knowledge of the impacts of *Bt* rice on functional arthropod guilds, including herbivores, predators, parasitoids and detritivores in laboratory and field studies. Laboratory results indicate *Bt* rice did not influence survival rate and developmental duration of herbivores, although exposure to *Bt* rice led to reduced egg laying, which correctly predicted their reduced abundance in *Bt* rice agroecosystems. Similarly, consuming prey exposed to *Bt* protein did not influence survival, development or fecundity of predators, indicating constant abundances of predators in *Bt* rice fields. Compared to control agroecosystems, parasitoid populations decreased slightly in *Bt* rice cropping systems, while detritivores increased. We draw two inferences. One, laboratory studies of *Bt* rice showing effects on ecological functional groups are mainly either consistent with or more conservative than results of field studies, and two, *Bt* rice will pose negligible risks to the non-target functional guilds in future large-scale *Bt* rice agroecosystems in China.

Introduction

Genetically modified (GM) crops expressing *cry* genes derived from *Bacillus thuringiensis* Berliner (*Bt*) have been grown commercially since 1996 to control target insect pests worldwide (Cohen *et al.*, 2008). GM crops production have been increasing, amounting to 179.7 million ha over 28 countries in 2015 (James, 2015). Despite the substantial economic and environmental benefits of deploying *Bt* crops (Klumper and Qaim, 2014; Raymond Park *et al.*, 2011), real concerns about their ecological risks continue (Brookes and Barfoot, 2015). These concerns drive contemporary ecological risk assessments designed to guide current and future risk management.

Meta-analyses have been applied widely in ecological risk assessments of *Bt* crops on non-target arthropods (Comas *et al.*, 2014; Naranjo, 2009; Peterson *et al.*, 2011). Marvier *et al.* (2007) conducted a meta-analysis to assess the ecological risks of *Bt* maize and *Bt* cotton on non-target invertebrates, and reported that certain non-target taxa were less abundant in *Bt* cotton and *Bt* maize fields compared with insecticide-free control fields. Naranjo (2009) concluded that negative effects of *Bt* crops on predators and parasitoids in laboratory tests coincided with lower abundance in fields except for *Bt* rice and *Bt* eggplant cropping systems. They detected no differences in the abundances of

non-target organisms in *Bt* rice fields, while another study revealed reduced spider abundances in *Bt* rice cropping systems relative to controls (Peterson *et al.*, 2011). These results highlight the need for a comprehensive analysis of the current state of our understanding of ecological risks associated with deploying *Bt* rice lines on a large scale.

Rice, *Oryza sativa* L., is one of the world's most important food crops (Zeigler and Barclay, 2008). Since development of the first transgenic *Bt* rice plant in 1989 (Yang *et al.*, 1989), a series of *Bt* rice lines, expressing Cry1A, Cry1C, Cry2A or Cry1Ab/Vip3H, had been generated in China (Chen *et al.*, 2011; Li *et al.*, 2016). These lines effectively suppress stem borer, leaf folder and other lepidopteran pests (Chen *et al.*, 2010; Wang *et al.*, 2016; Ye *et al.*, 2001a,b). Their environmental risks towards non-target arthropods have been fully studied in laboratory and field settings (Chen *et al.*, 2011; Cohen *et al.*, 2008; Li *et al.*, 2016). However, the results varied with transgenic rice lines and/or non-target taxa, and accurate predictions of the influence of *Bt* rice on rice agroecosystems remain problematic. In this report, we addressed this issue by developing and presenting an analysis of the accessible literature on the influence of *Bt* rice on non-target arthropod ecological functional guilds from laboratory to field conditions in China.

Results

Our database in this analysis contained 282 observations from 40 papers reporting laboratory studies and 585 observations from 27 papers reporting field studies (Tables S1 and S2). Details of the literature search are shown in Figure 1. Most of the data had no publication bias in our database (Figures S1–S3).

Laboratory studies

Among the three laboratory parameters we assessed, the survival or development of the indicated ecological functional groups were not significantly affected by Bt rice (Figure 2; Table 1). Also, the reproduction of predators, parasitoids and detritivores were similar between Bt and control rice. The effect size of herbivore reproduction was lower following Bt treatments, indicating herbivores on Bt rice laid fewer eggs (Table 1), although the herbivore data including thrips and aphids were highly heterogeneous (I^2 = 93.7%, $P_{heterogeneity}$ < 0.001). We re-analysed the data after removing the information on thrips and aphids, which, again, revealed a significant reduction in herbivore reproduction (E = −0.449, P < 0.001, I^2 = 35.8%, $P_{heterogeneity}$ = 0.050; Figure S4). We conducted a subgroup analysis to understand the high heterogeneity of the development data on predators. The developmental rates of predators was restrained after feeding on lepidopteran prey from Bt rice (E = 1.754, P < 0.001; Figure S5), while other subgroups did not significantly affect by Bt rice.

Field studies

The abundance of non-target herbivores (E = −0.286, 95% CI = −0.389 to −0.182, P < 0.001; Figure 3), including plant-feeding thrips on Bt rice (E = −0.591, P < 0.001; Figure 4a), was significantly reduced, compared to controls. Predator abundances on Bt rice did not differ from controls (E = −0.028, 95% CI = −0.140 to 0.085, P = 0.629; Figure 3). The densities of three main predatory orders, Araneae, Hemiptera and Coleoptera, were similar between Bt rice and the control (Figure 4b). There was a slight reduction for the abundance of parasitoids in

Bt rice paddies (E = −0.444, 95% CI = −0.882 to −0.005, P = 0.048). Populations of detritivores increased in Bt rice fields relative to controls (E = 0.309, 95% CI = 0.026 to 0.592, P = 0.032), possibly due to higher abundance of Collembola (E = 0.280, P = 0.016; Figure 4c).

Discussion

Many arthropods provide numerous ecosystem services in rice agricultural systems, such as biological control, pollination, crop residue decomposition and soil health improvement (Hao et al., 1998; Wolfenbarger et al., 2008). They are classified into ecological functional groups, herbivores, predators, parasitoids and detritivores, based on ecosystem services. Herbivores serve as prey for predators and hosts for parasitoids (Norris and Kogan, 2005). Predators and parasitoids are important natural enemies of crop pests in agroecosystems (Naranjo, 2005), and detritivores contribute to degrading plant litter and microorganisms (Rusek, 1998). Thus, our meta-analysis focused on impacts of Bt rice on these ecological functional guilds. We interpret the overarching results to show that the results of laboratory studies are reasonable predictors of the outcomes of field studies.

The fecundity and field abundances of herbivores were significantly suppressed after consuming Bt rice. Thrips and aphids did not account for these differences, because removing the thrips and aphids data did not change the outcome of our analysis. This is reasonable because several studies reported that aphids preferred Bt maize or Bt cotton compared to controls (Faria et al., 2007; Liu et al., 2005). Bt maize and Bt cotton did not impact thrips (Li et al., 2007; Obrist et al., 2005), although they could accumulate Cry protein from Bt cotton (Kumar et al., 2014). Other studies also reported that the brown planthopper, a non-target sucking pest of Bt rice, laid fewer eggs on Bt rice lines (KMD1 and KMD2) expressing the Cry1Ab protein in laboratory and field surveys (Chen et al., 2007, 2012; Gao et al., 2011). The results of an additional meta-analysis on planthoppers were consistent with these reports (Figure S6). Our analyses predicate

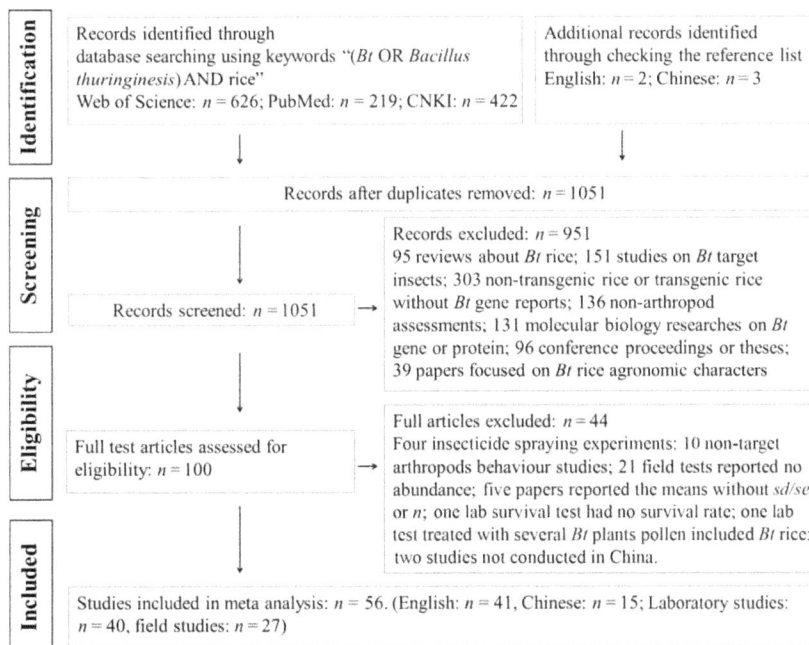

Figure 1 PRISMA flow diagram showing the procedure used for selection of studies for meta-analysis.

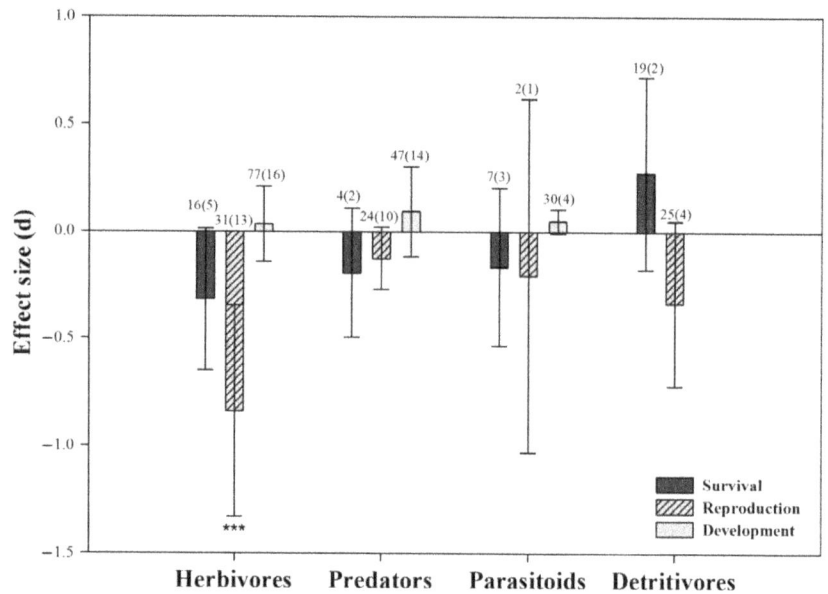

Figure 2 Meta-analysis of laboratory studies examining the influence of *Bt* rice on non-target ecological functional guilds biological parameters. For the development of detritivores, no data could be collected to conduct the analysis. Effect size (*E*) is Hedges'*d*, and error bars represent bias-corrected 95% *CI* (confidence interval). Values above each bar indicate the total number of studies for each group (number of papers). Asterisks denote significant differences in the observed effect sizes among the comparisons (*$P < 0.05$; **$P < 0.01$; ***$P < 0.001$).

Table 1 Influence of *Bt* rice on survival, development and reproduction of the ecological functional groups in laboratory

Functional guilds	Survival		Reproduction		Development	
	E (95% CI)	P	E (95% CI)	P	E (95% CI)	P
Herbivores	−0.318 (−0.648 to 0.012)	0.059	−0.838 (−1.331 to 0.345)	<0.001	0.034 (−0.141 to 0.208)	0.707
Parasitoids	−0.166 (−0.534 to 0.202)	0.376	−0.207 (−1.031 to 0.617)	0.622	0.047 (−0.009 to 0.102)	0.098
Predators	−0.194 (−0.494 to 0.106)	0.205	−0.125 (−0.270 to 0.019)	0.089	0.092 (−0.117 to 0.301)	0.386
Detritivores	0.272 (−0.175 to 0.719)	0.233	−0.336 (−0.719 to 0.048)	0.086	–	–

E = effect sizes, *P* = significance level, *CI* = confidence interval. '–' means that no data could be collected to conduct the analysis for the development of detritivores.

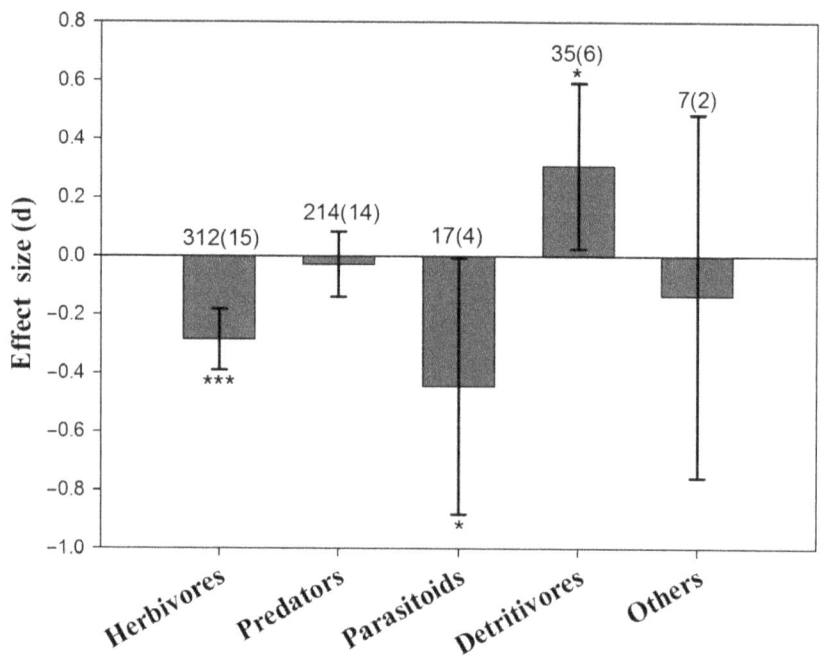

Figure 3 Meta-analysis of field studies examining the influence of *Bt* rice on non-target ecological functional guilds abundances. Asterisks denote significant differences in the observed effect sizes among the comparisons (*$P < 0.05$; **$P < 0.01$; ***$P < 0.001$).

that populations of the non-target herbivores including thrips and planthoppers in *Bt* rice fields are no more likely to achieve outbreak levels than insects in non-*Bt* rice cropping systems. We speculate that thrips and planthoppers laid fewer eggs on *Bt* rice because of the direct actions of *Bt* insecticidal proteins and the indirect impacts of alterations in nutritional qualities of *Bt* rice plants.

Two recent meta-analyses reported that predators developed more slowly in tri-trophic tests involving *Bt* crop–herbivore–predator systems (Duan *et al.*, 2010; Naranjo, 2009). This pattern

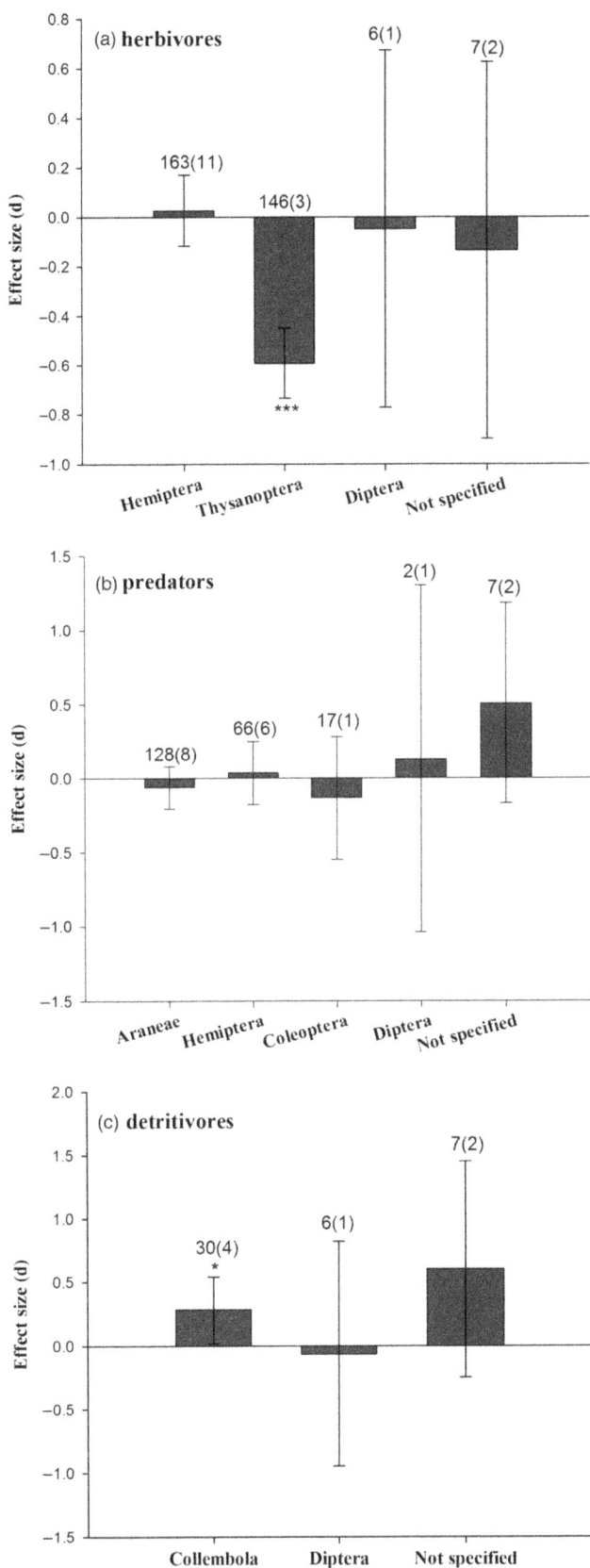

Figure 4 Meta-analyses of field studies examining the influence of *Bt* rice on the taxa abundance of herbivores (a), predators (b) and detritivores (c). Asterisks denote significant differences in the observed effect sizes among the comparisons (*P < 0.05; **P < 0.01; ***P < 0.001).

may be attributed to the varied types of prey consumed by predators. After feeding on Cry1Ab-containing rice leafrollers, *Cnaphalocrocis medinalis*, a *Bt*-targeted pest, the *Bt* protein was present in its predator, the wolf spider, *Pirata subpiraticus*. However, there was neither a binding protein in the brush border membrane vesicles nor an accumulation with longer feeding time for Cry1Ab in the spider. The authors inferred the spiders suffered elongated development times, possibly due to reduced prey quality (Chen *et al.*, 2009). Romeis *et al.* (2006) similarly concluded adverse *Bt* effects on predators were due to the reduced prey quality. Given that there are many prey types in rice fields, we infer that *Bt* rice does not influence predator abundance in field studies, in agreement with previous studies (Chen *et al.*, 2009; Han *et al.*, 2011; Lu *et al.*, 2016; Tian *et al.*, 2010).

Bt treatments led to small reductions in braconid parasitoid abundances in field, but not laboratory studies (Lu *et al.*, 2014, 2016; Tian *et al.*, 2008). Similarly, *Bt* maize led to negative effects on the parasitoid *Macrocentrus cingulum* Brischke (Hymenoptera: Braconidae) in previous meta-analyses (Duan *et al.*, 2010; Wolfenbarger *et al.*, 2008). These findings indicate that the biological performance of some parasitoid species may be adversely affected by *Bt* crops after parasitizing *Bt*-targeted lepidopteran pests. In our analysis, parasitoids were mainly composed of two families, larval parasitoids, Braconids, which parasitize *Bt*-targeted pests and egg parasitoids, Mymaridae, which parasitize non-targeted pests, such as planthoppers. The pooled effect size was significantly lower in *Bt* rice fields. The limited number of studies and the different sampling methods may help understand the variance among studies. Our assessment is based on 17 observations from four papers that met our selection criteria (Bai *et al.*, 2012; Lu *et al.*, 2014, 2016; Tian *et al.*, 2008). This small data set may lead to inaccurate assessments of the effect size. Our interpretation is that combining collection methods in the field, specifically, vacuum suction and sticky cards, may lead to a more complete appreciation of parasitoid populations in future.

We found that the number of surviving springtails, juveniles and adults, did not decrease after feeding on *Bt* rice plant materials or artificial diets containing *Bt* protein in the laboratory. Contrarily, the number of springtails in *Bt* rice cropping systems significantly increased. This effect differed from results with other transgenic crops, because *Bt* maize and *Bt* cotton did not impact detritivore abundance (Wolfenbarger *et al.*, 2008). A higher abundance of Collembola was collected in *Bt* rice crops using a vacuum suction method, with no difference in pitfall trap surveys (Lu *et al.*, 2014). We infer the ecological services due to detritivores, decomposing and recycling plant residue, would not be affected by *Bt* protein in rice agroecosystems.

In total, our results indicated the *Bt* rice effects on the functional groups obtained from laboratory were not always consistent with that in field trials for the complicated factors existed in paddy agroecosystem. A tiered approach is indispensable to assess the ecological risks of transgenic *Bt* rice in future (Romeis *et al.*, 2008). There is no doubt that China is playing a leading role in the development and risk assessment of *Bt* rice (Li *et al.*, 2014), and biosafety certificates for commercial planting of two *Bt* rice lines (Huahui 1 and *Bt* Shanyou 63) in Hubei Province have been issued twice by the government in 2009 and 2014 (Li *et al.*, 2016). Quantitative syntheses of the risk assessments of *Bt* rice would provide a strong evidence for the Chinese policy makers to avoid some disputes caused by German government in 2009 (Marvier, 2011; Ricroch *et al.*, 2010).

Despite that the effects of *Bt* rice in China might be not similar with them in other countries because of the ecology of insects/plants and climate zones varied around the world, it would offer some lessons to the country in which *Bt* rice is in urgent need to be developed.

Moreover, although we tried our best to collect data and conduct the analysis, there were still some limitations. Firstly, we failed to make some comparisons between the treatments of *Bt* insecticide protein and chemical insecticide for few cases with chemical insecticide. Also, the data of some functional guilds (especially parasitoids and detritivores) seemed to be not enough. With the knowledge accumulated of risk assessment, these aspects will be gradually improved in future.

Experimental procedures

Data search and database production

The database was created by searching the Web of Science (http://isiknowledge.com), PubMed (http://www.ncbi.nlm.nih.gov/pubmed) and China National Knowledge Infrastructure (CNKI, http://www.cnki.net) using the key words '(Bt OR *Bacillus thuringinesis*) AND rice'. The following criteria were applied to screen the studies: (i) transgenic rice expresses one or more Cry proteins and targets lepidopteran pests; (ii) non-transgenic rice was used as control plants in laboratory and field research; (iii) one or more non-target arthropods were assessed; (iv) the papers reported data on development, survival and reproduction as response variables for non-target taxa in laboratory studies, and species abundances in field studies; (v) each study reported means accompanied by standard deviations (SD) or standard error (SE) and sample size (*n*); (vi) all studies were conducted in China and published in English or Chinese up to September 2016.

To build the database, we followed the formulation described by Marvier *et al.* (2007). For each study, we recorded authors and journal information, details about the *Bt* rice (Cry protein, transgenic event and its control), the non-target group (taxonomy, functional guild and stage) and the experimental treatment with its control. We also recorded study location, cultivation, plot size, exposure method, sampling method and other methodological details. If a study reported figures without numerical data, we used ImageJ software version 1.45 to measure means and its variance (Abràmoff *et al.*, 2004). We contacted authors to obtain details when needed.

We developed data eligibility criteria for laboratory and field studies. For laboratory studies, we used (i) total eggs per female for reproduction, immature stage survival rate, adult emergence and pupation rate, but not egg-hatching data for survival; (ii) we selected the separately recorded male and female development time and the entire time span of larval or nymphal development; (iii) if studies reported multiple generation effects, we used the last generation for analyses. The criteria for field studies were (i) we chose the seasonal means and, when they were not available, the peak abundance; (ii) when multiple life-stage data was reported in one study, we chose the larval or nymphal abundance; (iii) we used the ecological functional groups, not individual species. However, when analysing by taxonomic group, we used the individual species.

We identified the ecological functional guilds specified in the original papers. If a paper did not identify the guilds, we classified the non-target taxa into five functional guilds: parasitoids, predators, herbivores, detritivores and others, following the description of Liu *et al.* (2003; Table S3).

Data analysis

Before the quantitative data synthesis, publication bias of the database was conducted by a funnel plot with the Begg–Mazumdar rank correlation test and Egger's regression test (Begg and Mazumdar, 1994; Egger *et al.*, 1997; Haworth *et al.*, 2016). A symmetrical funnel plot showed no publication bias of the data (Field and Gillett, 2010). Trim-and-fill method was used to balance the asymmetrical funnel plot (Duval and Tweedie, 2000).

A weighted mean effect size (*E*), Hedges'*d*, was used to calculate the difference between an experimental (*Bt*) and the control (non-*Bt*) mean divided by the pooled standard deviation and weighted by the reciprocal of the sampling variance (Hedges and Olkin, 1985). A negative effect size value indicates the *Bt* group had lower abundance, fecundity, survival or shorter development time compared with the non-*Bt* group, and a positive value indicates these parameters were higher or longer than controls. For hypothesis testing, we used the parametric 95% confidence interval (*CI*) to test the results. If the interval enclosed zero, we took the effect size as not significantly different from zero. We assessed heterogeneity with the Q test and I^2 statistic. When the I^2 value was larger than 50% and $P_{heterogeneity} < 0.001$, we considered the data highly heterogeneous and conducted subgroup analysis to analyse the high heterogeneity. The random-effect model was the more appropriate method to carry out all the analysis for the data included in the analysis came from different research groups (Borenstein *et al.*, 2009; Schmidt *et al.*, 2009). All the analyses were conducted using the STATA software version 12.0 (STATA Crop, College Station, TX).

Acknowledgements

We greatly thank Dr. Mao Chen (Monsanto Company, 700 Chesterfield Parkway W., GG3I Chesterfield, MO) and Dr. Junce Tian (Zhejiang Academy of Agricultural Sciences, Hangzhou, Zhejiang, 310021, China) for manuscript revision. This work was supported by the National Special Transgenic Project from Chinese Ministry of Agriculture (2016ZX08011-001 and 2014ZX08011-001) and China National Science Fund for Innovative Research Group of Biological Control (Grant No. 31321063). Mention of trade names or commercial products in this article is solely for the purpose of providing specific information and does not imply recommendation or endorsement by the US Department of Agriculture. All programs and services of the US Department of Agriculture are offered on a non-discriminatory basis without regard to race, colour, national origin, religion, sex, age, marital status or handicap.

References

Abràmoff, M.D., Magalhães, P.J. and Ram, S.J. (2004) Image processing with ImageJ. *Biophotonics Int.* **11**, 36–42.

Bai, Y.Y., Yan, R.H., Ye, G.Y., Huang, F.N., Wangila, D.S., Wang, J.J. and Cheng, J.A. (2012) Field response of aboveground non-target arthropod community to transgenic *Bt-Cry1Ab* rice plant residues in postharvest seasons. *Transgenic Res.* **21**, 1023–1032.

Begg, C.B. and Mazumdar, M. (1994) Operating characteristics of a rank correlation test for publication bias. *Biometrics*, **50**, 1088–1101.

Borenstein, M., Hedges, L.V., Higgins, J. and Rothstein, H.R. (2009) *Introduction to Meta-analysis*. Chichester, UK: John Wiley & Sons, Ltd. https://www.meta-analysis-workshops.com/download/bookChapterSample.pdf [Accessed 2 September 2016]

Brookes, G. and Barfoot, P. (2015) Environmental impacts of genetically modified (GM) crop use 1996–2013: impacts on pesticide use and carbon emissions. *GM Crops Food* **6**, 103–133.

Chen, M., Liu, Z.C., Ye, G.Y., Shen, Z.C., Hu, C., Peng, Y.F., Altosaar, I. *et al.* (2007) Impacts of transgenic *cry1Ab* rice on non-target planthoppers and their main predator *Cyrtorhinus lividipennis* (Hemiptera: Miridae) – a case study of the compatibility of *Bt* rice with biological control. *Biol. Control* **42**, 242–250.

Chen, M., Ye, G.Y., Liu, Z.C., Fang, Q., Hu, C., Peng, Y.F. and Shelton, A.M. (2009) Analysis of Cry1Ab toxin bioaccumulation in a food chain of *Bt* rice, an herbivore and a predator. *Ecotoxicology*, **18**, 230–238.

Chen, Y., Tian, J.C., Shen, Z.C., Peng, Y.F., Hu, C., Guo, Y.Y. and Ye, G.Y. (2010) Transgenic rice plants expressing a fused protein of Cry1Ab/Vip3H has resistance to rice stem borers under laboratory and field conditions. *J. Econ. Entomol.* **103**, 1444–1453.

Chen, M., Shelton, A. and Ye, G.Y. (2011) Insect-resistant genetically modified rice in China: from research to commercialization. *Annu. Rev. Entomol.* **56**, 81–101.

Chen, Y., Tian, J.C., Wang, W., Fang, Q., Akhtar, Z.R., Peng, Y.F., Hu, C. *et al.* (2012) *Bt* rice expressing Cry1Ab does not stimulate an outbreak of its non-target herbivore, *Nilaparvata lugens*. *Transgenic Res.* **21**, 279–291.

Cohen, M.B., Chen, M., Bentur, J.S., Heong, K.L. and Ye, G. (2008) *Bt* rice in Asia: potential benefits, impact, and sustainability. In: *Integration of Insect-resistant Genetically Modified Crops within IPM Programs* (Romeis, J., Shelton, A.M. and Kennedy, G.G., eds), pp. 223–248. Dordrecht: Springer Science + Business Media B. V.

Comas, C., Lumbierres, B., Pons, X. and Albajes, R. (2014) No effects of *Bacillus thuringiensis* maize on nontarget organisms in the field in southern Europe: a meta-analysis of 26 arthropod taxa. *Transgenic Res.* **23**, 135–143.

Duan, J.J., Lundgren, J.G., Naranjo, S. and Marvier, M. (2010) Extrapolating non-target risk of *Bt* crops from laboratory to field. *Biol. Lett.* **6**, 74–77.

Duval, S. and Tweedie, R. (2000) Trim and fill: a simple funnel-plot-based method of testing and adjusting for publication bias in meta-analysis. *Biometrics*, **56**, 455–463.

Egger, M., Smith, G.D., Schneider, M. and Minder, C. (1997) Bias in meta-analysis detected by a simple, graphical test. *Br. Med. J.* **315**, 629–634.

Faria, C.A., Wackers, F.L., Pritchard, J., Barrett, D.A. and Turlings, T.C. (2007) High susceptibility of *Bt* maize to aphids enhances the performance of parasitoids of lepidopteran pests. *PLoS ONE*, **2**, e600.

Field, A.P. and Gillett, R. (2010) How to do a meta-analysis. *Br. J. Math. Stat. Psychol.* **63**, 665–694.

Gao, M.Q., Hou, S.P., Pu, D.Q., Shi, M., Ye, G.Y., Peng, Y.F. and Chen, X.X. (2011) Effects of *Bt* rice on the number and hatch rate of planthopper eggs and their attack by natural enemies in paddy fields. *Acta Entomol. Sin.* **54**, 467–476.

Han, Y., Xu, X.L., Ma, W.H., Yuan, B.Q., Wang, H., Liu, F.Z., Wang, M.Q. *et al.* (2011) The influence of transgenic *cry1Ab/cry1Ac*, *cry1C* and *cry2A* rice on non-target planthoppers and their main predators under field conditions. *Agric. Sci. China* **10**, 1739–1747.

Hao, S.H., Zhang, X.X., Cheng, X.N., Luo, Y.J. and Tian, X.Z. (1998) The dynamics of biodiversity and the composition of nutrition classes and dominant guilds of arthropod community in paddy field. *Acta Entomol. Sin.* **41**, 343–353.

Haworth, M., Hoshika, Y. and Killi, D. (2016) Has the impact of rising CO_2 on plants been exaggerated by meta-analysis of free air CO_2 enrichment studies? *Front. Plant Sci.* **7**, 1153.

Hedges, L.V. and Olkin, I. (1985) *Statistical Method for Meta-analysis*. San Diego, CA: Academic Press. http://files.eric.ed.gov/fulltext/ED227133.pdf [Accessed 30 August 2015]

James, C. (2015) *Global Status of Commercialized Biotech/GM Crops: 2015*. ISAAA Brief, No. 51, Ithaca, NY: ASAAA. http://www.isaaa.org/resources/publications/briefs/51/ [Accessed 20 May 2016]

Klumper, W. and Qaim, M. (2014) A meta-analysis of the impacts of genetically modified crops. *PLoS ONE*, **9**, e111629.

Kumar, R., Tian, J.C., Naranjo, S.E. and Shelton, A.M. (2014) Effects of *Bt* cotton on *Thrips tabaci* (Thysanoptera: Thripidae) and its predator, *Orius insidiosus* (Hemiptera: Anthocoridae). *J. Econ. Entomol.* **107**, 927–932.

Li, H.B., Wu, K.M., Xu, Y., Yang, X.R., Yao, J. and Wang, F. (2007) Studies on population density dynamic of onion thrips in cotton field in the south of Xinjiang. *Xinjiang Agric. Sci.* **44**, 583–586.

Li, Y.H., Peng, Y.F., Hallerman, E.M. and Wu, K.M. (2014) Biosafety management and commercial use of genetically modified crops in China. *Plant Cell Rep.* **33**, 565–573.

Li, Y.H., Hallerman, E.M., Liu, Q.S., Wu, K.M. and Peng, Y.F. (2016) The development and status of *Bt* rice in China. *Plant Biotechnol. J.* **14**, 839–848.

Liu, Z.C., Ye, G.Y., Hu, C. and Datta, K.S. (2003) Impact of transgenic *indica* rice with a fused gene of *cry1Ab/cry1Ac* on the rice paddy arthropod community. *Acta Entomol. Sin.* **46**, 454–465.

Liu, X.D., Zhai, B.P., Zhang, X.X. and Zong, J.M. (2005) Impact of transgenic cotton plants on a non-target pest, *Aphis gossypii* Glover. *Ecol. Entomol.* **30**, 307–315.

Lu, Z.B., Tian, J.C., Han, N.S., Hu, C., Peng, Y.F., Stanley, D. and Ye, G.Y. (2014) No direct effects of two transgenic *Bt* rice lines, T1C-19 and T2A-1, on the arthropod communities. *Environ. Entomol.* **43**, 1453–1463.

Lu, Z.B., Dang, C., Han, N.S., Shen, Z.C., Peng, Y.F., Stanley, D. and Ye, G.Y. (2016) The new transgenic *cry1Ab/vip3H* rice poses no unexpected ecological risks to arthropod communities in rice agroecosystems. *Environ. Entomol.* **45**, 518–525.

Marvier, M. (2011) Using meta-analysis to inform risk assessment and risk management. *J. Verbrauch. Lebensm.* **6**, 113–118.

Marvier, M., McCreedy, C., Regetz, J. and Kareiva, P. (2007) A meta-analysis of effects of *Bt* cotton and maize on nontarget invertebrates. *Science*, **316**, 1475–1477.

Naranjo, S.E. (2005) Long-term assessment of the effects of transgenic *Bt* cotton on the abundance of nontarget arthropod natural enemies. *Environ. Entomol.* **34**, 1193–1210.

Naranjo, S.E. (2009) Impacts of *Bt* crops on non-target invertebrates and insecticide use patterns. *Perspect. Agr. Vet. Sci. Nutr. Nat. Resour.* **4**, 1–23.

Norris, R.F. and Kogan, M. (2005) Ecology of interactions between weeds and arthropods. *Annu. Rev. Entomol.* **50**, 479–503.

Obrist, L.B., Klein, H., Dutton, A. and Bigler, F. (2005) Effects of *Bt* maize on *Frankliniella tenuicornis* and exposure of thrips predators to prey-mediated *Bt* toxin. *Entomol. Exp. Appl.* **115**, 409–416.

Peterson, J.A., Lundgren, J.G. and Harwood, J.D. (2011) Interactions of transgenic *Bacillus thuringiensis* insecticidal crops with spiders (Araneae). *J. Arachnol.* **39**, 1–21.

Raymond Park, J., McFarlane, I., Hartley Phipps, R. and Ceddia, G. (2011) The role of transgenic crops in sustainable development. *Plant Biotechnol. J.* **9**, 2–21.

Ricroch, A., Berge, J.B. and Kuntz, M. (2010) Is the German suspension of MON810 maize cultivation scientifically justified? *Transgenic Res.* **19**, 1–12.

Romeis, J., Meissle, M. and Bigler, F. (2006) Transgenic crops expressing *Bacillus thuringiensis* toxins and biological control. *Nat. Biotechnol.* **24**, 63–71.

Romeis, J., Bartsch, D., Bigler, F., Candolfi, M.P., Gielkens, M.M.C., Hartley, S.E., Hellmich, R.L. *et al.* (2008) Assessment of risk of insect-resistant transgenic crops to nontarget arthropods. *Nat. Biotechnol.* **26**, 203–208.

Rusek, J. (1998) Biodiversity of Collembola and their functional role in the ecosystem. *Biodivers. Conserv.* **7**, 1207–1219.

Schmidt, F.L., Oh, I.S. and Hayes, T.L. (2009) Fixed-versus random-effects models in meta-analysis: model properties and an empirical comparison of differences in results. *Br. J. Math. Stat. Psychol.* **62**, 97–128.

Tian, J.C., Liu, Z.C., Yao, H.W., Ye, G.Y. and Peng, Y.F. (2008) Impact of transgenic rice with a *cry1Ab* gene on parasitoid subcommunity structure and the dominant population dynamics of parasitoid wasps in rice paddy. *J. Environ. Entomol.* **30**, 1–7.

Tian, J.C., Liu, Z.C., Chen, M., Chen, Y., Chen, X.X., Peng, Y.F., Hu, C. *et al.* (2010) Laboratory and field assessments of prey-mediated effects of transgenic *Bt* rice on *Ummeliata insecticeps* (Araneida: Linyphiidae). *Environ. Entomol.* **39**, 1369–1377.

Wang, Y.N., Ke, K.Q., Li, Y.H., Han, L.Z., Liu, Y.M., Hua, H.X. and Peng, Y.F. (2016) Comparison of three transgenic *Bt* rice lines for insecticidal protein expression and resistance against a target pest, *Chilo suppressalis* (Lepidoptera: Crambidae). *Insect Sci.* **23**, 78–87.

Wolfenbarger, L.L., Naranjo, S.E., Lundgren, J.G., Bitzer, R.J. and Watrud, L.S. (2008) *Bt* crop effects on functional guilds of non-target arthropods: a meta-analysis. *PLoS ONE*, **3**, e2118.

Yang, H., Li, J., Guo, S., Chen, X. and Fan, Y. (1989) Transgenic rice plants produced by direct uptake of δ-endotoxin protein gene from *Bacillus thuringenesis* into rice protoplasts. *Sci. Agric. Sin.* **22**, 1–5.

Ye, G.Y., Shu, Q.Y., Yao, H.W., Cui, H.R., Cheng, X.Y., Hu, C., Xia, Y.W. *et al.* (2001a) Field evaluation of resistance of transgenic rice containing a synthetic *cry1Ab* gene from *Bacillus thuringiensis* Berliner to two stem borers. *J. Econ. Entomol.* **94**, 271–276.

Ye, G.Y., Tu, J.M., Hu, C., Datta, K. and Datta, S.K. (2001b) Transgenic IR72 with fused *Bt* gene *cry1Ab/cry1Ac* from *Bacillus thuringiensis* is resistant against four lepidopteran species under field conditions. *Plant Biotechnol.* **18**, 125–133.

Zeigler, R.S. and Barclay, A. (2008) The relevance of rice. *Rice*, **1**, 3–10.

17

Sequencing of Australian wild rice genomes reveals ancestral relationships with domesticated rice

Marta Brozynska[1], Dario Copetti[2,3], Agnelo Furtado[1], Rod A. Wing[2,3], Darren Crayn[4], Glen Fox[5], Ryuji Ishikawa[6] and Robert J. Henry[1,*]

[1]Queensland Alliance for Agriculture and Food Innovation, University of Queensland, Brisbane, QLD, Australia

[2]Arizona Genomics Institute, School of Plant Sciences, University of Arizona, Tucson, AZ, USA

[3]International Rice Research Institute, T.T. Chang Genetic Resources Center, Los Baños, Laguna, Philippines

[4]Australian Tropical Herbarium, James Cook University, Cairns, QLD, Australia

[5]Queensland Alliance for Agriculture and Food Innovation, University of Queensland, Toowoomba, QLD, Australia

[6]Faculty of Agriculture and Life Science, Hirosaki University, Hirosaki, Aomori, Japan

*Correspondence
email robert.henry@uq.edu.au
This whole genome shotgun project has been deposited at DDBJ/EMBL/GenBank under the accessions LONB00000000 and LONC00000000. The version described in this paper is LONB01000000 and LONC01000000.

Keywords: assembly, molecular clock, sequencing, Oryza, phylogeny, wild rice.

Summary

The related A genome species of the *Oryza* genus are the effective gene pool for rice. Here, we report draft genomes for two Australian wild A genome taxa: *O. rufipogon*-like population, referred to as Taxon A, and *O. meridionalis*-like population, referred to as Taxon B. These two taxa were sequenced and assembled by integration of short- and long-read next-generation sequencing (NGS) data to create a genomic platform for a wider rice gene pool. Here, we report that, despite the distinct chloroplast genome, the nuclear genome of the Australian Taxon A has a sequence that is much closer to that of domesticated rice (*O. sativa*) than to the other Australian wild populations. Analysis of 4643 genes in the A genome clade showed that the Australian annual, *O. meridionalis*, and related perennial taxa have the most divergent (around 3 million years) genome sequences relative to domesticated rice. A test for admixture showed possible introgression into the Australian Taxon A (diverged around 1.6 million years ago) especially from the wild *indica/O. nivara* clade in Asia. These results demonstrate that northern Australia may be the centre of diversity of the A genome *Oryza* and suggest the possibility that this might also be the centre of origin of this group and represent an important resource for rice improvement.

Introduction

Rice is a pantropical crop that is a staple food consumed by over half of the world's population This crop has a long history of domestication and its cultivation dates back around 10 000 years in Asia (*O. sativa*) and over 3000 years in Africa (*O. glaberrima*). The *Oryza* genus diversified into six diploid (A–C and E–G) and five tetraploid (BC, CD, HJ, HK and KL) genome groups. The phylogeny of the *Oryza* genome groups has been widely studied and is now well known (Ammiraju *et al.*, 2010; Ge *et al.*, 1999; Lu *et al.*, 2009). The relationships between the most recently diverged A genome diploids, that include domesticated rice, have been more challenging and only lately have their phylogeny been more fully described, using both chloroplast (Wambugu *et al.*, 2015) and nuclear genomes [International Oryza Map Alignment Project (I-OMAP), unpublished].

Rice food security requires continued increases in rice productivity and relies on ongoing genetic improvement. Climate change adds to the difficulty of achieving the necessary rates of genetic gain (Abberton *et al.*, 2016). The wild relatives of rice provide a gene pool that allows for the expansion of diversity (Krishnan *et al.*, 2014) in domesticated rice for the creation of new high yielding genotypes, with new nutritional and functional traits (Kharabian-Masouleh *et al.*, 2012) and adaptation to new environments (Brozynska *et al.*, 2016). The A genome

species of *Oryza*, which include the species that are readily interfertile with rice, represent the effective primary gene pool for rice. Recent investigations of large and widespread wild populations in tropical Australia (Henry *et al.*, 2010) suggest the presence of two distinct and possibly novel perennial wild A genome taxa [Figure 1; Waters *et al.* (2012); Sotowa *et al.* (2013); Brozynska *et al.* (2014)]. Of these, Taxon A has plant and seed morphology similar to that of *O. rufipogon* and Taxon B appears to be similar to the annual *O. meridionalis*. Here, we report draft genomes for these two Australian wild rice taxa: Taxon A and Taxon B, which are likely to be novel and different species. The two taxa were sequenced and assembled by integration of two distinct next-generation sequencing (NGS) data, namely Illumina and Pacific Biosciences. The draft nuclear genome sequences of 384.8 Mb (Taxon A) and 354.9 Mb (Taxon B) were placed on 12 pseudochromosomes based on available rice reference sequences. Taken together, this study creates a new genomic platform for investigating the gene pool and agriculturally important traits potentially present in Australian taxa.

The Australian wild rices have been isolated from the impact of gene transfer from domesticated rice that may complicate interpretation of the genetics of wild rice populations in Asia where rice has been cultivated on a large scale for thousands of years. An understanding of genetic relationships and diversity between and within these Asian and Australian populations will guide the effective use of wild genetic resources for global rice improvement. The phylogenetic relationships between all of the A genome taxa have recently been estimated using whole chloroplast genome sequences (Brozynska *et al.*, 2014; Wambugu *et al.*, 2013). In this phylogeny, the Australian A genome taxa form a distinct clade, which is a sister to the Asian domesticated rice clade (Figure 2a). We now report a phylogenetic analysis of the corresponding nuclear genomes.

Results

Genome sequencing

The statistics of sequencing reads obtained in this study are shown in Table S1. Total data produced by the Illumina platform

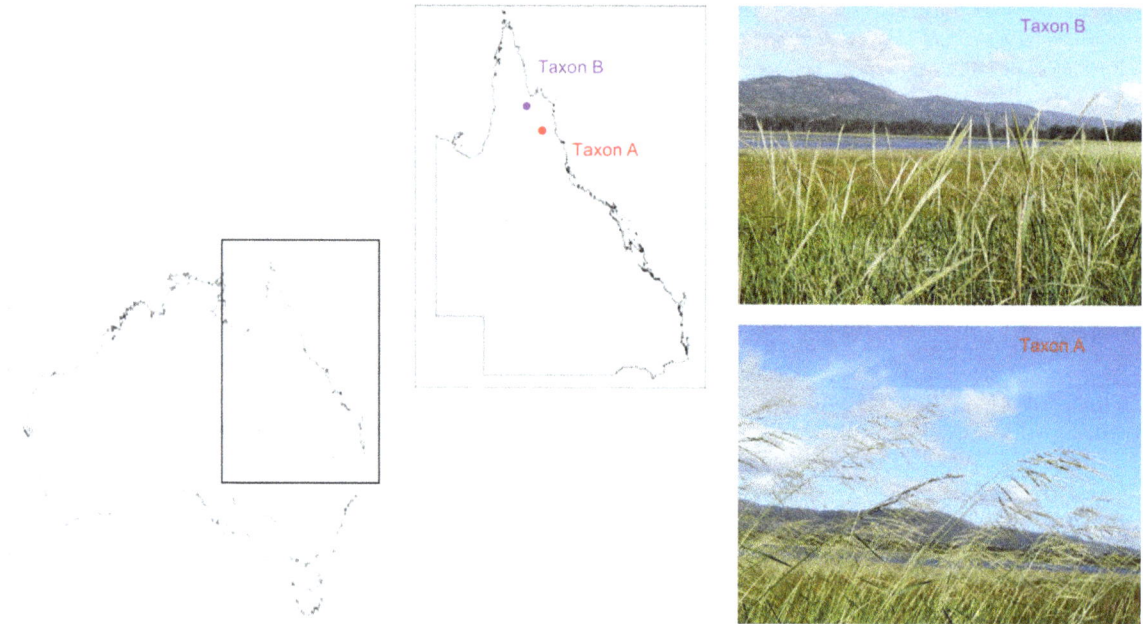

Figure 1 Australian perennial A genome taxa from northern Australia. Taxon A is characterized by open panicles, while Taxon B has closed panicles.

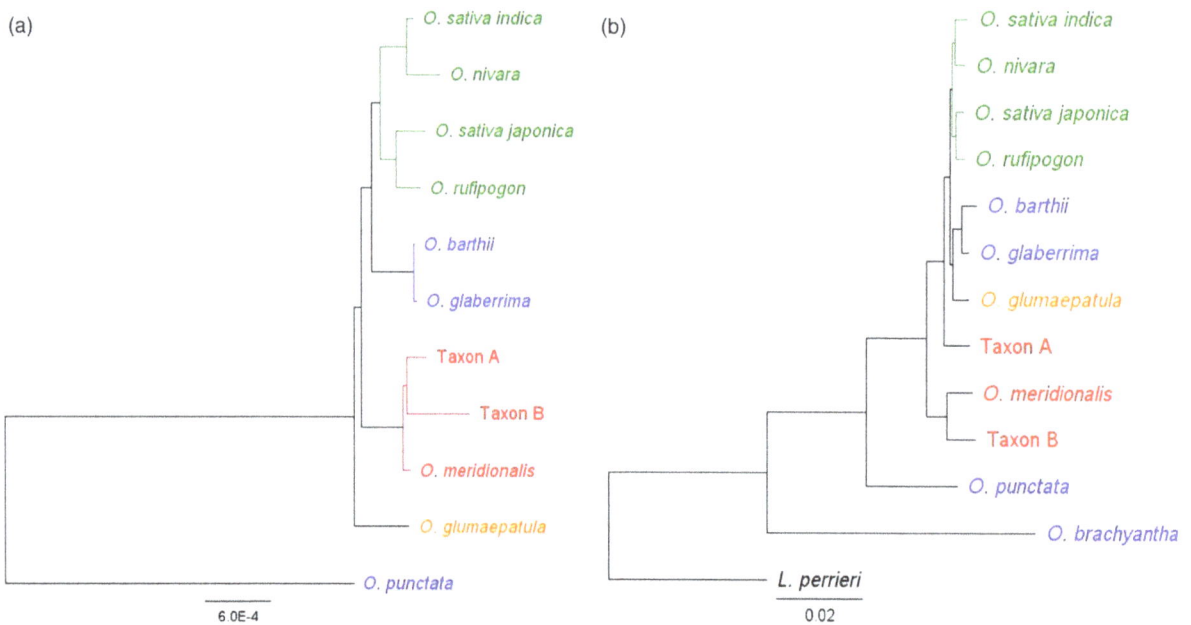

Figure 2 Phylogenetic relationships between A genome rice species; (a) tree topology based upon analysis of supermatrix of 4643 nuclear genes; (b) tree topology based on whole chloroplast genome sequences. Figure adapted and modified from Wambugu *et al.* (2015). Taxa marked in green represent Asian rice species, in blue: African, in orange: South American and in red: Australian. *L. perrieri* and *O. punctata* were used as outgroups in nuclear and chloroplast studies, respectively.

were 47.1 Gb and 41.4 Gb for Taxon A and Taxon B, respectively. The data generated on the PacBio instrument were long reads with an average length of 7693 bp and 8140 bp for Taxon A and Taxon B, respectively, with 14.8 Gb and 15.0 Gb of overall data for those taxa. The minimum and maximum read lengths for Taxon A were 50 bp and 49 742 bp, respectively, and for Taxon B: 50 bp and 50 242 bp.

The genome sizes were estimated *in silico* to be about 390 Mb and 370 Mb for Taxon A and Taxon B, in turn. These estimates were similar to other A genome rice species which fall between 341 and 413 Mb in size (Zhang *et al.*, 2014). Considering the estimated genome sizes, we also assessed the genome coverage of each of the data sets (Table S2). Furthermore, we used these estimations in evaluating completeness of genome assemblies.

Genome assemblies and evaluation

A total of 384.8 Mb (PacBio assembly) and 382.7 Mb (hybrid assembly) of the Taxon A, and 354.9 Mb (PacBio) and 446.4 Mb (hybrid) of the Taxon B genome sequences were assembled (Table 1). PacBio assemblies slightly outperformed hybrid assemblies in terms of standard assembly metrics, that is lower number of scaffolds, longest contig size, higher N50 and mean scaffold size. Both Taxon A and Taxon B assemblies exhibited high total lengths, as percentage of known genome sizes, with an unexpected high length of the Taxon B hybrid assembly that accounted for around 120% of estimated genome size (370 Mb). The high percentage of estimated genome size for Taxon B hybrid assembly may be due to the heterozygous and repetitive nature of this taxon's genome or to a nonprecise genome size estimation. Predicted heterozygous sites measured as the rate of variant branches caused by allelic differences in a de Bruijn graph, (1 in 400) and repeat content (1 in 300) rates were slightly higher for Taxon B compared to Taxon A (1 in 800 and 1 in 400, respectively; data not shown). Those traits might impact the assembly quality and completeness especially using short-read

data (Illumina) resulting in a more fragmented assembly with a higher number of repeated contigs.

The number of scaffolds obtained as a result of the assemblies was between 2585 and 3252 for Taxon A and Taxon B, respectively, for PacBio assemblies and between 3359 and 4718 for Taxon A and Taxon B, respectively, for hybrid assemblies. The number of scaffolds was slightly higher for hybrid assemblies for both taxa. Higher number of scaffolds in Taxon B assemblies might also be a result of the more heterozygous and repetitive nature of this taxon in comparison with Taxon A.

Both of the core gene presence evaluation methods, CEGMA and BUSCO, indicated PacBio assemblies to be more complete than hybrid assemblies for both taxa (Table S3 and Table S4). Normalized values ranged from 94.8% to 99.1% of completeness using CEGMA and from 89% to 98% using BUSCO. Neither wild rice assembly was found to have a higher number of mapped genes than the Nipponbare reference.

A high fraction of the Nipponbare reference genome was aligned to Taxon A assemblies, 70.7% and 71.5% to hybrid and PacBio, respectively (Table S5). However, a significantly lower percentage was aligned to Taxon B assemblies, 42.4% and 37.3% to hybrid and PacBio, respectively. These values were not high enough to use the Nipponbare genome as a reference for Taxon B in orienting and ordering contigs into chromosome pseudomolecules using Genome Puzzle Master [(GPM; Zhang *et al.* (2016)]. Based on the previous finding that Taxon B shares numerous molecular markers with *O. meridionalis* (Sotowa *et al.*, 2013) and that they descended from a common ancestor (this study), this genome was evaluated as well. QUAST results showed higher *O. meridionalis* genome fraction aligning to Taxon B assemblies than Nipponbare genome, 62.4% and 56.3% to hybrid and PacBio, respectively. These values were satisfactory allowing the use of *O. meridionalis* instead of Nipponbare sequences as a guide in GPM for Taxon B assembly.

Table 1 Taxon A and Taxon B hybrid and PacBio assembly statistics. The metrics were calculated for scaffolds and contigs for hybrid assembly and for scaffolds only for PacBio assembly

	Taxon A		Taxon B	
Assembly	Hybrid	PacBio-only	Hybrid	PacBio-only
Assembler	Sparse Assembler + DBG2OLC	Celera Assembler	Sparse Assembler + DBG2OLC	Celera Assembler
Scaffolds				
Number of scaffolds	3359	2585	4718	3252
Total size of scaffolds	382 655 312	384 759 810	446 369 637	354 906 376
Total scaffold length as percentage of known genome size	98.1	98.7	120.6	95.9
Longest scaffold	1 305 248	1 692 155	2 079 733	3 232 522
Shortest scaffold	2297	9523	2425	12 563
Mean scaffold size	113 919	148 843	94 610	109 135
Median scaffold size	61 996	97 803	54 787	61 207
N50 scaffold length	217 336	219 409	163 003	159 640
Contigs				
Number of contigs	3425	–	4808	–
Total size of contigs	382 644 322	–	446 351 110	–
Longest contig	1 158 569	–	1 449 836	–
Shortest contig	1139	–	790	–
Mean contig size	111 721	–	92 835	–
Median contig size	61 459	–	54 495	–
N50 contig length	211 599	–	159 759	–

Rice pseudomolecules

After the preliminary evaluation of the assemblies (basic assembly statistics, core gene presence and alignment to the reference genomes), the PacBio-only scaffolds were chosen for further analysis and investigation. Ordering and orientation of the contigs with GPM resulted in 12 pseudochromosomes for both taxa and 386 unordered contigs for Taxon A and 1080 for Taxon B (Table S6). 94.9% and 83.1% of the assemblies' length were anchored and oriented to chromosomes for Taxon A and Taxon B, respectively. Alignment of wild rice pseudomolecules to their reference genomes revealed better coverage and less ambiguity between Taxon A and *O. sativa japonica* than between Taxon B and *O. meridionalis* genome (Figure S1).

Genome annotation

Repetitive elements, RNAs and protein coding genes were annotated in Taxon A and Taxon B draft genomes. The sequences subjected to the annotation were the 12 pseudomolecules and remaining unordered contigs for each of the taxa.

Total repeats found in Australian wild rices made up 36.5% and 46.4% of the Taxon A and Taxon B genomes, respectively (Table S7 and Table S8). The most abundant class of transposable elements found were retrotransposons from the Gypsy superfamily. These represented 57.8% and 39.7% of all repeats described in Taxon A and Taxon B, respectively, followed by the Copia superfamily in Taxon A (8.3%) and Mutator in Taxon B (9.3%). The classes and fractions of other repetitive elements were similar in both taxa; however, the numbers and lengths were significantly higher in the Taxon B genome.

Noncoding RNAs annotated in the wild rice genomes included tRNA, miRNA, snoRNA, sRNA, rRNA and other (Table S9). Overall, RNAs consisted of approximately 0.25% of both genomes which corresponded to length of 960 496 bp in Taxon A and 883 779 bp in Taxon B. 675 and 558 tRNAs models were predicted by tRNAscan in Taxon A and Taxon B, respectively, whereas 629 and 581 tRNAs models were predicted by Infernal, respectively. Additional models predicted by Infernal but not by tRNAscan were added to the final annotations resulting in 677 and 615 tRNAs for Taxon A and Taxon B, respectively, of the combined length of 50 687 and 46 301 bp.

The number of gene models annotated in wild rice genomes is listed in Table S10. Slightly more genes were found in the Taxon A genome which was probably associated with the longer total assembly of this taxon (384.8 Mb as opposed to 354.9 Mb of Taxon B). In comparison with other wild rice species (I-OMAP, unpublished), these taxa showed a considerably lower number of genes. In the previous study, the lowest number of annotated loci was found in *O. brachyantha* (24 208), which also carries the smallest genome described so far in the genus *Oryza* (261 Mb). Similar numbers of InterPro protein domains, KEGG pathways or GO terms were found in the two genomes (Table S10). Overall, just over 61% of annotated models had matches in the InterPro database, about 9% in KEGG and around 40% in GO.

Phylogenetic analysis

Sequences of 4643 genes were extracted from nuclear genome sequences generated by whole genome sequencing (I-OMAP, unpublished; Table S11). The alignment of these 4643 gene sequences had a total length of 6 272 851 bp. The sequence similarity between the *Oryza* species was very high (Table 2) ranging from 86.6% (between *O. meridionalis and O. brachyantha*) to up to 98.4% (between *O. rufipogon* and *O. sativa* ssp. *japonica*). Of the bases that were subjected to the maximum parsimony (MP) analysis, 5 229 706 were constant, 741 498 were variable and parsimony uninformative, and 301 647 were parsimony informative. Both phylogenetic inference methods used in this study, maximum parsimony (MP) and Bayesian inference (BI), recovered the same optimal tree topology (Figure 2b) with the following values for the MP tree: length = 1 264 556 steps, consistency index CI = 0.90, retention index RI = 0.74, CI excluding uninformative characters = 0.72. The nodes on this topology were all strongly supported with MP bootstrap values of 100% and the posterior probabilities of all nodes in the BI equal to 1.

Our results showed that *O. meridionalis* and Taxon B in Australia are sister to all other A genome species including the Australian Taxon A (Figure 2b). Taxon A is in turn a sister to the clade that includes the Asian and African domesticated species. African domesticated rice, *O. glaberrima*, and its wild progenitor, *O. barthii*, together with *O. glumaepatula* from South America are a clade distinct from the Asian species (Wambugu *et al.*,

Table 2 Sequence similarities between rice taxa in the supermatrix used for phylogenetic inference

	Oryza species	1	2	3	4	5	6	7	8	9	10	11	12
1	*O. sativa indica*		97.7	97.6	97.5	96.7	97.2	97.2	96	94.8	94.7	93	87.1
2	*O. nivara*			98.1	98.2	97.4	97.9	98	96.6	95.3	95.2	93.5	87.6
3	*O. sativa japonica*				98.4	97.3	98	97.9	96.8	95.4	95.4	93.6	87.7
4	*O. rufipogon*					97.3	97.8	97.8	96.5	95.3	95.2	93.5	87.6
5	*O. barthii*						97.9	97.3	95.8	94.8	94.6	93	87.3
6	*O. glaberrima*							97.8	96.4	95.3	95.2	93.5	87.7
7	*O. glumaepatula*								96.4	95.4	95.2	93.5	87.7
8	Taxon A									94.4	95.2	92.7	87.1
9	*O. meridionalis*										95.1	92.5	86.6
10	Taxon B											92.4	86.7
11	*O. punctata*												87.6
12	*O. brachyantha*												

The panel represents percentage of bases that are identical in the supermatrix alignment between corresponding species. The heat map shows individual values in a matrix as colours: red cells indicate high similarities; orange, yellow and green cells show gradually lower similarities.

2013). *Indica* and *japonica* rice are represented by two well-resolved clades. *Japonica* and *O. rufipogon* show a close relationship which is consistent with the long-accepted view that *O. rufipogon* is the progenitor of *japonica* (Wei *et al.*, 2012) while *indica* rice was found in a clade with the Asian annual *O. nivara*. Recent SNP analysis of genomic regions under selection suggests the independent domestication of *indica* rice from wild rice in an area from southern Indochina to the Brahmaputra valley (Civáň *et al.*, 2015).

The relationship of the Australian A genome populations was greatly clarified by this study. *O. meridionalis* and the morphologically similar perennial populations (Taxon B) are sister to all other A genome species. The Australian wild populations with morphology similar to *O. rufipogon* (Taxon A) were found to be sister to the clades including all other A genome species. The Australian (Taxon A) population has a large anther like that of *O. rufipogon* in Asia but is morphologically distinct from the other Australian A genome species and domesticated rice with small anthers. Taxon A and Taxon B can be most readily distinguished in the field by the open panicles of Taxon A and closed panicles of Taxon B (Figure 1). Taxon B generally has longer awns, but the ranges of awn length for the two taxa overlap. The presence of these diverse A genome taxa makes northern Australia a key centre of diversity for rice and indicates the need for more collections from this poorly explored area and the need to ensure in situ conservation of these resources.

The nuclear phylogenies presented here showed a different relationship to those deduced from the chloroplast genomes suggesting different evolutionary histories for the maternally inherited plastid genetic material and the nuclear genome (Figure 2). This difference in evolutionary path for these genomes is commonly observed in recently diverged plant taxa (Tsitrone *et al.*, 2003).

Analysis of the timing of the evolutionary events in this study (Figure 3) agrees well with that reported in analysis of the *Oryza* genomes (I-OMAP, unpublished) despite the use of a different method of analysis. The average rate of evolution was estimated to be 3.53E-3 ± 1.85E-6 (Table S12) and the root age (divergence between *O. brachyantha* and *O. punctata*) to be 14.98 ± 0.97 mya (Figure 3). Given this root age, the A genome group diverged in the last 3 million years and the divergence of the *japonica* and *indica* clades dated at about 990 000 years ago (Figure 3). The chloroplast genomes appear to have diverged more recently (Wambugu *et al.*, 2015) possibly

due to some degree to the sharing of maternal genomes across this group.

Analysis of the phylogenetic relationship for each chromosome separately revealed some discordant results for chromosomes 5, 7, 10 and 11 (Figure S2, Table S13) that is likely due to introgression rather than incomplete lineage sorting. To test for potential recombination events between Australian Taxon A and other *Oryza* species, we performed a four-taxon test, also known as the D-statistic (Durand *et al.*, 2011; Green *et al.*, 2010) separately for each chromosome. This test screens the aligned data for two biallelic mutation patterns: ABBA and BABA. The first species set we used was ((*O. rufipogon*, *O. barthii*) Taxon A, *O. punctata*) where *O. punctata* was used as the outgroup. In this analysis, a negative D-statistic value would suggest introgression between Taxon A and the Asian species, whereas a positive value would mean an introgression between Taxon A and the African species (*O. barthii*). In this set, all but two chromosomes showed negative D-statistic values, indicating that introgression occurred between Australian Taxon A and Asian *O. rufipogon* (Figure 4, Table S14). Eight of these statistics were significant (chromosomes 1, 3, 5, 6, 8, 9, 11 and 12). Among chromosomes with significant results was chromosome 5 for which Taxon A was found to be closer on the phylogenetic tree to the Asian clade (Figure S2) than it was on the consensus tree (Figure 2b). The D-statistic calculated for chromosome 7 was positive and statistically significant, thereby providing evidence for introgression between Taxon A and *O. barthii*.

To investigate the relationship between Australian and Asian species, we used the set ((*O. rufipogon*, *O. nivara*), Taxon A, *O. punctata*). For this set, a negative D-statistic value would suggest introgression between Taxon A and the *japonica/O. rufipogon* clade, whereas a positive value would suggest introgression between Taxon A and the *indica/O. nivara* clade. Four chromosomes (1, 2, 4 and 12) returned significant positive values, whereas no negative value was meaningful. Introgression between the Asian and Australian populations was suggested with evidence for greater introgression between Taxon A and the *indica/O. nivara* clade than between Taxon A and the *japonica/O. rufipogon* clade.

Discussion

The discovery of two novel Australian wild rice taxa expands the understanding of the genetic diversity within the genus *Oryza*.

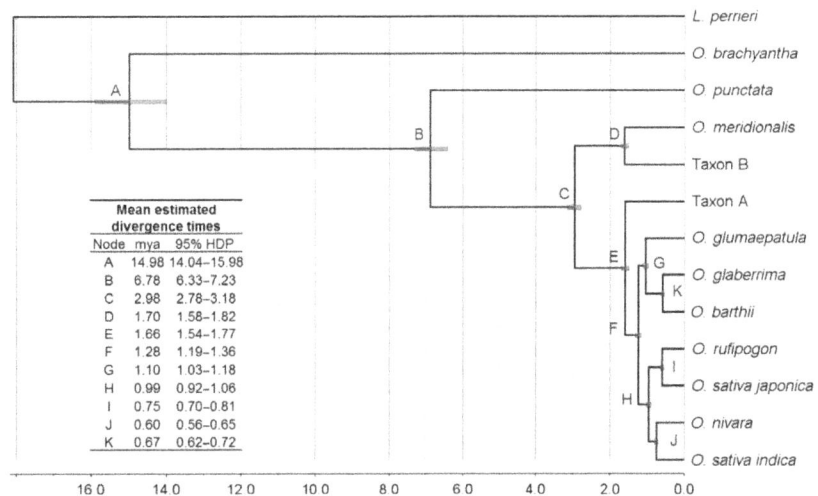

Figure 3 Molecular clock analyses for A genome rice evolution. The most frequent tree topology retrieved in analyses of alignments of separate chromosomes, inferred for 8 out of 12 chromosomes. Scale axis represents age in million years (mya). Node bars display 95% highest posterior density (HDP) interval.

Mean estimated divergence times		
Node	mya	95% HDP
A	14.98	14.04–15.98
B	6.78	6.33–7.23
C	2.98	2.78–3.18
D	1.70	1.58–1.82
E	1.66	1.54–1.77
F	1.28	1.19–1.36
G	1.10	1.03–1.18
H	0.99	0.92–1.06
I	0.75	0.70–0.81
J	0.60	0.56–0.65
K	0.67	0.62–0.72

(a)

Chromosome	1	2	3	4	5	6	7	8	9	10	11	12
(((*O. rufipogon*, *O. barthii*), Taxon A), *O. punctata*)												
D-statistic	−0.167	−0.08	−0.42	−0.06	−0.33	−0.25	0.29	−0.67	−0.33	0.05	−0.37	−0.54
Z-score	−1.96*	−0.73	−3.74*	−0.65	−3.43*	−1.98*	2.56*	−10.03*	−2.86*	0.10	−2.51*	−3.66*
(((*O. rufipogon*, *O. nivara*), Taxon A), *O. punctata*)												
D-statistic	0.26	0.25	−0.12	0.37	−0.21	0.09	0.17	0.12	−0.09	−0.20	−0.13	0.24
Z-score	2.59*	2.38*	−1.00	2.86*	−1.79	0.57	1.15	0.67	−0.75	−0.50	−1.51	2.74*

(b)

D < 0 * chromosomes: 1, 3, 5, 6, 8, 9, 11, 12

(c)

D > 0 * chromosomes: 1, 2, 4, 12

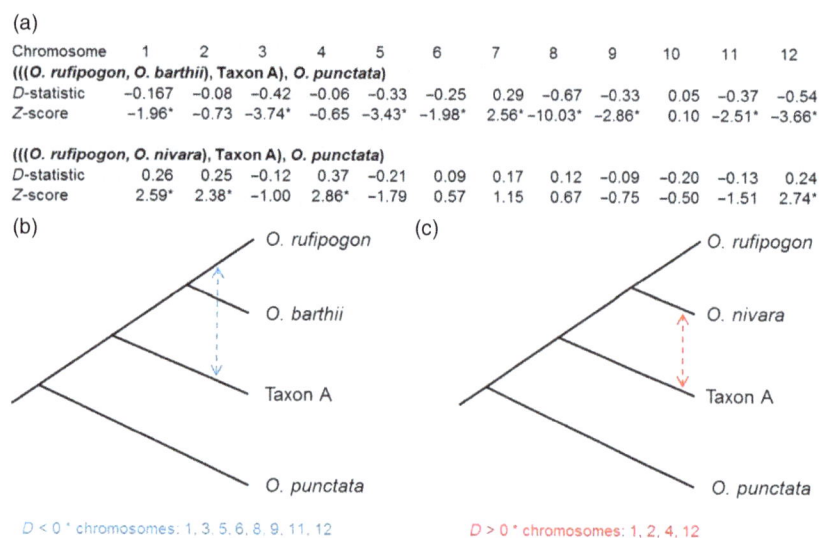

Figure 4 Results of four-taxon test for Taxon A and selected *Oryza* species; (a) D-statistics and Z-scores calculated for two sets of selected species per chromosome: (((*O. rufipogon*, *O. barthii*), Taxon A), *O. punctata*) and (((*O. rufipogon*, *O. nivara*), Taxon A), *O. punctata*). Z-scores marked with asterisk (*) indicate statistically significant values; (b) four-taxon tree used in the first test. Bidirectional arrow shows inferred introgression for chromosomes with significant D-statistics (listed below the tree); (c) four-taxon tree used in the second test. Bidirectional arrows show inferred introgression for chromosomes with significant D-statistics (listed below the tree).

The two draft genomes generated in this study produce an excellent platform for exploring the potential of Australian wild rices. These genomic resources will complement the already extensive study of numerous wild and cultivated species within *Oryza* (I-OMAP, unpublished) providing unique data for comparative genomics, evolutionary studies of the entire genus and widening the species pool for rice improvement.

The sequencing and assembly performed in this study show the particular challenges of generating genome assemblies for wild heterozygous plants. It also shows that the replacement of short-read data (Illumina) with long-read data (PacBio) can improve the overall completeness of a draft genome sequence. The results suggest that the best representation of the Taxon A and Taxon B genomes was obtained using the PacBio-only data. As a consequence, the PacBio-only assemblies for both taxa have been selected for further analysis. PacBio-only assembly improved almost all metrics, such as number of scaffolds, total size, the longest and the shortest scaffolds, mean and median lengths as well as completeness measured as presence or absence of orthologous genes.

The evolution of rice has been the subject of ongoing debate in particular regarding whether there was a single or multiple domestication in Asia (Bouchenak-Khelladi *et al.*, 2010; Kellogg, 2009; Vaughan *et al.*, 2008). The distinctness of the *indica* and *japonica* genomes suggests separate origins for most of each genome (Wei *et al.*, 2012). However, the presence of many shared domestication-related alleles has led to suggestions that some level of introgression between the two genomes has also been a feature of their domestication history (Civáň *et al.*, 2015; Fuller *et al.*, 2010; Huang *et al.*, 2012a,b; Molina *et al.*, 2011). Geographic separation may have allowed early populations to diverge resulting in distinct *O. rufipogon*-like populations in Asia and Australia. Chloroplast transfer or capture is common between recently diverged plant taxa which may explain the distinct Asian and Australian chloroplast genomes in the wild populations descended from the taxa that were domesticated in Asia.

Wild rice populations are a key genetic resource for rice improvement (Anacleto *et al.*, 2015). The Australian populations may provide an especially useful resource for evaluation of rice domestication due to their isolation from significant impact of gene flow from domesticated rice. An improved understanding of the geographic variation in A genome wild rice species provided by genome sequencing should guide the search for useful alleles in wild populations (Krishnan *et al.*, 2014). The taxonomy of the A genome wild rice species in Australia and Asia needs to be re-evaluated in the light of the molecular data now available to determine whether distinct wild taxa need to be recognized. Movement of flora between Sahul (New Guinea and Australia) and Sunda (Malay Peninsula, Sumatra, Borneo, Java) in both directions may be an important part of the evolutionary history of the A genome *Oryza* species (Crayn *et al.*, 2015; Prasad *et al.*, 2011; Tang *et al.*, 2010). The current diversity of A genome *Oryza* in northern Australia suggests the possibility of an Australian and/or South-East Asian origin for the A genome clade, but further historical biogeographical analyses based on more extensive data sets are required to evaluate this hypothesis.

Experimental procedures

Plant material

The wild rice plants used in the study came from perennial wild rice populations in North Queensland, Australia. The first individual, referred to here as Taxon A, was collected from Abattoir Swamp Environmental Park near Julatten and was described by Sotowa *et al.* (2013) as *Oryza rufipogon*-like taxon (r-type) collected from Jpn1 site. The second individual, referred to here as Taxon B, was collected from a small wetland beside the Peninsula Developmental Road and in Sotowa *et al.* (2013) was called *Oryza meridionalis*-like (m-type) taxon collected from Jpn2 site. Specimens of these wild rice populations were collected from their natural habitats and are now kept and maintained in glasshouse conditions at The University of Queensland in Brisbane, Australia.

DNA extraction and sequencing

DNA from leaf tissue of Taxon A and Taxon B individuals was extracted using a modification of the CTAB method (Furtado, 2014) and subsequently subjected to whole genome shotgun sequencing. Next-generation sequencing platforms used were Illumina HiSeq2000 (Illumina, San Diego, CA) and Pacific Biosciences RSII with P6-C4 chemistry (PacBio, Menlo Park, CA). The data generated on Illumina instrument were 101-bp reads with an average library insert of 550 bp (paired end reads, PE),

3000 bp (mate pair reads, 3 Kb MP) and 5000 bp (mate pair reads, 5 Kb MP). Samples for paired end sequencing were generated using TruSeq DNA PCR-free library preparation kit, whereas mate pair libraries were prepared using the Nextera Mate Pair protocol. Illumina sequencing was performed by Macrogen (Seoul, Korea) and PacBio sequencing by The University of Queensland Diamantina Institute (Brisbane, Australia). The SMRTbell template libraries were prepared following the standard protocol for long-insert libraries according to the manufacturer's instructions (PacBio) with an insert size of 20 kbp. 20 SMRT Cells per taxon were sequenced resulting in approximately 40-fold genome coverage for each sample.

Data processing and genome assembly

Raw reads from both platforms were assessed using FastQC (www.bioinformatics.babraham.ac.uk/projects/fastqc), a tool for evaluating the quality of sequencing reads in FASTQ files. Illumina reads were additionally used to estimate the genome size of the taxa. A preqc module (Simpson, 2014) from the SGA de novo genome assembler package (Simpson and Durbin, 2012) was used for this estimation. This utility also enabled an estimation of heterozygosity and repeat content in the genome.

Taxon A and Taxon B genomes were assembled *de novo* using two strategies. The first one utilized Illumina and PacBio sequencing reads together (hybrid assembly) and the second one – PacBio data only (PacBio-only assembly). The software used to accomplish the hybrid assembly was DBG2OLC package (Ye *et al.*, 2016). This assembly included Illumina PE raw reads and raw PacBio reads. The first step within the analysis involved two rounds of Illumina read error correction and subsequent assembly with SparseAssembler [beta version; Ye *et al.* (2012)]. The coverage threshold for both an error and for a correct sparse *k*-mer candidate in the correction process was set to 5. In the first round, the *k*-mer length was set to 15 and reads were trimmed at the ends, whereas in the second one, the *k*-mer length used was 31 and the trimming option was disabled. *K*-mer size in the assembly step was 31. The next step was 'overlap and layout' with the output contigs from the first step and PacBio reads with the following parameters: *k*-mer = 17, k-mer coverage threshold = 2, adaptive threshold = 0.001, minimum overlap = 20 and removal of chimeric reads in the data set. The last step called the consensus contigs from output files from two previous steps and raw PacBio reads. In PacBio-only assembly, first, the raw reads from the PacBio platform were corrected using the PBcR pipeline (Berlin *et al.*, 2015) with the self-correction feature enabled and the minimum length of PacBio fragment to keep set to 500. The assembly was done using The Celera Assembler (CA) version 8.3rc2 (Myers *et al.*, 2000) leaving the parameters as default. The primary contigs were filtered in order to keep only the unique contigs.

Evaluation of genome assemblies

First, each of the assemblies was assessed using Assembly Stats (assemblathon tool) from the Assemblathon project (Earl *et al.*, 2011) accessed through the iPlant Collaborative platform [iPlant; Goff *et al.* (2011)]. The expected genome sizes were set based on the estimations from this study. Second, core gene presence was assessed in the assemblies. This was done using both CEGMA [Core Eukaryotic Genes Mapping Approach; Parra *et al.* (2007)] and BUSCO [Benchmarking Universal Single-Copy Orthologs; Simao *et al.* (2015)], keeping the default cut-offs for genes. CEGMA uses a set of 248 CEGs (core eukaryotic genes) which are

very highly conserved in eukaryotes and are present in low copy number (Parra *et al.*, 2009). BUSCO also uses a set of universal single-copy orthologs and provides a set to evaluate plant genomes in particular with a set of 956 plant orthologs. Additionally, both evaluation tools were run for the rice reference sequence of *O. sativa japonica* and the values for completeness of this high quality rice genome were used for normalization.

The assemblies were also aligned to rice reference genomes using QUAST [Quality Assessment Tool for Genome Assemblies; Gurevich *et al.* (2013)] with the default parameters and the minimum alignment length of 1000 bp. The genome of *Oryza sativa* ssp. *japonica* var. Nipponbare [IRGSP_MSU.v7; Kawahara *et al.* (2013)] and *O. meridionalis* (GenBank assembly accession: GCA_000338895.2) was used as the reference sequences in this study.

Rice pseudomolecules

Assembled PacBio contigs were assigned to chromosome pseudomolecules using Genome Puzzle Master [GPM; Zhang *et al.* (2016)]. Rice reference genome sequences were used to guide the process: *Oryza sativa* ssp. *japonica* var. Nipponbare genome was used for Taxon A and *O. meridionalis* was used for Taxon B. The twelve pseudomolecules and unanchored contigs were then annotated for genes and other features.

Genome annotation

Protein coding genes were annotated using the MAKER-P v.2.3 annotation pipeline (Campbell *et al.*, 2014). Within the pipeline, the repeat elements were masked using the RepeatMasker (www.repeatmasker.org, v. 3.3.0). Due to the lack of expression data for these specific taxa, expression evidence included available expressed tags (ESTs) and full-length cDNA from other *Oryza* taxa. These CDS and their corresponding protein sequences (used as the protein homology evidence) consisted of annotated genes models of *O. sativa* ssp. *japonica* var. Nipponbare RefSeq, *O. glaberrima* (Wang *et al.*, 2014) and *Brachypodium distachyon* (The International Brachypodium Initiative 2010). ESTs comprised of *O. sativa* ssp. *japonica* var. Nipponbare (for Taxon A annotation) and *O. meridionalis* (for Taxon B annotation) transcripts generated by The International *Oryza* Map Alignment Project (I-OMAP, unpublished) and clustered at 95% similarity. The *ab initio* gene predictors run within MAKER-P were SNAP (Korf, 2004) using O. sativa.hmm parameter and AUGUSTUS 3.1 (Stanke and Waack, 2003) with rice as the gene prediction species model. Resulting gene models were filtered removing noncomplete models, that is without valid start and/or stop codons and with internal terminator codons, followed by removing transposable elements (TE) based on the specific rice- and Australian taxa-related libraries used at the genome masking step. Removed TE consisted of hits above the e-value threshold of 1e-5, with more than 40% of query coverage and longer than 100 nt. The gene models predicted by MAKER-P were functionally analysed using InterProScan version 5.16.55 (Jones *et al.*, 2014) including annotation with Gene Ontology (GO) and biological pathway information. The InterProScan results were further parsed for additional functional evidence (GO terms and KEGG pathway) using interproscanParser script available at iPlant.

The repeat annotation was obtained by merging the output of RepeatMasker and Blaster, a component of the REPET package (Flutre *et al.*, 2011), using nucleotide libraries (PReDa and RepeatExplorer) from RiTE-db (Copetti *et al.*, 2015) and an in-house curated collection of transposable element (TE) proteins. Additionally, for each of the two species, a custom repeat library was developed with RepeatExplorer and curated as described previously (Copetti *et al.*, 2015) using short-read data (Illumina sequencing reads). Infernal (Nawrocki and Eddy, 2013) was adopted to identify noncoding RNAs (ncRNAs) using the Rfam library Rfam.cm.1_1. Hits above the e-value threshold of 1e-5 were filtered, as well as results with scores lower than the family-specific gathering threshold. When loci on both strands were predicted, only the hit with the highest score was kept. Transfer RNAs were also predicted using tRNAscan-SE v. 1.23 (Schattner *et al.*, 2005) with default parameters.

Phylogenetic analysis

Phylogenetic analysis was undertaken using data from twelve fully sequenced diploid *Oryza* genomes including two taxa investigated in this study (Taxon A and Taxon B), and ten other species downloaded from GenBank (Table S11). Eleven of these species were A genome-type rice relatives. Moreover, we used *O. punctata*, which belongs to BB genome group, and *O. brachyantha*, which has FF genome type. *Leersia perrieri* was used as the outgroup species.

From these diploid genomes, a set of putatively single-copy orthologs was selected by blasting (BLASTn) the initial collection of 6,015 genes (representing the set of genes that could be identified in all *Oryza* taxa) used in a previous study (I-OMAP, unpublished) against the Australian *Oryza* genomes (Taxon A and Taxon B) applying the following thresholds: e-value of 1e-5, 40% of query coverage and 100 nt of hit length. The initial set of clusters of single-copy orthologous loci was identified by BLAST-Overlap-Synteny (BOS) filtering using a protocol described by Zwickl *et al.* (2014). A final subset of 4,643 genes sequences present in all *Oryza* assemblies was extracted from the genomes and used for further investigation. Nucleotide sequences of the 4,643 genes selected from each genome were separately aligned using CLUSTALW multiple sequence alignment program (Thompson *et al.*, 1994) with default parameters. Then, single gene alignments were concatenated to create a supermatrix of 6 272 851 base pairs, which was used in the following phylogenetic inference.

Phylogenetic tree reconstruction was conducted using maximum parsimony (MP) and Bayesian inference (BI) methods. MP was performed using PAUP* 4.0 software (Swofford, 2003). The following tree search settings were enabled in MP reconstruction: heuristic search with tree bisection–reconnection branch swapping and 200 random addition sequence replications. The group support was assessed using 2000 bootstrap pseudoreplications. Alignment gaps were treated as missing data. All characters were treated as unordered and weighted equally.

For the model-based approach (BI), jModelTest2 software (Darriba *et al.*, 2012) was used to determine the model of nucleotide substitution that best fits the data based on the Akaike information criterion. The Bayesian analyses used the general time reversible model with gamma-shaped among-site rate variation with an estimated proportion of invariable sites (GTR+I+G; p-inv = 0.3730, four gamma categories and gamma shape = 0.8890). The BI analysis was performed using MrBayes version 3.2 (Ronquist *et al.*, 2012). The branch length prior was set to exponential with parameter 10.0. Two independent and simultaneous analyses starting from distinct random trees were performed. Three heated (heating coefficient = 0.2) and one cold Monte Carlo Markov chains (MCMC) were run for 1×10^6

generations, with a tree sampled every 200 generations. The first 10% of trees were discarded as burn-in and a 50% majority rule consensus tree was constructed and rooted using the outgroup method.

Divergence time estimates

Divergence times were estimated using the Bayesian evolutionary method implemented in the software package BEAST 2 version 2.3.1 (Bouckaert *et al.*, 2014). A secondary clock calibration was used based on the estimated divergence time for the *Oryza* crown group (*O. brachyantha*–*O. punctata*) of 15 ± 0.5 mya (I-OMAP, unpublished). The 4643 genes used were divided according to the chromosome they were found on and aligned using CLUSTALW. The number of genes and the alignment length is shown in Table S15. The best fit evolutionary model, determined by jModelTest2, was the general time reversible model (GTR+I+G) for each of the alignments. Evolutionary rates were modelled under a strict molecular clock and speciation was modelled employing the Yule model. Posterior probabilities were estimated using MCMC algorithm with chain length 5 000 000 and a tree sampled every 1000th generation. The first 10% of sampled trees was discarded as burn-in. The output from BEAST 2 was analysed in Tracer version 1.63 (www.beast.bio.ed.ac.uk/Tracer). The best supported tree with the highest product of the posterior probability of all its nodes (maximum clade credibility tree) and the mean heights of each node was summarized using TreeAnnotator distributed in the BEAST 2 package. The final tree estimates were visualized in FigTree version 1.4.2 (www.tree.bio.ed.ac.uk/softwa re/figtree).

Genetic introgression

To test for genetic introgression between Australian and other *Oryza* species, we used the software package HYBRIDCHECK (Ward and van Oosterhout, 2016). We performed two tests and in each run four aligned sequences were analysed. In the first one, we tested for introgression between Taxon A and either *O. rufipogon* or *O. barthii,* and in the second, we tested for introgression between Taxon A and either *O. rufipogon* or *O. nivara.* Block jackknife was used to calculate the statistic with the block size of 20 000 as well as *Z*-score to measure the statistical significance.

Acknowledgements

This work was supported by the Australian Research Council and the Rural Industries Research and Development Corporation. We thank Mike Sanderson for advice on phylogenetic analysis. The authors declare no conflict of interest.

References

Abberton, M., Batley, J., Bentley, A., Bryant, J., Cai, H., Cockram, J., Costa de Oliveira, A. *et al.* (2016) Global agricultural intensification during climate change: a role for genomics. *Plant Biotechnol. J.* **14**, 1095–1098.

Ammiraju, J.S.S., Fan, C., Yu, Y., Song, X., Cranston, K.A., Pontaroli, A.C., Lu, F. *et al.* (2010) Spatio-temporal patterns of genome evolution in allotetraploid species of the genus Oryza. *Plant J.* **63**, 430–442.

Anacleto, R., Cuevas, R.P., Jimenez, R., Llorente, C., Nissila, E., Henry, R. and Sreenivasulu, N. (2015) Prospects of breeding high-quality rice using post-genomic tools. *Theor. Appl. Genet.* **128**, 1449–1466.

Berlin, K., Koren, S., Chin, C.S., Drake, J.P., Landolin, J.M. and Phillippy, A.M. (2015) Assembling large genomes with single-molecule sequencing and locality-sensitive hashing. *Nat. Biotechnol.* **33**, 623–630.

Bouchenak-Khelladi, Y., Verboom, G.A., Savolainen, V. and Hodkinson, T.R. (2010) Biogeography of the grasses (Poaceae): a phylogenetic approach to reveal evolutionary history in geographical space and geological time. *Bot. J. Linn. Soc.* **162**, 543–557.

Bouckaert, R., Heled, J., Kühnert, D., Vaughan, T., Wu, C.H., Xie, D., Suchard, M.A. *et al.* (2014) BEAST 2: a software platform for Bayesian evolutionary analysis. *PLoS. Computat. Biol.* **10**, e1003537.

Brozynska, M., Omar, E.S., Furtado, A., Crayn, D., Simon, B., Ishikawa, R. and Henry, R. (2014) Chloroplast genome of novel rice Germplasm identified in Northern Australia. *Tropical Plant Biol.* **7**, 111–120.

Brozynska, M., Furtado, A. and Henry, R.J. (2016) Genomics of crop wild relatives: expanding the gene pool for crop improvement. *Plant Biotechnol. J.* **14**, 1070–1085.

Campbell, M.S., Law, M., Holt, C., Stein, J.C., Moghe, G.D., Hufnagel, D.E., Lei, J. *et al.* (2014) MAKER-P: a tool kit for the rapid creation, management, and quality control of plant genome annotations. *Plant Physiol.* **164**, 513–524.

Civáň, P., Craig, H., Cox, C.J. and Brown, T.A. (2015) Three geographically separate domestications of Asian rice. *Nature Plants*, **1**, 15164.

Copetti, D., Zhang, J., El Baidouri, M., Gao, D., Wang, J., Barghini, E., Cossu, R.M. *et al.* (2015) RiTE database: a resource database for genus-wide rice genomics and evolutionary biology. *BMC Genom.*, **16**, 1–10.

Crayn, D.M., Costion, C., Harrington, M.G. and Richardson, J. (2015) The Sahul-Sunda floristic exchange: dated molecular phylogenies document Cenozoic intercontinental dispersal dynamics. *J. Biogeograp.* **42**, 11–24.

Darriba, D., Taboada, G.L., Doallo, R. and Posada, D. (2012) jModelTest 2: more models, new heuristics and parallel computing. *Nat. Methods*, **9**, 772.

Durand, E.Y., Patterson, N., Reich, D. and Slatkin, M. (2011) Testing for ancient admixture between closely related populations. *Mol. Biol. Evol.* **28**, 2239–2252.

Earl, D., Bradnam, K., St John, J., Darling, A., Lin, D., Fass, J., Yu, H.O.K. *et al.* (2011) Assemblathon 1: a competitive assessment of de novo short read assembly methods. *Genome Res.* **21**, 2224–2241.

Flutre, T., Duprat, E., Feuillet, C. and Quesneville, H. (2011) Considering transposable element diversification in De Novo annotation approaches. *PLoS ONE*, **6**, e16526.

Fuller, D.Q., Sato, Y.I., Castillo, C., Qin, L., Weisskopf, A.R., Kingwell-Banham, E.J., Song, J. *et al.* (2010) Consilience of genetics and archaeobotany in the entangled history of rice. *Archaeol. Anthropol. Sci.* **2**, 115–131.

Furtado, A. (2014) DNA extraction from vegetative tissue for next-generation sequencing. In: *Cereal Genomics. Methods in Molecular Biology* (Henry, R.J. and Furtado, A. eds), p. 1. New York: Springer Science+Business Media.

Ge, S., Sang, T., Lu, B.R. and Hong, D.Y. (1999) Phylogeny of rice genomes with emphasis on origins of allotetraploid species. *P Nat. Acad. Sci. USA*, **96**, 14400–14405.

Goff, S.A., Vaughn, M., McKay, S., Lyons, E., Stapleton, A.E., Gessler, D., Matasci, N. *et al.* (2011) The iPlant collaborative: cyberinfrastructure for plant biology. *Front Plant Sci.* **2**, 34.

Green, R.E., Krause, J., Briggs, A.W., Maricic, T., Stenzel, U., Kircher, M., Patterson, N. *et al.* (2010) A draft sequence of the neandertal genome. *Science*, **328**, 710–722.

Gurevich, A., Saveliev, V., Vyahhi, N. and Tesler, G. (2013) QUAST: quality assessment tool for genome assemblies. *Bioinformatics*, **29**, 1072–1075.

Henry, R.J., Rice, N., Waters, D.L.E., Kasem, S., Ishikawa, R., Hao, Y., Dillon, S. *et al.* (2010) Australian oryza: utility and conservation. *Rice*, **3**, 235–241.

Huang, P.U., Molina, J., Flowers, J.M., Rubinstein, S., Jackson, S.A., Purugganan, M.D. and Schaal, B.A. (2012a) Phylogeography of Asian wild rice, Oryza rufipogon: a genome-wide view. *Mol. Ecol.* **21**, 4593–4604.

Huang, X., Kurata, N., Wei, X., Wang, Z.X., Wang, A., Zhao, Q., Zhao, Y. *et al.* (2012b) A map of rice genome variation reveals the origin of cultivated rice. *Nature*, **490**, 497–501.

Jones, P., Binns, D., Chang, H., Fraser, M., Li, W., McAnulla, C., McWilliam, H. *et al.* (2014) InterProScan 5: genome-scale protein function classification. *Bioinformatics*, **30**, 1236–1240.

Kawahara, Y., de la Bastide, M., Hamilton, J., Kanamori, H., McCombie, W.R., Ouyang, S., Schwartz, D. *et al.* (2013) Improvement of the Oryza sativa Nipponbare reference genome using next generation sequence and optical map data. *Rice*, **6**, 4.

Kellogg, E.A. (2009) The evolutionary history of ehrhartoideae, oryzeae, and oryza. *Rice*, **2**, 1–14.

Kharabian-Masouleh, A., Waters, D.L.E., Reinke, R.F., Ward, R. and Henry, R.J. (2012) SNP in starch biosynthesis genes associated with nutritional and functional properties of rice. *Sci. Rep.* **2**, 557.

Korf, I. (2004) Gene finding in novel genomes. *BMC Bioinform.* **5**, 59.

Krishnan, S.G., Daniel, L.E.W. and Henry, R.J. (2014) Australian wild rice reveals pre-domestication origin of polymorphism deserts in rice genome. *PLoS ONE*, **9**, e98843.

Lu, F., Ammiraju, J.S.S., Sanyal, A., Zhang, S., Song, R., Chen, J., Li, G. *et al.* (2009) Comparative sequence analysis of MONOCULM1-orthologous regions in 14 Oryza genomes. *P Nat. Acad. Sci. USA*, **106**, 2071–2076.

Molina, J., Sikora, M., Garud, N., Flowers, J.M., Rubinstein, S., Reynolds, A., Huang, P. *et al.* (2011) Molecular evidence for a single evolutionary origin of domesticated rice. *P Nat. Acad. Sci. USA*, **108**, 8351–8356.

Myers, E.W., Remington, K.A., Anson, E.L., Bolanos, R.A., Chou, H.H., Jordan, C.M., Halpern, A.L. *et al.* (2000) A whole-genome assembly of Drosophila. *Science*, **287**, 2196–2204.

Nawrocki, E.P. and Eddy, S.R. (2013) Infernal 1.1: 100-fold faster RNA homology searches. *Bioinformatics*, **29**, 2933–2935.

Parra, G., Bradnam, K. and Korf, I. (2007) CEGMA: a pipeline to accurately annotate core genes in eukaryotic genomes. *Bioinformatics*, **23**, 1061–1067.

Parra, G., Bradnam, K., Ning, Z., Keane, T. and Korf, I. (2009) Assessing the gene space in draft genomes. *Nucleic Acids Res.* **37**, 289–297.

Prasad, V., Strömberg, C.A.E., Leaché, A.D., Samant, B., Patnaik, R., Tang, L., Mohabey, D.M. *et al.* (2011) Late Cretaceous origin of the rice tribe provides evidence for early diversification in Poaceae. *Nature Commun.* **2**, 480.

Ronquist, F., Teslenko, M., van der Mark, P., Ayres, D.L., Darling, A., Hohna, S., Larget, B. *et al.* (2012) MrBayes 3.2: efficient Bayesian phylogenetic inference and model choice across a large model space. *Sys. Biol.* **61**, 539–542.

Schattner, P., Brooks, A.N. and Lowe, T.M. (2005) The tRNAscan-SE, snoscan and snoGPS web servers for the detection of tRNAs and snoRNAs. *Nucleic Acids Res.* **33**, W686–W689.

Simao, F.A., Waterhouse, R.M., Ioannidis, P., Kriventseva, E.V. and Zdobnov, E.M. (2015) BUSCO: assessing genome assembly and annotation completeness with single-copy orthologs. *Bioinformatics*, **31**, 3210–3212.

Simpson, J.T. (2014) Exploring genome characteristics and sequence quality without a reference. *Bioinformatics*, **30**, 1228–1235.

Simpson, J.T. and Durbin, R. (2012) Efficient de novo assembly of large genomes using compressed data structures. *Genome Res.* **22**, 549–556.

Sotowa, M., Ootsuka, K., Kobayashi, Y., Hao, Y., Tanaka, K., Ichitani, K., Flowers, J. *et al.* (2013) Molecular relationships between Australian annual wild rice, Oryza meridionalis, and two related perennial forms. *Rice*, **6**, 26.

Stanke, M. and Waack, S. (2003) Gene prediction with a hidden Markov model and a new intron submodel. *Bioinformatics*, **19**, 215–225.

Swofford, D.L. (2003) *PAUP*. Phylogenetic Analysis Using Parsimony (*and Other Methods). Version 4.* Sunderland, Massachusetts: Sinauer Associates.

Tang, L., Zou, X.H., Achoundong, G., Potgieter, C., Second, G., Zhang, D.Y. and Ge, S. (2010) Phylogeny and biogeography of the rice tribe (Oryzeae): evidence from combined analysis of 20 chloroplast fragments. *Mol. Phylogenet. Evol.* **54**, 266–277.

The International Brachypodium Initiative. (2010) Genome sequencing and analysis of the model grass Brachypodium distachyon. *Nature*, **463**, 763–768.

Thompson, J.D., Higgins, D.G. and Gibson, T.J. (1994) CLUSTAL W: improving the sensitivity of progressive multiple sequence alignment through sequence weighting, position-specific gap penalties and weight matrix choice. *Nucleic Acids Res.* **22**, 4673–4680.

Tsitrone, A., Kirkpatrick, M. and Levin, D.A. (2003) A model for chloroplast capture. *Evolution*, **57**, 1776–1782.

Vaughan, D.A., Lu, B.R. and Tomooka, N. (2008) The evolving story of rice evolution. *Plant Sci.* **174**, 394–408.

Wambugu, P.W., Furtado, A., Waters, D.L.E., Nyamongo, D.O. and Henry, R.J. (2013) Conservation and utilization of African Oryza genetic resources. *Rice*, **6**, 1–13.

Wambugu, P.W., Brozynska, M., Furtado, A., Waters, D.L. and Henry, R.J. (2015) Relationships of wild and domesticated rices (Oryza AA genome species) based upon whole chloroplast genome sequences. *Sci. Rep.* **5**, 13957.

Wang, M., Yu, Y., Haberer, G., Marri, P.R., Fan, C., Goicoechea, J.L., Zuccolo, A. *et al.* (2014) The genome sequence of African rice (Oryza glaberrima) and evidence for independent domestication. *Nat. Genet.* **46**, 982–988.

Ward, B.J. and van Oosterhout, C. (2016) Hybridcheck: software for the rapid detection, visualization and dating of recombinant regions in genome sequence data. *Molecular Ecol. Res.* **16**, 534–539.

Waters, D.L., Nock, C.J., Ishikawa, R., Rice, N. and Henry, R.J. (2012) Chloroplast genome sequence confirms distinctness of Australian and Asian wild rice. *Ecol. Evol.* **2**, 211–217.

Wei, X., Qiao, W.H., Chen, Y.T., Wang, R.S., Cao, L.R., Zhang, W.X., Yuan, N.N. *et al.* (2012) Domestication and geographic origin of Oryza sativa in China: insights from multilocus analysis of nucleotide variation of *O. sativa* and *O. rufipogon*. *Mol. Ecol.* **21**, 5073–5087.

Ye, C.X., Ma, Z.S.S., Cannon, C.H., Pop, M. and Yu, D.W. (2012) Exploiting sparseness in de novo genome assembly. *BMC Bioinformatics*, **13**, S1.

Ye, C.X., Hill, C., Wu, S., Ruan, J. and Ma, Z.S.S. (2016) DBG2OLC: efficient assembly of large genomes using long erroneous reads of the third generation sequencing technologies. *Sci. Rep.* **6**, 31900.

Zhang, Q.J., Zhu, T., Xia, E.H., Shi, C., Liu, Y.L., Zhang, Y., Liu, Y. *et al.* (2014) Rapid diversification of five Oryza AA genomes associated with rice adaptation. *P Nat. Acad. Sci. USA*, **111**, E4954–E4962.

Zhang, J., Kudrna, D., Mu, T., Li, W., Copetti, D., Yu, Y., Goicoechea, J.L. *et al.* (2016) Genome Puzzle Master (GPM) – an integrated pipeline for building and editing pseudomolecules from fragmented sequences. *Bioinformatics*, **32**, 3058–3064.

Zwickl, D.J., Stein, J.C., Wing, R.A., Ware, D. and Sanderson, M.J. (2014) Disentangling methodological and biological sources of gene tree discordance on *Oryza* (Poaceae) chromosome 3. *Sys. Biol.* **63**, 645–659.

pOsNAR2.1:OsNAR2.1 expression enhances nitrogen uptake efficiency and grain yield in transgenic rice plants

Jingguang Chen[1,2], Xiaoru Fan[1,2], Kaiyun Qian[1,2], Yong Zhang[1,2], Miaoquan Song[1,2], Yu Liu[3], Guohua Xu[1,2] and Xiaorong Fan[1,2,*]

[1]*State Key Laboratory of Crop Genetics and Germplasm Enhancement, Nanjing Agricultural University, Nanjing, China*
[2]*Key Laboratory of Plant Nutrition and Fertilization in Low-Middle Reaches of the Yangtze River, Ministry of Agriculture, Nanjing Agricultural University, Nanjing, China*
[3]*State Key Laboratory of Plant Physiology and Biochemistry, College of Life Science, Zhejiang University, Hangzhou, China*

Correspondence
email xiaorongfan@njau.edu.cn

Keywords: *OsNAR2.1* promoter, *OsNAR2.1*, *Oryza sativa*, Nitrogen uptake efficiency.

Summary

The nitrate (NO_3^-) transporter has been selected as an important gene maker in the process of environmental adoption in rice cultivars. In this work, we transferred another native *OsNAR2.1* promoter with driving *OsNAR2.1* gene into rice plants. The transgenic lines with exogenous *pOsNAR2.1:OsNAR2.1* constructs showed enhanced *OsNAR2.1* expression level, compared with wild type (WT), and ^{15}N influx in roots increased 21%–32% in response to 0.2 mM and 2.5 mM $^{15}NO_3^-$ and 1.25 mM $^{15}NH_4^{15}NO_3$. Under these three N conditions, the biomass of the *pOsNAR2.1:OsNAR2.1* transgenic lines increased 143%, 129% and 51%, and total N content increased 161%, 242% and 69%, respectively, compared to WT. Furthermore in field experiments we found the grain yield, agricultural nitrogen use efficiency (ANUE), and dry matter transfer of *pOsNAR2.1:OsNAR2.1* plants increased by about 21%, 22% and 21%, compared to WT. We also compared the phenotypes of *pOsNAR2.1:OsNAR2.1* and *pOsNAR2.1:OsNRT2.1* transgenic lines in the field, found that postanthesis N uptake differed significantly between them, and in comparison with the WT. Postanthesis N uptake (PANU) increased approximately 39% and 85%, in the *pOsNAR2.1:OsNAR2.1* and *pOsNAR2.1:OsNRT2.1* transgenic lines, respectively, possibly because *OsNRT2.1* expression was less in the *pOsNAR2.1:OsNAR2.1* lines than in the *pOsNAR2.1:OsNRT2.1* lines during the late growth stage. These results show that rice NO_3^- uptake, yield and NUE were improved by increased *OsNAR2.1* expression *via* its native promoter.

Introduction

Nitrogen (N) is an essential macronutrient for plant growth and crop productivity. NO_3^- is the main inorganic N nutrient for plants in aerobic uplands, and NH_4^+ is the main form in anaerobic paddy fields (Foyer *et al.*, 1998; Scheible *et al.*, 2004; Stitt, 1999). In upland cultivation system, NO_3^- is readily dissolved in soil water and very mobile in soil and therefore it was very easily lost into environment (Jin *et al.*, 2015; Zarabi and Jalali, 2012). NO_3^- is acquired by roots through NO_3^- transporters and then transported throughout the plant, or it can be assimilated with carbon into amino acids before being redistributed (Katayama *et al.*, 2009; Miller *et al.*, 2007; Xu *et al.*, 2012). In plants, seed dormancy can be broken by NO_3^- as a signalling molecule (Alboresi *et al.*, 2005; Matakiadis *et al.*, 2009), regulating lateral root development (Zhang and Forde, 1998; Zhang *et al.*, 1999) and leaf growth (Hsu and Tsay, 2013; Rahayu *et al.*, 2005), integrating the expression of nitrate-induced genes for growth and development (Dechorgnat *et al.*, 2012; Ho and Tsay, 2010; Huang *et al.*, 2015; O'Brien *et al.*, 2016; Wang *et al.*, 2012) and altering flowering time (Castro Marin *et al.*, 2011).

As for adapting to the low and high NO_3^- concentrations in soil, the plants have developed two different absorption systems (Léran *et al.*, 2014; Miller *et al.*, 2007; Siddiqi *et al.*, 1990), including the low NO_3^- affinity system (LATS) and high NO_3^- affinity system (HATS) (Crawford and Glass, 1998). As we know the NPF (NRT1/PTR) and NRT2 families contribute to LATS and HATS responding the NO_3^- uptake and translocation in plants (Fan *et al.*, 2005; Léran *et al.*, 2014; Miller *et al.*, 2007; Orsel *et al.*, 2006; Szczerba *et al.*, 2006).

Some NRT2 family members in plant are needed NAR2 partners in transporting nitrate crossing cell membrane (Galván *et al.*, 1996; Liu *et al.*, 2014; Okamoto *et al.*, 2006; Orsel *et al.*, 2006; Quesada *et al.*, 1994; Tong *et al.*, 2005; Zhuo *et al.*, 1999). In *Chlamydomonas reinhardtii* Quesada *et al.* (1994) firstly found that CrNar2 and CrNar3 can restore NO_3^- absorption of the NO_3^- uptake-defective mutants. Zhou *et al.* (2000) further demonstrated that CrNar2 was a partner protein of CrNRT2.1 in NO_3^- transporting cross the oocyte cell membrane. Okamoto *et al.* (2006) reported that, based on NAR2-type gene expression, both NAR2s and NRT2s constitute the NO_3^- inducible HATS, but not the LATS in Arabidopsis, such as AtNRT3, although the protein had no known transport activity. Yong *et al.* (2010) reported that *in vivo* NAR2.1 and NRT2.1 forming a complex on plasma membrane and played the role in absorbing low concentration of nitrate in Arabidopsis roots. Orsel *et al.* (2006) used oocyte expression and yeast split-ubiquitin systems to show that AtNAR2.1 and AtNRT2.1 are partners in a two-component HATS.

Two-component NRT2-NAR2 system also exists in rice NO_3^- transport process. Feng *et al.* (2011) used an oocyte expression

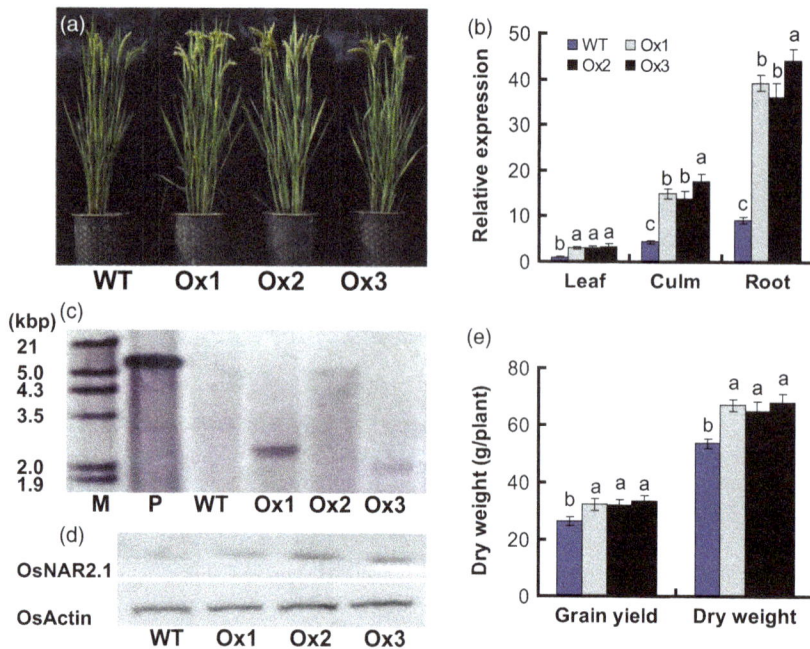

Figure 1 Characterization of *pOsNAR2.1: OsNAR2.1* transgenic lines. (a) Phenotype of wild-type and *pOsNAR2.1:OsNAR2.1* transgenic plants (Ox1, Ox2 and Ox3). (b) qRT-PCR analysis the expression of *OsNAR2.1*. RNA was extracted from root, culm and Leaf blade I. Error bars: SE ($n = 3$ plants). (c) Southern blot of genomic DNA isolated from T2 generation transgenic plants and WT. Hybridization using a hygromycin gene probe. P, positive control, M, marker. (d) Western blot of total proteins from shoots of T2 generation transgenic plants and WT. Hybridization with an OsActin-specific antibody and an OsNAR2.1-specific antibody. Each lane was loaded with equal quantity of protein (50 µg). (e) Biomass and grain yield per plant in the field. Error bars: SE ($n = 5$ plants). The different letters indicate a significant difference between the transgenic line and the WT ($P < 0.05$, one-way ANOVA).

system to show that only OsNAR2.1, but not OsNAR2.2, interacts with OsNRT2.3a or OsNRT2.1/2.2 to promote NO_3^- uptake. Katayama *et al.* (2009) reported that overexpression of *OsNRT2.1* improved the growth of rice seedlings, but did not increase nitrogen uptake. Tang *et al.* (2012) showed that rice *OsNRT2.3a* gene is involved in root transport of NO_3^- to shoots. The OsNRT2.3a or OsNRT2.1/2.2 and OsNAR2.1 interaction at the protein level was demonstrated using bimolecular fluorescence complementation, the yeast two-hybrid system and Western blot analysis (Liu *et al.*, 2014; Yan *et al.*, 2011). Furthermore Yan *et al.* (2011) also reported that knockdown of *OsNAR2.1* by RNA interference (RNAi) can suppress expression of *OsNRT2.3a*, *OsNRT2.2* and *OsNRT2.1* in mutants roots and demonstrated that *OsNAR2.1* does a key function in both high and low NO_3^- uptake.

Chen *et al.* (2016) showed that using *OsNAR2.1* promoter instead of ubiquitin promoter driving *OsNRT2.1* can improve the ANUE and yield in rice. In this study, we created new construct of *OsNAR2.1* promoter to drive the open reading frame (ORF) of the *OsNAR2.1*, investigated the transformation effects of *pOsNAR2.1:OsNAR2.1* on rice NO_3^- uptake, yield and NUE and also presented many different characteristics of *pOsNAR2.1: OsNAR2.1* from *pOsNAR2.1:OsNRT2.1* transgenic plants.

Results

Generation of transgenic rice expressing *pOsNAR2.1: OsNAR2.1*

We used the *Agrobacterium tumefaciens*-mediated method to introduce the *pOsNAR2.1:OsNAR2.1* expression construct (Figure S1) into Wuyunjing 7 (*O. sativa* L. ssp. Japonica cv., the wild type for this experiment, WT), a high yield rice cultivar used in Jiangsu, China. We obtained 10 lines with increased the expression of *OsNAR2.1* (Figure S2a) and analysed biomass and yield of the transgenic plants in the T1 generation. Compared to WT, biomass and yield of the 10 lines of T1 generation increased

by approximately 13% and 20%, respectively (Figure S2b). Based on a Southern blot analysis of T2 generation and the data of RNA expression for the T1 and T2 generations (Figures 1c, S2a and 1b), we selected three independent lines of *pOsNAR2.1: OsNAR2.1* designated Ox1, Ox2 and Ox3 (Figure 1a).

The expression of *OsNAR2.1* in roots was increased four- to fivefold in the Ox1, Ox2 and Ox3 lines. *OsNAR2.1* expression increased approximately 3.5-fold in culms and increased approximately 2.6-fold in leaf blades of the *pOsNAR2.1:OsNAR2.1* transgenic plants (Figure 1b). The Western blot showed that the protein level of OsNAR2.1 was increased in shoots of Ox1, Ox2 and Ox3 lines compared with WT (Figure 1d). The field data showed that the transgenic lines exhibited increased grain yield and dry weight, compared with the WT (Figures 1e and S2b). Field data of the T2, T3 and T4 generation lines showed that total aboveground biomass, increased by as much as 23%; yields of T3 transgenic plants grown at Sanya were enhanced by approximately 20%, and the yields of T2 and T4 plants grown at Nanjing increased by 21%–23%, relative to the WT (Table S3).

For the T4 transgenic plants at harvest, height increased 5%, total tiller number per plant increased 26%, panicle length increased approximately 12%, grain weight per panicle increased 25%, seed setting rate increased 13%, grain number per panicle increased 16%, and grain yields increased by 23% relative to the WT; however, 1000-grain weight had no difference between WT and the transgenic lines (Table 1).

Effects of *pOsNAR2.1:OsNAR2.1* expression on plant seedling growth and total nitrogen content

As previous data showed that knockdown of *OsNAR2.1* in rice affects N uptake and growth (Yan *et al.*, 2011). We further analysed the effect of *pOsNAR2.1:OsNAR2.1* expression on plant seedling growth and nitrogen content by planting WT and transgenic rice seedlings in the solution containing 1 mM NH_4^+ of IRRI for 2 weeks and then in 2.5 mM NH_4^+, 0.2 mM NO_3^-, 2.5 mM NO_3^- or 1.25 mM NH_4NO_3 for 3 more weeks (Figure 2a–d). While

Table 1 Comparison of agronomic traits of *pOsNAR2.1:OsNAR2.1* transgenic lines

Genotype	WT	Ox1	Ox2	Ox3
Plant height (cm)	83.81b	87.74a	87.22a	88.15a
Total tiller number per plant	20.48b	26.78a	25.14a	25.46a
Panicle length (cm)	13.78b	15.67a	15.24a	15.56a
Grain number per panicle	130.67b	153.80a	149.56a	152.66a
Seed setting rate (%)	72.67b	83.04a	80.33a	82.45a
Grain weight (g/panicle)	2.32b	3.01a	2.77a	2.89a
1000-grain weight (g)	25.79a	25.65a	25.87a	25.74a
Grain yield (g/plant)	26.37b	32.14a	31.81a	33.38a

Statistical analysis of data from T4 generation; $n = 3$ plots for each mean. The different letters indicate a significant difference between the transgenic line and the WT. ($P < 0.05$, one-way ANOVA).

the dry weight of roots, leaf sheaths and leaves of the *pOsNAR2.1:OsNAR2.1* transgenic line were not affected by growth in 2.5 mM NH_4^+ (Figure 2e), they increased, respectively, by 152%, 149% and 151% in 0.2 mM NO_3^- (Figure 2f); by 124%, 181% and 95% in 2.5 mM NO_3^- (Figure 2g); and by 62%, 51% and 47% in 1.25 mM NH_4NO_3, compared with WT after harvest (Figure 2h).

Total N concentrations of roots, leaf sheaths and leaves in *pOsNAR2.1:OsNAR2.1* were not affected by 2.5 mM NH_4^+ (Figure 3a), but were increased by 19%, 10% and 14%, in 0.2 mM NO_3^- (Figure 3b); by 62%, 25% and 60% in 2.5 mM NO_3^- (Figure 3c); and by 15%, 15% and 8% in 1.25 mM NH_4NO_3 (Figure 3d), respectively. Total N contents of roots, leaf sheaths and leaves in *pOsNAR2.1:OsNAR2.1* were not affected by 2.5 mM

NH_4^+ (Figure 3e), but were increased by 199%, 174% and 72%, in 0.2 mM NO_3^- (Figure 3f); by 263%, 251% and 212% in 2.5 mM NO_3^- (Figure 3g); and by 87%, 74 and 60% in 1.25 mM NH_4NO_3, compared with WT (Figure 3h), respectively.

Yan *et al.* (2011) reported that *OsNAR2.1* RNAi affects the expression of interacting proteins with the *OsNAR2.1* including *OsNRT2.1*, *OsNRT2.2* and *OsNRT2.3a* genes. We further analysed whether *OsNAR2.1* and *OsNRT2s* expression in transgenic rice roots was altered at differing N supply rates. Transcription of *OsNRT2.3a*, *OsNRT2.2* and *OsNRT2.1* in transgenic plant roots was not affected by growth in 2.5 mM NH_4^+ (Figure 4a); but was increased, respectively, by 117, 121 and 129% in 0.2 mM NO_3^- (Figure 4b); by 105%, 118% and 110%, in 2.5 mM NO_3^- (Figure 4c); and by 76%, 68% and 73% in 1.25 mM NH_4NO_3 (Figure 4d), compared with WT.

Rates of NO_3^- and NH_4^+ influx in WT and transgenic plants

We analysed short-term NO_3^- and NH_4^+ uptake in same-size seedlings of the *pOsNAR2.1:OsNAR2.1* transgenic lines and WT by exposing the plants to 2.5 mM $^{15}NH_4^+$, 0.2 mM $^{15}NO_3^-$, 2.5 mM $^{15}NO_3^-$, 1.25 mM $^{15}NH_4^{15}NO_3$, 1.25 mM $^{15}NH_4NO_3$ or 1.25 mM $NH_4^{15}NO_3$ for 5 min to determine the effect of *pOsNAR2.1: OsNAR2.1* expression on root NO_3^- and NH_4^+ influx into intact plants. The influx rate of $^{15}NH_4^+$ in the Ox1, Ox2 and Ox3 transgenic lines did not change compared with that of WT (Figure 5a); however, the influx rate of $^{15}NO_3^-$ increased 32% and 26% in response to 0.2 mM $^{15}NO_3^-$ and 2.5 mM $^{15}NO_3^-$, respectively, in the *pOsNAR2.1:OsNAR2.1* transgenic lines (Figure 5b, c). The influx rate of $^{15}NH_4^{15}NO_3$ in the transgenic lines increased about 20% in 1.25 mM $^{15}NH_4^{15}NO_3$ (Figure 5d), and the influx rates of $^{15}NH_4^+$ and $^{15}NO_3^-$ increased by 21% and 22% in 1.25 mM $^{15}NH_4NO_3$ and 1.25 mM $NH_4^{15}NO_3$, respectively

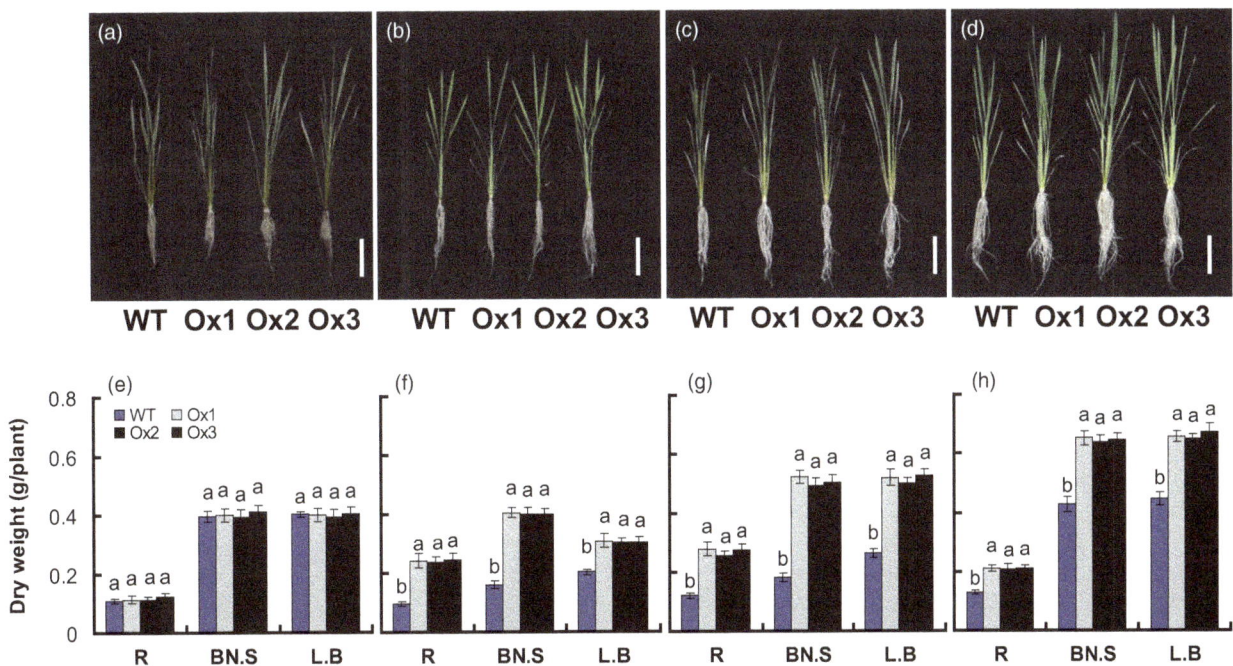

Figure 2 Comparison of growth of *pOsNAR2.1:OsNAR2.1* transgenic lines at different nitrogen supply levels. WT and transgenic rice seedlings in the solution containing 1 mM NH_4^+ of IRRI for 2 weeks and then in different forms of nitrogen for 3 additional weeks. Phenotype of the *pOsNAR2.1:OsNAR2.1* lines (Ox1, Ox2 and Ox3) grown with (a) 2.5 mM NH_4^+, (b) 0.2 mM NO_3^-, (c) 2.5 mM NO_3^- and (d) 1.25 mM NH_4NO_3; bar = 10 mm; dry weight of seedlings treated with (e) 2.5 mM NH_4^+, (f) 0.2 mM NO_3^-, (g) 2.5 mM NO_3^- and (h) 1.25 mM NH_4NO_3. L.B, leaf blade; BN.S, basal node and sheath; R, root. Error bars: SE ($n = 4$ plants). The different letters indicate a significant difference between the transgenic line and the WT ($P < 0.05$, one-way ANOVA).

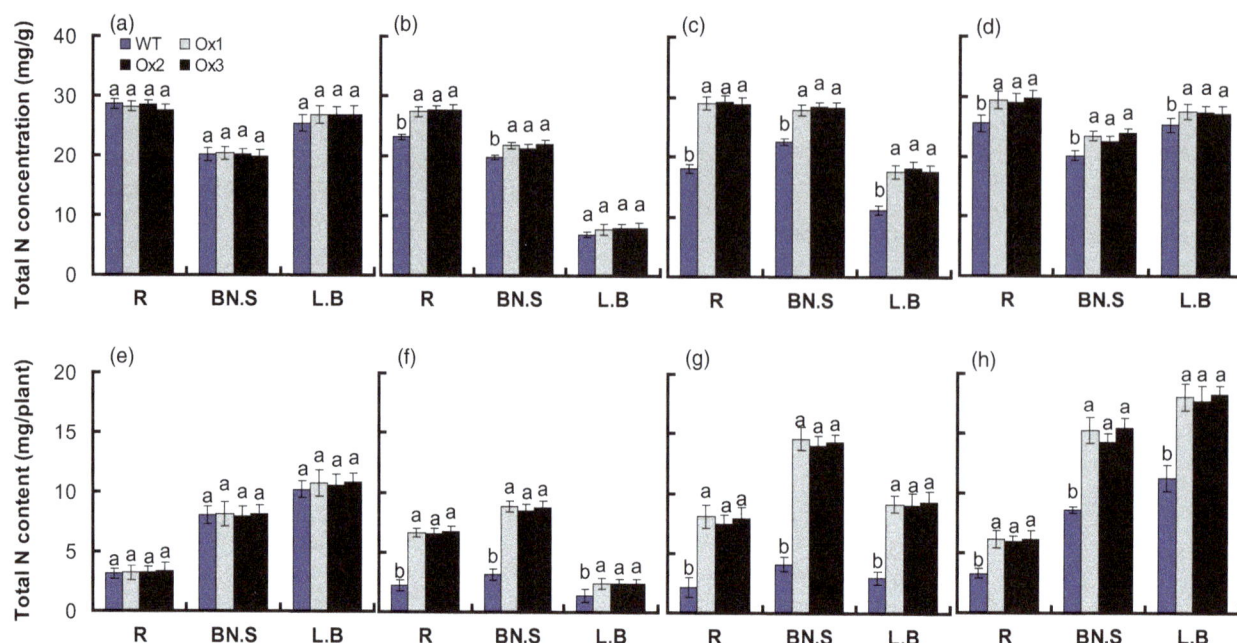

Figure 3 Comparison of total nitrogen concentration and total nitrogen content of *pOsNAR2.1:OsNAR2.1* transgenic plants at different nitrogen supply levels. WT and transgenic rice seedlings in the solution containing 1 mm NH_4^+ of IRRI for 2 weeks, and in different forms of nitrogen for 3 additional weeks. Total nitrogen concentration of seedlings treated with (a) 2.5 mm NH_4^+, (b) 0.2 mm NO_3^-, (c) 2.5 mm NO_3^- and (d) 1.25 mm NH_4NO_3; Total N content of seedlings grown with (e) 2.5 mm NH_4^+, (f) 0.2 mm NO_3^-, (g) 2.5 mm NO_3^- and (h) 1.25 mm NH_4NO_3. L.B, leaf blade; BN.S, basal node and sheath; R, root. Error bars: SE (n = 4 plants). The different letters indicate a significant difference between the transgenic line and the WT ($P < 0.05$, one-way ANOVA).

(Figure S3a, b). The ratio of $^{15}NH_4^+$ to $^{15}NO_3^-$ influx in *pOsNAR2.1: OsNAR2.1* transgenic and WT plants did not differ in response to 1.25 mm $^{15}NH_4NO_3$ and 1.25 mm $NH_4^{15}NO_3$ (Figure S3c).

Translocation of dry matter and nitrogen in WT and transgenic plants

Methods to measure NUE usually depend on calculating plant biomass production per unit of applied N, regardless of the crop and whether the root, leaf, fruit, or seed is measured, the transfer of N to plant organs and yield is known as "nutrient utilization efficiency" (Good et al., 2004; Xu et al., 2012). We analysed the dry matter, total nitrogen concentration and the total nitrogen content of the T4 generation of the *pOsNAR2.1:OsNAR2.1* transgenic lines in the anthesis and maturity stages. The result showed that the biomass of panicles, leaves and culms in the transgenic lines increased 26%, 20% and 28%, respectively, in the anthesis stage (Figure 6b), and increased 23%, 29% and 25% in the maturity stage compared to those of WT (Figure 6c). Total nitrogen concentration in leaves of the transgenic lines increased approximately 10% in the anthesis stage, but did not change in panicles or culms compared to those of WT. Total nitrogen concentration of panicles, leaves and culms was not different at the maturity stage compared to that in WT (Figure 6e); total nitrogen content of panicles, leaves and culms in the transgenic lines increased by approximately 34%, 33% and 33%, respectively, during the anthesis stage (Figure 6f), and by 35%, 33% and 34% in the maturity stage, respectively, compared to those in the WT (Figure 6f).

We calculated nitrogen and dry matter translocation in plants by determining dry matter at maturity (DMM), dry matter at anthesis (DMA), grain yield (GY), total nitrogen accumulation at

maturity (TNAM), grain nitrogen accumulation at maturity (GNAM) and total nitrogen accumulation at anthesis (TNAA). Compared to the WT, DMA, DMM, GY, TNAA, TNAM and GNAM increased by approximately 25%, 25%, 24%, 33%, 34% and 35%, respectively, in *pOsNAR2.1:OsNAR2.1* transgenic plants (Table 2).

We also calculated the harvest index (HI), dry matter translocation (DMT), dry matter translocation efficiency (DMTE) and the contribution of pre-anthesis assimilates to grain yield (CPAY), based on a method described by Chen et al. (2016). DMT increased by approximately 21%, whereas DMTE, CPAY and HI had no difference in *pOsNAR2.1:OsNAR2.1* transgenic plants from WT (Table 3). We investigated nitrogen translocation (NT), contribution of pre-anthesis nitrogen to grain nitrogen accumulation (CPNGN) and NT efficiency (NTE), based on a method described by Chen et al. (2016). NTE and CPNGN did not differ *pOsNAR2.1:OsNAR2.1* transgenic plants from WT, whereas NT increased by approximately 33%, relative to that in WT (Table 3).

NUE of *pOsNAR2.1:OsNAR2.1* transgenic lines

NUE is inherently compound and can be further defined with component parts, including NUpE, NUtE, ANR, AE NTE, NRE (Xu et al., 2012). Because both yield and biomass were increased in the *pOsNAR2.1:OsNAR2.1* transgenic lines, in the meanwhile, we also investigated ANUE in transgenic plants of T2–T4 generations, and nitrogen recovery efficiency (NRE), PANU, nitrogen harvest index (NHI), and physiological nitrogen use efficiency (PNUE) traits in the T4 transgenic plants to determine whether nitrogen use was changed in these lines, using the method described by Chen et al. (2016). Compared to WT, the ANUE of the *pOsNAR2.1: OsNAR2.1* transgenic lines was enhanced by approximately 22%

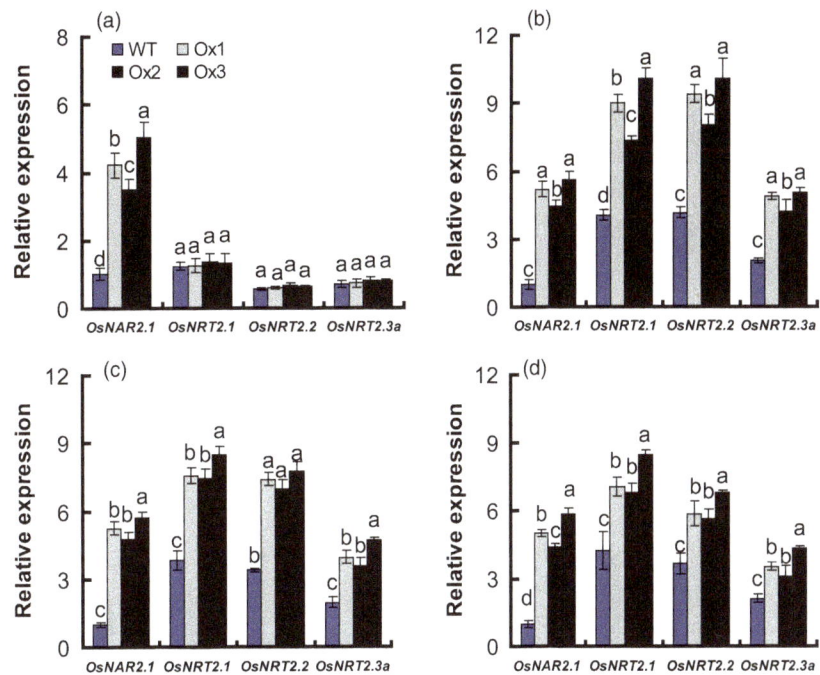

Figure 4 The expression of *OsNRT2s* and *OsNAR2.1* in *pOsNAR2.1:OsNAR2.1* transgenic lines. Extraction of total RNA from roots of WT and transgenic lines as showing in Figure 2 and qRT-PCR result under (a) 2.5 mM NH_4^+, (b) 0.2 mM NO_3^-, (c) 2.5 mM NO_3^- and (d) 1.25 mM NH_4NO_3 conditions. Error bars: SE (*n* = 3 plants). The different letters indicate a significant difference between the transgenic line and the WT (*P* < 0.05, one-way ANOVA).

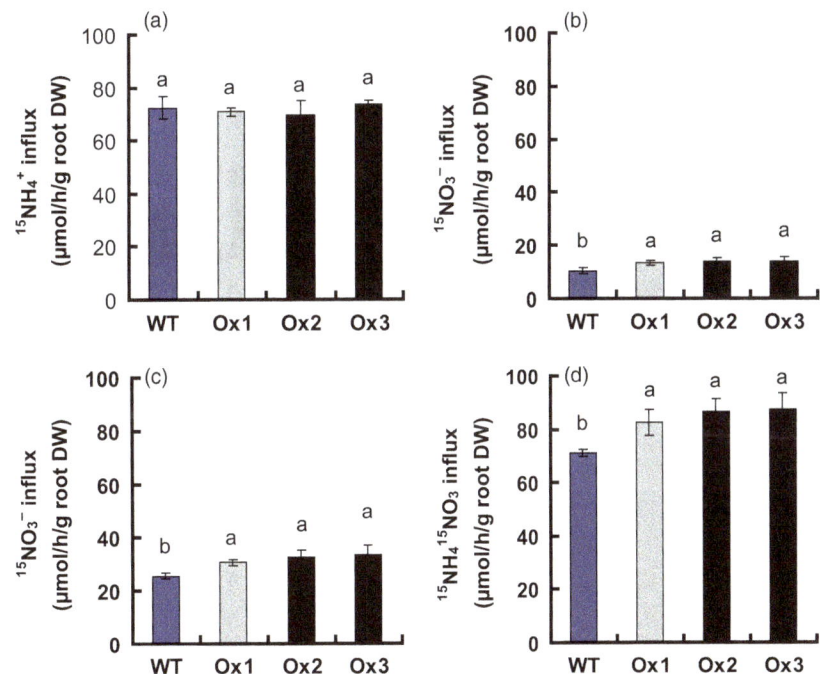

Figure 5 NH_4^+ and NO_3^- influx rates of *pOsNAR2.1:OsNAR2.1* transgenic lines measured using ^{15}N-enriched sources. WT and transgenic seedlings were grown in 1 mM NH_4^+ for 3 weeks and nitrogen starved for 1 week. ^{15}N influx rates were then measured at (a) 2.5 mM $^{15}NH_4^+$, (b) 0.2 mM $^{15}NO_3^-$, (c) 2.5 mM $^{15}NO_3^-$ and (d) 1.25 mM $^{15}NH_4{}^{15}NO_3$ during 5 min. DW, dry weight. Error bars: SE (*n* = 4 plants). The different letters indicate a significant difference between the transgenic line and the WT (*P* < 0.05, one-way ANOVA).

in T3 generation grown at Sanya under the tropical climate condition and by 21%–24% in the T2 and T4 plants grown at Nanjing under semi-tropical condition (Table S3, Figure 7a). NRE and PANU increased approximately 125% and 39% in the T4 generations, compared to those in WT (Figure 7b, c), but PNUE and NHI values had no different between those and WT (Table 3).

Comparison between NUEs in *pOsNAR2.1:OsNAR2.1* and *pOsNAR2.1:OsNRT2.1* transgenic plants

In the field, PNUE, NHI, DMTE, CPAY, HI and NTE values of *pOsNAR2.1:OsNAR2.1* and *pOsNAR2.1:OsNRT2.1* transgenic lines were the same as those of WT plants (Table S4). Compared

to WT, ANUE increased approximately 22% and 31%, in the *pOsNAR2.1:OsNAR2.1* and *pOsNAR2.1:OsNRT2.1* transgenic lines, respectively. NRE increased approximately 25% and 36%, and PANU increased approximately 39 and 85% (Figure 7d–f). The CPNGN of *pOsNAR2.1:OsNAR2.1* transgenic lines showed no difference compared with WT, but the CPNGN decreased about 15% in *pOsNAR2.1:OsNRT2.1* lines (Table S4).

The expression of *OsNRT2.1* and *OsNAR2.1* in transgenic lines

The expression levels of *OsNRT2.1* and *OsNAR2.1* in culms were significantly increased in all transgenic lines, compared to the WT plants (Figure 8a, b). The expression of *OsNRT2.1* was about 32%

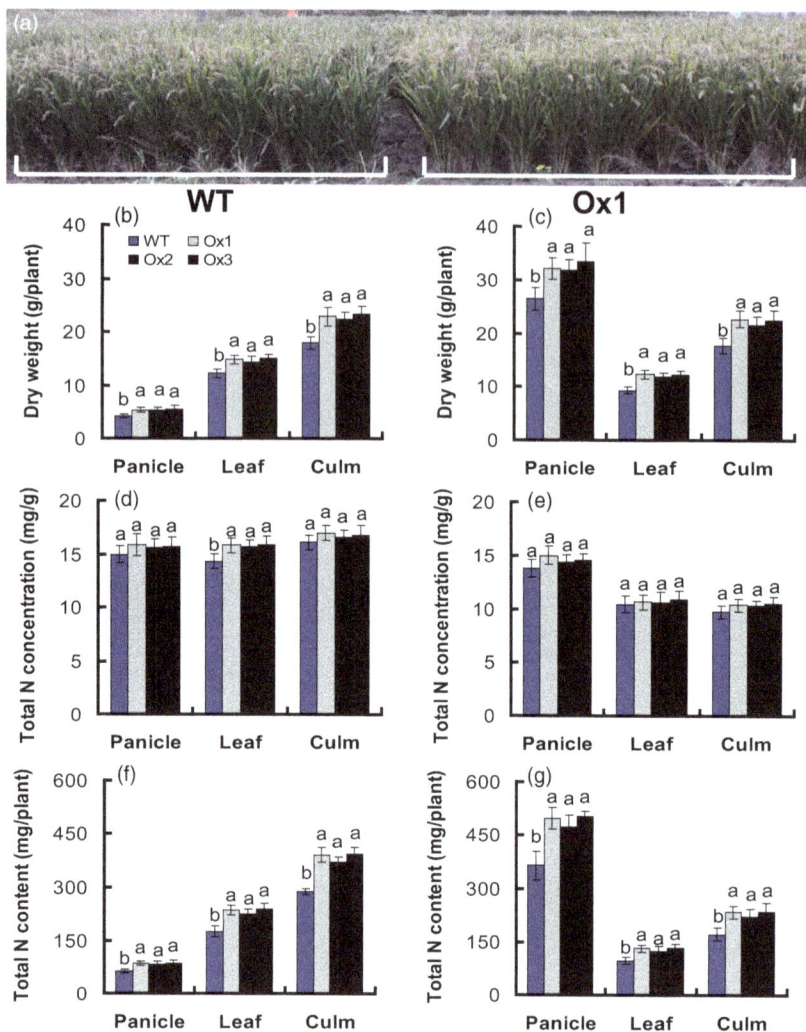

Figure 6 Biomass and nitrogen content in different parts of *pOsNAR2.1:OsNAR2.1* transgenic lines at the anthesis stage and maturity stage. (a) Photograph of WT and T4 generation Ox1 in the field experiment. Biomass in various parts of WT and T4 generation transgenic plants at (b) the anthesis stage and (c) maturity stage. Nitrogen concentration in different parts of transgenic lines and WT at the (d) anthesis stage and (e) maturity stage. Nitrogen content in different parts of transgenic lines and WT at the (f) anthesis stage and (g) maturity stage. Error bars: SE (*n* = 5 plants). The different letters indicate a significant difference between the transgenic line and the WT (*P* < 0.05, one-way ANOVA).

Table 2 Biomass and nitrogen content of *pOsNAR2.1:OsNAR2.1* transgenic lines

Dry matter and nitrogen components:	WT	Ox1	Ox2	Ox3
DMA (kg/m²)	0.86b	1.08a	1.05a	1.09a
DMM (kg/m²)	1.34b	1.68a	1.63a	1.70a
GY (kg/m²)	0.66b	0.81a	0.80a	0.84a
TNAA (g/m²)	13.17b	17.76a	16.90a	17.90a
TNAM (g/m²)	15.87b	21.61a	20.46a	21.74a
GNAM (g/m²)	9.11b	12.44a	11.79a	12.55a

Statistical analysis of data from T4 generation; *n* = 3 plots for each mean. The different letters indicate a significant difference between the transgenic line and the WT (*P* < 0.05, one-way ANOVA).

and 38% higher in the *pOsNAR2.1:OsNAR2.1* and *pOsNAR2.1: OsNRT2.1* lines than in WT (Figure 8a). The expression of *OsNAR2.1* was 4.1–6.4-fold higher in the *pOsNAR2.1:OsNAR2.1* lines than in WT and 2.3–3.6-fold higher in the *pOsNAR2.1: OsNRT2.1* lines than in WT (Figure 8b). And the expression of *OsNRT2.1* showed no difference between the *pOsNAR2.1: OsNAR2.1* and *pOsNAR2.1:OsNRT2.1* lines throughout the experimental growth period, except at 75 days (Figure 8a). The expression of *OsNAR2.1* was significantly higher in the culms of

the *pOsNAR2.1:OsNAR2.1* lines than of the *pOsNAR2.1: OsNRT2.1* lines (Figure 8b).

We also calculated the expression ratio of *OsNRT2.1* and *OsNAR2.1* in different plants, which was approximately 2.2:1 in the *pOsNAR2.1:OsNAR2.1* lines, 4.6:1 in the *pOsNAR2.1: OsNRT2.1* lines and 10.6:1 in WT (Figure S4).

Discussion

All levels of plant function were affected by nitrogen nutrition, from metabolism to growth, development and resource allocation (Crawford, 1995; Scheible *et al.*, 1997). NO_3^- is a main available form of nitrogen for plants and is absorbed in the roots by active transport processes and passive transport ion channels, stored in vacuoles of rice shoots (Fan *et al.*, 2007; Kucera, 2003; Li *et al.*, 2008; Pouliquin *et al.*, 2000). OsNRT2.1/2 and OsNRT2.3a need to be combined with OsNAR2.1 protein for uptake and transport of NO_3^- in rice (Liu *et al.*, 2014; Tang *et al.*, 2012; Yan *et al.*, 2011). The expression of *OsNAR2.1* is up-regulated by NO_3^- and down-regulated by NH_4^+ (Feng *et al.*, 2011; Yan *et al.*, 2011).

Feng *et al.* (2011) reported that the native *OsNAR2.1* promoter has strong activities in roots and basal nodes in seedlings. In this study, *OsNAR2.1* expression was up-regulated significantly in both roots and shoots of *pOsNAR2.1:OsNAR2.1* transgenic lines

Table 3 DMT and NT of the *pOsNAR2.1:OsNAR2.1* transgenic lines

	WT	Ox1	Ox2	Ox3
PNUE (g/g)	51.77a	50.16a	51.71a	52.15a
NHI (%)	57.36a	57.58a	57.62a	57.70a
DMT (g/m^2)	182.32b	223.39a	213.80a	226.97a
DMTE (%)	21.20a	18.86a	20.43a	20.74a
CPAGY (%)	27.62a	27.34a	26.89a	27.18a
HI (%)	49.35a	48.84a	48.85a	49.05a
NT (g/m^2)	6.40b	8.59a	8.23a	8.69a
NTE (%)	48.61a	48.39a	48.70a	48.57a
CPNGN (%)	70.32a	69.38a	69.82a	69.24a

PNUE (kg/kg) = (GY – GY of zero-N plot)/TNAM; NHI (%) = (GNAM/TNAM) × 100%; DMT (kg/ha) = DMA – (DMM – GY); DMTE (%) = (DMT/DMA) × 100%; CPAY (%) = (DMT/GY) × 100%; HI (%) = (GY/DMM) × 100%; NT (kg/ha) = TNAA – (TNAM – GNAM); NTE (%) = (NT/TNAA) × 100%; CPNGN (%) = (NT/GNAM) × 100%. Statistical analysis of data from T4 generation; $n = 3$ plots for each mean. The different letters indicate a significant difference between the transgenic line and the WT ($P < 0.05$, one-way ANOVA).

(Figure 1b). Previous report had addressed the *OsNAR2.1* promoter induction by NO_3^- in rice based on GUS fusion data (Feng et al., 2011); Yan et al. (2011) reported the effect of rice seedling stage and nitrogen uptake on growth after OsNAR2.1 knockdown; moreover Chen et al. (2016) reported the gain function of *pOsNAR2.1:OsNRT2.1* expression on rice growth and nitrogen use. Here, we focused on nitrogen uptake and growth at the seedling stage, field yield and NUE in *pOsNAR2.1:OsNAR2.1* transgenic lines.

pOsNAR2.1:OsNAR2.1 expression increases NO_3^- uptake of transgenic rice plants

Feng et al. (2011) had proved that *OsNAR2.1* interacts with *OsNRT2.3a* and *OsNRT2.1/2.2* in an oocyte expression system to take up NO_3^-. The OsNAR2.1 and OsNRT2.3a or OsNRT2.1/2.2 interaction in the protein level was demonstrated using bimolecular fluorescence complementation, Western blot analysis and a yeast two-hybrid assay (Liu et al., 2014; Yan et al., 2011). Tang et al. (2012) showed that *OsNRT2.3a* gene is important in NO_3^- root transport to shoots. Katayama et al. (2009) reported that increased *OsNRT2.1* expression slightly improved the growth of rice seedling in hydroponic condition, but did not affect the nitrogen uptake. In this study, we demonstrated that *OsNAR2.1* driven by the native *OsNAR2.1* promoter increased NO_3^- uptake by rice roots.

As we know, the native *OsNAR2.1* was expressed in all parts in rice plant, and mainly expressed roots and leaf sheaths (Chen et al., 2016; Feng et al., 2011; Liu et al., 2014; Yan et al., 2011), but we do not know why one more native promoter driving *OsNAR2.1* can increase the expression level of *OsNAR2.1* more than one time and in different organs, the increase patterns were different. The possible reason about this was that the methylation level was different in the transferred homologous exogenous promoter sequence compared with the endogenous promoter sequence (Matzke et al., 1989). However more experiments are needed for this understanding.

Rice dry weight and total nitrogen content of *pOsNAR2.1: OsNAR2.1* transgenic plants differed clearly from WT plants when the plants were supplied with 1.25 mM NH_4NO_3, 0.2 mM NO_3^- or 2.5 mM NO_3^- (Figures 2 and 3). *OsNRT2.3a, OsNRT2.2 and*

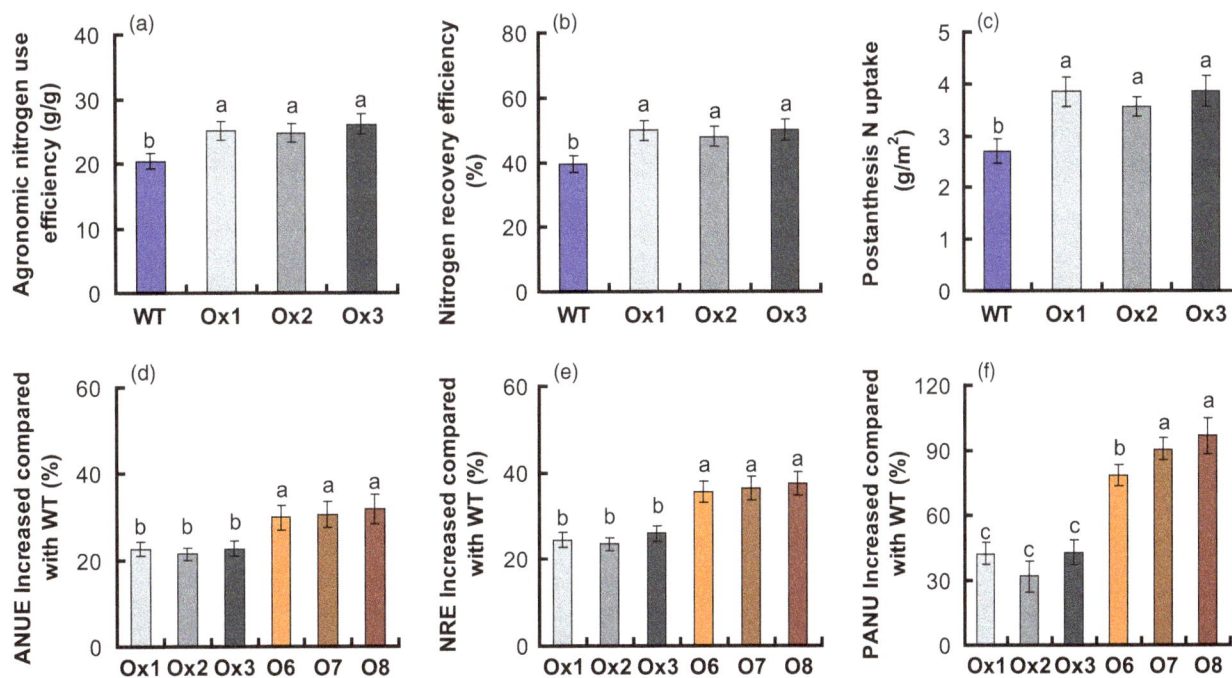

Figure 7 Increased nitrogen use efficiency (NUE) in *pOsNAR2.1:OsNAR2.1* and *pOsNAR2.1:OsNRT2.1* transgenic plants relative to wild type. Comparison of (a) agronomic nitrogen use efficiency (ANUE), (b) nitrogen recovery efficiency (NRE) and (c) postanthesis N uptake (PANU) between *pOsNAR2.1: OsNAR2.1* transgenic lines and wild type. Enhanced percentage of (d) ANUE, (e) NRE and (f) PANU of *pOsNAR2.1:OsNAR2.1* and *pOsNAR2.1:OsNRT2.1* relative to wild type. $n = 3$ plots for each mean. The different letters indicate a significant difference between the transgenic line and the WT ($P < 0.05$, one-way ANOVA).

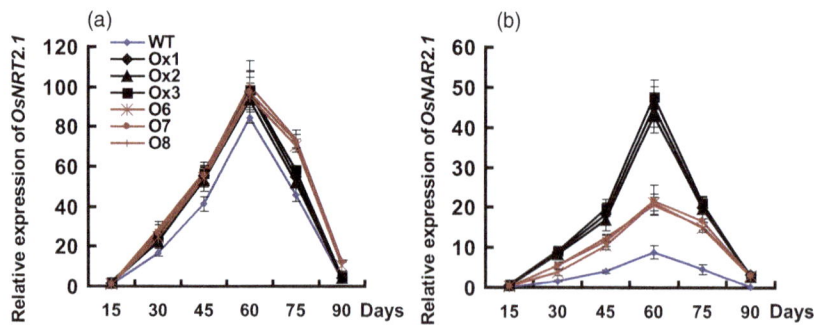

Figure 8 Expression of *OsNRT2.1* and *OsNAR2.1* in transgenic lines and wild type during the experimental growth period. RNA Samples of T4 generation plant culms were collected every 15 days, from the beginning of rice transplant to mature stage. Error bars: SE (*n* = 3 plants). D in *x*-axis means the day after transplanting.

OsNRT2.1 expression which encode OsNAR2.1-interacting proteins, increased significantly in 1.25 mM NH_4NO_3, 0.2 mM NO_3^- and 2.5 mM NO_3^-, compared with WT (Figure 4). The up-regulated expression of *OsNAR2.1* and *OsNRT2.3a*, *OsNRT2.2* and *OsNRT2.1* caused the $^{15}NO_3^-$ influx rates of *pOsNAR2.1: OsNAR2.1* transgenic lines in 0.2 mM $^{15}NO_3^-$, 2.5 mM $^{15}NO_3^-$ and 1.25 mM $NH_4^{15}NO_3$ to increase 32%, 26% and 22%, respectively (Figures 5 and S3).

Enhanced NO_3^- uptake promotes NH_4^+ uptake in rice

Kronzucker *et al.* (2000) used ^{13}N to show that the presence of NO_3^- promotes NH_4^+ uptake, accumulation and metabolism in rice. Duan *et al.* (2006) found that increasing NO_3^- uptake promotes dry weight and NO_3^- accumulation and assimilation of NH_4^+ and NO_3^- by 'Nanguang', which is an N-efficient rice cultivar, during the entire growth period. Li *et al.* (2006) showed that supplying NH_4^+ and NO_3^- enhances *OsAMT1;3*, *OsAMT1;2* and *OsAMT1;1* expression compared with supplying only NH_4^+ or NO_3^-, thereby enhancing NH_4^+ uptake by rice.

High expression of *OsNRT2.3b* in rice improves the pH-buffering capacity of the rice resulting in less ^{15}N-$NH_4^{15}NO_3$ uptake in 5-min uptake experiment, and more ^{15}N-$^{15}NH_4NO_3$ increased uptake at pH 4 and pH 6 (Fan *et al.*, 2016).

Our results showed that the influx rates of $^{15}NO_3^-$ and $^{15}NH_4^+$ increased 22% and 21%, respectively, in *pOsNAR2.1:OsNAR2.1* transgenic lines in 1.25 mM $NH_4^{15}NO_3$ or 1.25 mM $^{15}NH_4NO_3$ (Figure S3), and that the ratio of $^{15}NO_3^-$ to $^{15}NH_4^+$ influx into *pOsNAR2.1:OsNAR2.1* transgenic plants was not different from WT in 1.25 mM $NH_4^{15}NO_3$ or 1.25 mM $^{15}NH_4NO_3$ (Figure S3). Eventually, the biomass and total nitrogen content of *pOsNAR2.1: OsNAR2.1* transgenic lines increased by 50.7% and 68.9% after 3 weeks in 1.25 mM NH_4NO_3 (Figures 2d, 2h and 3h).

Exogenous of *pOsNAR2.1:OsNAR2.1* transformation in rice enhances ANUE and NRE

During recent years, NO_3^- transporter gene as a target gene was applied in crop high NUE breeding (Fan *et al.*, 2017). For examples, the *OsNRT1.1B* low-affinity NO_3^- transporter can increase the *indica* rice NUE by approximately 30% (Hu *et al.*, 2015). Fan *et al.* (2016) showed that increased *OsNRT2.3b* expression improved NUE and grain yield by up to 40% in Japonica cultivars. Chen *et al.* (2016) reported the ANUE of *pOsNAR2.1:OsNRT2.1* transgenic plants increased by 28% of in the same background cultivar (Wuyunjing 7) as this experiment. Our present data show that *OsNAR2.1* driven by the native *OsNAR2.1* promoter can produce a relatively higher yield and ANUE in rice plants (Figure 7, Figure S2b and Table S3).

Nitrogen redistribution can be altered by the expression change of some nitrogen use relative gene, such as the autophagy gene

ATG8c (Islam *et al.*, 2016) and also presents different patterns in different genotypes (Sanchez-Bragado *et al.*, 2017; Souza *et al.*, 1998). During rice grain filling, 70-90% of the nitrogen was redistributed from the vegetative organs to the panicles (Yoneyama *et al.*, 2016). Dry matter and nitrogen content of *pOsNAR2.1: OsNAR2.1* lines were more than WT plants in the anthesis and maturity stages (Figure 6). Although DMT and NT increased by approximately 21% and 33%, compared to that of WT, DMTE and NTE of *pOsNAR2.1:OsNAR2.1* transgenic plants and WT were not different (Table 3), suggested that dry matter and nitrogen transfer from shoots to grains did not change significantly between *pOsNAR2.1:OsNAR2.1* transgenic plants and WT; thus, the physiological NUE and NHI of *pOsNAR2.1:OsNAR2.1* transgenic plants did not increase (Table 3). NRE and ANUE increased 25% and 22% due to the increase in nitrogen accumulation and grain yield, respectively, at maturity in *pOsNAR2.1:OsNAR2.1* transgenic lines (Table 2; Figure 7a, b).

Comparison of growth and NUE of *pOsNAR2.1:OsNRT2.1* and *pOsNAR2.1:OsNAR2.1* transgenic lines

Chen *et al.* (2016) reported that co-expressing *OsNAR2.1* and *OsNRT2.1* in *pOsNAR2.1:OsNRT2.1* transgenic lines increased rice grain yield and ANUE. We also compared field growth between *pOsNAR2.1:OsNRT2.1* and *pOsNAR2.1:OsNAR2.1* transgenic lines. *OsNAR2.1* expressed significantly higher in the culms in the *pOsNAR2.1:OsNAR2.1* lines than in the *pOsNAR2.1: OsNRT2.1* lines, but there was no difference in *OsNRT2.1* expression between them, except at 75 days (Figure 8), which is a key period for grain filling and a critical transition period between rice vegetative and reproductive growth (Zhang *et al.*, 2009). *OsNRT2.1* expression and N uptake decreased in the *pOsNAR2.1:OsNAR2.1* lines during the grain filling stage. Postanthesis N uptake decreased in the *pOsNAR2.1:OsNAR2.1* lines compared with the *pOsNAR2.1:OsNRT2.1* plants (Figure 7f). N for rice grain filling comes mainly from accumulation before flowering (Table S4). Although DMTE and NTE did not differ between *pOsNAR2.1:OsNRT2.1* and *pOsNAR2.1:OsNAR2.1* transgenic lines (Table S4), NRE and ANUE of *pOsNAR2.1:OsNAR2.1* transgenic lines were significantly lower than those of *pOsNAR2.1:OsNRT2.1* transgenic lines (Figure 7d, e).

Designing a genetically modified crop using tissue-specific expression conferred by selected promoters

Although using either the ubiquitin promoter (*pUbi*) or *OsNAR2.1* promoter (*pOsNAR2.1*) to drive *OsNRT2.1* expression could significantly increase total biomass and grain yield compared with those in WT, ANUE was decreased 17% by *pUbi:OsNRT2.1* expression and increased 28% by *pOsNAR2.1:OsNRT2.1* expression (Chen *et al.*, 2016). These opposite effects of different promoters driving *OsNRT2.1* expression on ANUE were caused mainly by altered tissue localization and abundance of *OsNRT2.1* transcripts which may be linked to postflowering transfer of dry matter into grains (Chen *et al.*, 2016). Another transformation example of native promoter driving its ORF is *pOsPTR9:OsPTR9* transgene in rice with improving on growth, grain yield and NUE (Fang *et al.*, 2013). Fang *et al.* (2013) investigated the expression pattern of *OsPTR9* and found that it is regulated by nitrogen sources and light. Although OsPTR9 does not appear to directly transport NO_3^-, its overexpression results in enhanced NH_4^+ uptake, increased grain yield and promoted lateral root formation (Fang *et al.*, 2013). These results indicate that expression of genes using specific promoters may be a good approach for plant breeding.

Several phloem NO_3^- transporters, such as NPF2.13, NPF1.1 and NPF1.2, are responsible for redistributing xylem-borne NO_3^- into developing leaves to increase shoot growth (Fan *et al.*, 2009; Hsu and Tsay, 2013). Therefore, selecting and applying the promoters of genes specifically expressed in senescing leaves or other source organs could be used to drive phloem-expressed NO_3^-, transporters, which would decrease residual N in old vegetative organs and increase growth and NUE.

In this experiment, we demonstrated that rice NO_3^- uptake, yield and NUE of rice were ameliorated by increasing *OsNAR2.1* expression using its native promoter.

Experimental procedures

Construction of transgenic rice with *pOsNAR2.1: OsNAR2.1*

The primers in Table S1 amplified the *OsNAR2.1* ORF sequence from the cDNA of *Oryza sativa* L. ssp. Japonica cv. 'Nipponbare'. The *OsNAR2.1* promoter was amplified from the *pOsNAR2.1-(1,698 bp):GUS* constructs (Feng *et al.*, 2011). The PCR amplification products were ligated into pMD19-T vector (TaKaRa Bio, Shiga, Japan) independently and after sequencing check we construct the *pOsNAR2.1:OsNAR2.1* plasmid by subcloning. The constructed *pOsNAR2.1:OsNAR2.1* vector is shown in Figure S1 and was transformed into callus of Wuyunjing 7 (*O. sativa* L. ssp. Japonica cv.) by *Agrobacterium tumefaciens* strain EHA105 (Chen *et al.*, 2016).

qRT-PCR and Southern blot analysis

Total RNA was extracted using TRIzol reagent (Vazyme Biotech Co., Ltd, http://www.vazyme.com). DNase I-treated total RNAs were subjected to reverse transcription (RT) with HiScript Q RT SuperMix for qPCR (+gDNA wiper) kit (Vazyme Biotech Co.). Triplicate quantitative assays were performed using the AceQ qPCR SYBR Green Master Mix kit (Vazyme Biotech Co.) and a Step One Plus Real-Time PCR System (Applied Biosystems, Foster City, CA). The relative quantitative calculation of real-time PCR was described in Chen *et al.* (2016). The primers for PCR are shown in Table S2.

The Southern blot was carried to identify the T-DNA insertion. The genomic DNA exaction of T2 plant shoots, DNA digestion and hybridization were followed the previous report (Chen *et al.*, 2016)

Western blot

OsNAR2.1 antibody and Western blot process was described in Yan *et al.* (2011). The total protein of 10 g shoots were sampled and 50 µg of each protein was analysed in gel-loaded buffer and boiled in 10% SDS-PAGE. Protein transfer to PVDF membrane and incubated with OsActin (1 : 5000), or OsNAR2.1 (1 : 2000) overnight at 4 °C. The membrane was then incubated with the appropriate secondary antibody (1 : 20 000; Pierce), then carries on the chemiluminescence detection (Tang *et al.*, 2012; Yan *et al.*, 2011).

Field experiments for harvest yield

The rice plants of T0 to T4 generations, except T3 generation, were cultivated in plots at the Experimental Site of Nanjing Agricultural University, Nanjing, with subtropical climate from May to October in a year. For T3 generation, transgenic lines were tested in plots of Experiment Site of Sanya Nanjing Agricultural University with tropical climate from December to April. Soil properties in Nanjing field experiment were described as before (Chen *et al.*, 2016).

T2–T4 generation *pOsNAR2.1:OsNAR2.1* and wild-type plants were planted in three plots with 300 kg N/ha and without nitrogen fertilizer as blank control. The plots were 2 × 2 m in size, and the seedlings were planted in a 10 × 10 array. During rice flowering and mature stages, we collected samples from each plot for further analysis. Random four replicates (each replicate with four individual plants) from each plot were selected within the plots free from the edges, and therefore, the data of total 16 individual plants were pulled into mean value of each plot (Chen *et al.*, 2016).

The agronomic characters of T4 generation plant height, total tiller number per plant, grain weight per panicle, grain number per panicle, seed setting rate, panicle length, 1000-grain weight, yield and biomass per plant were measured at the maturity stage under 300 kg N/ha N fertilizer condition.

Dry weight, total nitrogen measurement and calculation of nitrogen use efficiency

We harvested T4 generation shoot samples from the field to analyse biomass and nitrogen under 300 kg N/ha fertilizer condition according to our previous method (Chen *et al.*, 2016) DMTE and NTE were calculated according to Chen *et al.* (2016). DMT (kg/ha) = DMA–(DMM–GY); CPAY (%) = (DMT/GY) × 100%; DMTE (%) = (DMT/DMA) × 100%; HI (%) = (GY/DMM) × 100%; The NUE method was used for the calculation as described by Chen *et al.* (2016). ANUE (kg/kg) = (GY–GY of zero-N plot)/N supply; PNUE (kg/kg) = (GY–GY of zero-N plot)/TNAM; NRE (%) = (TNAM–TNAM of zero-N plot)/N supply; PANU (kg/ha) = TNAM–TNAA; NHI (%) = (GNAM/TNAM) × 100%; NT (kg/ha) = TNAA–(TNAM–GNAM); CPNGN (%) = (NT/GNAM) × 100%; NTE (%) = (NT/TNAA) × 100%.

Determination of total N content, root ^{15}N-NO_3^- influx rate and ^{15}N-NH_4^+ influx rate in WT and transgenic seedlings

WT and transgenic rice seedlings were grown in the solution containing 1 mм NH_4^+ in IRRI solution for 2 weeks and then transferred in different forms of nitrogen for 3 additional weeks. The nitrogen treatments in this experiment included in 2.5 mм NH_4^+, 0.2 mм NO_3^-, 2.5 mм NO_3^- and 1.25 mм NH_4NO_3. The biomass and nitrogen concentration were measured for each line (n = 4 plants) under each N treatment after 3-week treatment.

For root ^{15}N uptake experiment, new rice seedlings were grown in 1 mм NH_4^+ for 3 weeks and then were nitrogen starved for 1 week before ^{15}N uptake. 2.5 mм $^{15}NH_4^+$, 0.2 mм $^{15}NO_3^-$, 2.5 mм $^{15}NO_3^-$, 1.25 mм $^{15}NH_4NO_3$, 1.25 mм $NH_4^{15}NO_3$ or 1.25 mм $^{15}NH_4^{15}NO_3$ (atom % ^{15}N: $^{15}NO_3^-$, 99%; $^{15}NH_4^+$, 99%)

was used, and the ^{15}N influx rate was calculated following the method in Tang *et al.* (2012).

Statistical analysis

The single-factor analysis of variance (ANOVA) and Tukey's test data analysis were applied in our data statistical analysis (Chen *et al.* (2016).

Acknowledgements

This work was supported by China National Key Program for Research and Development (2016YFD0100700), National Natural Science Foundation (Grant 31372122), Jiangsu Science Fund for Distinguished Young Scholars (Grant BK20160030) and the Transgenic Project (Grant 2016ZX08001003-008). The authors declare no conflict of interest.

References

Alboresi, A., Gestin, C., Leydecker, M.T., Bedu, M., Meyer, C. and Truong, H.N. (2005) Nitrate, a signal relieving seed dormancy in Arabidopsis. *Plant, Cell Environ.* **28**, 500–512.

Castro Marin, I., Loef, I., Bartetzko, L., Searle, I., Coupland, G., Stitt, M. and Osuna, D. (2011) Nitrate regulates floral induction in Arabidopsis, acting independently of light, gibberellin and autonomous pathways. *Planta*, **233**, 539–552.

Chen, J., Zhang, Y., Tan, Y., Zhang, M., Zhu, L., Xu, G. and Fan, X. (2016) Agronomic nitrogen-use efficiency of rice can be increased by driving OsNRT2.1 expression with the OsNAR2.1 promoter. *Plant Biotechnol. J.* **14**, 1705–1715.

Crawford, N.M. (1995) Nitrate: nutrient and signal for plant growth. *Plant Cell*, **7**, 859–868.

Crawford, N.M. and Glass, A.D.M. (1998) Molecular and physiological aspects of nitrate uptake in plants. *Trends Plant Sci.* **3**, 389–395.

Dechorgnat, J., Patrit, O., Krapp, A., Fagard, M. and Daniel-Vedele, F. (2012) Characterization of the Nrt2.6 gene in Arabidopsis thaliana: a link with plant response to biotic and abiotic stress. *PLoS ONE*, **7**, e42491.

Duan, Y.H., Zhang, Y.L., Shen, Q.R. and Wang, S.W. (2006) Nitrate effect on rice growth and nitrogen absorption and assimilation at different growth stages. *Pedosphere*, **16**, 707–717.

Fan, X., Shen, Q., Ma, Z., Zhu, H., Yin, X. and Miller, A.J. (2005) A comparison of nitrate transport in four different rice (Oryza sativa L.) cultivars. *Sci China C Life Sci.* **48**, 897–911.

Fan, X., Jia, L., Li, Y., Smith, S.J., Miller, A.J. and Shen, Q. (2007) Comparing nitrate storage and remobilization in two rice cultivars that differ in their nitrogen use efficiency. *J. Exp. Bot.* **58**, 1729–1740.

Fan, S.C., Lin, C.S., Hsu, P.K., Lin, S.H. and Tsay, Y.F. (2009) The Arabidopsis nitrate transporter NRT1.7, expressed in phloem, is responsible for source to sink remobilization of nitrate. *Plant Cell*, **9**, 2750–2761.

Fan, X., Tang, Z., Tan, Y., Zhang, Y., Luo, B., Yang, M., Lian, X. *et al.* (2016) Over expression of a pH sensitive nitrate transporter in rice increases crop yields. *Proc. Natl Acad. Sci. USA*, **113**, 7118–7123.

Fan, X., Naz, M., Fan, X., Xuan, W., Miller, A.J. and Xu, G. (2017) Plant nitrate transporters: from gene function to application. *J. Exp. Bot.* doi:10.1093/jxb/erx011 (online).

Fang, Z., Xia, K., Yang, X., Grotemeyer, M.S., Meier, S., Rentsch, D., Xu, X. *et al.* (2013) Altered expression of the PTR/NRT1 homologue OsPTR9 affects nitrogen utilization efficiency, growth and grain yield in rice. *Plant Biotechnol. J.* **11**, 446–458.

Feng, H., Yan, M., Fan, X., Li, B., Shen, Q., Miller, A.J. and Xu, G. (2011) Spatial expression and regulation of rice high-affinity nitrate transporters by nitrogen and carbon status. *J. Exp. Bot.* **62**, 2319–2332.

Foyer, C.H., Valadier, M.H., Migge, A. and Becker, T.W. (1998) Drought-induced effects on nitrate reductase activity and mRNA and on the coordination of nitrogen and carbon metabolism in maize leaves. *Plant Physiol.* **117**, 283–292.

Galván, A., Quesada, A. and Fernández, E. (1996) Nitrate and nitrate are transported by different specific transport systems and by a bispecific transporter in Chlamydomonas reinhardtii. *J. Biol. Chem.* **271**, 2088–2092.

Good, A.G., Shrawat, A.K. and Muench, D.G. (2004) Can less yield more? Is reducing nutrient input into the environment compatible with maintaining crop production? *Trends Plant Sci.* **12**, 597–605.

Ho, C.H. and Tsay, Y.F. (2010) Nitrate, ammonium, and potassium sensing and signaling. *Curr. Opin. Plant Biol.* **13**, 604–610.

Hsu, P.K. and Tsay, Y.F. (2013) Two phloem nitrate transporters, NRT1.11 and NRT1.12, are important for redistributing xylem-borne nitrate to enhance plant growth. *Plant Physiol.* **163**, 844–856.

Hu, B., Wang, W., Ou, S., Tang, J., Li, H., Che, R., Zhang, Z. *et al.* (2015) Variation in NRT1.1B contributes to nitrate-use divergence between rice subspecies. *Nat. Genet.* **47**, 834–838.

Huang, S., Chen, S., Liang, Z., Zhang, C., Yan, M., Chen, J., Xu, G. *et al.* (2015) Knockdown of the partner protein OsNAR2.1 for high-affinity nitrate transport represses lateral root formation in a nitrate-dependent manner. *Sci. Rep.* **5**, 18192.

Islam, M.M., Ishibashi, Y., Nakagawa, A.C., Tomita, Y., Iwaya-Inoue, M., Arima, S. and Zheng, S.H. (2016) Nitrogen redistribution and its relationship with the expression of GmATG8c during seed filling in soybean. *J. Plant Physiol.* **192**, 71–74.

Jin, Z., Zhu, Y., Li, X., Dong, Y. and An, Z. (2015) Soil N retention and nitrate leaching in three types of dunes in the Mu Us desert of China. *Sci. Rep.* **5**, 14222.

Katayama, H., Mori, M., Kawamura, Y., Tanaka, T., Mori, M. and Hasegawa, H. (2009) Production and characterization of transgenic rice plants carrying a high-affinity nitrate transporter gene (OsNRT2.1). *Breeding Sci.* **59**, 237–243.

Kronzucker, H.J., Glass, A.D.M., Siddiqi, M.Y. and Kirk, G.J.D. (2000) Comparative kinetic analysis of ammonium and nitrate acquisition by tropical lowland rice: implications for rice cultivation and yield potential. *New Phytol.* **145**, 471–476.

Kucera, I. (2003) Passive penetration of nitrate through the plasma membrane of Paracoccus denitrificans and its potentiation by the lipophilic tetraphenylphosphonium cation. *Biochim. Biophys. Acta*, **1557**, 119–124.

Léran, S., Varala, K., Boyer, J.C., Chiurazzi, M., Crawford, N., Daniel-Vedele, F., David, L. *et al.* (2014) A unified nomenclature of NITRATE TRANSPORTER 1/PEPTIDE TRANSPORTER family members in plants. *Trends Plant Sci.* **19**, 5–9.

Li, B., Xin, W., Sun, S., Shen, Q. and Xu, G. (2006) Physiological and molecular responses of nitrogen-starved rice plants to re-supply of different nitrogen sources. *Plant Soil*, **287**, 145–159.

Li, Y.L., Fan, X.R. and Shen, Q.R. (2008) The relationship between rhizosphere nitrification and nitrogen-use efficiency in rice plants. *Plant, Cell Environ.* **31**, 73–85.

Liu, X., Huang, D., Tao, J., Miller, A.J., Fan, X. and Xu, G. (2014) Identification and functional assay of the interaction motifs in the partner protein OsNAR2.1 of the two-component system for high-affinity nitrate transport. *New Phytol.* **204**, 74–80.

Matakiadis, T., Alboresi, A., Jikumaru, Y., Tatematsu, K., Pichon, O., Renou, J.P., Kamiya, Y. *et al.* (2009) The Arabidopsis abscisic acid catabolic gene CYP707A2 plays a key role in nitrate control of seed dormancy. *Plant Physiol.* **149**, 949–960.

Matzke, M.A., Primig, M., Trnovsky, J. and Matzke, A.J.M. (1989) Reversible methylation and inactivation of marker genes in sequentially transformed tobacco plants. *EMBO J.* **8**, 643–649.

Miller, A.J., Fan, X., Orsel, M., Smith, S.J. and Wells, D.M. (2007) Nitrate transport and signalling. *J. Exp. Bot.* **58**, 2297–2306.

O'Brien, J.A., Vega, A., Bouguyon, E., Krouk, G., Gojon, A., Coruzzi, G. and Gutierrez, R.A. (2016) Nitrate transport, sensing, and responses in plants. *Mol. Plant.* **9**, 837–856.

Okamoto, M., Kumar, A., Li, W., Wang, Y., Siddiqi, M.Y., Crawford, N.M. and Glass, A.D. (2006) High-affinity nitrate transport in roots of Arabidopsis depends on expression of the NAR2-like gene AtNRT3.1. *Plant Physiol.* **140**, 1036–1046.

Orsel, M., Chopin, F., Leleu, O., Smith, S.J., Krapp, A., Daniel-Vedele, F. and Miller, A.J. (2006) Characterization of a two-component high-affinity nitrate uptake system in Arabidopsis. Physiology and protein-protein interaction. *Plant Physiol.* **142**, 1304–1317.

Pouliquin, P., Boyer, J.C., Grouzis, J.P. and Gibrat, R. (2000) Passive nitrate transport by root plasma membrane vesicles exhibits an acidic optimal pH like the H(+)-ATPase. *Plant Physiol.* **122**, 265–274.

Quesada, A., Galvan, A. and Fernandez, E. (1994) Identification of nitrate transporter genes in Chlamydomonas reinhardtii. *Plant J.* **5**, 407–419.

Rahayu, Y.S., Walch-Liu, P., Neumann, G., Römheld, V., von Wiren, N. and Bangerth, F. (2005) Root-derived cytokinins as long-distance signals for NO$_3^-$-induced stimulation of leaf growth. *J. Exp. Bot.* **56**, 1143–1152.

Sanchez-Bragado, R., Serret, M.D. and Araus, J.L. (2017) The nitrogen contribution of different plant parts to wheat grains: exploring genotype, water, and nitrogen effects. *Front Plant Sci.* **7**, 1986.

Scheible, W.R., Gonzalez-Fontes, A., Lauerer, M., Muller-Rober, B., Caboche, M. and Stitt, M. (1997) Nitrate acts as a signal to induce organic acid metabolism and repress starch metabolism in tobacco. *Plant Cell*, **9**, 783–798.

Scheible, W.R., Morcuende, R., Czechowski, T., Fritz, C., Osuna, D., Palacios-Rojas, N., Schindelasch, D. *et al.* (2004) Genome-wide reprogramming of primary and secondary metabolism, protein synthesis, cellular growth processes, and the regulatory infrastructure of Arabidopsis in response to nitrogen. *Plant Physiol.* **136**, 2483–2499.

Siddiqi, M.Y., Glass, A.D., Ruth, T.J. and Rufty, T.W. (1990) Studies of the uptake of nitrate in barley: I. Kinetics of NO(3) influx. *Plant Physiol.* **93**, 1426–1432.

Souza, S.R., Stark, E.M.L.M. and Fernandes, M.S. (1998) Nitrogen remobilization during the reproductive period in two Brazilian rice varieties. *J. Plant Nutr.* **21**, 2049–2063.

Stitt, M. (1999) Nitrate regulation of metabolism and growth. *Curr. Opin. Plant Biol.* **2**, 178–186.

Szczerba, M.W., Britto, D.T. and Kronzucker, H.J. (2006) The face value of ion fluxes: the challenge of determining influx in the low-affinity transport range. *J. Exp. Bot.* **57**, 3293–3300.

Tang, Z., Fan, X., Li, Q., Feng, H., Miller, A.J., Shen, Q. and Xu, G. (2012) Knockdown of a rice stelar nitrate transporter alters long-distance translocation but not root influx. *Plant Physiol.* **160**, 2052–2063.

Tong, Y., Zhou, J.J., Li, Z. and Miller, A.J. (2005) A two-component high-affinity nitrate uptake system in barley. *Plant J.* **41**, 442–450.

Wang, Y.Y., Hsu, P.K. and Tsay, Y.F. (2012) Uptake, allocation and signaling of nitrate. *Trends Plant Sci.* **17**, 458–467.

Xu, G., Fan, X. and Miller, A.J. (2012) Plant nitrogen assimilation and use efficiency. *Annu. Rev. Plant Biol.* **63**, 153–182.

Yan, M., Fan, X., Feng, H., Miller, A.J., Sheng, Q. and Xu, G. (2011) Rice OsNAR2.1 interacts with OsNRT2.1, OsNRT2.2 and OsNRT2.3a nitrate transporters to provide uptake over high and low concentration ranges. *Plant, Cell Environ.* **34**, 1360–1372.

Yoneyama, T., Tanno, F., Tatsumi, J. and Mae, T. (2016) Whole-plant dynamic system of nitrogen use for vegetative growth and grain filling in rice plants (Oryza sativa L.) as revealed through the production of 350 Grains from a germinated seed over 150 days: a review and synthesis. *Front Plant Sci.* **7**, 1151.

Yong, Z., Kotur, Z. and Glass, A.D. (2010) Characterization of an intact two-component high-affinity nitrate transporter from Arabidopsis roots. *Plant J.* **63**, 739–748.

Zarabi, M. and Jalali, M. (2012) Leaching of nitrogen from calcareous soils in western Iran: a soil leaching column study. *Environ. Monit. Assess.* **184**, 7607–7622.

Zhang, H. and Forde, B.G. (1998) An Arabidopsis MADS box gene that controls nutrient-induced changes in root architecture. *Science*, **279**, 407–409.

Zhang, H., Jennings, A., Barlow, P.W. and Forde, B.G. (1999) Dual pathways for regulation of root branching by nitrate. *Proc. Natl Acad. Sci. USA*, **96**, 6529–6534.

Zhang, Y.L., Fan, J.B., Wang, D.S. and Shen, Q.R. (2009) Genotypic differences in grain yield and physiological nitrogen use efficiency among rice cultivars. *Pedosphere*, **19**, 681–691.

Zhou, J.J., Fernández, E., Galván, A. and Miller, A.J. (2000) A high affinity nitrate transport system from Chlamydomonas requires two gene products. *FEBS Lett.* **466**, 225–227.

Zhuo, D., Okamoto, M., Vidmar, J.J. and Glass, A.D. (1999) Regulation of a putative high-affinity nitrate transporter (Nrt2;1At) in roots of Arabidopsis thaliana. *Plant J.* **17**, 563–568.

19

The expression of heterologous Fe (III) phytosiderophore transporter *HvYS1* in rice increases Fe uptake, translocation and seed loading and excludes heavy metals by selective Fe transport

Raviraj Banakar[1], Ána Alvarez Fernández[2], Javier Abadía[2], Teresa Capell[1] and Paul Christou[1,3,*]

[1]Departament de Producció Vegetal i Ciència Forestal, Universitat de Lleida-Agrotecnio Center Lleida, Lleida, Spain
[2]Department of Plant Nutrition, Aula Dei Experimental Station, Consejo Superior de Investigaciones Científicas (CSIC), Zaragoza, Spain
[3]ICREA, Catalan Institute for Research and Advanced Studies, Barcelona, Spain

*Correspondence
email christou@pvcf. udl.es

Keywords: Rice, metal transporters, iron, toxic metals, barley YS1 transporter, 2' deoxymugenic acid.

Summary

Many metal transporters in plants are promiscuous, accommodating multiple divalent cations including some which are toxic to humans. Previous attempts to increase the iron (Fe) and zinc (Zn) content of rice endosperm by overexpressing different metal transporters have therefore led unintentionally to the accumulation of copper (Cu), manganese (Mn) and cadmium (Cd). Unlike other metal transporters, barley Yellow Stripe 1 (HvYS1) is specific for Fe. We investigated the mechanistic basis of this preference by constitutively expressing *HvYS1* in rice under the control of the *maize ubiquitin1* promoter and comparing the mobilization and loading of different metals. Plants expressing *HvYS1* showed modest increases in Fe uptake, root-to-shoot translocation, seed accumulation and endosperm loading, but without any change in the uptake and root-to-shoot translocation of Zn, Mn or Cu, confirming the selective transport of Fe. The concentrations of Zn and Mn in the endosperm did not differ significantly between the wild-type and *HvYS1* lines, but the transgenic endosperm contained significantly lower concentrations of Cu. Furthermore, the transgenic lines showed a significantly reduced Cd uptake, root-to-shoot translocation and accumulation in the seeds. The underlying mechanism of metal uptake and translocation reflects the down-regulation of promiscuous endogenous metal transporters revealing an internal feedback mechanism that limits seed loading with Fe. This promotes the preferential mobilization and loading of Fe, therefore displacing Cu and Cd in the seed.

Introduction

Iron (Fe) is an important micronutrient for all living organisms (Winterbourn, 1995). Plants acquire Fe from the soil and mobilize it from the roots to the aerial organs to support essential processes such as photosynthesis, electron transport and respiration (Morrissey and Guerinot, 2009). Fe is also loaded into the seed endosperm to support germination (Lanquar *et al.*, 2005) and thus becomes available as a micronutrient for humans. Rice is an important staple food crop, particularly in the developing world, but rice grains do not accumulate high levels of Fe, leading to severe Fe deficiency in populations that rely mostly on rice for their nutritional needs (Gómez-Galera *et al.*, 2010; Pérez-Massot *et al.*, 2013).

Metal acquisition and mobilization in plants are controlled by several families of membrane-bound metal transporters (Hall and Williams, 2003; Vert *et al.*, 2002) including the Fe-regulated transporter (IRT), natural resistance-associated macrophage protein (NRAMP), cation diffusion facilitator (CDF), yellow stripe-like (YSL) and heavy metal ATPase (HMA) transporter families, as well as other Fe transporters in the chloroplast and vacuolar membranes (Duy *et al.*, 2007; Hall and Williams, 2003; Vert *et al.*, 2002; Zhang *et al.*, 2012). Iron acquisition in rice involves different strategies for Fe^{2+} and Fe^{3+} (Ishimaru *et al.*, 2006; Kobayashi and Nishizawa, 2012; Sperotto *et al.*, 2012). In strategy I, Fe^{2+} ions are taken up into the root epidermis by OsIRT1/OsIRT2 in the plasma membrane (Ishimaru *et al.*, 2006; Lee and An, 2009; Vert *et al.*, 2002) and are then transported via the phloem and xylem to accumulate in the seeds (Ishimaru *et al.*, 2010; Takahashi *et al.*, 2011). Phloem transport involves the Fe^{2+} chelator nicotianamine (NA) and the YSL family transporters YSL2 and YSL16, whereas xylem transport involves NRAMP1 (Ishimaru *et al.*, 2010; Takahashi *et al.*, 2011) and the citrate efflux transporter FRD3 (Durrett *et al.*, 2007). In strategy II, phytosiderophores (PS) such as mugineic acid (MA) and deoxymugenic acid (DMA) are secreted to the rhizosphere (Ma *et al.*, 1999) where they solubilize Fe^{3+} by forming DMA-Fe^{3+} complexes (Ma *et al.*, 1999). The complex is taken up into the roots by YSL15 in the plasma membrane (Inoue *et al.*, 2009). The DMA-Fe^{3+} complex is transported through the phloem by YSL18 and accumulates in the seeds in the same form (Ayoma *et al.*, 2009), whereas translocation through the xylem is also mediated by the citrate efflux transporter FRDL1 (Yokosho *et al.*, 2009). Rice, which is adapted for growing in anaerobic soils where Fe is more soluble, produces much less PS than barley, which is adapted to alkaline soils. In fact, rice is the only cereal species that combines components of strategy I plants (OsIRT1 and OsIRT2; Ishimaru *et al.*, 2006) with PS production and Fe-PS uptake (OsYSL15; Inoue *et al.*, 2009).

Although Fe is abundant in the soil, rice has only a limited ability to acquire and mobilize Fe and load it into the endosperm

(Lee and An, 2009; Lee et al., 2009) most likely due to the weak expression of Fe transporters in the root (Inoue et al., 2009; Lee and An, 2009; Lee et al., 2009; Tan et al., 2015). Previous efforts to increase the uptake of Fe into rice plants have therefore focused on the overexpression of metal transporters (Bashir et al., 2013). However, most transporters are promiscuous and those responsible for the mobilization of Fe may also transport Zn (another important micronutrient) and other metals such as Cu, Mn, Ni and Cd, some of which are toxic even at low levels (Hall and Williams, 2003; Ishimaru et al., 2010; Takahashi et al., 2011; Thomine and Vert, 2013; Vert et al., 2002). The overexpression of OsIRT1, OsIRT2, MxIRT1, AtIRT1, OsYSL15 and OsYSL2 in rice therefore increased the levels of Zn, Cu, Mn, Cd and Ni mobilized from the soil and this was shown to be detrimental to plant health (Lee and An, 2009; Nishida et al., 2011; Tan et al., 2015; Uraguchi and Fujiwara, 2012).

One approach that can address this challenge is the overexpression of heterologous metal transporters that are selective for Fe, with no affinity for other divalent cations (Clemens et al., 2013; Slamet-Loedin et al., 2015). The barley (Hordeum vulgare) YS1 protein (HvYS1) is an Fe-selective metal transporter expressed in the root epidermal cells (Murata et al., 2006, 2008, 2015). HvYS1 expression is induced by Fe deficiency but not by the depletion of other metals (Ueno et al., 2009). Yeast complementation studies have shown that HvYS1 is a strict DMA-Fe^{3+} transporter that does not interact with Zn, Cu, Mn or Cd complexed with DMA or metals complexed with NA (Murata et al., 2006). Hence, this selectivity is attributed to an Fe-specific outer membrane loop between the sixth and seventh transmembrane domains (Murata et al., 2008).

Here, we investigated the mechanism by which HvYS1 promotes the selective transport of Fe using rice as a model. The heterologous expression of HvYS1 improved Fe uptake and root-to-shoot translocation, and achieved a moderate increase in Fe seed loading, without increasing the uptake and root-to-shoot translocation of Zn, Cu or Mn. The concentrations of Zn and Mn in the seed were unaffected by HvYS1 expression, whereas the concentration of Cu declined. Cadmium uptake, root-to-shoot translocation and seed loading were also inhibited in these plants. The preferential mobilization of Fe at the expense of other metals reflects the inhibition of heavy metal seed loading due to the selective transport of Fe by HvYS1.

Results

The constitutive overexpression of HvYS1 in rice improves Fe uptake, translocation and seed loading

We co-transformed 7-day-old mature seed-derived zygotic rice embryos with a plasmid containing HvYS1 driven by the constitutive maize ubiquitin 1 (ubi-1) promoter and another plasmid carrying the selectable marker hpt driven by the CaMV35S promoter and regenerated transgenic plants under hygromycin selection. HvYS1 expression in 15 independent transgenic lines was confirmed by RNA blot analysis (Figure 1). These lines and corresponding wild-type plants were grown to maturity and T_1 seeds were collected. The five transgenic lines with the highest levels of HvYS1 expression were bred to homozygosity for detailed analysis.

We hypothesized that constitutive HvYS1 expression might improve Fe uptake, root-to-shoot translocation and seed loading in the transgenic lines because HvYS1 is a specific Fe transporter in barley expressed in root epidermal cells and achieves Fe (III)-PS

translocation when expressed in yeast (Murata et al., 2006), X. laevis oocytes (Murata et al., 2008) and petunia (Murata et al., 2015). Accordingly, the T_2 HvYS1 transgenic lines contained up to 1.6-fold more Fe in the roots than wild-type controls, that is 566 ± 38 vs 345 ± 10 µg Fe/g dry weight (DW) (Figure 2a). This in turn enhanced the root-to-shoot translocation of Fe in the transgenic lines, resulting in up to 2.2-fold more Fe in the leaves, that is 231 ± 10 vs 104 ± 5 µg Fe/g DW (Figure 2b). This increase in Fe uptake and root-to-shoot translocation also had an impact on Fe seed loading. The husks of the transgenic seeds contained up to 2.1-fold more Fe than wild-type seeds: 216 ± 3 vs 102 ± 4 µg Fe/g DW (Figure 2c). The unpolished transgenic seeds contained up to 1.6-fold more Fe than wild-type seeds: 24.0 ± 0.5 vs 15.4 ± 0.4 µg Fe/g DW (Figure 2d), whereas the polished transgenic seeds (the endosperm) contained 2.1-fold more Fe than wild-type endosperm: 8.7 ± 0.3 vs 4.0 ± 0.1 µg/g DW Fe (Figure 2e). These results suggest that HvYS1 expression in the transgenic lines improved Fe mobilization from the soil to the roots, root-to-shoot translocation and seed loading, with loading of Fe occurred preferentially into the endosperm rather than into the bran.

DMA synthesis and accumulation are enhanced in the HvYS1 transgenic plants

Rice produces DMA (Araki et al., 2015), and HvYS1 transports Fe^{3+} as a complex with DMA and MA with the same efficiency (Murata et al., 2008). We therefore hypothesized that the higher levels of Fe in the transgenic lines should be accompanied by higher levels of DMA. We measured the amount of DMA in the roots, leaves and seeds of selected T_2 HvYS1 transgenic lines and observed significantly higher levels of DMA in all three tissues compared to wild-type plants (Figure 3a, b, c). These data confirm that the increased mobilization of Fe in the transgenic plants coincides with higher levels of DMA, indicating that the additional Fe is likely to be mobilized as an Fe^{3+}-DMA complex. We then measured the levels of NA in the tissues where we measured DMA to investigate whether the expression of HvYS1 followed by Fe^{3+}-DMA transport influences NA levels. Although the quantification of NA was not possible in roots as the levels were below the detection limit, transgenic lines did not differ significantly from wild type for NA levels in leaves and seeds (Figure 3d, e, f). The data indicate that endogenous NA synthesis and accumulation were not influenced due to Fe^{3+}-DMA transport by HvYS1.

The selective mobilization of Fe by HvYS1 does not affect the uptake or root-to-shoot translocation of Zn, Cu and Mn

As many Fe transporters can also transport Zn, Cu and Mn (Lee et al., 2009), we investigated the distribution of these three metals in the HvYS1 transgenic lines to confirm the specificity of the transporter in its heterologous environment. We found no difference in the distribution of these three metals when comparing transgenic and wild-type roots (Figure 4a) and leaves (Figure 4b) suggesting that HvYS1 achieves the selective uptake and root-to-shoot translocation of Fe and excludes Zn, Cu and Mn.

The selective mobilization of Fe by HvYS1 does not affect seed loading with Mn but influences the distribution of Zn in the husk and Cu in the endosperm

In contrast to the straightforward metal distribution profile in the vegetative tissues, the impact of HvYS1 on metal distribution in

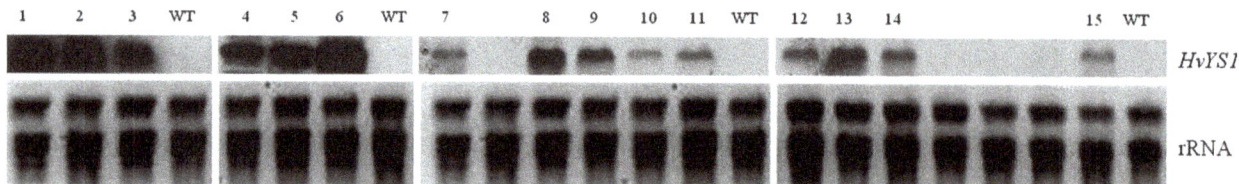

Figure 1 RNA blot analysis showing transgene expression in the leaf tissue of wild-type (WT) and transgenic lines expressing *HvYS1*. rRNA: ribosomal RNA; *HvYS1*: barley yellow stripe 1 transporter.

Figure 2 Concentrations of Fe (μg/g DW) in roots (a), leaves (b), husks (c), unpolished seeds (d) and polished seeds (e) of wild-type (WT) and T$_2$ generation transgenic lines expressing *HvYS1* (lines 1, 2, 3, 4, 5). Asterisks indicate a statistically significant difference between wild-type and transgenic plants as determined by Student's *t*-test (*P* < 0.05; *n* = 6). DW: dry weight. Iron measurements in husk were taken from two representative transgenic lines.

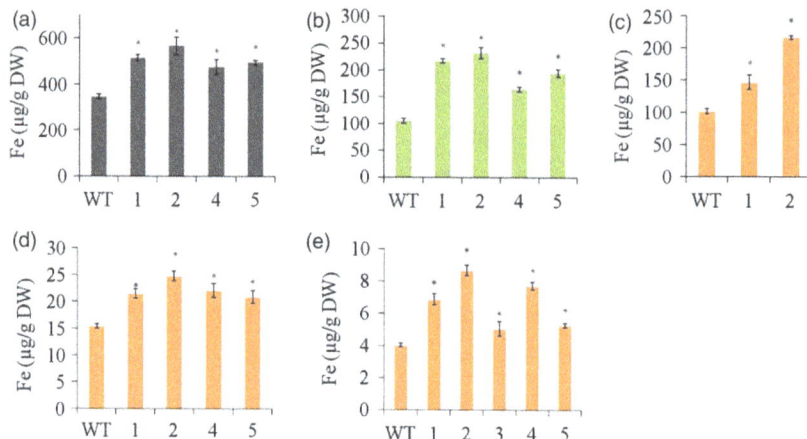

Figure 3 Concentration of 2'-deoxymugenic acid (DMA) and nicotianamine (NA) (μg/g FW) in roots, leaves and polished seeds of wild-type (WT) and two selected T$_2$ generation transgenic lines expressing *HvYS1* (lines 1, 2). Asterisks indicate a statistically significant difference between wild-type and transgenic plants as determined by Student's *t*-test (*P* < 0.05; *n* = 3). NA levels in the roots were below the detection limit. FW: fresh weight.

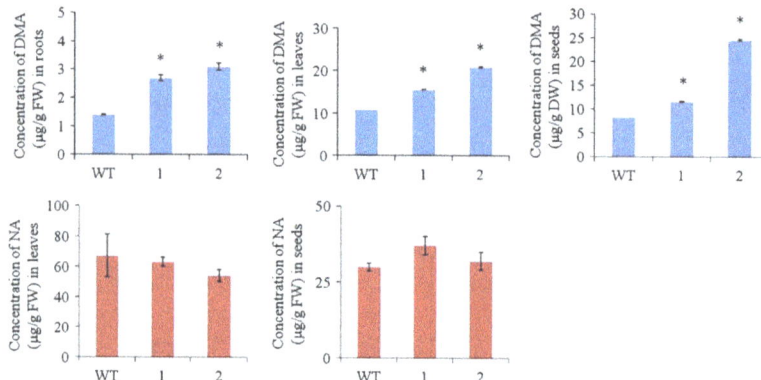

the seeds was more complex. We compared the distribution of metals in the husk (Figure 4c), unpolished (Figure 4d) and polished seeds (Figure 4e). We found no difference between the transgenic and wild-type seeds in terms of Mn loading, suggesting the distribution of Mn among the different seed tissues was unaffected by the moderate increase in Fe loading caused by the expression of *HvYS1*. However, Zn was specifically displaced from the husk in the transgenic lines, resulting in a 1.7-fold depletion, from 45 ± 1 down to 26 ± 2 μg Zn/g DW (Figure 4c), although there was no significant difference in Zn levels when we compared the unpolished or polished transgenic and wild-type seed. In contrast to the situation for Mn and Zn, we found that Cu was depleted in all three seed tissues in the transgenic lines. The transgenic husk contained 7.8 ± 0.3 μg Cu/g DW compared to 20 ± 0.6 μg Cu/g DW in the wild-type husk, reflecting a 2.5-fold decrease in Cu (Figure 4c). The unpolished transgenic seed contained 3.03 ± 0.1 μg Cu/g DW, 3.7-fold lower than the wild-type level of 11.5 ± 0.1 μg Cu/g DW (Figure 4d). Finally, the polished transgenic seed contained 2.5 ± 0.1 μg/g DW Cu, 3.8-fold lower than the wild-type level

of 9 ± 0.1 μg Cu/g DW (Figure 4e). The lower levels of Zn and Cu in the transgenic seeds suggest that the increase in the delivery of Fe selectively suppresses Zn accumulation in the husk and Cu accumulation in all seed tissues, with the effect being particularly intense in the endosperm.

The selective mobilization of Fe in the transgenic lines suppresses the mobilization of Cd at all steps along the translocation pathway

Many Fe transporters not only transport other divalent cations such as Zn, Mn and Cu, but also toxic metals such as Cd (Lee *et al.*, 2009; Takahashi *et al.*, 2011). We therefore compared the distribution of Fe and Cd in the transgenic lines and wild-type controls when Cd was added to the soil to gain more insight into the selective mobilization of different metals by HvYS1 in its heterologous environment. Unlike Zn, Mn and Cu, whose distribution in vegetative tissues was unaffected, we found that the transgenic lines contained significantly lower levels of Cd than wild-type plants in the roots and leaves as well as the seeds (Figure 5). When compared to the wild type, the transgenic lines

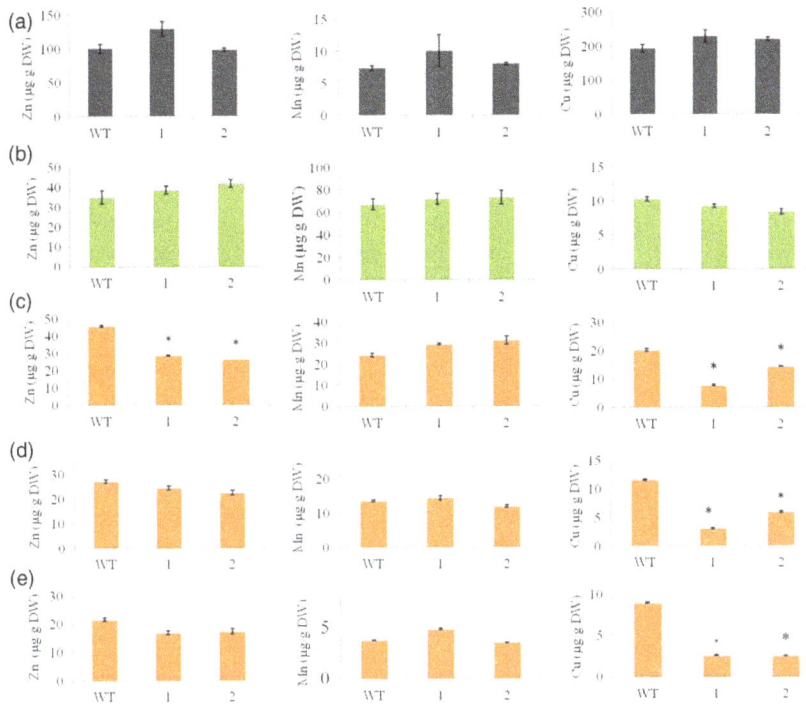

Figure 4 Concentrations of Zn (left), Mn (middle) and Cu (right), all in µg metal per g DW, in roots (a), leaves (b), husks (c), unpolished seeds (d) and polished seeds (e) of wild-type (WT) and two selected T_2 generation transgenic lines expressing HvYS1 (lines 1, 2). Asterisks indicate a statistically significant difference between wild-type and transgenic plants as determined by Student's t-test ($P < 0.05$; $n = 6$). DW: dry weight.

accumulated 2.3-fold less Cd in roots (Figure 5a), fivefold less Cd in leaves (Figure 5b) and 2.4-fold less Cd in unpolished seeds (Figure 5c). In contrast, when compared to the wild type, the transgenic lines contained 2.4-fold more Fe in roots (Figure 5d), 1.8-fold more Fe in leaves (Figure 5e) and 1.9-fold more Fe in seeds (Figure 5f). These data suggest that Fe mobilized by HvYS1 in the roots and shoots suppresses the uptake and translocation of Cd and that Fe delivery to the seeds also prevents seed loading with Cd and/or displaces Cd that is already in situ.

Homeostasis mechanisms limit Fe seed loading in the transgenic lines

The selective mobilization of Fe by HvYS1 in the transgenic lines leads to a moderate increase in Fe levels in the seeds. A possible explanation for the modest increase in Fe levels in transgenic lines is that an Fe homeostasis mechanism imposes limitations on Fe accumulation. Fe homeostasis in rice involves a number of genes controlling uptake, root-to-shoot translocation, remobilization from the flag leaf and deposition in seeds (Table S1) suggesting that these endogenous genes may be modulated by the heterologous expression of HvYS1. To investigate the influence of HvYS1 on the expression of endogenous Fe homeostasis

genes, we measured the expression of genes controlling Fe uptake (OsIRT1, OsYSL15 and OsNRAMP5), long-distance transport (OsFRDL1, OsYSL2, OsYSL16, OsYSL18 and OsNRAMP1), vacuolar sequestration (OsVIT1), storage (OsFERRITIN1), endogenous phytosiderophore synthesis pathway (OsSAMS1, OsNAS2, OsNAS3, OsNAAT1, OsDMAS1) and transcription factor (OsIDEF1) in the roots, leaves and seeds of the HvYS1 transgenic lines and wild-type controls (Table S1).

In the roots, OsIRT1 and OsYSL15 (controlling Fe uptake) were down-regulated by 3-fold and 2.2-fold, respectively, in the transgenic lines (Table 1; Figure S1). Furthermore, the Fe^{2+}-NA transporter OsYSL16 was down-regulated by 3.7-fold, the Fe^{3+}-citrate transporter OsFRDL1 was down-regulated by 5.1-fold, the vacuolar transporter OsVIT1 was down-regulated by 7.7-fold, iron storage OsFERRITIN1 was down-regulated by 5.6-fold, and the transcription factor regulating metal homeostasis OsIDEF1 was down-regulated by 3-fold (Table 1; Figure S1). Our results suggest that endogenous Fe uptake and root-to-shoot translocation are down-regulated by HvYS1 expression. The expression of OsNRAMP5 and OsNRAMP1 was up-regulated by 3.2-fold and 4.4-fold, respectively, in the transgenic lines compared to wild-type controls (Table 1; Figure S1). This suggests that these genes

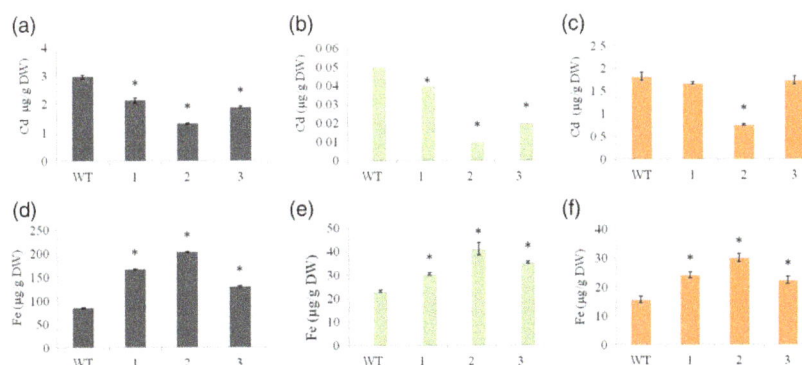

Figure 5 Concentrations of Cd (top row) and Fe (bottom row), both in µg/g DW, in (a and d) roots, (b and e) leaves and (c and f) unpolished seeds of wild-type (WT) and T_3 generation transgenic lines expressing HvYS1 (lines 1, 2, 3) supplied with 10 µM $CdCl_2$. Asterisks indicate a statistically significant difference between wild-type and transgenic plants as determined by Student's t-test ($P < 0.05$; $n = 6$). DW: dry weight.

Table 1 Fold change in the relative expression level of *OsIRT1*, *OsYSL15*, *OsNRAMP5*, *OsVIT1*, *OsYSL2*, *OsYSL16*, *OsFRDL1*, *OsYSL18*, *OsNRAMP1*, *OsFERRITIN1*, *OsSAMS1*, *OsNAS2*, *OsNAS3*, *OsNAAT1*, *OsDMAS1 and OsIDEF1* in roots (left), flag leaf (centre) and seeds (right) at grain filling stage in wild-type (WT) and T$_2$ generation transgenic lines expressing *HvYS1* (Line 1 and Line 2). Arrows show up-regulation and down-regulation. Gene-specific primers are listed in Table S2. NC, no change; ND, not determined

Genes		Roots	Flag leaf	Seeds
Metal uptake	*OsIRT1*	↓3	1.8↑	2.8↓
	OsYSL15	↓2.2	ND	7.2↓↓
	OsNRAMP5	↑3.2	2.6↑	NC
Vacuolar sequestration	*OsVIT1*	↓↓7.7	NC	9↓↓
Long-distance transport	*OsYSL2*	NC	NC	3.4↓
	OsYSL16	↓3.7	3↑	NC
	OsFRDL1	↓↓5.1	NC	2.7↓
	OsYSL18	NC	2.2↑	7.5↓↓
	OsNRAMP1	↑4.4	NC	NC
Iron storage	*OsFERRITIN1*	↓5.6	3.1↑	NC
Endogenous phytosiderophore synthesis pathway	*OsSAMS1*	6.4↑	2.2↓	NC
	OsNAS2	17↓↓↓↓	3.2↑	1.8↑
	OsNAS3	1.7↓	9↑↑	3.4↑
	OsNAAT1	2.5↓	2.2↑	2.1↓
	OsDMAS1	2↑	2.2↑	2.5↓
Transcription factor	*OsIDEF1*	3↓	1.8↑	2.5↓

were up-regulated to balance Fe uptake and translocation. Among the endogenous PS synthesis genes, expression of *OsSAMS1* (6.4-fold), *OsDMAS1* (2-fold) was up-regulated, whereas the expression of *OsNAS2* (17-fold), *OsNAS3* (1.7-fold) and *OsNAAT1* (2.5-fold) was down-regulated. This suggests that the expression of *HvYS1* modulates expression of genes for the conversion of L-methionine to S-adenosyl methionine (*OsSAMS1*) and 3′-keto intermediate to DMA (*OsDMAS1*) but such an alteration in endogenous PS pathway suppresses the expression of genes for the conversion of S-adenosyl methionine to NA (*OsNAS2*, *OsNAS3*) and NA to 3′-keto intermediate (*OsNAAT1*).

In the leaves, *OsIRT1*, *OsNRAMP5*, *OsYSL16*, *OsYSL18*, *OsFERRITIN1* and *OsIDEF1* were all up-regulated in the transgenic lines by between 1.8-fold and 3.1-fold (Table 1; Figure S1), indicating that Fe remobilization from leaves and storage was enhanced in the transgenic lines. Similarly, expression of *OsNAS2*, *OsNAS3*, *OsNAAT1 and OsDMAS1* was up-regulated by 3.2-, 9-, 2.2- and 2.2-fold, respectively, in the transgenic lines compared to the wild type (Table 1; Figure S1). In contrast, expression of *OsSAMS1* was down-regulated by 2.2-fold in the transgenic lines compared to wild type (Table 1; Figure S1). These results suggest that generally the expression of endogenous PS pathway genes was modulated to enhance Fe mobilization. In the seeds, *OsIRT1*, *OsYSL15*, *OsYSL2*, *OsYSL18*, *OsFRDL1*, *OsVIT1* and *OsIDEF1* were down-regulated by 2.8-fold, 7.2-fold, 3.4-fold, 7.5-fold, 2.7-fold, ninefold and 2.5-fold, respectively (Table 1; Figure S1), suggesting that endogenous genes that promote Fe accumulation are down-regulated to limit Fe accumulation in the seeds. In contrast to the general down-regulation of metal transporters, the expression of *OsNAS2* and *OsNAS3* was up-regulated by 1.8- and 3.4-fold, respectively, whereas *OsNAAT1* and *OsDMAS1* expressions were down-regulated by 2.1- and 2.5-fold, respectively (Table 1; Figure S1). These results suggest that the

conversion of NA to the 3′-keto intermediate (by *OsNAAT1*) followed by the latter's conversion to DMA (by *OsDMAS1*) was down-regulated to limit Fe accumulation in seeds.

Discussion

Rice plants secrete DMA from the root surface (Suzuki *et al.*, 2008), which chelates Fe^{3+} in the soil allowing the resulting Fe^{3+}-DMA complex to be taken up by Fe^{3+}-DMA transporters. The complexes are translocated internally and ultimately accumulate in the seeds (Inoue *et al.*, 2009). One strategy to enhance Fe uptake, translocation and accumulation is therefore to overexpress appropriate metal transporters. However, by and large metal transporters are promiscuous and they can transport toxic metals such as Cd, along with metals that are essential nutrients. The broad impact of heterologous metal transporter overexpression on metal accumulation in seeds, and the expression of endogenous genes involved in metal homeostasis, is thus still unclear because the mechanisms of metal homeostasis in plants are complex and they depend on many different factors.

To address these issues in more detail, we generated transgenic rice plants overexpressing the barley Fe^{3+}-DMA transporter HvYS1, which is strictly specific for Fe and therefore allows studying the impact on Fe levels. The constitutive expression of *HvYS1* increased Fe uptake from the soil, root-to-shoot translocation and seed loading, resulting in concentration increases of 1.6-, 2.2- and 2.1-fold, respectively, in the roots, leaves and endosperm of the T$_2$ transgenic plants. The transgenic lines also accumulated significantly higher levels of DMA in the roots, leaves and seeds. Similar results were reported by others when the Fe^{2+} transporter genes *OsIRT1*, *MxIRT1* and *AtIRT1* were expressed in rice, as well as the Fe^{3+}-DMA transporter gene *OsYSL15* and the promiscuous metal transporter gene *OsNRAMP5* (whose product can transfer Fe, Mn and Cd), but the increase in endosperm Fe levels was more moderate, leading to concentration increases of 1.2- to 1.3-fold when compared to wild-type seeds (Boonyaves *et al.*, 2016; Ishimaru *et al.*, 2012; Lee and An, 2009; Lee *et al.*, 2009; Tan *et al.*, 2015). This suggests that the overexpression of *HvYS1* enhances Fe^{3+}-DMA uptake, root-to-shoot translocation and seed loading more efficiently than the other genes, resulting in a 2.1-fold increase in Fe levels in the endosperm (i.e. from 4 µg Fe/g DW in wild-type plants to 8.7 µg Fe/g DW in the transgenic lines). Compared to other cereals, barley is highly tolerant to Fe deficiency and the presence of the efficient Fe transporter YS1 in the plasma membrane may explain this phenomenon (Murata *et al.*, 2006, 2008).

Next, we investigated the impact of heterologous *HvYS1* expression on Zn, Mn and Cu uptake, root-to-shoot translocation and seed accumulation. The *HvYS1* lines did not differ significantly from wild-type plants in terms of the concentration of Zn, Mn and Cu in the roots and leaves. Similarly, *HvYS1* expression in *Xenopus laevis* oocytes revealed that HvYS1 has the ability to transport Fe^{3+}-MA complexes but not complexes with other metals (Murata *et al.*, 2006, 2008). In contrast, genes encoding the promiscuous metal transporters OsIRT1, MxIRT1, OsNRAMP5, OsHMA3 and AtIRT1 increased the levels of Zn, Mn and Cu, respectively, by 1.3-, 1.2- and 1.4-fold in rice roots, and by 1.4-, 1.2- and 1.3-fold in rice leaves (Boonyaves *et al.*, 2016; Ishimaru *et al.*, 2012; Lee and An, 2009; Lee *et al.*, 2009; Tan *et al.*, 2015; Ueno *et al.*, 2010). There was no difference in the distribution of Zn and Mn in the unpolished and polished seeds of the transgenic lines compared to wild-type seeds, but the concentration of Cu was 3.8-fold lower in the transgenic seeds. The overexpression of *OsIRT1*, *MxIRT1*, *OsHMA3*, *OsNRAMP5* and *AtIRT1* increased the concentrations of Zn, Mn and Cu in rice seeds by 1.5-, 1.3- and 1.6-fold, respectively (Boonyaves *et al.*, 2016; Ishimaru *et al.*, 2012; Lee and An, 2009; Lee *et al.*, 2009; Tan *et al.*, 2015; Ueno *et al.*, 2010). The promiscuous transporters increase seed loading with Zn, Mn and Cu by directly transporting these metals into the seed, whereas the specificity of HvYS1 means that only Fe is loaded and any differences in other metals must be attributed to passive effects, that is Cu being passively displaced by Fe in the *HvYS1* transgenic rice plants. Zinc is nutritionally important for human health, whereas Mn and Cu are toxic even at moderate levels (Alimba *et al.*, 2016). The selective loading of Fe into the endosperm of the *HvYS1* lines is therefore advantageous over the general increase in metal levels previously achieved by the overexpression of promiscuous transporters (Boonyaves *et al.*, 2016; Ishimaru *et al.*, 2012; Lee and An, 2009; Lee *et al.*, 2009; Tan *et al.*, 2015; Ueno *et al.*, 2010).

The expression of *HvYS1* doubled the concentration of Fe in the transgenic seeds compared to wild-type seeds, an effect similar to those achieved by expressing *OsIRT1*, *AtIRT1*, *MxIRT1*, *OsYSL15* or *OsNRAMP5* (Boonyaves *et al.*, 2016; Ishimaru *et al.*, 2012; Lee and An, 2009; Lee *et al.*, 2009; Slamet-Loedin *et al.*, 2015; Tan *et al.*, 2015; Ueno *et al.*, 2010). This suggests there may be a limit to the amount of Fe that can be deposited in the seed, because metal homeostasis mechanisms are tightly regulated and do not allow Fe accumulation beyond certain limits (Sperotto *et al.*, 2012; Wang *et al.*, 2013). Hence, we investigated the impact of *HvYS1* on the expression of endogenous genes controlling Fe mobilization, including the Fe-regulated metal uptake transporters encoded by *OsIRT1* (Fe-Zn-Mn), *OsYSL15* (Fe^{3+}-DMA) and *OsNRAMP5* (Fe-Mn); the vacuolar Fe-Zn transporter encoded by *OsVIT1*; long-distance transporters encoded by *OsYSL2* (Fe-Mn), *OsYSL16* (Fe), *OsNRAMP1* (Fe), *OsFRDL1* (Fe^{3+}-citrate) and *OsYSL18* (Fe^{3+}-DMA), Fe storage protein ferritin encoded by *OsFERRITIN1*, genes involved in PS synthesis such as *OsSAMS1*, *OsNAS2*, *OsNAS3*, *OsNAAT1*, *OsDMAS1* and finally *OsIDEF1* a transcription factor regulating Fe homeostasis. This allowed us to unravel facets of the mechanism through which Fe accumulation in the seeds is regulated and how the homeostasis mechanism operating in different tissues regulates Fe accumulation in roots, leaves and seeds.

In the roots of the *HvYS1* transgenic lines, *OsIRT1* and *OsYSL15* were slightly down-regulated. These encode Fe-regulated transporters and the corresponding genes are induced by Fe deficiency and repressed when Fe levels are sufficient (Inoue *et al.*, 2009; Lee and An, 2009). Therefore, the higher Fe levels in the transgenic lines appear to create an Fe-sufficient environment causing these two genes to be suppressed. OsIRT1 carries Zn and Mn in addition to Fe, so the down-regulation of *OsIRT1* may trigger the expression of the Fe-Mn transporter *OsNRAMP5* to increase the uptake of Mn. Iron mobilization from the roots through the xylem promotes Fe seed loading (Yoneyama *et al.*, 2015). The transporter OsNRAMP1 loads the xylem with Fe (Takahashi *et al.*, 2011), and OsNRAMP5 promotes both Fe uptake and xylem loading (Yang *et al.*, 2014). The up-regulation of these two transporters in the *HvYS1* lines therefore suggests an increase in Fe xylem loading and root-to-shoot translocation. DMA plays a major role in uptake and root-to-shoot translocation of Fe in rice (Bashir *et al.*, 2014). Synthesis of S-adenosyl methionine (SAM) from L-methionine is carried out by *OsSAMS1*, and NA is synthesized from SAM through expression of *OsNAS2 and OsNAS3*. NA is then converted to a 3′-keto

intermediate by *OsNAAT1*, and finally, *OsDMAS1* catalyses the formation of DMA through the 3′-keto intermediate precursor molecule (Bashir *et al.*, 2014). In *HvYS1* lines, expression of *OsSAMS1* and *OsDMAS1* was up-regulated, whereas expression of *OsNAS2, OsNAS3 and OsNAAT1* was down-regulated. Expression of *OsSAMS1, OsNAS2, OsNAS3, OsNAAT1 and OsDMAS1* was up-regulated in roots under Fe deficiency, while the reverse was true under Fe sufficiency conditions (Bashir and Nishizawa, 2006; Bashir *et al.*, 2014; Inoue *et al.*, 2003, 2008). Therefore, in *HvYS1* lines, up-regulation of *OsDMAS1* increased DMA levels due to increased Fe levels in roots. The down-regulation of *OsNAS2, OsNAS3* and *OsNAAT1* indicates that the Fe homeostasis mechanism operates to restrict Fe uptake and root-to-shoot translocation by limiting the synthesis of NA and its conversion into DMA.

The remobilization of Fe from the flag leaf through the phloem is important for seed loading (Curie *et al.*, 2009; Yoneyama *et al.*, 2015), and this is facilitated by the transporters encoded by *OsYSL16* (Kakei *et al.*, 2012) and *OsYSL18* (Ayoma *et al.*, 2009). *OsYSL16* and *OsYSL18* were up-regulated in the transgenic lines, suggesting an increase in phloem loading with Fe, resulting in higher Fe levels in the seeds. Similar to its role in uptake and root-to-shoot translocation of Fe, DMA is also important in the remobilization of Fe from flag leaf to seeds (Ayoma *et al.*, 2009; Masuda *et al.*, 2009). *OsNAS2, OsNAS3, OsNAAT1* and *OsDMAS1* were up-regulated in *HvYS1* lines. The up-regulation of *OsNAS2, OsNAS3, OsNAAT1* and *OsDMAS1* suggests increased synthesis and accumulation of DMA in flag leaf leading to enhanced Fe remobilization from flag leaf in transgenic lines compared to wild type. The Fe storage protein ferritin is also regulated by the amount of Fe present in the cell (Jain and Connolly, 2013). The induction of *OsFERRITIN1* in the flag leaf suggests that Fe in the flag leaf was not freely available for remobilization through the phloem because Fe is diverted to the chloroplast (Long *et al.*, 2008). Iron storage as a complex with ferritin therefore appears to act as a buffer to control the remobilization of Fe through the phloem (Long *et al.*, 2008). *OsIRT1, OsYSL15, OsFRDL1* and *OsYSL18* were down-regulated in the transgenic seeds, which was surprising because all four corresponding proteins are known to contribute to Fe seed loading. Indeed, the suppression of *OsYSL15* and *OsFRDL1* expressions resulted in 1.5-fold and 1.3-fold lower levels of Fe in rice seeds, respectively (Lee *et al.*, 2009; Yokosho *et al.*, 2009), whereas the overexpression of *OsIRT1* increased Fe levels in the seed by 1.3-fold, with OsYSL18 proposed to facilitate Fe loading into the phloem (Ayoma *et al.*, 2009; Lee and An, 2009). Similar to the metal transporters, expression of *OsNAAT1* and *OsDMAS1* was down-regulated in the transgenic lines. DMA is important for Fe seed loading (Masuda *et al.*, 2009). Therefore, limited loading of Fe in the transgenic seeds suggests that homeostasis is triggered once a certain threshold is reached, which involves the down-regulation of genes encoding endogenous transporters and DMA synthesis responsible for the mobilization of Fe. This mechanism operates in the roots, flag leaf and seeds. Similarly, rice engineered to produce higher levels of phytosiderophores increased only fourfold the wild-type level of Fe in the seeds, due to the modulation of genes controlling metal uptake, translocation and seed loading (Wang *et al.*, 2013; Banakar *et al.* under review).

Increasing the loading of seeds with Fe decreased the seed concentrations of Cd. Previous reports have shown that Fe-specific transporters limit the uptake of Cd in yeast (Lee *et al.*, 2009; Murata *et al.*, 2006, 2008), but this is the first time that a Cd decrease has been observed directly in the seeds of plants exposed to high levels of Cd in the environment. We investigated

Cd uptake, translocation and seed loading in *HvYS1* lines with Cd supplied in the soil. The expression of *HvYS1* reduced Cd levels by 2.3-fold in roots, 5-fold in leaves and 2.3-fold in seeds. The decrease in Cd seed concentration is particularly important given the simultaneous 2-fold increase in Fe levels, because such an approach would simultaneously address the issues of Fe deficiency and Cd toxicity in rice fields with low-Fe/high-Cd soils (Clemens *et al.*, 2013; Slamet-Loedin *et al.*, 2015). Our results show that plants can take up more Fe in the presence of Cd, and Fe acquisition in the presence of Cd may thus act as a defence mechanism to mitigate Cd-induced stress (Astolfi *et al.*, 2014; Meda *et al.*, 2007). Similarly, overexpression of the plastid Fe transporter gene *NtPIC1* in tobacco boosted the Fe/Cd ratio in leaves and improved Cd tolerance (Gong *et al.*, 2015), and rice expressing *HvNAS1* and *OsNAS1* + *HvNAATb* also accumulated more Fe but less Cd in the seeds compared to wild-type plants (Masuda *et al.*, 2012; Banakar *et al.*, under review). In contrast, rice plants exposed to Fe deficiency in the presence of excess Cd accumulated more Cd in the seeds (Nakanishi *et al.*, 2006). The specific uptake, translocation and seed loading of Fe by the *HvYS1* transgenic plants therefore appear to inhibit the uptake, translocation and loading of Cd.

Our findings can be summarized in the mechanistic model presented in Figure 6, which shows that the constitutive expression of *HvYS1* in rice selectively increases the uptake of Fe leading to higher levels of Fe in the roots, followed by selective root-to-shoot translocation increasing the Fe concentration in the leaves, promoting the remobilization of Fe from flag leaves and ultimately causing the selective accumulation of Fe in seeds. Iron homeostasis in the roots, leaves and seeds imposes a limit on the concentration of Fe in the seeds (2-fold when compared with the wild-type level) through the modulation of endogenous metal transporters, PS synthesis and the Fe storage protein ferritin. The selective mobilization of Fe by HvYS1 has no impact on Zn, Mn and Cu in most tissues, but displaces Cu and Cd from the seeds

and Cd from other tissues, providing a strategy for the selective modulation of different metal ions.

In conclusion, we have shown that the heterologous expression of *HvYS1* in rice increases Fe uptake, translocation and seed loading without affecting the uptake, translocation or seed loading of Zn and Mn, without affecting the uptake and translocation of Cu but nevertheless displacing this metal from the endosperm. The concentration of Fe in the seeds of the *HvYS1* transgenic plants is limited to double the normal level, reflecting feedback from the endogenous Fe homeostasis machinery as demonstrated by the modulation of genes controlling endogenous metal transporters and the Fe storage protein ferritin. In contrast to Zn, Mn and Cu, all of which are micronutrients required for the biological activity of certain enzymes and other proteins, Cd is robustly excluded in the transgenic plants during uptake, translocation and seed loading. Our data provide insight into the molecular basis of ion-selective metal mobilization in plants, which may have evolved to reduce the impact of stress caused by exposure to toxic heavy metals.

Materials and methods

Gene cloning and transformation vectors

The *HvYS1* cDNA (GenBank ID AB214183.1) was cloned from the roots of 2-week-old barley plants (*Hordeum vulgare* L. cv. Ordalie) growing in vitro on MS medium without Fe (Murashige and Skoog, 1962). Total RNA was extracted using the RNeasy Plant Mini Kit (Qiagen, Hilden, Germany) and 1 mg of total RNA was reverse-transcribed using the Omniscript RT Kit (Qiagen). The full-size cDNA (2037 bp) was amplified by PCR using forward primer HvYS1-BamHI-FOR (5′-AGG ATC CAT GGA CAT CGT CGC CCC GGA CCG CA-3′) and reverse primer HvYS1-HindIII-REV (5′-AAA GCT TTT AGG CAG CAG GTA GAA ACTTCA TG-3′). The product was transferred to the pGEM®-T Easy vector (Promega, Madison, WI) for sequencing and verification. The *HvYS1* cDNA was then subcloned using the BamHI and HindIII sites and inserted into the

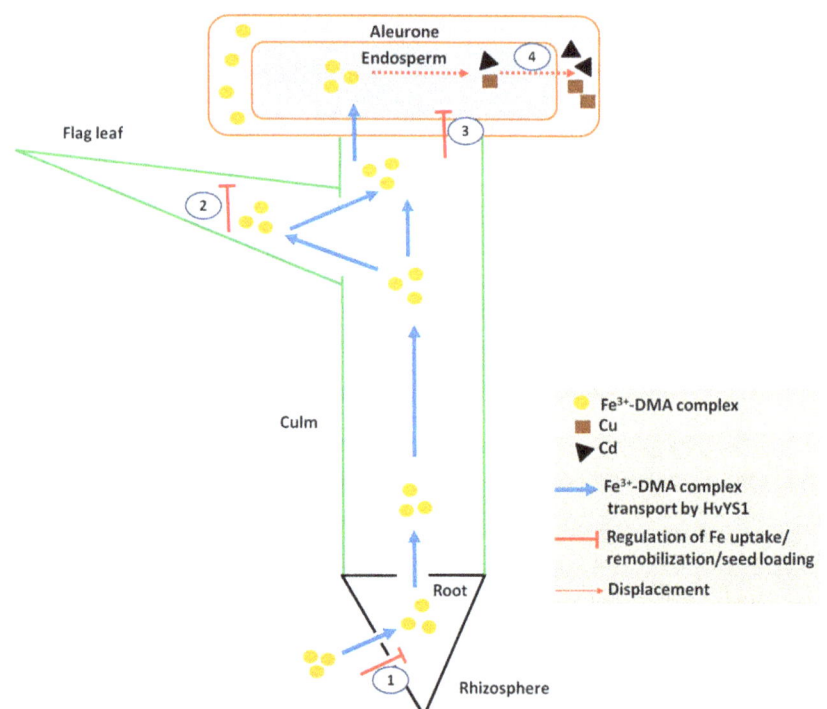

Figure 6 The mechanistic basis of selective Fe transport by HvYS1. Heterologous expression of HvYS1 results in the selective uptake, translocation, remobilization and seed loading of Fe. Endogenous Fe homeostasis limits Fe accumulation in seeds to rather modest levels (i.e. twofold) by modulating the expression of endogenous genes controlling Fe uptake (1), remobilization (2) and seed loading (3), but this is sufficient to displace the toxic heavy metals Cd and Cu from the endosperm (4).

expression vector pAL76 (Christensen and Quail, 1996), which contains the maize ubiquitin-1 (ubi-1) promoter and first intron, and an *Agrobacterium tumefaciens nos* transcriptional terminator. The hygromycin phosphotransferase selectable marker gene was controlled by the *CamV35S* promoter and carried a nos terminator for transcriptional termination.

Rice transformation

Mature rice seed-derived embryos (*Oryza sativa* L. cv EYI 105) were cultured and excised as previously described (Sudhakar et al., 1998; Valdez et al., 1998). After 7 days, the embryos were bombarded with gold particles carrying the *HvYS1* transgene and *hpt* selectable marker on separate vectors, with a 3 : 1 molar ratio (Christou et al., 1991). The rice embryos were incubated on high-osmoticum medium (0.2 M mannitol, 0.2 M sorbitol) for 4 h prior to bombardment. Bombarded embryos were selected on MS medium supplemented with 30 mg/L hygromycin, and callus pieces were transferred sequentially to shooting and rooting medium containing hygromycin as above. Regenerated plantlets were transferred to pots containing Traysubstract soil (Klasmann-Deilmann GmbH, Geeste, Germany) and were grown under flooded conditions in a chamber at 26 ± 2 °C, with a 12-h photoperiod (900 μmol/m^2/s photosynthetically active radiation) and 80% relative humidity. Plants were irrigated with a solution of 100 μM Fe provided as Fe (III)-EDDHA in the form of Sequestrene 138 Fe G-100 (Syngenta Agro SA, Madrid, Spain).

RNA blot analysis

Total leaf RNA was isolated using the RNeasy Plant Mini Kit (Qiagen) and 20-μg aliquots were fractionated on a denaturing 1.2% agarose gel containing formaldehyde before blotting. The membranes were probed with digoxigenin-labelled partial *HvYS1* cDNA at 50 °C overnight using DIG Easy Hyb (Roche Diagnostics, Mannheim, Germany). After washing and immunological detection with anti-DIG-AP (Roche Diagnostics) according to the manufacturer's instructions, CSPD chemiluminescence (Roche Diagnostics) was detected on Kodak BioMax light film (Sigma-Aldrich, St Louis, MO).

Cadmium uptake studies

Seeds from three representative transgenic rice lines (1, 2 and 3) were germinated on ½ MS medium supplemented with 50 mg/L hygromycin, and wild-type seeds were germinated on ½ MS medium without hygromycin. After 7 days, 15 uniform seedlings from wild-type and transgenic lines were transferred to nutrient solution (Kobayashi et al., 2005) containing 10 μM CdCl$_2$. The pH of the solution was adjusted to 5.3 with 0.1 M KOH and the plants were maintained as above until seed maturity. Roots, leaves and seeds were harvested from all plants and metal concentrations were quantified by inductively coupled plasma mass spectrometry (ICP-MS).

Measurement of metal concentrations by ICP-MS

Roots and leaves were collected in plastic containers prewashed with 6.5% HNO$_3$ to avoid metal contamination. Metals were also removed from the surface of each sample by washing three times in double-deionized water followed by 100 μM Na$_2$EDTA, and EDTA was then removed with two further washes in double-deionized water. To avoid metal contamination during polishing, dehusked wild-type and transgenic seeds were polished using a noncontaminating polisher (Kett, Villa Park, CA) and ground using a mortar and pestle prewashed with 6.5% HNO$_3$. Roots,

leaves and seeds were dried at 70 °C for 2 days and 300-mg portions were digested with 4.4 M HNO$_3$, 6.5 M H$_2$O$_2$ and double-deionized water (3 : 2 : 2) for 20 min at 230 °C using a MarsXpress oven (CEM Corp, Matthews, NC). Metal concentrations were determined in diluted samples by ICP-MS using an Agilent 7700X instrument (Agilent Technologies, Santa Clara, CA).

Quantitation of NA and DMA

NA (98% purity) was obtained from Hasegawa Co. Ltd. (Kawasaki, Japan), and DMA (98% purity) was obtained from Toronto Research Chemicals Inc. (Toronto, Canada). Nicotyl-lysine was synthesized as described by Wada et al. (2007). Stock solutions were prepared at concentrations of 1–10 mM and stored in darkness at −80 °C. Working solutions were prepared by diluting the stock solutions with double-deionized water. Each 5-μL standard solution was diluted with 5 μL of 50 mM EDTA, 5 μL nicotyl-lysine and 30 μL of a 1 : 9 ratio mixture of 10 mM ammonium acetate and acetonitrile (pH 7.3), and the mixture was filtered through polyvinylidene fluoride (Durapore® PVDF) 0.45-μm ultrafree-MC centrifugal filter devices (Merck KGaA, Darmstadt, Germany) before injection into the HPLC-ESI-TOF-MS system (see below). Fresh root and leaf tissues were extracted as described by Schmidt et al. (2011) with some modifications. Samples stored as 200-mg aliquots at −80 °C prior to extraction were homogenized in 200 μL (roots) or 400 μL (leaves) double-deionized water containing 36 μL 1 mM nicotyl-lysine. The homogenate was vortexed for 30 s, sonicated for 5 min and centrifuged at 15 000 *g* for 10 min at 4 °C before the supernatant was passed through a 3-kDa centrifugal filter (cellulose Amicon® Ultra filter units, Merck KGaA). The filtrate was centrifuged as above for 30 min and dried under vacuum. Seeds were ground to a fine powder under liquid N$_2$ and extracted three times as described by Wada et al. (2007) with some modifications. Aliquots of 50 mg seed powder were extracted in 300 μL double-deionized water containing 18 μL of 1 mM nicotyl-lysine. The supernatant was recovered by centrifugation at 15 000 *g* for 15 min at 4 °C and stored at −20 °C, and the pellet was extracted twice as above. The three supernatant fractions were pooled and the total extract was passed through the centrifugal filter, centrifuged again and concentrated under vacuum as described above. The dry residues from the leaf/root and seed extracts were dissolved in 20 and 10 μL of type I water, respectively. Then, 5-μL aliquots of extracts were diluted with 10 μL of 50 mM EDTA, 15 μL type I water and 30 μL of a 1 : 9 ratio mixture of 10 mM ammonium acetate and acetonitrile (pH 7.3), and the mixture was filtered through 0.45-μm polyvinylidene fluoride (PVDF) ultrafree-MC centrifugal filter devices (Merck KGaA, Darmstadt, Germany) before analysis.

NA and DMA levels were determined by high-performance liquid chromatography electrospray ionization time-of-flight mass spectrometry (HPLC-ESI-TOF-MS) as described by Xuan et al. (2006), with modifications. Details of HPLC conditions are described in SI Materials and Methods, and the details of TOF-MS operating conditions are listed in Table S2.

Quantitation of endogenous gene expression

Quantitative real-time RT-PCR was carried out to measure steady state mRNA levels in roots, flag leaf and immature seeds, representing the endogenous genes listed in Table S1. Due to its stable expression, actin is a reliable reference gene for qRT-PCR studies (Cheng et al., 2007; Lee et al., 2011). Hence, *OsActin1* was used as a reference gene (details of PCR conditions are described in SI Materials and Methods).

Acknowledgements

We acknowledge support from the European Research Council IDEAS Advanced Grant Program (BIOFORCE) to P.C., and the Spanish Ministry of Economy and Competitivity (MINECO; projects AGL2013-42175-R, co-financed with FEDER) and the Aragón Government (Group A03) to J.A. R.B was supported by a PhD fellowship from the University of Lleida, Spain.

Author contributions

R.B., A.A.F. and P.C. designed the research; R.B. performed the research; R.B and A.A.F. analysed the data; R.B., A.A.F., J.A., T.C and P.C. wrote the manuscript.

References

Alimba, G.C., Dhillon, V., Bakare, A.A. and Fenech, M. (2016) Genotoxicity and cytotoxicity of chromium, copper, manganese and lead, and their mixture in WIL2-NS human B lymphoblastoid cells is enhanced by folate depletion. *Mut. Res.-Gen. Tox. En.* **798**, 35–47.

Araki, R., Kousaka, K., Namba, K., Murata, Y. and Murata, J. (2015) 2′-Deoxymugineic acid promotes growth of rice (*Oryza sativa* L.) by orchestrating iron and nitrate uptake processes under high pH conditions. *Plant J.* **81**, 233–246.

Astolfi, S., Ortolani, M.R., Catarcione, G., Paolacci, A.R., Cesco, S., Pinton, R. and Ciaffi, M. (2014) Cadmium exposure affects iron acquisition in barley (*Hordeum vulgare*) seedlings. *Physiol. Plant.* **152**, 646–659.

Ayoma, T., Kobayashi, T., Takahashi, M., Nagasaka, S., Usada, K., Kakei, Y., Ishimaru, Y. *et al.* (2009) OsYSL18 is a rice iron (III)-deoxymugineic acid transporter specifically expressed in reproductive organs and phloem of lamina joints. *Plant Mol. Biol.* **70**, 681–692.

Bashir, K. and Nishizawa, N.K. (2006) Deoxymugineic acid synthase: a gene important for Fe acquisition and Homeostasis. *Plant Signal. Behav.* **1**, 290–292.

Bashir, K., Takahashi, R., Nakanishi, H. and Nishizawa, N.K. (2013) The road to micronutrient biofortification of rice: progress and prospects. *Front Plant Sci.* **4**, 1–7.

Bashir, K., Hanada, K., Shimizu, M., Seki, M., Nakanishi, H. and Nishizawa, N.K. (2014) Transcriptomic analysis of rice in response to iron deficiency and excess. *Rice*, **7**, 18–33.

Boonyaves, K., Gruissem, W. and Bhullar, N. (2016) NOD promoter controlled AtIRT1 expression functions synergistically with NAS and FERRITIN genes to increase iron in rice grains. *Plant Mol. Biol.* **90**, 207–215.

Cheng, L., Wang, F., Shou, H., Huang, F., Zheng, L., He, F., Li, J. *et al.* (2007) Mutation in nicotianamine aminotransferase stimulated the Fe (II) acquisition system and led to iron accumulation in rice. *Plant Physiol.* **145**, 1647–1657.

Christensen, A.H. and Quail, P.H. (1996) Ubiquitin promoter based vectors for high-level expression of selectable and/or screenable marker genes in monocotyledonous plants. *Transgenic Res.* **5**, 213–218.

Christou, P., Ford, T.L. and Kofron, M. (1991) Production of transgenic rice (*Oryza sativa* L.) plants from agronomically important indica and japonica varieties via electric discharge particle acceleration of exogenous DNA immature zygotic embryos. *Biotechnology*, **9**, 957–962.

Clemens, S., Aarts, M.G., Thomine, S. and Verbruggen, N. (2013) Plant science: the key to preventing slow cadmium poisoning. *Trends Plant Sci.* **18**, 92–99.

Curie, C., Cassin, G., Couch, D., Divol, F., Higuchi, K., Jean, M.L., Mission, J., Schikora, A., Czernic, P. and Mari, S. (2009) Metal movement within the plant: contribution of nicotianamine and yellow stripe 1-like transporters. *Ann. Bot.* **103**, 1–11.

Durrett, T.P., Gassmann, W. and Rogers, E.E. (2007) The FRD3-mediated efflux of citrate into the root vasculature is necessary for efficient iron translocation. *Plant Physiol.* **144**, 197–205.

Duy, D., Wanner, G., Meda, A.R., von Wirén, N., Soll, J. and Philippar, K. (2007) PIC1, an ancient permease in Arabidopsis chloroplasts, mediates iron transport. *Plant Cell*, **19**, 986–1006.

Gómez-Galera, S., Rojas, E., Sudhakar, D., Zhu, C., Pelacho, A.M., Capell, T. and Christou, P. (2010) Critical evaluation of strategies for mineral fortification of staple food crops. *Transgenic Res.* **19**, 165–180.

Gong, X., Yin, L., Chen, J. and Guo, C. (2015) Overexpression of the iron transporter NtPIC1 in tobacco mediates tolerance to cadmium. *Plant Cell Rep.* **34**, 1963–1973.

Hall, J.L. and Williams, L.E. (2003) Transition metal transporter in plants. *J. Exp. Bot.* **54**, 2601–2613.

Inoue, H., Higuchi, K., Takahashi, M., Nakanishi, H., Mori, S. and Nishizawa, N.K. (2003) Three rice nicotianamine synthase genes OsNAS1, OsNAS2 and OsNAS3 are expressed in cells involved in long distance transport of iron and differentially regulated by iron. *Plant J.* **36**, 366–381.

Inoue, H., Takahashi, M., Kobayashi, T., Suzuki, M., Nakanishi, H., Mori, S. and Nishizawa, N.K. (2008) Identification and localization of rice nicotianamine aminotransferase OsNAAT1 expression suggests the site of phytosiderophore synthesis in rice. *Plant Mol. Biol.* **66**, 193–203.

Inoue, H., Kobayashi, T., Nozoye, T., Takhashi, M., Kakei, Y., Suzuki, K., Nakazono, M. *et al.* (2009) Rice OsYSL15 is an iron-regulated iron (III)-deoxymugenic acid transporter expressed in the roots and is essential for iron uptake in early growth of the seedlings. *J. Biol. Chem.* **284**, 3470–3479.

Ishimaru, Y., Suzuki, M., Tsukamoto, T., Suzuki, K., Nakazono, M., Kobayashi, T., Wada, Y. *et al.* (2006) Rice plants take up iron as an Fe^{3+}-phytosiderophore and as Fe^{2+}. *Plant J.* **45**, 335–346.

Ishimaru, Y., Masuda, H., Bashir, K., Inoue, H., Tsukamoto, T., Takahashi, M., Nakanishi, H. *et al.* (2010) Rice metal-nicotianamine transporter, OsYSL2, is required for the long-distance transport of iron and manganese. *Plant J.* **62**, 379–390.

Ishimaru, Y., Takahashi, R., Bashir, K., Shimo, H., Senoura, T., Sugimoto, K., Ono, K., Yano, M., Ishikawa, S., Arao, T., Nakanishi, H. and Nishizawa, N.K. (2012) Charecterizing the role of NRAMP5 in manganese, iron and cadmium transport. *Sci Rep.* **2**, 286–294.

Jain, A. and Connolly, E.L. (2013) Mitochondrial iron transport and homeostasis in plants. *Front Plant Sci.* **4**, 348–354.

Kakei, Y., Ishimaru, Y., Kobayashi, T., Yamakawa, T., Nakanishi, H. and Nishizawa, N.K. (2012) OsYSL16 plays a role in the allocation of iron. *Plant Mol. Biol.* **79**, 583–594.

Kobayashi, T. and Nishizawa, N.K. (2012) Iron uptake, translocation and regulation in higher plants. *Ann. Rev. Plant Biol.* **63**, 131–152.

Kobayashi, T., Suzuki, M., Inoue, H., Itai, R.N., Takahashi, M., Nakanishi, H., Mori, S. *et al.* (2005) Expression of iron-acquisition-related genes in iron-deficient rice is coordinately induced by partially conserved iron-deficiency responsive elements. *J. Exp. Bot.* **56**, 1305–1316.

Lanquar, V., Lelièvre, F., Bolte, S., Hamès, C., Alcon, C., Neumann, D., Vansuyt, G. *et al.* (2005) Mobilization of vacuolar iron by AtNRAMP3 and AtNRAMP4 is essential for seed germination on low iron. *EMBO J.* **7**, 4041–4051.

Lee, S. and An, G. (2009) Over-expression of OsIRT1 leads to increased iron and zinc accumulations in rice. *Plant, Cell Environ.* **32**, 408–416.

Lee, S., Chiecko, J.C., Kim, S.A., Walker, E.L., Lee, Y., Guerinot, M.L. and An, G. (2009) Disruption of OsYSL15 leads to iron inefficiency in rice plants. *Plant Physiol.* **150**, 786–800.

Lee, S., Person, D.P., Hansen, T.H., Husted, S., Schjoerring, J.K., Kim, S.-Y., Jeon, U.S. *et al.* (2011) Bio-available zinc in rice seeds is increased by activation tagging of nicotianamine synthase. *Plant Biotech. J.* **9**, 865–873.

Long, J.C., Sommer, K., Allen, M.D., Lu, F.-S. and Merchant, S.S. (2008) *FER1* and *FER2* encoding two ferritin complexes in *Chlamydomonas reinhardtii* chloroplasts are regulated by iron. *Genetics*, **179**, 137–147.

Ma, J.F., Taketa, S., Chang, Y.C., Takeda, K. and Matsumoto, H. (1999) Biosynthesis of phytosiderophores in several Triticeae species with different genomes. *J. Exp. Bot.* **50**, 723–726.

Masuda, H., Usada, K., Kobayashi, T., Ishimaru, Y., Kakei, Y., Takahashi, M., Higuchi, K. et al. (2009) Overexpression of barley nicotianamine synthase gene HvNAS1 increases iron and zinc concentration in rice grains. Rice, **2**, 155–166.

Masuda, H., Ishimaru, Y., Aung, M.S., Kobayashi, T., Kakei, Y., Takahashi, M., Higuchi, K. et al. (2012) Iron biofortification in rice by the introduction of multiple genes involved in iron nutrition. Sci. Rep. **2**, 543–550.

Meda, A.R., Scheuermann, E.B., Prechs, U.E., Erenoglu, B., Schaaf, G., Hayen, H., Weber, G. et al. (2007) Iron acquisition by phytosiderophores contributes to cadmium tolerance. Plant Physiol. **143**, 1761–1773.

Morrissey, J. and Guerinot, M.L. (2009) Iron uptake and transport in plants: the good, the bad, and the ionome. Chem. Rev. **109**, 4553–4567.

Murashige, T. and Skoog, F. (1962) A revised medium for rapid growth and bioassays with tobacco tissue cultures. Physiol. Plant. **15**, 473–497.

Murata, Y., Ma, J.F., Yamaji, N., Ueno, D., Nomoto, K. and Iwashita, T. (2006) A specific transporter of iron (III)-phytosiderophore in barley roots. Plant J. **46**, 563–572.

Murata, Y., Harada, E., Sugase, K., Namba, K., Horikawa, M., Ma, J.F., Yamaji, N. et al. (2008) Specific transporter for iron (III) phytosiderophore complex involved in iron uptake by barley roots. Pure Appl. Chem. **80**, 2689–2697.

Murata, H., Itoh, Y., Iwashita, T. and Namba, K. (2015) Transgenic petunia with the Iron (III)-phytosiderophore transporter gene acquires tolerance to iron deficiency alkaline environments. PLoS ONE, **10**, e0120227.

Nakanishi, H., Ogawa, I., Ishimaru, Y., Mori, S. and Nishizawa, N.K. (2006) Iron deficiency enhances cadmium uptake and translocation mediated by the Fe^{2+} transporters OsIRT1 and OsIRT2 in rice. J. Soil. Sci. Plant Nutr. **52**, 464–469.

Nishida, S., Tsuzuki, C., Kato, A., Aisu, A., Yoshida, J. and Mizuno, T. (2011) AtIRT1, the Primary iron-uptake transporter in the root, mediates excess nickel accumulation in Arabidopsis thaliana. Plant Cell Physiol. **52**, 1433–1442.

Pérez-Massot, E., Banakar, R., Gómez-Galera, S., Zorrilla-López, U., Sanahuja, G., Arjó, G., Miralpeix, B. et al. (2013) The contribution of transgenic plants to better health through improved nutrition: opportunities and constraints. Genes Nutr. **29**, 29–41.

Schmidt, H., Böttcher, C., Trampczynska, A. and Clemens, S. (2011) Use of recombinantly produced ^{15}N3-labelled nicotianamine for fast and sensitive stable isotope dilution ultra-performance liquid chromatography/electrospray ionisation time-of-flight mass spectrometry. Anal. Bioanal. Chem. **399**, 1355–1361.

Slamet-Loedin, I.H., Johnson-Beebout, S.E., Impa, S. and Tsakirpaloglou, N. (2015) Enriching rice with Zn and Fe while minimizing Cd risk. Front Plant Sci. **6**, 1–9.

Sperotto, R.A., Ricachenevsky, F.K., Waldow, V.A. and Fett, J.P. (2012) Iron biofortification in rice: it's a long way to the top. Plant Sci. **190**, 24–39.

Sudhakar, D., Duc, L.T., Bong, B.B., Tinjuangjun, P., Maqbool, S.B., Valdez, M., Jefferson, R. et al. (1998) An efficient rice transformation system utilizing mature seed-derived explants and a portable, inexpensive particle bombardment device. Transgenic Res. **7**, 289–294.

Suzuki, M., Tsukamoto, T., Inoue, H., Watanabe, S., Matsuhashi, S., Takahashi, M., Nakanishi, H. et al. (2008) Deoxymugineic acid increases Zn translocation in Zn-deficient rice plants. Plant Mol. Biol. **66**, 609–617.

Takahashi, R., Ishimaru, Y., Senoura, T., Shimo, H., Ishikawa, S., Arao, T., Nakanishi, H. et al. (2011) The OsNRAMP1 iron transporter is involved in Cd accumulation in rice. J. Exp. Bot. **62**, 4843–4850.

Tan, S., Han, R., Li, P., Yang, G., Li, S., Zhang, P., Wang, W.B. et al. (2015) Over-expression ofthe MxIRT1 gene increases iron and zinc content in rice seeds. Transgenic Res. **24**, 109–122.

Thomine, S. and Vert, G. (2013) Iron transport in plants: better be safe than sorry. Curr. Opin. Plant Biol. **16**, 322–327.

Ueno, D., Yamaji, N. and Ma, J.F. (2009) Further characterization of ferric-phytosiderophore transporters ZmYS1 and HvYS1 in maize and barley. J. Exp. Bot. **60**, 3513–3520.

Ueno, D., Yamaji, N., Kono, I., Huang, F.C., Ando, T., Yano, M. and Ma, F.J. (2010) Gene limiting cadmium accumulation in rice. Proc Natl Acad Sci USA, **21**, 16500–16505.

Uraguchi, S. and Fujiwara, T. (2012) Cadmium transport and tolerance in rice: perspectives for reducing grain cadmium accumulation. Rice, **5**, 5–12.

Valdez, M., Cabrera-Ponce, J.L., Sudhakhar, D., Herrera-Estrella, L. and Christou, P. (1998) Transgenic Central American, West African and Asian elite rice varieties resulting from particle bombardment of foreign DNA into mature seed-derived explants utilizing three different bombardment devices. Ann. Bot. **82**, 795–801.

Vert, G., Grotz, N., Dedaldechamp, F., Gaymard, F., Guerinot, M.L., Briat, J.F. and Curie, C. (2002) IRT1, an Arabidopsis transporter essential for iron uptake from the soil and for plant growth. Plant Cell. **14**, 1223–1233.

Wada, Y., Yamaguchi, I., Takahashi, M., Nakanishi, H., Mori, S. and Nishizawa, N.K. (2007) Highly sensitive quantitative analysis of nicotianamine using LC/ESI-TOF-MS with and internal standard. Biosci. Biotechnol. Biochem. **71**, 435–441.

Wang, M., Gruissem, W. and Bhullar, N.K. (2013) Nicotianamine synthase overexpression positively modulates iron homeostasis-related genes in high iron rice. Front Plant Sci. **29**, 156–171.

Winterbourn, C.C. (1995) Toxicity of iron and hydrogen peroxide the Fenton reaction. Toxicol. Lett. **82–83**, 969–974.

Xuan, Y., Scheuermann, E.B., Meda, A.R., Hayen, H., vonWiren, N. and Weber, G. (2006) Separation and identification of phytosiderophores and their metal complexes in plants by zwitterionic hydrophilic interaction liquid chromatography coupled to electrospray ionization mass spectrometry. J. Chromatgr. A. **1136**, 73–81.

Yang, M., Zhang, Y., Zhang, L., Hu, J., Zhang, X., Lu, K., Dong, H. et al. (2014) OsNRAMP5 contributes to manganese translocation and distribution in rice shoots. J. Exp. Bot. **65**, 4849–4861.

Yokosho, K., Yamaji, N., Ueno, D., Mitani, N. and Ma, J.F. (2009) OsFRDL1 is a citrate transporter required for efficient translocation of iron in rice. Plant Physiol. **149**, 297–305.

Yoneyama, T., Ishikawa, S. and Fujimaki, S. (2015) Route and regulation of zinc, cadmium, and iron transport in rice plants (Oryza sativa L.) during vegetative growth and grain filling: metal transporters, metal speciation, grain Cd reduction and Zn and Fe biofortification. Int. J. Mol. Sci. **16**, 19111–19129.

Zhang, Y., Xu, Y.H., Yi, H.Y. and Gong, J.M. (2012) Vacuolar membrane transporters OsVIT1 and OsVIT2 modulate iron translocation between flag leaves and seeds in rice. Plant J. **72**, 400–410.

Permissions

List of Contributors

Oropeza-Aburto Araceli, Mora-Macias Javier and Herrera-Estrella Luis
Metabolic Engineering Laboratory, Unidad de Genomica Avanzada – LANGEBIO CINVESTAV, Irapuato, Guanajuato, Mexico

Cruz-Ramirez Alfredo
Molecular and Developmental Complexity Laboratory, Unidad de Genomica Avanzada – LANGEBIO CINVESTAV, Irapuato, Guanajuato, Mexico

Min Deng, Dongqin Li, Jingyun Luo, Yingjie Xiao, Haijun Liu, Qingchun Pan, Xuehai Zhang, Minliang Jin, Mingchao Zhao and Jianbing Yan
National Key Laboratory of Crop Genetic Improvement, Huazhong Agricultural University, Wuhan, China

Tim Fox, Jason DeBruin, Kristin Haug Collet, Mary Trimnell, Joshua Clapp, April Leonard, Bailin Li, Eric Scolaro, Sarah Collinson, Kimberly Glassman, Michael Miller, Jeff Schussler, Dennis Dolan, Lu Liu, Carla Gho, Marc Albertsen, Dale Loussaert and Bo Shen
DuPont Pioneer, Johnston, IA, USA

Zhesi He, Lihong Wang, Andrea L. Harper, Ian Bancroft and Lenka Havlickova
Department of Biology, University of York, Heslington, York, UK

Akshay K. Pradhan
Department of Genetics and Centre for Genetic Manipulation of Crop Plants, University of Delhi, New Delhi, India

Isobel A. P. Parkin
Agriculture and Agri-Food Canada, Saskatoon, SK, Canada

Rod A. Herman, Brandon J. Fast, Peter N. Scherer, Alyssa M. Brune, Barry W. Schafer, Ricardo D. Ekmay, George G. Harrigan and Greg A. Bradfisch
Dow AgroSciences LLC, Indianapolis, IN, USA

Denise T. de Cerqueira
Dow AgroSciencies Sementes e Biotecnologia Brasil LTDA, Cravinhos, SP, Brazil

Inger Bæksted Holme, Giuseppe Dionisio, Claus Krogh Madsen and Henrik Brinch-Pedersen
Department of Molecular Biology and Genetics, Faculty of Science and Technology, Research Centre Flakkebjerg, Aarhus University, Slagelse, Denmark

Junfang Kang, Jianmin Li, Shuang Gao, Chao Tian and Xiaojun Zha
College of Chemistry and Life Sciences, Zhejiang Normal University, Jinhua, China

Dong-Keun Lee, Pil Joong Chung, Jin Seo Jeong, Seung Woon Bang, Harin Jung, Youn Shic Kim and Ju-Kon Kim
Graduate School of International Agricultural Technology and Crop Biotechnology Institute/GreenBio Science and Technology, Seoul National University, Pyeongchang, Korea

Geupil Jang and Yang Do Choi
Department of Agricultural Biotechnology, Seoul National University, Seoul, Korea

Sun-Hwa Ha
Department of Genetic Engineering and Graduate School of Biotechnology, Kyung Hee University, Yongin, Korea

Yunhe Li, Qingsong Liu, Yan Yang, Yanan Wang, Xiuping Chen and Yufa Peng
State Key Laboratory for Plant Diseases and Insect Pests, Institute of Plant Protection, Chinese Academy of Agricultural Sciences, Beijing, China

Hongxia Hua
College of Plant Science & Technology, Huazhong Agricultural University, Wuhan, China

Qingling Zhang
State Key Laboratory for Plant Diseases and Insect Pests, Institute of Plant Protection, Chinese Academy of Agricultural Sciences, Beijing, China
College of Plant Science & Technology, Huazhong Agricultural University, Wuhan, China

Jörg Romeis
State Key Laboratory for Plant Diseases and Insect Pests, Institute of Plant Protection, Chinese Academy of Agricultural Sciences, Beijing, China

Agroscope, Biosafety Research Group, Zurich, Switzerland

Michael Meissle
Agroscope, Biosafety Research Group, Zurich, Switzerland

Guosheng Xie, Jiangfeng Huang, Ran Zhang, Yu Li, Yanting Wang, Ao Li, Xukai Li and Liangcai Peng
Biomass and Bioenergy Research Centre, Huazhong Agricultural University, Wuhan, China
National Key Laboratory of Crop Genetic Improvement, Huazhong Agricultural University, Wuhan, China
College of Plant Science and Technology, Huazhong Agricultural University, Wuhan, China

Fengcheng Li
Biomass and Bioenergy Research Centre, Huazhong Agricultural University, Wuhan, China
National Key Laboratory of Crop Genetic Improvement, Huazhong Agricultural University, Wuhan, China
College of Plant Science and Technology, Huazhong Agricultural University, Wuhan, China
Key Laboratory of Crop Physiology, Ecology, Genetics and Breeding, Ministry of Agriculture, Rice Research Institute, Shenyang Agricultural University, Shenyang, China

Miaomiao Zhang and Tao Xia
Biomass and Bioenergy Research Centre, Huazhong Agricultural University, Wuhan, China
National Key Laboratory of Crop Genetic Improvement, Huazhong Agricultural University, Wuhan, China
College of Life Science and Technology, Huazhong Agricultural University, Wuhan, China

Chengcheng Qu
State Key Laboratory of Agricultural Microbiology, Huazhong Agricultural University, Wuhan, China

Fan Hu and Arthur J. Ragauskas
Department of Chemical and Biomolecular Engineering, The University of Tennessee- Knoxville, Knoxville, TN, USA
Department of Forestry, The University of Tennessee-Knoxville, Knoxville, TN, USA

Jingyu Lin, Xinlu Chen, Mitra Mazarei, Vincent R. Pantalone, Charles Neal Stewart Jr and Feng Chen
Department of Plant Sciences, University of Tennessee, Knoxville, TN, USA

Dan Wang and Ningning Wang
Department of Plant Biology and Ecology, College of Life Sciences, Nankai University, Tianjin, China

Tobias G. Köllner
Department of Biochemistry, Max Planck Institute for Chemical Ecology, Jena, Germany

Hong Guo
Department of Biochemistry, Cellular and Molecular Biology, University of Tennessee, Knoxville, TN, USA

Prakash Arelli
Crop Genetics Research Unit, USDA-ARS, Jackson, TN, USA

Ku-Ting Chen and Lin-Chih Yu
Institute of Molecular Biology, Academia Sinica, Nankang, Taipei, Taiwan, ROC

Shuen-Fang Lo, Yi-Lun Liu and Mirng-Jier Jiang
Institute of Molecular Biology, Academia Sinica, Nankang, Taipei, Taiwan, ROC
Agricultural Biotechnology Center, National Chung Hsing University, Taichung, Taiwan, ROC

Su-May Yu
Institute of Molecular Biology, Academia Sinica, Nankang, Taipei, Taiwan, ROC
Agricultural Biotechnology Center, National Chung Hsing University, Taichung, Taiwan, ROC
Department of Life Sciences, National Chung Hsing University, Taichung, Taiwan, ROC

Tuan-Hua David Ho
Agricultural Biotechnology Center, National Chung Hsing University, Taichung, Taiwan, ROC
Institute of Plant and Microbial Biology, Academia Sinica, Taipei, Taiwan, ROC
Department of Life Sciences, National Chung Hsing University, Taichung, Taiwan, ROC

Liang-Jwu Chen
Agricultural Biotechnology Center, National Chung Hsing University, Taichung, Taiwan, ROC
Institute of Molecular Biology, National Chung Hsing University, Taichung, Taiwan, ROC

Kun-Ting Hsieh
Institute of Molecular Biology, National Chung Hsing University, Taichung, Taiwan, ROC

Miin-Huey Lee, Chi-yu Chen and Tzu-Pi Huang
Department of Plant Pathology, National Chung Hsing University, Taichung, Taiwan, ROC

Mikiko Kojima and Hitoshi Sakakibara
RIKEN Center for Sustainable Resource Science, Yokohama, Kanagawa, Japan

Poonam Mehra, Bipin Kumar Pandey and Jitender Giri
National Institute of Plant Genome Research, New Delhi, India

Timothy L. Fitzgerald, Jiri Stiller, Paul J. Berkman and Donald M. Gardiner
Commonwealth Scientific a nd I ndustrial Research Organisation Agriculture, St Lucia, Queensland, Australia

Kemal Kazan and Jonathan J. Powell
Commonwealth Scientific and Industrial Research Organisation Agriculture, St Lucia, Queensland, Australia
Queensland Alliance for Agriculture and Food Innovation, University of Queensland, St Lucia, Queensland, Australia

Robert J. Henry
Queensland Alliance for Agriculture and Food Innovation, University of Queensland, St Lucia, Queensland, Australia

John M. Manners
Commonwealth Scientific a nd I ndustrial Research Organisation Agriculture, Black Mountain, Australian Capital Territory, Austral

Amy Rinaldo, Brian Gilbert, Evans Lagudah and Michael Ayliffe
CSIRO Agriculture, Canberra, ACT, Australia

Rainer Boni and Simon G. Krattinger
Department of Plant and Microbial Biology, University of Zurich, Zurich, Switzerland

Davinder Singh and Robert F. Park
Plant Breeding Institute, University of Sydney, Narellan, NSW, Australia

Cong Dang, Long Wang, Xuefei Chang, Fang Wang, Hongwei Yao and Gongyin Ye
State Key Laboratory of Rice Biology & Key Laboratory of Agricultural Entomology of Ministry of Agriculture, Institute of Insect Sciences, Zhejiang University, Hangzhou, China

Zengbin Lu
State Key Laboratory of Rice Biology & Key Laboratory of Agricultural Entomology of Ministry of Agriculture, Institute of Insect Sciences, Zhejiang University, Hangzhou, China
Institute of Plant Protection, Shandong Academy of Agricultural Sciences, Jinan, China

Yufa Peng
State Key Laboratory for Biology of Plant Diseases and Insect Pests, Institute of Plant Protection, Chinese Academy of Agricultural Sciences, Beijing, China

David Stanley
Biological Control of Insects Research Laboratory, USDA/Agricultural Research Service, Columbia, MO, USA

Marta Brozynska, Robert J. Henry and Agnelo Furtado
Queensland Alliance for Agriculture and Food Innovation, University of Queensland, Brisbane, QLD, Australia

Dario Copetti and Rod A. Wing
Arizona Genomics Institute, School of Plant Sciences, University of Arizona, Tucson, AZ, USA
International Rice Research Institute, T.T. Chang Genetic Resources Center, Los Baños, Laguna, Philippines

Darren Crayn
Australian Tropical Herbarium, James Cook University, Cairns, QLD, Australia

Glen Fox
Queensland Alliance for Agriculture and Food Innovation, University of Queensland, Toowoomba, QLD, Australia

Ryuji Ishikawa
Faculty of Agriculture and Life Science, Hirosaki University, Hirosaki, Aomori, Japan

Jingguang Chen, Xiaoru Fan, Kaiyun Qian, Yong Zhang, Miaoquan Song, Guohua Xu and Xiaorong Fan
State Key Laboratory of Crop Genetics and Germplasm Enhancement, Nanjing Agricultural University, Nanjing, China
Key Laboratory of Plant Nutrition and Fertilization in Low-Middle Reaches of the Yangtze River, Ministry of Agriculture, Nanjing Agricultural University, Nanjing, China

Yu Liu
State Key Laboratory of Plant Physiology and Biochemistry, College of Life Science, Zhejiang University, Hangzhou, China

Raviraj Banakar and Teresa Capell
Departament de Producció Vegetal i Ciència Forestal, Universitat de Lleida-Agrotecnio Center Lleida, Lleida, Spain

Àna Alvarez Fernández and Javier Abadía
Department of Plant Nutrition, Aula Dei Experimental
Station, Consejo Superior de Investigaciones Científicas
(CSIC), Zaragoza, Spain

Paul Christou
ICREA, Catalan Institute for Research and Advanced
Studies, Barcelona, Spain

Index

www.ingramcontent.com/pod-product-compliance
Lightning Source LLC
Chambersburg PA
CBHW080642200326
41458CB00013B/4713